ESSAYS IN INDEX NUMBER THEORY
VOLUME 1

CONTRIBUTIONS TO ECONOMIC ANALYSIS

217

ESSAYS IN INDEX NUMBER THEORY VOLUME 1

Edited by

W. Erwin DIEWERT
University of British Columbia
Vancouver, Canada

Alice O. NAKAMURA
University of Alberta
Edmonton, Canada

Emerald

United Kingdom – North America – Japan
India – Malaysia – China

Emerald Group Publishing Limited
Howard House, Wagon Lane, Bingley BD16 1WA, UK

British Library Cataloguing in Publication Data
A catalogue record for this book is available from the British Library

ISBN: 978-0-44-487471-9
ISSN: 0573-8555 (Series)

ISOQAR QUALITY ASSURED ISO 9001
UKAS QUALITY MANAGEMENT 026
Certificate number 1985

Awarded in recognition of Emerald's production department's adherence to quality systems and processes when preparing scholarly journals for print

INVESTOR IN PEOPLE

INTRODUCTION TO THE SERIES

This series consists of a number of studies, which are introduced by the editors in the belief that they represent valuable contributions to economic science.
The term "economic analysis" as used in the title of the series has been adopted because it covers both the activities of the theoretical economist and the research worker.
Although the analytical methods used by the various contributors are not the same, they are nevertheless conditioned by the common origin of their studies, namely theoretical problems encountered in practical research. Since for this reason, business cycle research and national accounting, research work on behalf of economic policy, and problems of planning are the main sources of the subjects dealt with, they necessarily determine the manner of approach adopted by the authors. Their methods tend to be "practical" in the sense of not being too far remote from application to actual economic conditions. In additon they are quantitative.
It is the hope of the editors that the publication of these studies will help to stimulate the exchange of scientific information and to reinforce international cooperation in the field of economics.

The Editors

CONTENTS FOR VOLUME I

Part Four

SUPERLATIVE INDEX NUMBERS

Part Five

THE AXIOMATIC APPROACH TO INDEX NUMBER THEORY

Part Six

AGGREGATION USING PRICE PROPORTIONALITY

Essays in Index Number Theory, Volume I
W.E. Diewert and A.O. Nakamura (Editors)

Chapter 1
OVERVIEW OF VOLUME I

W. Erwin Diewert*

1. Introduction

This volume and a subsequent one gather together most of my published papers on index number and aggregation theory. In addition, these volumes contain some new papers which have not been published before. In Volume I, Chapters 1, 2 and 14 are new. Also proofs have been added to the previously published version of Chapter 6. The previously published papers have not been changed in content but minor errors have been corrected and the references to unpublished papers have been replaced by references to the published versions in most cases.

This volume is divided into six parts.

Part One provides an introduction to this volume.

Part Two is devoted to the early history of index number theory along with biographical comments on two of the giants of index number theory: A.A. Konüs (who sadly died recently) and E.L.E. Laspeyres.

Part Three presents two surveys of index number theory. The first one is Chapter 5 and the second is Chapter 7. As the second survey is more mathematical in nature, the underlying economic theory that is used in Chapter 7 is presented in Chapter 6.

Part Four is concerned mainly with the concept of a superlative index number formula. This concept will be defined below in Section 4 of this overview.

Part Five is devoted to axiomatic or test approaches to index number theory and some closely related measurement problems involving the axiomatic characterization of symmetric means. The axiomatic approach to index number theory will be explained in Section 2 of this overview.

Part Six is concerned with one particular mechanism for aggregating over commodities, namely aggregation when prices move proportionally. The original result in this area is known as Hicks' [1946] Composite Commodity Theorem.

---

*This research was supported by a Strategic Grant from the Social Science and Humanities Research Council of Canada. Thanks are due to Shelley Hey and Louise Hebert for typing assistance.

Looking ahead to Volume II in this series, it will deal with five topics: (i) the measurement of input, output and productivity; (ii) the measurement of price and welfare change using consumer theory; (iii) alternative approaches to the measurement of change (using differences instead of ratios); (iv) multilateral comparisons; and (v) nonparametric approaches to measurement.

This chapter provides a brief overview of index number theory. In addition to serving as an introduction to the subject, this overview should be helpful to readers who are interested in locating material on specific topics covered in this volume. Those readers who want a short course on index number theory should read this overview plus Chapters 2 and 5. For a more in depth course, the reader should add Chapters 7 and 8 and Sections 1–3 of Chapter 13 (the axiomatic part). Readers who are primarily interested in the measurement of inequality or in functional forms for social welfare functions or in the theory of choice under uncertainty can restrict themselves to Chapter 14.

Several chapters could be used as readings for a topics course in economic theory which would cover duality theory and the economic theory of index numbers. The relevant chapters are 6 (which provides the duality theory background for Chapter 7), 7, 8, 11, 14, 15 and 16.

In these volumes, we shall study aggregation problems in economics. Economic theory is for the most part concerned with modeling the demand and supply for *individual* goods and services (commodities) by *individual* economic agents (producers or consumers). However, due to the truly enormous numbers of both commodities and agents in real life economies, empirical economics uses data that are always aggregated over commodities and often aggregated over agents. How should this aggregation over goods and agents be accomplished?

More specifically, the aggregation over goods problem asks: how can we aggregate or summarize individual microeconomic data on prices into a single aggregate price level and individual data on quantities into a single aggregate quantity level so that the product of the price level times the quantity level equals the sum of the individual prices times the quantities for the commodities to be aggregated? How exactly to construct these aggregate levels is the index number problem in economics.

Although these volumes are primarily concerned with the aggregation over goods problem, some of the chapters touch on aspects of the aggregation over agents problem. The aggregation over agents problem asks the following question: under what conditions will price and quantity data, which are constructed by summing over economic agents, behave as if the aggregate data were the solution to a microeconomic optimization problem involving a single consumer or producer? On the aggregation over producers problem, the reader is referred to Gorman [1968b], F.M. Fisher [1965], Diewert [1980] and Blackorby and Schworm [1984]. On the aggregation over consumers problem, see Gorman [1953], Muellbauer [1975] [1976], Berndt, Darrough and Diewert [1977],

Lau [1977a] [1977b] [1982], Diewert [1983a], Jorgenson and Slesnick [1983] and Chapters 6 (Section 10) and 11 in this volume.

Returning to the aggregation over goods problem, it should be noted that this problem also encompasses two other aggregation problems: (i) the aggregation over time problem and (ii) the aggregation over space problem.[1] As Debreu [1959] noted many years ago, the definition of a commodity is flexible enough to encompass not only the "physical" characteristics of a good or service, but also its time and spatial characteristics; i.e., the same good sold at a different place or time can be regarded as a distinct commodity.[2] Triplett [1990a; 11–13] also stressed that different terms of sale can serve to make the same physical good into different commodities. However, for practical measurement purposes, we cannot take the "fundamental" unit of time or space to be too small, since the smaller we make the unit of time or space within which production or consumption takes place, the less actual production or consumption there will be to observe, and comparisons between these tiny units will become meaningless.[3] Thus for normal economic data, the time period under consideration is usually: (i) a shift (a part of a working day), (ii) a day, (iii) a week, (iv) a month, (v) a quarter, or (vi) a year. A normal "spatial" unit is usually: (i) an enterprise[4] or a household at a specific address or (ii) an aggregate of enterprises or households over a region. The region could be: (i) a county or municipality, (ii) a metropolitan region, (iii) a state or province, (iv) a country, or (v) a group of countries.

Once the fundamental units of time and space have been chosen, we typically aggregate production or consumption data on an individual "physical"

---

[1]This problem could be an aggregation over commodities problem or an aggregation over agents problem. If the same physical good is being sold by a single firm at several locations, then we have an aggregation over commodities problem.

[2]Alfred Marshall [1887; 373–374] seems to have been the first to appreciate that strawberries being made available at different times of the year or at different locations were separate commodities; see Chapter 2, Section 10 below.

[3]Thus if the fundamental unit of time or space is too small, production or consumption of most goods will be zero and comparisons between quantities and prices of the same good between adjacent periods will not be informative (this can be viewed as an example of the "new good" problem to be discussed later).

[4]An enterprise is usually defined to be the smallest production unit with a specific geographical address that can provide basic statistics on its inputs and outputs. The Standard Industrial Classification (SIC) attempts to develop criteria for grouping together enterprises. Triplett [1990a] [1991] criticized the "theory" behind these grouping attempts and provided his own criteria for grouping based on economic approaches to aggregation theory.

commodity i as follows: calculate the aggregate value V_i^t and the total number of units x_i^t of the good produced or consumed within period t. Then the microeconomic quantity of this good is taken to be x_i^t and the corresponding microeconomic price is defined to be $p_i^t \equiv V_i^t / x_i^t$. If we are given microeconomic price and quantity data (p_i^t, x_i^t) for T periods and the N commodities, so that $t = 1, 2, \ldots, T$ and $i = 1, 2, \ldots, N$, then to solve this particular aggregation over goods problem, we want T aggregate prices $P^1, P^2, \ldots, P^T$ and T aggregate quantities $Q^1, Q^2, \ldots, Q^T$ such that

$$(1) \qquad \sum_{i=1}^{N} p_i^t x_i^t = P^t Q^t \quad \text{for } t = 1, 2, \ldots, T.$$

Thus we want the aggregate value in period t, $P^t Q^t$, to equal the corresponding microeconomic value, $\sum_{i=1}^{N} p_i^t x_i^t$, for each time period t.[5] The aggregate period t price P^t is supposed to represent all of the period t microeconomic prices $p_1^t, p_2^t, \ldots, p_N^t$ in some sense and the aggregate period t quantity Q^t is supposed to represent all of the period t microeconomic quantities $x_1^t, x_2^t, \ldots, x_N^t$ in some sense. The index number problem is: how exactly are these aggregates P^t and Q^t to be constructed?[6]

There are two main approaches to index number theory: (i) the axiomatic approach and (ii) the economic approach. The difference between the two approaches can be explained as follows. Referring back to equation (1), denote the period t microeconomic price and quantity vectors as $p^t \equiv (p_1^t, \ldots, p_N^t)$ and $x^t \equiv (x_1^t, \ldots, x_N^t)$ respectively. In the axiomatic or test approach, the period t price and quantity levels, P^t and Q^t, are regarded as functions of the microeconomic price and quantity vectors, p^t and x^t, where p^t and x^t are *both* free to vary independently. Thus we have

$$(2) \qquad P^t \equiv P(p^t, x^t); \; Q^t \equiv Q(p^t, x^t) \quad \text{for } t = 1, 2, \ldots, T$$

where $P(p, x)$ and $Q(p, x)$ are each functions of $2N$ variables and $(p, x) \equiv (p_1, \ldots, p_N, \; x_1, \ldots, x_N)$. In the economic approach, the microeconomic price vectors p^t are regarded as *independent* variables, but the quantity vectors x^t are regarded as dependent variables; i.e., x^t is determined as a solution to some microeconomic optimization problem involving the observed price vector p^t.

---

[5]Of course, t could index locations instead of time periods, in which case we are attempting to construct comparable aggregates P^t, Q^t over space rather than time.

[6]Note that we do not attempt to determine how the goods to be aggregated are chosen. Triplett [1990a] [1991] addresses this question and suggests a number of economic approaches that could be used to choose the goods that are to be aggregated. We shall discuss three of his suggested approaches below (separability, Leontief aggregation, and Hicksian aggregation).

As a concrete example of the economic approach, suppose a consumer has preferences over differing amounts of N goods that can be represented by the linearly homogeneous and increasing utility function f where $u = f(x)$ is the utility or satisfaction level associated with the nonnegative consumption vector $x \equiv (x_1, \ldots, x_N) \geq 0_N \equiv (0, \ldots, 0)$. Given a strictly positive vector of consumer prices $p^t \gg 0_N$ for period t and an observed period t consumption vector $x^t \geq 0_N$, if the consumer is maximizing utility subject to a budget constraint or is minimizing the cost of achieving the utility level $u^t \equiv f(x^t)$ then x^t will solve the following cost (or expenditure) minimization problem:

$$\min_x \{p^t \cdot x : f(x) \geq u^t\} = c(p^t)f(x^t) \quad \text{for } t = 1, 2, \ldots, T, \tag{3}$$

where $p^t \cdot x \equiv \sum_{i=1}^N p_i^t x_i$ and $c(p^t)$ is the minimum cost of achieving one unit of utility; i.e., c is the unit cost function that corresponds to the linearly homogeneous utility function f. The equality in (3) follows from the linear homogeneity of f; see equation (2.17) in Chapter 6 below. Since x^t solves the minimization problem in (3) by assumption, we have

$$p^t \cdot x^t = \sum_{i=1}^N p_i^t x_i^t = c(p^t)f(x^t) \quad \text{for } t = 1, 2, \ldots, T. \tag{4}$$

Comparing equations (1) and (4), it is reasonable to identify the period t unit cost $c(p^t)$ with the price level aggregate P^t and the period t level of utility $f(x^t)$ with the quantity level Q^t; i.e., in the economic approach, we have:

$$P^t \equiv c(p^t); \; Q^t \equiv f(x^t) \quad \text{for } t = 1, 2, \ldots, T. \tag{5}$$

We are now in a position to give an outline of this overview.

In Section 2, we pursue the axiomatic approach in a bit more depth. In particular, we indicate how the functional forms for the functions P and Q which occur in (2) are determined.

Section 3 explains the theory behind the economic approach in more detail. Unfortunately, as one can see by inspecting equations (3)–(5) above, the economic theory of index numbers is often of limited use due to the unobservable nature of the functions which crop up; e.g., in (5), the utility function f and the corresponding unit cost function c cannot be observed directly. Thus in Section 4, we discuss how the basic theory behind the economic approach is modified and extended to yield "good" observable approximations to the unobservable theoretical constructs. It is a rather remarkable fact that at this point, the two major approaches to index number theory converge; i.e., it will turn out that "good" functional forms for P and Q in the axiomatic approach are also "good" functional forms from the viewpoint of the economic approach to index number theory.

Section 5 gives a brief overview of other approaches to index number theory while Section 6 explains the new good problem.

Section 7 discusses the problem of choosing a functional form for taking an average. For example, Irving Fisher [1922] used arithmetic, geometric and harmonic means as well as some other types of averages to construct an average price ratio. Which type of average should we use and what are the mathematical properties of these different kinds of averages?

Section 8 discusses the economic and mathematical tools that the various chapters in the book require.

Section 9 concludes with some personal observations.

2. The Axiomatic or Test Approach

Recall from the previous section that the vectors of period t microeconomic prices and quantities are $p^t \equiv (p^t, \ldots, p^t_N)$ and $x^t \equiv (x^t_1, \ldots, x^t_N)$ for $t = 1, 2, \ldots, T$. In the axiomatic approach, the prices and quantities, p^t_i and x^t_i, are regarded as independent variables and our index number problem is to somehow construct period t price and quantity aggregates, P^t and Q^t, such that $p^t \cdot x^t \equiv \sum_{i=1}^N p^t_i x^t_i = P^t Q^t$ for $t = 1, 2, \ldots, T$; i.e., such that equations (1) hold. It is (at first sight) natural to require that the price level P^t be a function of the components of p^t and that the quantity level Q^t be a function of the components of x^t; i.e., we define P^t and Q^t by

$$(6) \qquad P^t \equiv P(p^t); \; Q^t \equiv Q(x^t) \qquad \text{for } t = 1, 2, \ldots, T$$

where $P(p_1, \ldots, p_N)$ and $Q(x_1, \ldots, x_N)$ are functions of N variables that are to be determined somehow. Note that in equations (6), the functions P and Q are assumed to be the same for all T periods (but, of course, the individual price and quantity vectors p^t and x^t usually change from period to period).

At this point, we have to distinguish a number of separate branches of the axiomatic approach to index number theory. The reader will note that equations (6) are different from our earlier equations (2); in (6), the price function $P(p^t)$ depends only on the period t price vector p^t while in (2), the price function $P(p^t, x^t)$ depended on the period t price and quantity vectors, (p^t, x^t). Thus equations (2) and (6) give rise to different branches of the axiomatic approach to index number theory.

Yet another branch is obtained by setting $T = 2$ and by asking that the price aggregates for periods 1 and 2, P^1 and P^2, and the quantity aggregates, Q^1 and Q^2, satisfy the following equations:

$$(7) \qquad P^2/P^1 = P(p^1, p^2, x^1, x^2); \quad Q^2/Q^1 = Q(p^1, p^2, x^1, x^2).$$

The functions P and Q which occur in (7) are now functions of the $4N$ micro prices and quantities that pertain to periods 1 and 2. Note that P^2/P^1 is to be interpreted as (one plus) the *aggregate growth rate in prices* going from period 1 to 2 and Q^2/Q^1 is to be interpreted as (one plus) the *aggregate growth rate in quantities* going from period 1 to 2. The forms for the functions P and Q which occur in (7) are to be determined by this branch of the axiomatic approach to index number theory. This branch is known as *bilateral index number theory* since it aggregates price and quantity data over only two periods.

Finally, in *multilateral index number theory,* we ask that the price aggregates P^t and quantity aggregates Q^t satisfy the following equations:

$$(8)\qquad \begin{aligned} P^t &= F^t(p^1,\ldots,p^T,x^1,\ldots,x^T) \qquad t=1,2,\ldots,T\\ Q^t &= G^t(p^1,\ldots,p^T,x^1,\ldots,x^T) \qquad t=1,2,\ldots,T \end{aligned}$$

where now the $2T$ functions F^t and G^t to be determined depend in principle on all of the micro prices and quantities over all of the periods.

Thus there are separate axiomatic approaches to index number theory that are centered around each of the equations (2), (6), (7) and (8). Most of the analysis of the axiomatic approach to index number theory to be presented in this book will be on the bilateral approach (7) with some coverage of the multilateral approach (8).

For material on the bilateral approach, see Chapter 2, Section 4; Chapter 5, Section 2; Chapter 12, Section 3; and Chapter 13, Sections 2 and 3.

For material on the multilateral approach, see Chapter 2, Sections 6.3 and 8; Chapter 5, Sections 6 and 9; and Chapter 12.

None of the chapters in this volume utilize the approaches given by equations (2) and (6). In this introduction, we explain why this is so.

First consider the case represented by equations (6). Recall that we want our aggregates P^t and Q^t to satisfy equations (1) as well. Substitution of (6) into (1) yields:

$$(9)\qquad P(p^t)Q(x^t) = \sum_{i=1}^{N} p_i^t x_i^t, \quad t=1,2,\ldots,T.$$

Simply by examining equations (9), it can be seen that if the number of goods N is equal to or greater than two, then it is *impossible* for functions P and Q to exist which will satisfy (9) for all $p^t \equiv (p_1^t,\ldots,p_N^t) \geq 0_N$ and $x^t \equiv (x_1^t,\ldots,x_N^t) \geq 0_N$.[7] Thus our analysis of this particular branch of the axiomatic approach to index number theory comes to an abrupt end.

Now consider the branch of the axiomatic approach represented by equations (1) and (2). Substitution of (2) into (1) yields:

$$(10)\qquad P(p^t,x^t)Q(p^t,x^t) = \sum_{i=1}^{N} p_i^t x_i^t \equiv p^t\cdot x^t \qquad \text{for } t=1,\ldots,T.$$

[7]This result is due to Eichhorn [1978b; 144].

What properties should the price level function $P(p, x)$ and the quantity level function $Q(p, x)$ have? It seems reasonable to demand that $P(p, x)$ be linearly homogeneous in the components of the price vector $p \equiv (p_1, \ldots, p_N)$ so that if all prices increase by a common factor λ, then so will the price level; i.e., we assume that the function P satisfies the following *test* or *axiom*:

$$(11) \qquad P(\lambda p, x) = \lambda P(p, x) \quad \text{for all } \lambda > 0,\ p \gg 0_N \text{ and } x \gg 0_N.$$

Similarly, it seems reasonable to demand that $Q(p, x)$ be linearly homogeneous in its quantity variables x; i.e., we assume that Q satisfies the following axiom:

$$(12) \qquad Q(p, \lambda x) = \lambda Q(p, x) \quad \text{for all } \lambda > 0,\ p \gg 0_N \text{ and } x \gg 0_N.$$

It is also reasonable to ask that the aggregate price and quantity levels be positive if all microeconomic prices and quantities are positive. Thus we assume that the functions P and Q satisfy:

$$(13) \qquad P(p, x) > 0;\ Q(p, x) > 0 \quad \text{if } p \gg 0_N \text{ and } x \gg 0_N.$$

In order that (10) hold for all possible price and quantity vectors p and x, we require that P and Q jointly satisfy the following axiom (the *product test*):

$$(14) \qquad P(p, x)Q(p, x) = p \cdot x \quad \text{for } p \gg 0_N \text{ and } x \gg 0_N.$$

Note that (13) and (14) imply that the functions P and Q cannot be determined independently; e.g., if we have determined Q, then P is determined as $P(p, x) \equiv p{\cdot}x/Q(p, x)$. Thus property (12) for Q implies that $P(p, x)$ will be homogeneous of degree 0 in the elements of x; i.e., if P and Q satisfy (12), (13) and (14), then for $p \gg 0_N$, $x \gg 0_N$ and $\lambda > 0$, we have:

$$(15) \qquad \begin{aligned} P(p, \lambda x) &= p \cdot \lambda x / Q(p, \lambda x) && \text{using (13) and (14)} \\ &= \lambda p \cdot x / \lambda Q(p, x) && \text{by (12)} \\ &= p \cdot x / Q(p, x) \\ &= P(p, x) && \text{using (13) and (14).} \end{aligned}$$

Since units of measurement in economics are quite arbitrary (e.g., purchases of gasoline could be measured in litres or gallons), it would be very useful if our price and quantity aggregates were *invariant* to changes in the units of measurement. Thus our last test that we impose on the price function $P(p, x)$ is the following one:

$$(16) \quad P(d_1 p_1, \ldots, d_N p_N,\ x_1/d_1, \ldots, x_N/d_N) = P(p_1, \ldots, p_N,\ x_1, \ldots, x_N)$$
$$\text{for all } p_i > 0,\ x_i > 0 \text{ and } d_i > 0 \text{ for } i = 1, \ldots, N.$$

Test or axiom (16) can be written more compactly using matrix notation as $P(Dp, D^{-1}x) = P(p, x)$ for all diagonal matrices D with positive elements on the main diagonal where D^{-1} denotes the inverse of the matrix D.

We can now show that the above properties on the functions P and Q are inconsistent.

PROPOSITION 1. *Properties (11), (13), (15) and (16) on P are inconsistent; i.e., there does not exist a function of $2N$ variables, $P(p_1, \ldots, p_N, x_1, \ldots, x_N)$, which satisfies these properties.*

Proof. Applying (16) with $d_i = x_i$ for $i = 1, \ldots, N$, we obtain the following equation:

$$(17) \qquad P(p_1, \ldots, p_N, x_1, \ldots, x_N) = P(p_1 x_1, \ldots, p_N x_N, 1, \ldots, 1).$$

If P satisfies the linear homogeneity in prices property (11) so that $P(\lambda p, x) = \lambda P(p, x)$, then (17) implies that P is also linearly homogeneous in x; i.e., (11) and (17) imply that $P(p, \lambda x) = \lambda P(p, x)$ and hence (15) cannot hold. Q.E.D.

An impossibility result similar to that given in Proposition 1 was obtained by Eichhorn [1978b; 144–145] who used monotonicity axioms on P and Q in place of our positivity axiom (13).[8]

In view of the impossibility theorem in Proposition 1, it does not seem fruitful to pursue the axiomatic model defined by equations (2) any further. Thus we have eliminated our first two axiomatic models from further consideration and can now turn to the bilateral model defined by equations (7).

In order for (7) to be consistent with (1), we shall require that the functions P and Q satisfy the following equation (the *product test* or the *weak factor reversal test*):

$$(18) \qquad P(p^1, p^2, x^1, x^2) Q(p^1, p^2, x^1, x^2) = p^2 \cdot x^2 / p^1 \cdot x^1.$$

If (18) holds (and if P and Q are nonzero), then it can be seen that once we have determined either P or Q, then the remaining function is determined by rearranging (18). For example, suppose we have pinned down the functional form for P; then $Q(p^1, p^2, x^1, x^2)$ is determined as the function $p^2 \cdot x^2 / p^1 \cdot x^1 P(p^1, p^2, x^1, x^2)$. Thus, given that the product test (18) holds, if we require that $P(p^1, p^2, x^1, x^2)$ satisfies enough tests or properties so that the functional form for P is determined, then the functional form for Q will also be determined.

How are these properties for P determined; i.e., what are "reasonable" properties that P should possess? One way of finding such reasonable properties is to consider what the mathematical properties of P are when $N = 1$; i.e., when there is only one good and there is no aggregation over commodities problem. In the one good case, $P(p_1^1, p_1^2, x_1^1, x_1^2)$ must equal the ratio of the period 2 price to the period 1 price, p_1^2 / p_1^1. Note that this last function has the

[8]Eichhorn assumed that $P(p, x)$ was increasing in each component of p and $Q(p, x)$ was increasing in each component of x. For more on the history of this axiomatic approach, see footnote 15 in Chapter 13.

following homogeneity properties: it is homogeneous of degree 1 in p_1^2, homogeneous of degree -1 in p_1^1, homogeneous of degree 0 in x_1^1, and homogeneous of degree 0 in x_1^2. Thus it seems "reasonable" to impose these same homogeneity properties on P when we are aggregating over N goods. This leads to the following homogeneity tests or axioms for P: for $p^1 \gg 0_N$, $p^2 \gg 0_N$, $x^1 \gg 0_N$, $x^2 \gg 0_N$ and $\lambda > 0$:

$$P(p^1, \lambda p^2, x^1, x^2) = \lambda P(p^1, p^2, x^1, x^2); \tag{19}$$
$$P(\lambda p^1, p^2, x^1, x^2) = \lambda^{-1} P(p^1, p^2, x^1, x^2); \tag{20}$$
$$P(p^1, p^2, \lambda x^1, x^2) = P(p^1, p^2, x^1, x^2); \tag{21}$$
$$P(p^1, p^2, x^1, \lambda x^2) = P(p^1, p^2, x^1, x^2). \tag{22}$$

Note that when $N = 1$, the single price ratio p_1^2/p_1^1 is invariant to changes in the units of measurement. In the case of a general N, it seems "reasonable" to ask that P also be invariant to changes in the units of measurement, so that

$$P(Dp^1, Dp^2, D^{-1}x^1, D^{-1}x^2) = P(p^1, p^2, x^1, x^2) \tag{23}$$

where D is any diagonal matrix with positive elements on the main diagonal.

At this point, the reader should be able to get the gist of the bilateral test approach to index number theory. The task of this approach is to specify a number of "reasonable" tests or axioms which are sufficient to determine a unique functional form for P. At the same time, we cannot be too ambitious and specify too many desirable properties so that an impossibility theorem results of the type that was derived for the axiomatic models (2) and (6). The longest set of mutually consistent axioms for the bilateral test approach that we have been able to find appears in Sections 2 and 3 of Chapter 13 below. There we find that the Fisher [1922] ideal price index defined as

$$P_F(p^1, p^2, x^1, x^2) \equiv (p^2 \cdot x^1 p^2 \cdot x^2 / p^1 \cdot x^1 p^1 \cdot x^2)^{\frac{1}{2}} \tag{24}$$

satisfies some 20 "reasonable" tests. However, before turning to these sections, the reader should read Section 4 of Chapter 2 which lays out the early history of the test approach.

The multilateral test approach has not been as well developed as the bilateral approach. A dominant set of tests which uniquely determines the functions F^t and G^t which appear in (8) has not yet been presented in the literature. In this volume, aspects of the multilateral approach to index number theory are developed in Chapter 5, Sections 6 and 9, and in Chapter 12.

3. The Economic Approach: Theory

The economic theory of index numbers is based on the definitions of two constrained optimization problems. One of these optimization problems leads to the Konüs [1924] price index and the other leads to the Malmquist [1953] quantity index.

We first consider the economic theory of price indexes. Consider a consumer who wants to minimize the cost of achieving a certain utility level or a producer who wants to minimize the cost of achieving a certain output level. For the sake of definiteness, we will use the language of utility theory in what follows.[9] Thus let $x \equiv (x_1, \ldots, x_N) \geq 0_N$ be a nonnegative consumption vector, let $y \equiv (y_1, \ldots, y_M)$ be a vector of "other" consumer variables that matter, and let F be the consumer's utility or preference function so that $u = F(x, y)$ denotes the level of utility that the consumer achieves if he or she consumes x and y. Let $p \equiv (p_1, \ldots, p_N) \gg 0_N$ be a positive vector of prices for the goods in the x vector and consider the following (conditional on $\bar{x}, \bar{y}$) *cost minimization problem*:

$$C(p, \bar{x}, \bar{y}) \equiv \min_{x \geq 0_N} \{p \cdot x : F(x, \bar{y}) \geq F(\bar{x}, \bar{y})\}. \tag{25}$$

Thus in (25), the consumer minimizes the commodity cost $p \cdot x \equiv \sum_{i=1}^{N} p_i x_i$ of the goods in the x vector required to achieve the reference utility level $\bar{u} \equiv F(\bar{x}, \bar{y})$, given that the consumer's y vector is fixed at $\bar{y}$. Note that $C(p, \bar{x}, \bar{y})$ is linearly homogeneous in the elements of p; i.e., for $\lambda > 0$, we have:

$$C(\lambda p, \bar{x}, \bar{y}) = \lambda C(p, \bar{x}, \bar{y}). \tag{26}$$

If $p^t \gg 0_N$ is the observed vector of prices for x goods that the consumer faces in period t for $t = 1, 2$, then the Konüs [1924] conditional *price index*[10] for the x goods for period 2 relative to period 1 is defined as

$$P_K(p^1, p^2, \bar{x}, \bar{y}) \equiv C(p^2, \bar{x}, \bar{y}) / C(p^1, \bar{x}, \bar{y}). \tag{27}$$

---

[9]To get the producer interpretation in what follows, let F be the producer's production function, let x be an input quantity vector, let y be a vector of "other" variables, and let u be the output target.

[10]In the original formulation of Konüs, the y vector did not appear in F, so that $F(x, y)$ was replaced by $F(x)$ and $C(p, \bar{x}, \bar{y})$ was replaced by $C(p, \bar{x})$. The more general formulation that we are outlining now is due to Pollak [1975] at this level of generality. However, a special case of the present theory was worked out by Shephard [1953; 61–71] [1970; 145–146] (the homogeneously separable case). See also Solow [1955–56], Green [1964], Geary and Morishima [1973] and Arrow [1974].

By using the homogeneity equation (26), it can be seen that $P_K(p^1, p^2, \bar{x}, \bar{y})$ defined by (27) is homogeneous of degree 1 in p^2 and homogeneous of degree -1 in p^1, which are counterparts to the homogeneity properties (19) and (20) in the test approach to index number theory.

If the y vector does not actually appear in the consumer's utility function, then the reference vectors $\bar{x}$ and $\bar{y}$ in (27) can be replaced by the reference utility level $\bar{u} \equiv F(\bar{x})$, $C(p, \bar{x}, \bar{y})$ can be replaced by $C^*(p, \bar{u}) \equiv \min_{x \geq 0_N}\{p \cdot x : F(x) \geq \bar{u}\}$,[11] and $P_K(p^1, p^2, \bar{x}, \bar{y})$ can be replaced by

$$(28) \qquad P_K^*(p^1, p^2, \bar{u}) \equiv C^*(p^2, \bar{u})/C^*(p^1, \bar{u}).$$

If, in addition, F is linearly homogeneous so that $F(\lambda x) = \lambda F(x)$ for $x \geq 0_N$ and $\lambda > 0$, then $C^*(p^t, \bar{u}) = C^*(p^t, 1)\bar{u} \equiv c(p^t)\bar{u}$ and (28) reduces to $P_K^*(p^1, p^2, \bar{u}) = c(p^2)/c(p^1)$. Thus in this case, we end up with the same economic indexes that we had earlier in equations (4) and (5).

Another way that we can end up with (4) and (5) from the general model (27) is to assume that the consumer's utility function has the following structure:

$$(29) \qquad F(x, y) = F^*[f(x), y] \quad \text{where } f(\lambda x) = \lambda f(x) \text{ for } \lambda > 0.$$

We also require that F^* be increasing in its first argument. If F has the structure given by (29), then we say that there is a ***homogeneously separable aggregate*** in x; note that the micro aggregator function f is linearly homogeneous. With F having the structure given by (29), the conditional cost minimization problem (25) becomes:

$$\begin{aligned} C(p, \bar{x}, \bar{y}) &\equiv \min_{x \geq 0_N}\{p \cdot x : F^*[f(x), \bar{y}] \geq F^*[f(\bar{x}), \bar{y}]\} \\ &= \min_{x \geq 0_N}\{p \cdot x : f(x) \geq f(\bar{x})\} \\ &\quad \text{since } F^* \text{ is increasing in its first variable} \\ (30) \qquad &= c(p)f(\bar{x}) \end{aligned}$$

where $c(p) \equiv \min_x\{p \cdot x : f(x) \geq 1\}$ is the unit cost function corresponding to f and (30) follows from the linear homogeneity of f. Thus under (29), the Konüs price index (27) again reduces to $c(p^2)/c(p^1)$; this follows by substituting (30) into (27). This implication of (29) is essentially due to Shephard [1953; 61–71] who made the same assumptions in the context of production theory.

---

[11]The mathematical properties of C^* are laid out in Chapter 6, Section 2. Since the conditional cost function C can be regarded as the negative of a conditional profit function, the mathematical properties of C can be determined by studying Section 11 of Chapter 6.

Thus under a variety of assumptions, the rather complicated theoretical price index defined by (27) reduces to the unit cost ratio $c(p^2)/c(p^1)$ which is consistent with our initial economic approach given by equations (4) and (5).

We turn now to the economic theory of quantity indexes. Again, let the consumer's utility be defined by $u = F(x,y)$. Consider first the following constrained maximization problem involving a single positive variable λ:

$$(31) \qquad D(x,\bar{x},\bar{y}) = \max_{\lambda \geq 0}\{\lambda : F(\lambda^{-1}x,\bar{y}) \geq F(\bar{x},\bar{y})\}.$$

In the maximization problem (31), we take a given quantity vector $x \equiv (x_1, \ldots, x_N)$ and deflate each component by the positive number λ, obtaining the λ deflated vector $\lambda^{-1}x \equiv (x_1/\lambda, x_2/\lambda, \ldots, x_N/\lambda)$. We choose this deflation factor λ to be as large as possible subject to the constraint that the deflated vector $\lambda^{-1}x$ when combined with the fixed reference vector $\bar{y}$ yields just as much utility as that yielded by the fixed reference vectors $\bar{x}$ and $\bar{y}$. This maximal deflation factor defines the *distance* or *deflation function* $D(x,\bar{x},\bar{y})$.

Note that $D(x,\bar{x},\bar{y})$ is linearly homogeneous in the components of x; i.e., we have

$$(32) \qquad D(\lambda x,\bar{x},\bar{y}) = \lambda D(x,\bar{x},\bar{y}) \quad \text{for } \lambda > 0.$$

If $F(x,y)$ is increasing in the components of x, then by inspecting (31), it can be seen that $D(x,\bar{x},\bar{y})$ will also be increasing in the components of x. This monotonicity property and the homogeneity property (32) lead us to use the deflation function D to define the following conditional Malmquist [1953] *quantity index* for the x goods:

$$(33) \qquad Q_M(x^1,x^2,\bar{x},\bar{y}) \equiv D(x^2,\bar{x},\bar{y})/D(x^1,\bar{x},\bar{y})$$

where x^t is the consumer's observed period t quantity vector for the x goods for $t = 1,2$.

By using the homogeneity property (32), it can be seen that $Q_M(\lambda x^1,x^2,\bar{x},\bar{y}) = \lambda^{-1}Q_M(x^1,x^2,\bar{x},\bar{y})$ and $Q_M(x^1,\lambda x^2,\bar{x},\bar{y}) = \lambda Q_M(x^1,x^2,\bar{x},\bar{y})$ for $\lambda > 0$. Thus our economic quantity index has homogeneity properties with respect to x^1 and x^2 that correspond to the homogeneity properties (21) and (22) in the test approach.

If the y vector does not actually appear in the consumer's utility function (this is the case considered by Malmquist [1953]), then the reference vectors $\bar{x}$ and $\bar{y}$ can be replaced by the reference utility level and $\bar{u} \equiv F(\bar{x})$, $D(x,\bar{x},\bar{y})$ can be replaced by $D^*(x,\bar{u}) \equiv \max_{\lambda \geq 0}\{\lambda : F(\lambda^{-1}x) \geq \bar{u}\}$, and $Q_M(x^1,x^2,\bar{x},\bar{y})$ can be replaced by

$$(34) \qquad Q_M^*(x^1,x^2,\bar{u}) \equiv D^*(x^2,\bar{u})/D^*(x^1,\bar{u}).$$

This is the definition of the Malmquist quantity index that we utilize in Chapters 5, 7 and 11 below.[12]

If, in addition to not depending on y, $F(x)$ is linearly homogeneous so that $F(\lambda x) = \lambda F(x)$ for $\lambda > 0$, then for $\bar{u} > 0$ we have:

$$\begin{aligned} D^*(x,\bar{u}) &\equiv \max_{\lambda\geq 0}\{\lambda : F(\lambda^{-1}x) \geq \bar{u}\} \\ &= \max_{\lambda\geq 0}\{\lambda : \lambda^{-1}F(x) \geq \bar{u}\} \\ &= \max_{\lambda\geq 0}\{\lambda : \lambda \leq F(x)/\bar{u}\} \\ &= F(x)/\bar{u}. \end{aligned} \tag{35}$$

Thus under the assumptions that $F(x,y)$ does not depend on y and $F(x)$ is linearly homogeneous in x, the Malmquist quantity index (33) or (34) reduces to

$$Q_K^*(x^1,x^2,\bar{u}) = [F(x^2)/\bar{u}]/[F(x^1)/\bar{u}] = F(x^2)/F(x^1) \tag{36}$$

which is again consistent with our earlier economic model defined by (4) and (5).

Another way that we can end up with (4) and (5) is to assume that $F(x,y)$ is homogeneously separable in x; i.e., assume F satisfies (29). Under these conditions, the distance function D defined by (31) becomes:

$$\begin{aligned} D^*(x,\bar{x},\bar{y}) &\equiv \max_{\lambda\geq 0}\{\lambda : F^*[f(\lambda^{-1}x),\bar{y}] \geq F^*[f(\bar{x}),\bar{y}]\} \\ &= \max_{\lambda\geq 0}\{\lambda : f(\lambda^{-1}x) \geq f(\bar{x})\} \\ &\quad \text{since } F^* \text{ is increasing in its first variable} \\ &= \max_{\lambda\geq 0}\{\lambda : \lambda^{-1}f(x) \geq f(\bar{x})\} \\ &\quad \text{using the linear homogeneity of } f \\ &= \max_{\lambda\geq 0}\{\lambda : \lambda \leq f(x)/f(\bar{x})\} \\ &\quad \text{assuming that } f(\bar{x}) > 0 \\ &= f(x)/f(\bar{x}). \end{aligned} \tag{37}$$

Substituting (37) into (33) yields $Q_M(x^1,x^2,\bar{x},\bar{y}) = f(x^2)/f(x^1)$ so that, in this homogenously separable case, the Malmquist quantity index is independent of the reference vectors $\bar{x}$ and $\bar{y}$ and is consistent with our preliminary economic model defined by equations (4) and (5) above.

[12]The mathematical properties of D^* are studied in Section 5 of Chapter 6.

Note that the optimization problem (31) that defined the distance function D did not involve consumer prices. However, prices soon make their appearance when we attempt to find observable bounds on the theoretical Malmquist indexes defined by (33) or (34); see Chapter 5, Section 4; Chapter 7, Section 3; and Chapter 11, Section 6.

We have not specified how the reference vectors $\bar{x}$ and $\bar{y}$, which appear in the definition of the Konüs price index (27) and the Malmquist quantity index (33), are chosen. In practical situations, we generally choose $\bar{x}, \bar{y}$ to be either (x^1, y^1) or (x^2, y^2) or an average of these two vectors which are the observed data pertaining to periods 1 and 2. In the following section, we shall indicate how these choices lead to observable bounds on the theoretical economic indexes.

Triplett [1990a] [1991] has recently suggested that economic theory could be helpful in deciding how industrial, commodity and occupational classifications could be set up. One of the methods he suggested was essentially based on (29), the assumption that a homogeneously separable aggregate exists on the producer side of the economy[13] (two of the other methods that he suggested, Hicksian and Leontief aggregation, we shall discuss shortly). In order to implement Triplett's suggestion, we need methods for determining whether separable aggregates exist; i.e., we need methods for testing whether assumption (29) is valid. In the early literature on testing for the existence of a separable aggregate,[14] researchers generally assumed that under the alternative hypothesis of nonseparability, the function $F(x, y)$ on the left hand side of (29) was a flexible functional form[15] and they then found restrictions on the parameters of F which collapsed $F(x, y)$ down into the null hypothesis form, $F^*[f(x), y]$. However, Blackorby, Primont and Russell [1977c] [1978] showed that for commonly used flexible functional forms for F (such as the translog), under the null hypothesis of homogeneous separability, the resulting F^* or f (or both) would necessarily be inflexible. This problem with the early separability tests has been overcome; see Hall [1973], Woodland [1978], Blackorby, Schworm and Fisher [1986], and Diewert and Wales [1991]. However, the results of this later

---

[13]Actually Triplett [1990a] assumed only a separable structure (not a homogeneously separable structure) so that (29) holds but $f(x)$ is not necessarily linearly homogeneous. In this case, the economic price index defined by (27) does simplify to (28) but it does *not* simplify to a ratio of unit cost functions, $c(p^2)/c(p^1)$. The type of separability assumed by Triplett is studied at great length in Blackorby, Primont and Russell [1978].

[14]See Berndt and Christensen [1973b] [1974], Berndt and Wood [1975], Jorgenson and Lau [1975], and Denny and Fuss [1977].

[15]A flexible functional form is one that can provide a second order approximation to an arbitrary twice continuously differentiable functional form; see Chapter 6, Section 10.

separability testing literature are not terribly encouraging from the viewpoint of finding an effective economic classification mechanism: the available empirical results generally *reject* the assumption that a separable aggregate exists.

There is one additional theoretical topic which arises in the economic approach to index numbers: namely the role of price or quantity proportionality as an aggregating mechanism.

Consider first the role of price proportionality. Suppose that the consumer or producer is solving a cost minimization problem like (25) for T periods. Assume that the observed x vector in period t, $x^t \equiv (x_1^t, \ldots, x_N^t)$, solves the following cost minimization problem:

$$(38) \qquad \min_x \{p^t \cdot x : F(x, y^t) \geq F(x^t, y^t) = u^t\} \equiv \widetilde{C}(p^t, y^t, u^t)$$

for $t = 1, 2, \ldots, T$ where (p^t, x^t, y^t) is the observed price and quantity data pertaining to the consumer or producer in period t. Suppose that in addition to the above assumption of cost minimizing behavior on the part of the consumer or producer, the prices $p^t \equiv (p_1^t, \ldots, p_N^t)$ vary in strict proportion over time so that

$$(39) \qquad p^t = \lambda^t \alpha, \qquad t = 1, 2, \ldots, T,$$

where $\lambda^t > 0$ is the period t proportionality factor and $\alpha \equiv (\alpha_1, \ldots, \alpha_N)$ is a fixed vector. Under these conditions, it is natural to choose the period t price level for the x goods P^t to be λ^t and, of course, the corresponding aggregate quantity level Q^t will have to equal $p^t \cdot x^t / \lambda^t$ so that the adding up relations (1) hold. Thus define

$$(40) \qquad P^t \equiv \lambda^t; \; Q^t \equiv p^t \cdot x^t / \lambda^t \quad \text{for } t = 1, 2, \ldots, T.$$

Substituting (39) into (38) yields for $t = 1, \ldots, T$:

$$(41) \qquad p^t \cdot x^t = \widetilde{C}(p^t, y^t, u^t) = \widetilde{C}(\lambda^t \alpha, y^t, u^t) = \lambda^t \widetilde{C}(\alpha, y^t, u^t)$$

where the last equality follows from the linear homogeneity of $\widetilde{C}(p, y, u)$ in p. Comparing (40) and (41), we have

$$(42) \qquad Q^t = \widetilde{C}(\alpha, y^t, u^t), \quad t = 1, 2, \ldots, T.$$

Now consider the following sequence of cost minimization problems involving the scalar aggregate Q:

$$(43) \qquad \min_Q \{P^t Q : Q = F(y^t, u^t)\}, \quad t = 1, 2, \ldots, T,$$

where the original micro function F has now been replaced by the aggregate commodity requirements function $\widetilde{F}$ defined by

$$\widetilde{F}(y,u) \equiv \widetilde{C}(\alpha,y,u). \tag{44}$$

Using (42) and (44), it can be seen that the Q^t defined in (40) solves the tth minimization problem in (43) for $t = 1,2,\ldots,T$ and we have $P^tQ^t = p^t \cdot x^t$ for $t = 1,\ldots,T$. Thus if the prices of a group of goods vary in strict proportion over time, then there exist aggregate prices and quantities, P^t and Q^t, defined by (40) and the aggregate quantity Q^t can be treated as if it were an actual microeconomic good. This is a rough and ready version of Hicks' [1946; 312–313] Aggregation Theorem. For a more detailed exposition of this result in both the consumer and producer contexts, see Chapter 15 below. Chapter 16 below spells out some implications of Hicks' Aggregation Theorem for elasticities of substitution. Chapter 16 also considers what happens if prices vary only *approximately* in fixed proportions rather than *exactly* proportionally.

Now consider the role of quantity proportionality. Suppose that $F(x,y)$ is the consumer's utility function and that the observed quantity data pertaining to the consumer for period t are (x^t,y^t) for $t = 1,2,\ldots,T$. The period t utility level is u^t defined as

$$u^t \equiv F(x^t,y^t), \quad t = 1,\ldots,T. \tag{45}$$

Suppose that the quantity vectors $x^t \equiv (x_1^t,\ldots,x_N^t)$ vary in strict proportion over time so that we have

$$x^t \equiv \lambda^t\beta, \quad t = 1,\ldots,T, \tag{46}$$

where $\beta \equiv (\beta_1,\ldots,\beta_N)$ is a fixed vector and $\lambda^t > 0$ is the period t factor of proportionality. Define an aggregate utility function $\widetilde{F}$ for an aggregate Q of the x goods and an arbitrary y vector as follows:

$$\widetilde{F}(Q,y) \equiv F(Q\beta,y). \tag{47}$$

Under suitable regularity conditions on F, the aggregate function $\widetilde{F}$ will inherit its properties. If we define the period t aggregate Q^t to be

$$Q^t \equiv \lambda^t, \quad t = 1,\ldots,T, \tag{48}$$

then the aggregated data (Q^t,y^t) will be consistent with the micro data (x^t,y^t) since

$$\begin{aligned} u^t &= F(x^t,y^t) && \text{by (45)} \\ &= F(\lambda^t\beta,y^t) && \text{by (46)} \\ &= \widetilde{F}(\lambda^t,y^t) && \text{by (47)} \\ &= \widetilde{F}(Q^t,y^t) && \text{by (48).} \end{aligned} \tag{49}$$

This is Leontief's [1936] Aggregation Theorem in the consumer context. Of course, a similar result holds in the producer context. If p^t is the micro period t price vector for the x goods, then the aggregate period t price P^t for the x aggregate can be defined as

$$P^t \equiv p^t \cdot x^t / Q^t, \quad t = 1, \ldots, T. \tag{50}$$

Leontief's Aggregation Theorem makes an appearance in Chapter 10 below. In that chapter, Allen and Diewert attempt to determine empirically whether proportional quantity variation or proportional price variation is more justified. They suggest that if quantities vary proportionally, then the aggregates should be constructed using equations (48) and (50), while if prices vary proportionally, then equations (40) should be used to construct the aggregates.

Leontief and Hicksian aggregation can be regarded as special cases of the homogeneous separability model (29) above. For Leontief type aggregation, the micro aggregator function f has the following functional form:

$$f(x_1, \ldots, x_N) \equiv \min_i \{x_i/\beta_i : i = 1, 2, \ldots, N\}. \tag{51}$$

For Hicksian aggregation, the micro aggregator function f has the following functional form:

$$f(x_1, \ldots, x_N) \equiv \alpha_1 x_1 + \cdots + \alpha_N x_N. \tag{52}$$

Thus the aggregation theorems of Leontief and Hicks are consistent with the assumption of homogeneous separability where the micro aggregator function is either a fixed coefficients, no substitution Leontief aggregator function or it is a linear, infinite substitution aggregator function. However, these aggregation theorems can be given a broader interpretation: (i) in planned economies, some groups of commodities may be produced in fixed proportions by central decree irrespective of the technology, and (ii) in market economies, some groups of prices may move proportionally due to regulation or due to union wage policies.

In this section, we have given a brief outline of how economic theory can be used in order to define a theoretical Konüs price index of the form (27) or (28) and to define a theoretical Malmquist quantity index of the form (33) or (34). However elegant the theory of these indexes may be, there is a practical problem associated with the empirical use of these indexes. The problem is that the theoretical price index depends on knowing the consumer's or producer's *cost function* and the theoretical quantity index depends on knowing the economic agent's *distance* or *deflation function* and we, as outside observers, do not have this knowledge. In the following section, we turn to a description of some of the methods which have been suggested to remedy this knowledge deficiency.

4. The Economic Approach: Empirical Approximations

There are at least three different ways to operationalize the theoretical indexes defined in the previous section: (i) econometric estimation; (ii) nonparametric bounds; and (iii) the theory of exact index numbers. We shall consider each of these possibilities in turn in this section.

Consider first an approach that relies on econometric estimation. This approach is relatively straightforward: given time series or cross section data on production units or households, we can postulate a functional form for the cost function C or the aggregator function F and estimate the unknown parameters which appear in the functional form by regression analysis. Typically, we choose functional forms for C or F that are flexible; i.e., the functional form has a sufficient number of free parameters so that it can provide a second order approximation to an arbitrary cost or aggregator function (with the appropriate regularity conditions).[16] This econometric approach is discussed in Section 6.4 of Chapter 2 and in Sections 10 and 11 of Chapter 6 below.

The problem with the econometric approach that relies on flexible functional forms is that it becomes unwieldy or impossible[17] as the number of goods N to be aggregated becomes large, since the number of unknown parameters to be estimated grows at a rate approximately equal to $(1/2)N^2$. This leads us to discuss the second approach for implementing the economic theory of index numbers, the method of bounds.

To show how the bounds method works, consider the tth cost minimization problem in (38) above and suppose that the observed x vector for period t, x^t, solves this problem for $t = 1, 2$. We suppose that we can observe the data (p^t, x^t, y^t) for $t = 1, 2$. We shall use the vectors (x^t, y^t) for $t = 1, 2$ as the reference vectors $(\bar{x}, \bar{y})$ in (27) which defines the theoretical Konüs price index $P_K(p^1, p^2, \bar{x}, \bar{y})$. Consider the following cost minimization problem:

$$(53) \qquad C(p^2, x^1, y^1) \equiv \min_x \{p^2 \cdot x : F(x, y^1) \geq F(x^1, y^1)\} \leq p^2 \cdot x^1$$

[16] Recent work on functional forms has moved beyond the local approximation idea that is inherent in the definition of a flexible functional form and has attempted to provide some global approximations; e.g., see the seminonparametric functional form literature started by Gallant [1981] [1982], Barnett and Jonas [1983], Barnett and Yue [1988], and Barnett, Geweke and Wolfe [1991]. See also the attempts by Diewert and Wales [1992a] [1992b] to provide functional forms that are flexible but at the same time can approximate arbitrary Engel curves (in the consumer context) or input expansion paths (in the producer context).

[17] This impossibility result will be discussed in more detail in Volume II; see Diewert [1992a].

where the inequality in (53) follows since x^1 is a feasible (but not necessarily optimal) solution to the cost minimization problem in (53). Similarly, we have:

$$(54) \qquad C(p^1, x^2, y^2) \equiv \min_x \{p^1 \cdot x : F(x, y^2) \geq F(x^2, y^2\} \leq p^1 \cdot x^2$$

where the inequality in (54) follows since x^2 is a feasible solution for the cost minimization problem in (54). If $C(p^1, x^2, y^2) > 0$ and $p^1 \cdot x^2 > 0$, then (54) may be rewritten as

$$(55) \qquad 1/C(p^1, x^2, y^2) \geq 1/p^1 \cdot x^2.$$

We can now substitute the inequalities (53) and (55) into (27) for $(\bar{x}, \bar{y}) \equiv (x^1, y^1)$ and $(\bar{x}, \bar{y}) \equiv (x^2, y^2)$ respectively to deduce inequalities or bounds on the following Konüs price indexes P_K (assuming that all costs are positive):

$$(56) \qquad \begin{aligned} P_K(p^1, p^2, x^1, y^1) &\equiv C(p^2, x^1, y^1)/C(p^1, x^1, y^1) \\ &= C(p^2, x^1, y^1)/p^1 \cdot x^1 \\ &\leq p^2 \cdot x^1/p^1 \cdot x^1; \end{aligned}$$

$$(57) \qquad \begin{aligned} P_K(p^1, p^2, x^2, y^2) &\equiv C(p^2, x^2, y^2)/C(p^1, x^2, y^2) \\ &= p^2 \cdot x^2/C(p^1, x^2, y^2) \\ &\geq p^2 \cdot x^2/p^1 \cdot x^2. \end{aligned}$$

Thus (56) says that the (unobservable) theoretical Konüs price index $P(p^1, p^2, x^1, y^1)$ (which uses the period 1 level of utility $u^1 \equiv F(x^1, y^1)$ as the reference indifference surface) is bounded from above by the (observable) Laspeyres price index $p^2 \cdot x^1/p^1 \cdot x^1$ and (57) says that the (unobservable) Konüs price index $P(p^1, p^2, x^2, y^2)$ (which uses the period 2 level of utility $u^2 \equiv F(x^2, y^2)$ as the reference indifference surface) is bounded from below by the (observable) Paasche price index $p^2 \cdot x^2/p^1 \cdot x^2$.

Konüs [1924] worked out a brilliant technique which enables us to go beyond the one sided bounds (56) and (57) to deduce two sided bounds for a theoretical Konüs price index: under suitable regularity conditions, we can take a weighted average of the two reference vectors (x^1, y^1) and (x^2, y^2) and deduce that the resulting theoretical price index $P[p^1, p^2, \lambda x^1 + (1-\lambda)x^2, \lambda y^1 + (1-\lambda)y^2]$ lies *between* the Laspeyres and Paasche price indexes for some λ such that $0 < \lambda < 1$. This technique of proof is used repeatedly in this volume; e.g., see Section 3 of Chapter 5, Sections 2 and 3 of Chapter 7 and Sections 3 and 4 of Chapter 11.

The above argument shows that the gap between the Paasche and Laspeyres price indexes will include the value of a theoretical economic index. This suggests that taking some sort of average or symmetric mean of the Paasche and Laspeyres price indexes should yield an empirically observable price index which is "close" to an unobservable theoretical price index. If we take the geometric mean of the Laspeyres and Paasche price indexes, we obtain the following index:

$$(58) \qquad (p^2 \cdot x^1/p^1 \cdot x^1)^{\frac{1}{2}}(p^2 \cdot x^2/p^1 \cdot x^2)^{\frac{1}{2}} = P_F(p^1, p^2, x^1, x^2)$$

where P_F is the Fisher [1922] ideal price index defined earlier by (24). Recall that the Fisher ideal index seemed best from the viewpoint of the test or axiomatic approach to index number theory. The above argument suggests that it is also very good from the viewpoint of the economic approach.

The theory of bounds can be extended to cover the case where there are more than two observations. This leads us into revealed preference theory and Afriat's [1967] nonparametric approach to preference estimation in the context of consumer theory. There is also a corresponding nonparametric approach to technology estimation in the context of producer theory; see the discussion at the end of Chapter 7. We pursue this topic in Volume II.

We turn now to the exact index number approach to approximating the unobservable economic indexes defined in the previous section. In this approach, an explicit functional form for the aggregator function or the dual cost function is chosen. With a strategic choice of functional form plus the assumption of optimizing behavior, it turns out that we can determine the value of an economic price or quantity index using only the observable price and quantity data (p^1, p^2, x^1, x^2) that pertain to the goods to be aggregated for observations 1 and 2. An example will illustrate how the method works.

Suppose that the cost function C^* which appears in (28) has the following explicit functional form:

$$(59) \qquad C^*(p, u) \equiv (p^T B p)^{\frac{1}{2}} u$$

where $B \equiv [b_{ij}]$ is an $N \times N$ symmetric matrix of parameters which satisfy certain regularity conditions.[18]

Assume that the consumer or producer is cost minimizing in periods 1 and 2. Using (59), we have:

$$(60) \qquad p^1 \cdot x^1 = C^*(p^1, u^1) = (p^{1T} B p^1)^{\frac{1}{2}} u^1;$$

$$(61) \qquad p^2 \cdot x^2 = C^*(p^2, u^2) = (p^{2T} B p^2)^{\frac{1}{2}} u^2.$$

Cost minimizing input demand functions can be obtained by differentiating $C^*(p^t, u^t)$ with respect to the components of the price vector (see Lemma 4

[18] These regularity conditions ensure that $C^*(p, u)$ is nondecreasing and concave in p over a domain of prices which includes the observed price vectors p^1 and p^2.

in Section 2 of Chapter 6, Shephard's Lemma). Thus differentiating (59), the cost minimizing observed vectors x^1 and x^2 satisfy:

$$x^1 = \nabla_p C^*(p^1, u^1) = (p^{1T} B p^1)^{-\frac{1}{2}} B p^1 u^1; \tag{62}$$

$$x^2 = \nabla_p C^*(p^2, u^2) = (p^{2T} B p^2)^{-\frac{1}{2}} B p^2 u^2 \tag{63}$$

where $\nabla_p C^*(p^t, u^t) \equiv [\partial C^*(p^t, u^t)/\partial p_1, \ldots, \partial C^*(p^t, u^t)/\partial p_N]^T$ for $t = 1, 2$. Now let us evaluate the Konüs price index (28) for an arbitrary $\bar{u} > 0$:

$$\begin{aligned} P_K^*(p^1, p^2, \bar{u}) &\equiv C^*(p^2, \bar{u})/C^*(p^1, \bar{u}) \\ &= (p^{2T} B p^2)^{\frac{1}{2}}/(p^{1T} B p^1)^{\frac{1}{2}} \quad \text{using (59).} \end{aligned} \tag{64}$$

Consider the formula for the Fisher ideal price index (24). Replace $p^t \cdot x^t$ by $(p^{tT} B p^t)^{\frac{1}{2}} u^t$ for $t = 1, 2$ (see (60) and (61) above). Then

$$\begin{aligned} P_F(p^1, p^2, x^1, x^2) &= \left[\frac{p^2 \cdot x^1 (p^{2T} B p^2)^{\frac{1}{2}} u^2}{p^1 \cdot x^2 (p^{1T} B p^1)^{\frac{1}{2}} u^1}\right]^{\frac{1}{2}} \\ &= \left[\frac{(p^{1T} B p^1)^{-\frac{1}{2}} p^{2T} B p^1 u^1 (p^{2T} B p^2)^{\frac{1}{2}} u^2}{(p^{2T} B p^2)^{-\frac{1}{2}} p^{1T} B p^2 u^2 (p^{1T} B p^1)^{\frac{1}{2}} u^1}\right]^{\frac{1}{2}} \\ &\qquad \text{using (62) and (63) to eliminate } x^1 \text{ and } x^2 \\ &= (p^{2T} B p^2)^{\frac{1}{2}}/(p^{1T} B p^1)^{\frac{1}{2}} \end{aligned} \tag{65}$$

where (65) follows from the line above using the symmetry of B which implies that $p^{2T} B p^1 = p^{1T} B p^2$. Since (65) equals (64), we have for any $\bar{u} > 0$:

$$P_K^*(p^1, p^2, \bar{u}) = P_F(p^1, p^2, x^1, x^2). \tag{66}$$

Thus the theoretical Konüs price index $P_K^*(p^1, p^2, \bar{u})$ is exactly equal to the empirically observable Fisher ideal price index $P_F(p^1, p^2, x^1, x^2)$, provided that there is cost minimizing behavior in the two periods and that the functional form for the cost function C^* is given by (59). In this case, the Fisher ideal price index P_F is said to be *exact* for the cost function defined by (59).

The theory of exact index numbers was started by Konüs and Byushgens [1926] and resurrected by Pollak [1971a], Afriat [1972b] and Samuelson and Swamy [1974]. Diewert noticed (see Chapter 8 below) that the cost function defined by (59) is *flexible*; i.e., it can provide a second order approximation to an arbitrary twice continuously differentiable cost function that is dual to a linearly homogeneous aggregator function. Thus Diewert called the Fisher price index P_F a *superlative* index number formula since it is exact for a flexible functional form. Many additional superlative index number formulae are derived in Chapter 8 below.

Note that we have now provided three separate justifications for the use of the Fisher ideal price index.

5. Other Approaches to Index Number Theory

The earliest approach to index number theory was the *fixed basket approach*. In this approach, a representative basket of quantities $x \equiv (x_1, \ldots, x_N)$ is chosen. The price level P^t for period t is defined to be the cost of purchasing this fixed basket at the period t prices, $p^t \equiv (p_1^t, \ldots, p_N^T)$:

$$(67) \qquad P^t \equiv p^t \cdot x, \qquad t = 1, \ldots, T.$$

This approach eventually evolved into the axiomatic approach to index number theory as researchers attempted to be more precise about the basket x; see Chapter 2, Section 2.

Another early approach to index number theory was the *statistical approach* initiated by Jevons [1865][1884] and pushed forward by Edgeworth [1888] [1901] [1923]. In this approach, ratios of the price of each good i in period 2 to its price in period 1, p_i^2/p_i^1, for $i = 1, \ldots, N$ are regarded as independent random variables which have a common mean: (one plus) the inflation rate going from period 1 to period 2. Thus by the law of large numbers, an average of these price ratios such as $\sum_{i=1}^N (1/N)(p_i^2/p_i^1)$ should approach (one plus) the inflation rate as N (the number of goods) becomes large. This approach was effectively criticized by Fisher [1911], Walsh [1924], Bowley [1928] and Keynes [1930; 71–81] and eventually fell out of favor; see Section 3 of Chapter 2. However, more recently "neostatistical" approaches that rely on minimizing some sort of statistical error have been suggested; see Section 10 of Chapter 5 and Section 4 of Chapter 7 for descriptions of the neostatistical approaches of Stuvel [1957], van Yzeren [1956], Theil [1960], Kloek and deWit [1961] and Banerjee [1975]. Finally, a few papers have looked at the statistical sampling problems involved in the construction of a conventional Laspeyres price index under conditions of incomplete information; see Szulc [1989] and Balk [1991].

Yet another approach to index number theory is the *continuous time approach* initiated by Divisia [1926]. Up to this point, we have regarded the basic microeconomic price and quantity data p_i^t, x_i^t as functions of a discrete time indicator, $t = 1, 2, \ldots, T$. In the Divisia approach, prices and quantities $p_i(t), x_i(t)$ are regarded as functions of a continuous time variable t where $0 \leq t \leq T$. The problem with this approach is that economic data are almost never available as continuous time variables. Hence for empirical purposes, it is necessary to approximate the continuous time Divisia price and quantity indexes by discrete time data. Since there are many ways of performing these approximations,[19] the Divisia approach does not seem to lead to a definitive result. The Divisia approach is discussed in Section 5 of Chapter 2 and Section 7 of Chapter 7; it will be discussed further in Volume II.

---

[19] See Diewert [1980; 443–446].

A final approach to index number theory might be termed the *difference approach*. The traditional (bilateral) approach to index number theory is based on a *ratio approach*. For example, the Konüs price index defined by (28) is a ratio of cost functions $C^*[p^2, F(x)]/C^*[p^1, F(x)]$ evaluated at different prices p^1 and p^2, holding the quantity vector x fixed. As another example, the Allen [1949] quantity index[20] Q_A is defined as a ratio of cost functions evaluated at the period 1 and 2 quantity vectors, x^1 and x^2, and at a common reference price vector p:

$$(68) \qquad Q_A(x^1, x^2, p) \equiv C^*[p, F(x^2)]/C^*[p, F(x^1)].$$

However, some early index number researchers such as Bennet [1920] and Montgomery [1929] [1937] used differences instead of ratios.[21] Thus the difference counterparts to the Konüs price index and the Allen quantity index can be defined as follows:

$$(69) \qquad P_D(p^1, p^2, x) \equiv C^*[p^2, F(x)] - C^*[p^1, F(x)];$$

$$(70) \qquad Q_D(x^1, x^2, p) \equiv C^*[p, F(x^2)] - C^*[p, F(x^1)].$$

The difference price index P_D has been forgotten in modern economic analysis but the difference quantity index Q_D lives on and is now known as a consumer surplus measure. In fact, if we set the reference prices in (70) equal to p^1 and p^2, we obtain the *equivalent* and *compensating variations* defined by Hicks [1941–42; 128] [1946; 331]; see Section 7 of Chapter 7.

For more material on Bennet's contributions, see Sections 5 and 6 of Chapter 2 and Section 7 of Chapter 7. The difference quantity index defined by (70) will be discussed more fully in Volume II; see Diewert [1992b].

6. The New Good Problem

In both the economic and axiomatic approaches to index number theory, we require price and quantity information on the *same* list of commodities for two or more observations. How can we apply these theories when the list of commodities changes?

---

[20]For more on the Allen quantity index, see Section 3 of Chapter 7.

[21]Bennet [1920] developed an economic approach while Montgomery [1929] [1937] developed a test approach using differences. Bennet [1920; 461] also developed Divisia's [1926] mechanical differentiation approach (before Divisia's publications appeared) as well as a mechanical differencing approach.

The list of commodities utilized or produced by a consumer or producer could change for a large number of reasons: (i) the consumer's *wealth* or *educational* status could change from period to period, inducing a demand for different goods and services; (ii) *advertising* could make consumers or producers aware of a product and trigger a demand for it; (iii) a *transportation improvement* could lower shipping costs and lead to an increased availability of goods along the route of the improvement; (iv) a *prohibitive tariff* could be lowered opening up the domestic market to previously unavailable imported goods; (v) *government regulations* prohibiting the consumption of certain goods could be relaxed;[22] (vi) *population growth* in a region could allow new firms offering specialized goods or services to locate in the region (e.g., a speciality bookstore, a used machinery dealer, the delivery of cable television, a sports franchise, etc.); and (vii) *technological progress* creates thousands of new products every year (e.g., video recorders, video cameras, more powerful computers, robots, thousands of new food products, etc.). Thus the "new" good problem is a tremendously pervasive phenomenon.

Obviously, the quantity of a "new" good produced or consumed in the period before its introduction is zero. However, the economic theory of index numbers requires a price to go along with this zero quantity — what should this price be? Hicks [1940; 114] provided a satisfactory theoretical solution to this problem many years ago: (i) from the viewpoint of a consumer or producer buying units of the "new" good in the first period that it makes its appearance, the price in the previous period should be that price which would have been just high enough to have driven the purchaser's demand down to zero and (ii) from the viewpoint of the producer of the "new" good, the price in the previous period should be that price which would have been just low enough to have induced the producer to supply zero units. Although this Hicksian solution to the pricing problem for new goods is theoretically satisfactory, it is not so satisfactory from a practical point of view since the determination of the appropriate shadow prices will generally require rather sophisticated econometric analysis and statistical agencies are not usually in a position to undertake this.[23] In view of the increasing pace of the introduction of new goods and services in most economies, we are led to conclude that the relative

[22]Examples of this phenomenon include: (i) the repeal of prohibition in the U.S.; (ii) the recent striking down of the German purity of beer laws which were used to exclude imports; (iii) countless local procurement rules which are gradually being whittled down by national and international agreements; and (iv) privatization of previous government monopolies which often leads to an increased choice set.

[23]Statistical agencies in Canada and the U.S. are currently doing some econometric analysis to deal with the new good problem (or the problem of quality change as it is sometimes called) in the areas of housing and computers.

neglect of the new good problem has probably led to an upward bias in the measurement of inflation in most market economies. This topic requires a great deal of additional research.

The new good problem is discussed in more detail below in Section 10 of Chapter 2 and in Section 11 of Chapter 5.

7. Symmetric Means

A mean or an average of N numbers $x_1, x_2, \ldots, x_N$ can be defined as a function $M(x_1, \ldots, x_N)$ of the N numbers which has the following property:

$$M(k, k, \ldots, k) = k; \tag{71}$$

i.e., if all N numbers are equal to the same number k, then the mean of these numbers is also equal to k. A symmetric mean $M(x_1, \ldots, x_N)$ has the additional property that the function treats each variable x_n in a symmetric manner, so that we can interchange the order in which the x_n appear in the function but the function value remains unchanged.

Symmetric means are used in index number theory fairly frequently: recall our earlier discussion around equation (58) where we took a particular symmetric mean (the geometric average) of the Paasche and Laspeyres price indexes to obtain an approximation to a theoretical price index. Symmetric means are used in this volume in Appendix 3 of Chapter 9.

In addition to being used in index number theory, symmetric means are used in two other areas of economics: (i) the measurement of welfare and inequality and (ii) in modeling choice under uncertainty.

In view of the importance of symmetric means in measurement economics, a systematic survey of symmetric means from an axiomatic perspective is presented in Chapter 14 below. In Chapter 14, we also develop a model of choice under uncertainty, due originally to Dekel [1986], Chew [1989] and Chew and Epstein [1989a], from a new axiomatic perspective. Our Implicit Expected Utility Model turns out to be much more flexible than the traditional Expected Utility Model. We illustrate this by applying our model to some simple problems involving insurance, gambling and investing.

8. Mathematical and Economic Prerequisites

A substantial portion of the proofs in this volume are quite elementary in nature. As in Section 4 of this overview, quite often proofs rest on feasibility arguments; i.e., we need only show that a certain set of quantities is feasible (but not necessarily optimal) for some optimization problem. Throughout the book, calculus and some matrix algebra will enable the reader to follow the majority of the proofs.

Occasionally, some specialized mathematical material is used which is not covered in the usual mathematics for economists course which is given to advanced undergraduates or beginning graduate students.

In Chapters 6 and 15, some use is made of the Maximum Theorem of Debreu [1952; 889-890][1959; 19] and Berge [1963; 116], and also of specialized material on convex and concave functions such as the Fenchel [1953] closure operation. These somewhat complex results are required only to establish various continuity properties; the reader who is primarily interested in economics can simply ignore the material on continuity problems. Good references for the results on convex sets and concave functions that are used in Chapters 6 and 15 are Rockafellar [1970] and Roberts and Varberg [1973]. However, we use only a small fraction of the material presented in these references.

The theory of functional equations is used when developing the axiomatic approach to index number theory but, as we saw in Section 2 of this chapter, the functional equations that arise are usually so simple that no specialized preparation is required. An exception to this rule occurs in Chapter 14 where Aczel's [1966] monograph on functional equations is used in a few places.

Chapter 14 also uses some material on two specialized topics which are not usually covered in standard mathematics for economists courses: (i) inequalities and (ii) generalized concave functions. The main inequality used is the Cauchy-Schwarz inequality. Proofs of it may be found in Hardy, Littlewood and Polya [1934; 16] and many other places. Two useful references for the material on generalized concave functions are Diewert, Avriel and Zang [1981] and Avriel, Diewert, Schaible and Zang [1988].

9. Personal Notes

It seems appropriate to conclude this introductory chapter by indicating how I got into the index number business in the first place.

As a student at Berkeley during the years 1964–1968, I had become interested in index number and aggregation problems by reading some of Richard Stone's work. However, I did not immediately pursue this interest. While I was an Assistant Professor at the University of Chicago during the years 1968–1970, I was asked to referee a paper by Sydney Afriat for the *Journal of Political Economy*. I had to work on this paper for two or three weeks before I finally understood it. I wrote up a lengthy referee's report with a recommendation to accept subject to the author revising the paper to make it more understandable. Sydney did not choose to revise the paper. Instead he resubmitted it to the *International Economic Review* where it was published as Afriat [1973d].[24] In any case, I had become acquainted with Afriat's research, and he eventually sent me his classic work on index numbers, Afriat [1972b], which was originally written in 1970.

Afriat [1972b] developed the theory of exact index numbers; i.e., he showed how certain functional forms for a utility function were consistent with certain index number formulae.[25] I became very interested in this approach to index number theory since my Ph.D. thesis was concerned with functional form problems for utility and production functions. Thus while visiting Stanford during the summer of 1973,[26] I decided to look up some of the more obscure articles that were listed in Afriat's [1972b] references, including Konüs [1924] and Byushgens [1925].[27] Fortunately, the Hoover Institution Library at Stanford had all of the old issues of the Russian journal *Voprosi Konyunkturi* and so I stumbled onto the classic article on exact index numbers and duality theory by Konüs and Byushgens [1926] which had not been translated into English or referred to in the English language literature on index numbers.[28] I studied the technique of proof used by Konüs and Byushgens, and I was able to utilize their techniques to establish a large class of exact index number formulae. This led me to write a Stanford Technical Report, "Homogeneous Weak Separability and Exact Index Numbers," during the summer of 1973. I submitted this paper to the *Review of Economic Statistics* and the *Quarterly Journal of Economics* but they both rejected the paper. I then submitted it to the *Journal of Econometrics* because I thought that I might get a better reception there, since I was an Associate Editor at the time. Dennis Aigner accepted the paper except that he asked me to drop the material on homogeneous weak separability (recall our discussion around equation (30) above) which was mostly due to

[24]Several years later, I dug out this old referee's report, revised it a bit and published it as Diewert [1973b]. I also gave a seminar on this material when I visited Harvard during the summer of 1970. I shared an office with Giora Hanoch who immediately saw that Afriat's revealed preference techniques could also be applied to production theory; see Hanoch and Rothschild [1972]. My thanks to Zvi Griliches and Dale Jorgenson for arranging that visit to Harvard.

[25]Pollak [1971a] picked up on Afriat's work and provided a beautiful exposition of the economic approach to index number theory.

[26]My thanks to Larry Lau for arranging this and other summer visits to Stanford.

[27]I believe Afriat found these references in Schultz [1939].

[28]Actually Schultz [1939; 9] quoted parts of a letter from Konüs in which Konüs alluded to his joint article with Byushgens (Buscheguennce) but Konüs did not provide a title or place of publication of his joint article.

Shephard [1953]. I agreed and the paper was published as Diewert [1976a] and is reprinted as Chapter 8 in the present volume.

Although the *Review of Economics and Statistics* had rejected my paper, they remembered that I had submitted a paper on index number theory to them. Thus in 1975, I was asked to referee Vartia [1976a]. I was very enthusiastic about the paper and recommended publication, but the *Review* rejected it despite my positive report. However, my refereeing experience proved fruitful since it led to a friendship with Vartia, and it also stimulated me to write a paper in the summer of 1975 entitled "Ideal Log Change Index Numbers and Consistency in Aggregation." This paper was eventually published as Diewert [1978b] and is reprinted as Chapter 9 in this volume.

My first two papers on index numbers, Chapters 8 and 9 below, were concerned with the properties of superlative index numbers. Irving Fisher [1922; 244] called an index number formula "superlative" if it agreed (numerically for his chosen data set of 36 commodities for the years 1913–1918) very closely with the Fisher ideal index numbers for the same data set. Since the Fisher ideal index is exact for a flexible functional form, I decided to use Fisher's term "superlative" to describe any index number formula that was exact for a flexible functional form (for either a unit cost function or an aggregator function). Somewhat surprisingly, my classification of superlative index number formulae turns out to be roughly equivalent to Fisher's classification since all known superlative index number formulae (according to my definition) closely approximate each other numerically; see Theorems 5 and 6 in Chapter 9.

Over the years, I have been gratified to see that the superlative index number concept has received some recognition in statistical agency circles; see the Bureau of Labor Statistics [1983], Hill [1988] and Triplett [1992].

It seems appropriate to note here that my attitude towards the test approach to index numbers has changed over the years. Initially, I was very much influenced by Frisch's [1936] survey article on index numbers (which can still be read with profit) where he criticized the test approach for its indeterminacy: no index number formula satisfied all "reasonable" tests and it was difficult to determine which tests should be dropped in order to end up with a consistent set of tests leading to a unique index number formula. Thus Frisch (and I) criticized the test approach for not leading to a definite result. My early papers on index number theory favored the economic approach (which I still generally favor).

My early enthusiasm for the economic approach to index number theory has been tempered by two considerations. The first is that it is sometimes very difficult to implement the economic approach empirically. For example, in regulated industries, it is usually unrealistic to assume that firms are competitively maximizing profits subject to the constraints of technology and fixed exogenous prices. Under these circumstances, the economic approach can be rescued if observed output prices are replaced by appropriate shadow prices,[29] which are usually equal or proportional to marginal costs. The problem with this solution is that many regulated firms (such as telecommunications firms) generally produce thousands of outputs and it is virtually impossible to determine the marginal cost of each product with any degree of accuracy. Thus the economic approach can break down from the viewpoint of practical measurement problems and hence must be replaced by another approach. The second consideration which increased my appreciation for the test approach was the discovery that the Fisher ideal index number formula seems to have emerged as the "best" alternative from the viewpoint of the axiomatic approach, since, as I show in Chapter 13, it satisfies some 20 "reasonable" tests. None of the other leading index number formulae satisfy as many "reasonable" tests. Thus the test approach does seem to lead to a determinate choice of functional form for an index number formula.

A final perspective on the test or axiomatic approach is due to Wolfgang Eichhorn. A decade ago, he pointed out to me that the economic approach is in fact an axiomatic approach: it just uses different axioms (optimizing behavior with prices being independent variables generating quantities as endogenous dependent variables whereas the traditional test approach to index number theory regards both prices and quantities as exogenous independent variables). Eichhorn's insightful comment led me to write the paper reprinted here as Chapter 11, which develops the axioms of the economic approach to index number theory. In Chapter 14, an additional axiomatic approach to another economic problem is developed: I attempt to axiomatize a decision maker's choices in an uncertain environment.

To conclude this review, I would like to thank my co-author, Robert Allen, for his permission to include our joint work in this volume. A very special thanks is due to my co-editor, Alice Nakamura, and also to Dale Jorgenson who made these volumes possible. As well, I would like to thank the following people for fruitful discussions or correspondence over the years on index number and related measurement problems: Sydney Afriat, Dennis Aigner, Maurice Allais, Bob Allen, Keir Armstrong, Ken Arrow, B.K. Atrostic, Bert Balk, Bill Barnett, Ernst Berndt, Jeff Bernstein, Chuck Blackorby (Blackie), Walter Bossert, John Bossons, Alexandra Cass, Doug Caves, Peter Chinloy, Lau Christensen, Dianne Cummings, Rob Danielson, Masako Darrough, Ed Dean, Angus Deaton, Gerard Debreu, Mike Denny, David Donaldson, Lorraine Eden, Wolfgang Eichhorn, Larry Epstein, Rolf Färe, Rob Feenstra, Ivan Fellegi, Frank Fisher, Kevin Fox, Mel Fuss, Frank Gallop, Terrence Gorman, Zvi Griliches, Shauna Grosskopf, Tom Gussman, Shirley Hahn, Peter Hammond, Diana Hancock, Arnold Harberger, Rick Harris, Peter Hill, Chuck Hulten, Dale Jorgenson, Frank Kiss, Ulrich Kohli, Serge Kolm, Alexander Konüs, Larry Lau, Dennis

[29]This point was originally made by Frisch [1936].

Lawrence, Peter Lawrence, Bernie Lefebvre, Tracy Lewis, Ramon Lopez, Marilyn Manser, Doug May, Jim Melvin, Nimfa Mendoza, Claude Montmarquette, Cathy Morrison, Alice Nakamura, Dale Orr, Tae Oum, Celik Parkan, Don Patterson, Andreas Pfingsten, Bob Pollak, Dan Primont, Craig Riddell, Alan Russell, Bob Russell (R Cubed), Tom Rymes, Jacob Ryten, Paul Samuelson, Bohdan Schultz (Szulc), Bill Schworm, Karl Shell, Ron Shephard, Margaret Slade, Barbara Slater, Dunc Smeaton, Spenser Star, Genio Staranczak, Frank Stehling, Leo Törnqvist, Mike Tretheway, Jack Triplett, Yasuko Tsurusaki, Arja Turunen-Red, Ralph Turvey, Dan Usher, Jan van Yzeren, Yrjö Vartia, Arthur Vogt, Terry Wales, Bill Waters, John Weymark, Ken White, Frank Wykoff, and Kim Zieschang. To all of you, thank you for the many hours of thought on measurement problems that you have stimulated.

Essays in Index Number Theory, Volume I
W.E. Diewert and A.O. Nakamura (Editors)

Chapter 2

THE EARLY HISTORY OF PRICE INDEX RESEARCH*

W.E. Diewert

1. Introduction

The purpose of this paper is to present a brief overview of the ancient history of price measurement. At least five distinct approaches to price and quantity measurement (or index number theory) can be distinguished in the early literature on the subject: (i) the tabular standard (or the commodity standard or the fixed basket approach); (ii) the statistical approach; (iii) the test approach; (iv) the Divisia index approach and (v) the economic approach. We shall discuss the history of each approach in turn in Sections 2–6 below.

In Sections 7 to 11 below, we shall discuss issues that are perhaps somewhat controversial. Section 7 briefly discusses the merits of the test approach to index number theory while Section 8 presents an extended discussion of the chain principle. Sections 9 and 10 discuss the possible magnitudes of the substitution bias and the new good bias respectively while Section 11 asks whether the theory of the cost of living index has been exhausted.

Section 12 concludes with a list of recommendations directed towards statistical agencies.

---

*This paper was originally a comment on "Price Index Research and its Influence on Data: A Historical Review," a paper presented by Jack E. Triplett in Washington, D.C. on May 12, 1988 at the 50th Anniversary Conference of the Conference on Research in Income and Wealth. The author is indebted to R.C. Allen, B.M. Balk and J.E. Triplett for valuable historical references and discussions over the years. This research was supported by a grant from the Social Sciences and Humanities Research Council of Canada and by the Summer Institute of the National Bureau of Economic Research on Productivity. Neither institution is to be held responsible for any opinions expressed by the author.

2. The Fixed Basket Approach

The essence of the fixed basket approach or the tabular standard may be explained as follows. Suppose that there are N goods that consumers in a location can purchase during two periods. In periods 1 and 2, the relevant price vectors are $p^1 \equiv (p_1^1, \ldots, p_N^1)$ and $p^2 \equiv (p_1^2, \ldots, p_N^2)$, respectively. Suppose further (unrealistically) that the quantities purchased of the N goods are constant during the two periods, with the constant vector of purchases being defined as $q \equiv (q_1, \ldots, q_N)$. Then a natural measure of the average level of prices in period 2 relative to period 1 is $p^2 \cdot q/p^1 \cdot q$ where $p^t \cdot q \equiv \sum_{n=1}^{N} p_n^t q_n$ is the inner product of the vectors p^t and q.

The above approach to price measurement has been independently proposed by many people. The earliest known proposer of the method was William Fleetwood, the Bishop of Ely, who wrote the book Chronicon Preciosum in 1707.[1] The constant basket of goods he used to compare the value of money (or, conversely, the level of prices) for an Oxford student of 1707 compared to an Oxford student of 1460 was 5 quarters of wheat, 4 hogsheads of beer and 6 yards of cloth.

Perhaps the next independent discovery of the tabular standard was made by the Legislature of Massachusetts in 1780. An account of this discovery is given by Willard Fisher [1913]. A tabular standard was used to index the pay of soldiers fighting in the Revolutionary War (a massive inflation had drastically reduced the real value of the fixed nominal pay of the soldiers). The constant quantity basket was 5 bushels of corn, 68 and 4/7 pounds of beef, 10 pounds of sheep's wool and 16 pounds of sole leather.[2]

Joseph Lowe [1823; 316] was not an independent discoverer of the constant basket index number formula

$$(1) \qquad p^2 \cdot q/p^1 \cdot q,$$

since he explicitly refers to Fleetwood's book. However, he developed the concept in such detail that he should be considered the father of the consumer price index. Lowe was well aware that the constant basket of commodities q could vary across demographic groups; on page 332, he presented some representative family budgets for cottagers and for the middle class. On page 97 of the Appendix, he noted that price indexes may be required for other classes of consumers or producers such as farmers and miners while on page 336 of the main text, he advocated the construction of separate "standards" for the laboring class, decomposed into unmarried laborers and married laborers with 2,

[1] A good account of Fleetwood's contributions with extensive quotations can be found in Ferger [1946].

[2] See W. Fisher [1913; 437].

3 or 4 children. Finally, Lowe [1823; 33] also envisaged a national "table of reference" which would price out a constant national consumption vector at the prices of each year t and on pages 94 and 95 of the Appendix, he constructed two such hypothetical tables.

How would the constant vector of commodities q in (1) be determined? Lowe answered this question as follows:

> As to quantity, a variation can take place only with increase of population or change of habits, and any alteration of that kind must be so gradual, that we run very little hazard in assuming a similarity of amount during a given period, which for the sake of precision, we shall suppose to be five years.
>
> Lowe [1823; Appendix 95]

Lowe [1823; 334] also proposed that the national government should fund the collection of the relevant price and quantity statistics, but if this was not done, then Lowe felt that government agencies should at least provide what data they had at their disposal "on the demand of any respectable association."

Lowe [1823; 335–343] listed a host of applications for his proposed tables of reference, including the following: (i) wages, salaries and rents could be indexed to eliminate the anomalies arising out of unforeseen fluctuations in the value of the country's currency, (ii) they would facilitate salary negotiations, (iii) they could be used to index long term agricultural leases and (iv) bond holders could be paid in real terms if they wanted that option.

Lowe [1823; 346] concluded with some pertinent observations on why his proposal had not been implemented up to his time:

> This has, we believe, been owing to two causes; the unfortunate neglect of political economy in the education of our public men; and the interest of government, the greatest of all debtors, to prevent the public from fixing its attention on the gradual depreciation of money that went on during the half century to the late peace.
>
> Lowe [1823; 346]

Scrope [1833; 406–407] followed in Lowe's footsteps but was the first to use the term ***tabular standard*** to describe the price index defined by (1). However, his treatment was not nearly as detailed as that of Lowe,[3] so we will pass on to list others who have endorsed the tabular standard.

[3]Scrope does not refer Lowe (or anybody else for that matter) since, as a Member of Parliament, he seemed to be writing an extended political speech. However, it seems likely that he knew of Lowe's work since his proposals and terminology were so similar to those of Lowe. Scrope [1833; 320] also proposed a comprehensive system of social security for workers "when rendered incapable of labor by illness, age or accident." Another early proposer of the tabular standard who does not refer to Fleetwood or Lowe is the American economist Willard Phillips who made his proposal in 1828; see Jastram [1951] for a description of the contributions of Phillips.

If quantities were to remain constant during the two periods under consideration, a whole host of authors endorsed formula (1) to measure price change, including Jevons [1865; 122] [1884; 122], Sidgwick [1883; 67–68], Edgeworth [1925; 212][4] (originally published in 1887), Marshall [1887; 363], Bowley [1899] [1901; 227] [1928; 223], Walsh [1901; 540] [1921; 543] [1924; 544] and Pigou [1912; 38]. During this period, the precise specification of the constant quantity vector q was a problem which was addressed. Thus Laspeyres [1871] proposed that q should equal $q^1 \equiv (q_1^1 \ldots, q_N^1)$, the base period quantity vector, while Paasche [1874] proposed that q should equal $q^2 \equiv (q_1^1, \ldots, q_N^2)$, the current period quantity vector. Thus (1) can be specialized to yield the famous Laspeyres and Paasche price indexes, P_L and P_P:

$$P_L(p^1, p^2, q^1, q^2) \equiv p^2 \cdot q^1 / p^1 \cdot q^1; \tag{2}$$

$$P_P(p^1, p^2, q^1, q^2) \equiv p^2 \cdot q^2 / p^1 \cdot q^2. \tag{3}$$

Given that quantities would not be exactly equal during the two periods under consideration, various authors started to argue that *averages* of (2) and (3) should be used to measure price change. Thus Sidgwick [1883; 68] and Bowley [1901; 227] proposed the use of $(1/2)P_L + (1/2)P_P$, while Edgeworth [1925; 214] (originally published in 1887) proposed that the q in (1) be set equal to the arithmetic average of the two quantity vectors, $(1/2)q^1 + (1/2)q^2$, (Edgeworth states that this variant was also independently proposed by Alfred Marshall). Bowley [1899] suggested the geometric mean of P_L and P_P, which later came to be known as Irving Fisher's [1922] ideal price index P_F defined as

$$P_F(p^1, p^2, q^1, q^2) \equiv (p^2 \cdot q^1 p^2 \cdot q^2 / p^1 \cdot q^1 p^1 \cdot q^2)^{1/2}. \tag{4}$$

Walsh [1901; 398] proposed that the components q_i of the quantity vector q in (1) should be set equal to the geometric means of the quantities in the two periods. Thus the Walsh price index is

$$P_W(p^1, p^2, q^1, q^2) \equiv \sum_{i=1}^{N} (q_i^1 q_i^2)^{1/2} p_i^2 / \sum_{j=1}^{N} (q_j^1 q_j^2)^{1/2} p_j^1. \tag{5}$$

Finally, Pigou [1912; 46] suggested $P_L P_P$ as a measure of price change. Since this price index has rather poor homogeneity properties, Pigou later modified his measure by taking the square root which yields P_F defined by (4); see Pigou [1932; 69].

At this stage, the fixed basket approach to index number theory merged into the test and economic approaches.

---

[4]Edgeworth [1925; 331] called the method the consumption standard rather than the tabular standard.

3. The Statistical Approach

This approach, which originated with Jevons [1865] [1884], assumed that increases in the supply of money increased all prices proportionately except for random fluctuations. Thus with additive errors and a sufficient number of independent observations, an appropriate price index could be obtained by taking the arithmetic mean of the price ratios p_i^2/p_i^1 while with multiplicative errors, an appropriate price index could be obtained by taking the geometric mean of the price ratios. This second alternative was advocated by Jevons, and thus we obtain the Jevons price index P_J:

$$P_J(p^1, p^2) \equiv \prod_{i=1}^{N} (p_i^2/p_i^1)^{1/N}. \tag{6}$$

In addition to Jevons, two other prominent economists who advocated the statistical approach to index numbers were Bowley [1901; 223–226] [1921; 202] [1928; 217–223] and Edgeworth [1888] [1896] [1901] [1923] [1925].[5] Edgeworth mainly advocated the median of the price ratios p_i^2/p_i^1 as the best estimator of price change.

The statistical approach was criticized by Irving Fisher [1911; 194–196] who explained in an absolutely convincing manner why all prices cannot move proportionately (due to the existence of fixed price contracts, for example). Fisher's criticisms were ignored by the profession as were those of Walsh [1924]. However, Keynes [1930; 71–81] effectively demolished the naive statistical approach by constructing various tables of index numbers which showed systematic differences over time and hence the hypothesis of approximate proportional change in all prices could not be maintained empirically.[6] Bowley [1928; 221] also criticized the approach on narrower statistical grounds by indicating that the price movements were not statistically independent.

---

[5]Bowley was not exclusively an advocate of the statistical approach as we shall see later. Also, initially Edgeworth took a broader view of the index number theory (recall his endorsement of the tabular standard if quantities remained constant), but after this initial broadness, he became a very strident defender of the statistical approach; in particular, Edgeworth's [1923] criticisms of Walsh's test approach became quite heated.

[6]Keynes [1930; 72] also used some rather colorful language to criticize the statistical approach as the following quotation indicates: "I have long believed that this is a will-o'-the-wisp, a circle-squaring expedition which has given an elusive taint, difficult to touch or catch, to the theory of price index numbers traditional in England. This is not equally true of America. Nevertheless, whilst the Americans have not worshipped the mythical creature, they have not (with the exception, perhaps, of Mr. Walsh) actively combatted him or dragged him out of the twilit cave where Edgeworth judiciously kept him."

Although Jevons' naive statistical approach is no longer advocated, statistical sampling of the prices of the various components of a price index is still done today. A problem with many of these sampling procedures is that prices are sampled independently of quantities. Pigou [1932; 77] was perhaps the first to propose that *values* should be sampled in the two periods under consideration, along with the corresponding prices and quantities, and then the Fisher ideal index P_F defined by (4) should be used to construct a measure of price change over the commodities in the sample of values. This *sample* price index could then be used to deflate the *population* value ratio over the two periods. Pigou's proposal deserves serious consideration by statistical agencies even today.

4. The Test Approach

The origins of the test approach are rooted in the more or less casual observations of the early workers in the index number field on their favorite index number formulae or those of their competitors.

Thus Jevons [1884; 152] (originally published in 1865) recognized that his unweighted geometric mean formula (6) gave index number comparisons between any two years that were independent of the base year. Edgeworth [1896; 137] gave a clear general treatment of this *base invariance test*[7] which we can phrase as follows. Let $P(p^0, p^t, q^0, q^t)$ be a generic index number formula of the type defined by (2) to (5) above which compares the level of prices in period t to the level of prices in period 0, the base year. Let p^t and q^t be the price and quantity vectors pertaining to year t for $t = 0, 1, \ldots, T$. Let i, s and t denote arbitrary years. With the base year equal to 0, the level of prices in year t relative to s is taken to be $P(p^0, p^t, q^0, q^t)/P(p^0, p^s, q^0, q^s)$. If we change the base to year i, then the level of prices in period t relative to s is $P(p^i, p^t, q^i, q^t)/P(p^i, p^s, q^i, q^s)$. The base invariance test demands that these two numbers be equal; i.e., that

$$(7) \quad P(p^0, p^t, q^0, q^t)/P(p^0, p^s, q^0, q^s) = P(p^i, p^t, q^i, q^t)/P(p^i, p^s, q^i, q^s).$$

Our next test was first proposed by Laspeyres [1871; 308], and has come to be known as the *strong identity test*: if prices in the two periods under

[7]Edgeworth mistakenly attributed this test to Westergaard [1890], but Westergaard's circular test is slightly different as we shall see.

consideration remain constant, then even if the quantities change,[8] the level of prices should remain unchanged; i.e., we should have

$$P(p, p, q^1, q^2) = 1 \tag{8}$$

where P denotes the index number formula or function, $p \equiv (p_1, \ldots, p_N)$ denotes the common price vector in both periods and $q^t \equiv (q_1^t, \ldots, q_N^t)$ denotes the quantity vector in period t for $t = 1, 2$.

The statistician Westergaard [1890; 218–219] formulated what later[9] became known as the *circularity test*: the bilateral index number formula should satisfy the following equation:

$$P(p^1, p^2, q^1, q^2)P(p^2, p^3, q^2, q^3) = P(p^1, p^3, q^1, q^3) \tag{9}$$

where p^t and q^t are the price and quantity vectors pertaining to periods t for $t = 1, 2, 3$. The right hand side of (9) computes the price level in period 3 relative to the price level in period 1 in one step, using the bilateral index number formula or function P. The left hand side of (9) computes the level of prices in period 3 relative to period 1 in two steps: in the first step, we use the bilateral formula $P(p^1, p^2, q^1, q^2)$ to compute the level of prices in period 2 relative to period 1 and then in the second step, we use the bilateral formula $P(p^2, p^3, q^2, q^3)$ to compute the level of prices in period 3 relative to period 2. The product of these two steps is supposed to yield the level of prices in period 3 relative to period 1.

The Dutch economist Pierson [1896] informally proposed two tests: (i) invariance to changes in the units of measurement (which Irving Fisher [1911] [1922; 420] first called the change of units test and later called the *commensurability test*) and (ii) the *time reversal test* which can be stated mathematically as follows:

$$P(p^2, p^1, q^2, q^1) = 1/P(p^1, p^2, q^1, q^2). \tag{10}$$

Up to this point in time, research on the test or axiomatic approach to index number theory was rather casual and unsystematic. The first systematic researcher on the axiomatic approach was Walsh [1901] [1921] [1924] who proposed a number of tests, including the *constant quantities test*;[10] i.e., if

---

[8] A weaker identity test would require both prices and quantities to remain constant. No one could object to this weaker identity test, but it is certainly possible to object to the strong identity test.

[9] See Fisher [1922; 413].

[10] Alternative names for this test might be the *constant basket test* or the *tabular standard test*. Walsh [1901; 540] [1921; 543] does not provide a name for this test.

quantities remain fixed at the vector q during the two periods under consideration, then the appropriate formula for the price index is

$$(11) \qquad P(p^1, p^2, q, q) = p^2 \cdot q / p^1 \cdot q.$$

Another test proposed by Walsh was the *strong proportionality* (in prices) *test*; i.e., if λ is a positive scalar and prices in period 2 are equal to λ times the corresponding prices in period 1, then[11]

$$(12) \qquad P(p^1, \lambda p^1, q^1, q^2) = \lambda.$$

A final test proposed by Walsh [1901; 389] [1921; 540] [1924; 506] was his *multiperiod identity test*;[12] i.e., the bilateral index number function P is to satisfy the following functional equation:

$$(13) \qquad P(p^1, p^2, q^1, q^2)P(p^2, p^3, q^2, q^3)P(p^3, p^1, q^3, q^1) = 1.$$

Note that the prices and quantities in period 4 are exactly equal to the prices and quantities in period 1, p^1 and q^1 respectively. As we shall see later, the test (13) will be useful in evaluating the usefulness of the chain principle.

The next major contributor to the test approach is Irving Fisher [1911] [1921] [1922]. Since Fisher's contributions to the test approach are quite well known (in fact, he is often credited with inventing the approach), we will not review his contributions in any detail. However, we do wish to make two comments about his work.

Our first comment is to note that Fisher [1911; 403] seems to have been the first to observe that the choice of a functional form $P(p^1, p^2, q^1, q^2)$ for a price index implicitly determines the functional form for the corresponding quantity index $Q(p^1, p^2, q^1, q^2)$; i.e., the product of the two indexes should equal the value ratio for the two periods under consideration. Thus given P, Q is implicitly determined by the following equation:

$$(14) \qquad P(p^1, p^2, q^1, q^2)Q(p^1, p^2, q^1, q^2) = p^2 \cdot q^2 / p^1 \cdot q^1.$$

---

[11]Walsh [1901; 385] noted that the strong proportionality test implies the strong identity test; i.e., (12) implies (8).

[12]Walsh [1924; 506] finally called this test the circular test but it seems best not to confuse Walsh's test (13) with Westergaard's test (9). If P satisfies the time reversal test (10), then (9) implies (13). However, if P satisfies (13), then P satisfies the weak identity test $P(p, p, q, q) = 1$ as well as (10) and (9). Upon noting that the index number formula $P(p^1, p^2, q^1, q^2) \equiv 0$ satisfies (9) but not (10), we see that (9) and (13) are not equivalent tests; i.e., Walsh's test (13) is more restrictive than Westergaard's test (9).

Frisch [1930; 399] called (14) the *product test* while Samuelson and Swamy [1974; 572] called it the *weak factor reversal test*.

Our second comment about Fisher's contributions to the test approach is relatively unknown. Fisher [1922; 140] explained how to "rectify" an arbitrary bilateral index number formula $P(p^1, p^2, q^1, q^2)$ so that the rectified formula P^* would satisfy the time reversal test (10). Simply define P^* as follows:

$$(15) \qquad P^*(p^1, p^2, q^1, q^2) \equiv [P(p^1, p^2, q^1, q^2)/P(p^2, p^1, q^2, q^1)]^{1/2}.$$

Fisher's time rectification procedure indeed does work as advertised. The P^* defined by (15) will satisfy (10). The only problem is that the procedure is clearly due to Walsh [1921; 542], who was a discussant for Fisher's [1921] paper which was a preview for Fisher [1922]. Unfortunately, Fisher's [1922; 183] historical comments on the rectification principle fail to mention Walsh at all. However, on the positive side, Fisher [1922; 396–398] generalized Walsh's basic idea by showing how an index number formula could also be rectified to satisfy Fisher's *factor reversal test*[13] or be rectified to simultaneously satisfy the factor and time reversal tests.

Frisch [1930] [1936; 5–7] effectively criticized the test approach to index number theory on the grounds that it could be shown that no bilateral index number formula $P(p^1, p^2, q^1, q^2)$ could satisfy all reasonable tests or axioms[14] and when some tests were dropped so as to achieve a consistent set of tests, there was no general agreement on which subset of tests should be dropped. Hence the test approach did not seem to lead anywhere.

In recent years, the test approach has sprung to life again, largely due to the efforts of Wolfgang Eichhorn [1973] [1976] and his students and colleagues.[15]

---

[13]This test postulates that the functional form that works for the price index $P(p^1, p^2, q^1, q^2)$ should also do the job as a quantity index, provided that the role of prices and quantities is interchanged. This transforms (14) into Fisher's *strong factor reversal test*: $P(p^1, p^2, q^1, q^2)P(q^1, q^2, p^1, p^2) = p^2 \cdot q^2/p^1 \cdot q^1$.

[14]Frisch [1930; 404–405] tried to show that a bilateral index number formula P could not simultaneously satisfy the base, commensurability, determinateness and factor reversal tests while Wald [1937; 180–182] and Samuelson [1974a; 18–20] showed that P could not simultaneously satisfy the proportionality, circular, and factor reversal tests. Many other impossibility theorems were obtained by Eichhorn [1976] and Eichhorn and Voeller [1976].

[15]In particular, see Eichhorn and Voeller [1976].

5. The Divisia Approach

Divisia's [1926; 39–40] derivation of the price and quantity indexes associated with his name can be summarized as follows. Let the prices $p_i(t)$ and the quantities $q_i(t)$, $i = 1, \ldots, N$, be functions of (continuous) time t and let expenditure at time t be the value $v(t) \equiv \sum_{i=1}^{N} p_i(t)q_i(t)$. Assuming differentiability, the rate of change of value at time t is:

$$dv(t)/dt = \sum_{i=1}^{N} p_i(dq_i/dt) + \sum_{i=1}^{N} q_i(dp_i/dt). \tag{16}$$

Divisia then divided both sides of (16) by $p(t) \cdot q(t) \equiv \sum_{i=1}^{N} p_i(t)q_i(t)$ and equated the right hand side of the resulting equation to $Q'(t)/Q(t) + P'(t)/P(t)$ where $Q(t)$ and $P(t)$ are aggregate quantity and price levels pertaining to period t and $Q'(t)$ and $P'(t)$ denote their time derivatives. Thus we have:

$$\sum_{i=1}^{N} \frac{p_i q_i'(t)}{v(t)} + \sum_{i=1}^{N} \frac{q_i p_i'(t)}{v(t)} = \frac{Q'(t)}{Q(t)} + \frac{P'(t)}{P(t)}. \tag{17}$$

Divisia then defined $Q(t)$ and $P(t)$ as solutions to the following differential equations:

$$\frac{Q'(t)}{Q(t)} \equiv \sum_{i=1}^{N} \frac{p_i(t)q_i'(t)}{p(t) \cdot q(t)}; \quad \frac{P'(t)}{P(t)} \equiv \sum_{i=1}^{N} \frac{q_i(t)p_i'(t)}{p(t) \cdot q(t)}. \tag{18}$$

Somewhat surprisingly, virtually the same derivation was made earlier by the English economist, T.L. Bennet [1920; 461], except that he did not divide (16) through by $v(t) = p(t) \cdot q(t)$.

The above derivation of the Divisia indexes is very mechanical and is unrelated to economics (i.e., choice under constraint). However, later both Ville [1951–52] and Hulten [1973] related the Divisia indexes to economic price and quantity indexes under the assumptions of optimizing behavior and a linearly homogenous aggregator function.[16]

The problem with the Divisia approach to price measurement is that we generally cannot observe prices and quantities continuously. Thus the continuous time Divisia indexes must be approximated using discrete time data and there are many ways of forming discrete time approximations to

---

[16]Ville assumed the maximization of a linearly homogeneous utility function subject to a budget constraint while Hulten assumed cost minimization subject to a linearly homogeneous production function constraint. Results equivalent to the Ville–Hulten results were also obtained by Samuelson and Swamy [1974; 578–580].

say $P(2)/P(1)$,[17] where $P(t)$ is the Divisia index for time period t defined by (18) (plus an initial normalization). Diewert [1980; 444–445] showed that the Laspeyres and Paasche indexes, P_L and P_P defined by (2) and (3) above, could be regarded as discrete time approximations to $P(2)/P(1)$ as could the Törnqvist-translog P_T defined by

$$\ln P_T(p^1, p^2, q^1, q^2) \equiv \sum_{i=1}^{N} (1/2)(s_i^1 + s_i^2) \ln(p_i^2/p_i^1) \tag{19}$$

where the shares s_i^t are defined as $s_i^t \equiv p_i^t q_i^t / p^t \cdot q^t$, $t = 1, 2$ and $i = 1, \ldots, N$. Since the indexes P_L, P_P and P_T can differ considerably, the Divisia approach does not lead to a practical resolution of the price measurement problem.[18]

To conclude this section on the Bennet–Divisia approach, we note that Bennet [1920; 457] suggested the following discrete approximations to measure *differences* (rather than the *ratios* of Divisia) in the aggregate price and quantity levels:

$$\Delta P \equiv P(2) - P(1) \equiv \sum_{i=1}^{N} (1/2)(q_i^1 + q_i^2)(p_i^2 - p_i^1); \tag{20}$$

$$\Delta Q \equiv Q(2) - Q(1) \equiv \sum_{i=1}^{N} (1/2)(p_i^1 + p_i^2)(q_i^2 - q_i^1). \tag{21}$$

Bennet also showed that the difference in expenditures for the two periods, $\sum_{i=1}^{N} p_i^2 q_i^2 - \sum_{i=1}^{N} p_i^1 q_i^1$, was exactly equal to $\Delta P + \Delta Q$, where ΔP and ΔQ are defined by the right hand sides of (20) and (21).[19]

6. The Economic Approach

The economic approach to index number theory[20] relies on the assumption of optimizing behavior on the part of economic agents: utility maximizing or expenditure minimizing behavior on the part of consumers and profit maximizing or cost minimizing behavior on the part of producers.

---

[17] Hofsten [1952] appears to be the first researcher who noticed this difficulty with the Divisia price index.

[18] See also Samuelson and Swamy [1974; 579] for alternative suggestions on how to approximate the continuous time Divisia indexes with discrete data.

[19] The right hand side of (21) can be regarded as an approximation to the arithmetic average of the compensating and equivalent variations defined by Hicks [1941–42]. Thus the measurement of *differences* in aggregate quantities led to consumer surplus theory while the measurement of *ratios* led to the economic theory of index numbers.

[20] For surveys, see Frisch [1936], Samuelson [1947; 146–163], Ruggles [1967], Pollak [1971a], Samuelson and Swamy [1974], Afriat [1977], Deaton and Muellbauer [1980; 167–184] and Diewert [1980] [1981a] [1983a] [1983b] [1987].

The first two papers to use an explicit utility maximizing framework appear to be by Bennet [1920] and Konüs [1924]. Bennet's paper drew on an earlier paper by Bowley [1919] (he used Bowley's notation and data) and may be regarded as an attempt to determine the approximate magnitude of the substitution bias using the assumption of a quadratic utility function. Bowley [1928; 226] [1938] was in turn influenced by Bennet and developed his own quadratic approximations. Bennet's paper was very short and sketchy and did not have the impact that the Konüs paper eventually had. Konüs [1924; 16–18] not only presented a very clear definition of the true cost of living for an individual optimizing consumer, he also developed the now well known Paasche and Laspeyres bounds.[21] Konüs [1924; 20–21] also showed that the Paasche and Laspeyres price indexes, (3) and (2) above, bound the true cost of living index even in the general nonhomothetic preferences case, provided that we evaluate the true index at a suitable utility level that is between the base and current period levels.

To complete our brief survey of the early history of the economic approach to index number theory, we shall review the economic approach under four subheadings: (i) basic theoretical definitions, (ii) the theory of bounds, (iii) exact index numbers and (iv) econometric approaches.

6.1 Basic theoretical definitions

There are three main branches of the economic approach. (i) For the *true cost of living index*, see Konüs [1924], Samuelson [1947; 156] and Pollak [1971a]. For related quantity indexes, see Bowley [1928; 230], Allen [1949], Malmquist [1953] and Pollak [1971a]. (ii) For theoretical definitions of the *output price index*, see Hicks [1940], Fisher and Shell [1972b], Samuelson and Swamy [1974; 588–592], Archibald [1977] and Diewert [1983b]. For related quantity indexes, see Bowley [1921; 203], Bergson [1961; 31–34], Moorsteen [1961], Fisher and Shell [1972b; 53], Samuelson and Swamy [1974, 588–591], Sato [1976b; 438] and Hicks [1981; 256]. (iii) *The input cost index* was defined by Court and Lewis [1942–43], Triplett [1983; 274] and Diewert [1980; 459] and corresponding quantity indexes were defined in Diewert [1980; 456–460].

There is a fourth branch of the economic approach that has received less attention: (iv) *constant utility income deflators.* On this last branch of theoretical index number theory, see Diewert and Bossons [1992].

---

[21]See, for example, Diewert [1983a; 171].

6.2 The theory of bounds

Observable bounds to the generally unobservable economic price and quantity indexes were first worked out by Pigou [1912; 44–46] [1932; 62–63] and Haberler [1927; 78–92] independently of Konüs [1924; 17–19], who established the Paasche and Laspeyres bounds for the true cost of living. For a generalization of these bounds to nonlinear budget constraints, see Frisch [1936; 18].[22] It is clear that a large portion of revealed preference theory that is often attributed to Hicks [1940] and Samuelson [1947; 157] had already been developed by Pigou, Konüs, Haberler and Frisch.

Other researchers who established bounds on true indexes in the two observation situation include Leontief [1936; 49], Friedman [1938; 125], Allen [1949], Malmquist [1953], Moorsteen [1961; 464], Pollak [1971a], Fisher and Shell [1972b; 57–62], Samuelson and Swamy [1974; 581–591], Archibald [1977] and Diewert [1981a; 167–179] [1983a; 173–210] [1983b; 1056–1090].

The above theory of bounds all pertains to the two observation situation. Afriat [1967] [1977] generalized the two observation theory to cover the many observation case.

6.3 Exact index numbers

Let an aggregator function[23] $f(q)$ be given where q is an N dimensional quantity vector; i.e., $q \equiv (q_1, \ldots, q_N)$. The *cost function* C which is generated by f may be defined as

$$C(u,p) \equiv \min_q \{p \cdot q : f(q) \geq u\}; \tag{22}$$

i.e., $C(u,p)$ is the solution to the problem of minimizing the cost $p \cdot q \equiv \sum_{i=1}^N p_i q_i$ of achieving at least the utility (or output) level u, where $p \equiv (p_1, \ldots, p_N)$ is an exogenous vector of prices facing the consumer (or producer).

An index number formula or function $P(p^1, p^2, q^1, q^2)$ of the type we considered in Section 4 is defined to be *exact*[24] for an aggregator function f if

$$P(p^1, p^2, q^1, q^2) = C(u, p^2)/C(u, p^1) \tag{23}$$

[22] The English translation of the original Russian article by Konüs [1924] did not become available until 1939. Thus Frisch [1936; 25] did not have access to the original Konüs article and he mistakenly attributed the Konüs limits or bounds on the true cost of living to Haberler [1927].

[23] Diewert [1976a; 115] introduced this term to cover both the production and utility function context.

[24] The first mention of the word "exact" in this index number context appears to be by Samuelson [1947; 155]. Afriat [1972b; 45] also uses the term exact.

for some utility or output level u where q^t solves (22) when $p = p^t$ for $t = 1, 2$; i.e., P is exact for f (or its dual cost function C) if P equals the relevant economic index under the assumption of optimizing behavior on the part of an economic agent using the aggregator function f. The right hand side of (23) is $P_K(p^1, p^2, u)$, the Konüs price index or true cost of living index for a consumer that has the utility function f and faces the vector of prices p^t in period t for $t = 1, 2$.

The English language literature on exact index numbers has its roots in the theory of quadratic approximations. As we indicated earlier, Bennet [1920; 460] attempted to determine an appropriate index number formula for the true cost of living of a single "satisfaction" maximizing consumer under the hypothesis that the underlying utility function $f(q)$ was a general quadratic function. Bowley [1928; 226] [1938] followed up on Bennet's approach and provided his own second order approximation. Frisch [1936; 27–29] criticized Bowley's index number formula and developed an alternative formula which he called the double expenditure method. Wald [1939; 329] and Balk [1981; 1556] correctly pointed out that Frisch's index number formula was *not* exact for a general quadratic utility function. However, Frisch [1936; 29–30] did correctly show[25] that his index number formula collapsed to the Fisher ideal price index P_F defined by (4) if one assumed homothetic quadratic preferences so that $f(q) \equiv \sum_{i=1}^{N}\sum_{j=1}^{N} a_{ij}q_iq_j \equiv q \cdot Aq$ where $A \equiv [a_{ij}]$ is a symmetric $N \times N$ matrix of parameters that characterize tastes. This is an early example of an exact index number formula.

Another early example was given by Wald [1939; 325] who assumed the following general quadratic aggregator function:

$$f(q) \equiv a_0 + a \cdot q + (1/2)q \cdot Aq, \qquad A = A^t, \tag{24}$$

where a_0, $a \equiv [d_1, \ldots, d_N]$ and $A \equiv [a_{ij}]$ are respectively a parameter, a vector of parameters and a symmetric $N \times N$ matrix of parameters.

Unfortunately, in order to evaluate Wald's general index number formula that is exact for (24), information on income elasticities is required (we shall not write out his general index number formula since it is rather complex). However, if we assume homothetic preferences again (i.e., the aggregator function is a monotonically increasing function of a linearly homogeneous function) so that $a_0 = 0$ and $a = 0_N$ in (24)[26] then all of the consumer's income elasticities equal unity and Wald's general index number formula collapses down to

[25]See Balk [1981; 1556] for a clear proof of Frisch's result.

[26]With these restrictions, $f(q) = q \cdot Aq = h[g(q)]$ where $h(x) \equiv x^2$ and $g(q) \equiv (q \cdot Aq)^{1/2}$. Thus in this case, f is a monotonically increasing function of the linearly homogeneous function $g(q)$ over the set of q such that $q \cdot Aq \geq 0$.

the Fisher price index (4) and again we obtain the exact index number result of Frisch.[27]

Unknown to the above authors, the Frisch–Wald exact index number result had already been obtained by Konüs and Byushgens [1926; 167–172] a decade earlier.[28] In this remarkable paper, they introduced duality theory into the economics literature; i.e., they expressed consumer preferences not only by the direct utility function $f(q)$ but also by the corresponding indirect utility function g defined as follows:

$$(25) \qquad g(p,y) \equiv \max_q \{f(q) : p \cdot q \leq y\};$$

i.e., the indirect utility function $g(p, y)$ gives the maximum utility attainable as a function of the prices faced by the consumer $p \equiv (p_1, \ldots, p_N)$ and the income or expenditure $y \geq 0$ to be spent on the N goods during the period under consideration. Konüs and Byushgens assumed that the direct utility function f was linearly homogeneous in which case the indirect utility function g can be expressed as follows in terms of the unit cost function $c(p) \equiv C(1,p)$ where $C(u,p)$ was defined by (22) above:

$$(26) \qquad g(p,y) = y/c(p).$$

Konüs and Byushgens considered three classes of homothetic preferences which were defined via the indirect utility function $g(p, y)$ or equivalently, using (26), via the unit cost function $c(p)$.

The first case they considered had the following unit cost function:

$$(27) \qquad c(p) \equiv \sum_{i=1}^{N} a_i p_i, \qquad a_i > 0, \quad i = 1, \ldots, N.$$

As is well known,[29] the dual direct utility function is the fixed coefficient or no substitution f defined as follows:

$$(28) \qquad f(q_1, \ldots, q_n) \equiv \min_i \{q_i/a_i : i = 1, \ldots, N\}.$$

Under these conditions, Konüs and Byushgens [1926; 162] showed that the Laspeyres and Paasche indexes, P_L and P_P defined by (2) and (3) above, will

[27]For a nice exposition and evaluation of the Frisch and Wald quadratic approximation results as well as of some proposals by Samuelson [1974a; 16], see Balk [1981].

[28]The Konüs and Byushgens [1926] paper was published in Russian and non Russian speaking economists seemed to be unaware of the paper although it was indirectly mentioned by Schultz [1939; 8].

[29]See Pollak [1971a; 105].

exactly equal the true cost of living (defined by the right hand side of (23) for any positive utility level u), provided that the consumer's direct utility function is defined by (26) and (27).

The second case they considered was the case of Cobb-Douglas preferences[30] which can be characterized by the following unit cost function:

$$c(p) \equiv \alpha \prod_{i=1}^{N} p_i^{\alpha_i}, \qquad \alpha_i > 0, \quad \sum_{i=1}^{N} \alpha_i = 1. \tag{29}$$

Konüs and Byushgens [1926; 165] showed that the generalized Jevons index defined by (6) (except that $1/N$ is replaced by α_i) is *exact* for these Wicksell-Cobb-Douglas preferences, where the unknown parameters α_i can be determined as follows: $\alpha_i = p_i^t q_i^t / p^t \cdot q^t$, the ith expenditure share, $i = 1, \ldots, N$, for any period t.

In the final case considered by Konüs and Byushgens [1926; 168], the consumer's preferences were characterized by the following unit cost function:

$$c(p) \equiv (p \cdot Bp)^{1/2} = \left(\sum_{i=1}^{N} \sum_{j=1}^{N} b_{ij} p_i p_j\right)^{1/2} \tag{30}$$

where $B \equiv [b_{ij}]$ is a symmetric $N \times N$ matrix of unknown parameters that characterize preferences. They showed that the Fisher ideal index P_F defined by (4) was *exact* for the preferences characterized by (26) and (30). Konüs and Byushgens [1926; 171] also showed that if the inverse of the matrix B existed, say $A = B^{-1}$, then the direct utility function corresponding to (26) and (30) was

$$f(q) \equiv (q \cdot Aq)^{1/2} \tag{31}$$

which is a monotonic transformation of the homogeneous quadratic utility function considered by Frisch and Wald.

Finally, Konüs and Byushgens [1926; 171] exhibited both the system of inverse demand functions, $p/y = Aq/q \cdot Aq$, and the system of ordinary demand functions, $q = yBp/p \cdot Bp$, that correspond to the homogeneous quadratic preferences defined by (31). They also suggested that the unknown parameters appearing in the A or B matrices could be determined given a sufficient number of price and quantity observations. However, note that a knowledge of A or B is *not* required in order to evaluate the Fisher price index P_F.

After the contributions of Bowley, Frisch and Wald to the theory of exact index numbers, the subject remained dormant until Afriat [1972b; 44–47], Pollak [1971a; 117–132] and Samuelson and Swamy [1974; 573–574] reexamined

---

[30]The Cobb-Douglas functional form was perhaps first used in the economics literature by Wicksell [1958; 98] in the nineteenth century.

the subject. All of these authors examined the three cases considered by Konüs and Byushgens and some other cases as well.

Diewert [1976a; 134] defined a price index function $P(p^1,p^2,q^1,q^2)$ to be *superlative* if P was exact for preferences which had a cost function $C(u,p) = uc(p)$ where $c(p)$ is a unit cost function that could provide a second order approximation to an arbitrary twice continuously differentiable linearly homogeneous function. The idea was that a superlative index number formula $P(p^1,p^2,q^1,q^2)$, which could be evaluated using only observable price and quantity data for the two periods under consideration, would correspond to a flexible functional form for a unit cost function $c(p)$. For example, P_F defined by (4) is a superlative price index since it is exact for the c defined by (30) and this c has the required second order approximation property. Another example of a superlative index is the Walsh index P_W defined by (5) since it is exact for the unit cost function which is dual to the aggregator function $f(q) \equiv \sum_{i=1}^N \sum_{j=1}^N a_{ij} {q_i}^{1/2} {q_j}^{1/2}$; see Diewert [1976a; 132]. A third example of a superlative price index is P_T defined by (19) which is exact for a translog unit cost function;[31] see Diewert [1976a; 121].

Unfortunately, Diewert [1976a] defined two (infinite) families of superlative index number formulae and this raised the question as to which formula should be used in empirical applications. However, Diewert [1978b] showed that all choices of a superlative formulae gave the same answer to the second order and hence the choice was usually immaterial. More precisely, Diewert showed that every known superlative index number formula $P(p^1,p^2,q^1,q^2)$ had the same first and second derivatives when evaluated at equal prices (i.e., $p^1 = p^2$) and equal quantities (i.e., $q^1 = q^2$).[32]

As an interesting footnote to the history of economic thought, it should be noted that Diewert was not the first to use the above second order approximation technique when evaluating index number formulas and their derivatives; Edgeworth [1901; 410–411] used a variant of it to show that the Walsh index P_W defined by (5) approximated to the second order the Edgeworth [1925; 213] – Marshall [1887; 372] index defined as follows:

(32)
$$P_{EM}(p^1,p^2,q^1,q^2) \equiv \left[\sum_{i=1}^N (1/2)(q_i^1 + q_i^2)p_i^2\right] \Big/ \left[\sum_{j=1}^N (1/2)(q_j^1 + q_j^2)p_j^2\right].$$

The Taylor series expansion technique around an equal price and quantity point was used again by Edgeworth [1923; 347] to show that the Laspeyres index P_L defined by (2) and the Fisher index P_F defined by (4) satisfy the

[31]The translog unit cost function was introduced by Christensen, Jorgenson and Lau [1971].

[32]Vartia [1978] provided an alternative derivation of this result.

circularity test (9) to the first order. Since this proposition seems to have been forgotten, we sketch a proof of it.

Let the index number function P be either P_L or P_F and define the functions f and g as follows:

$$(33)\qquad \begin{aligned} f(p^1,p^2,p^3,q^1,q^2,q^3) &\equiv P(p^1,p^2,q^1,q^2)P(p^2,p^3,q^2,q^3);\\ g(p^1,p^2,p^3,q^1,q^2,q^3) &\equiv P(p^1,p^3,q^1,q^3). \end{aligned}$$

For $P = P_L$ or for $P = P_F$, it can be verified that the following equalities hold:

$$(34)\qquad \begin{aligned} f(p^1,p^2,p^3,q^1,q^2,q^3) &= g(p^1,p^2,p^3,q^1,q^2,q^3) &&= 1;\\ \nabla_{p^1} f(p^1,p^2,p^3,q^1,q^2,q^3) &= \nabla_{p^1} g(p^1,p^2,p^3,q^1,q^2,q^3) &&= -q/p\cdot q;\\ \nabla_{p^2} f(p^1,p^2,p^3,q^1,q^2,q^3) &= \nabla_{p^2} g(p^1,p^2,p^3,q^1,q^2,q^3) &&= 0_N;\\ \nabla_{p^3} f(p^1,p^2,p^3,q^1,q^2,q^3) &= \nabla_{p^3} g(p^1,p^2,p^3,q^1,q^2,q^3) &&= q/p\cdot q;\\ \nabla_{q^i} f(p^1,p^2,p^3,q^1,q^2,q^3) &= \nabla_{q^i} g(p^1,p^2,p^3,q^1,q^2,q^3) &&= 0_N, \quad i=1,2,3, \end{aligned}$$

provided that the above functions are evaluated at equal prices (i.e., $p^1 = p^2 = p^3 \equiv p$) and equal quantities (i.e., $q^1 = q^2 = q^3 \equiv q$) where $\nabla_{p^i} f \equiv (df/dp_1^i, \ldots, df/dp_N^i)$ is the vector of first order partial derivatives of f with respect to the components of p^i, etc. The meaning of (34) is that the functions f and g approximate each other to the first order when evaluated at an equal price and quantity point. Thus the Laspeyres and Fisher price indexes satisfy the circularity test to the first order.

Using the results in Diewert [1978b; 898] and the results in the above paragraph, it can be shown that the Paasche, Laspeyres and all superlative index number formulae will similarly satisfy the circularity test to the first order.

Note that Edgeworth's proposition helps to explain Fisher's [1922; 280] empirical finding that P_F satisfied the circularity test to a very high degree of approximation.

6.4 Econometric estimation of preferences

In this variant of the economic approach to index numbers, the parameters that characterize consumer preferences are estimated. Preferences may be represented by: (i) the direct utility function; (ii) the indirect utility function; (iii) the distance function[33] or (iv) the cost or expenditure function. Once any one of these functions is known, the other three functions can be calculated, at least in principle. In particular, the cost function $C(u,p)$ can be calculated

[33] Malmquist [1953] was the first to use the distance function in index number theory.

and hence the Konüs price index defined by the right hand side of (23) can be calculated. Additional references to the recent literature on the econometric approach to the estimation of preferences can be found in Chapter 7 of Deaton and Muellbauer [1980].

It seems appropriate to add a few references to the early history of this approach to index number theory.

The earliest effort at a strategy for determining the parameters which characterize preferences was made by Bennet [1920; 462]. He assumed a quadratic direct utility function and showed how the consumer's system of demand functions could be obtained in the three good case (although he did not quite exhibit a closed form solution). He then made some comments on how many observations would be required in the general N good case in order to determine all $1+N+(1/2)N(N+1)$ of the parameters of the quadratic utility function $f(q)$ defined by (24).[34]

We have already seen that Konüs and Byushgens [1926] were contributors to the econometric approach, since they derived the demand functions corresponding to Cobb–Douglas preferences and homogeneous quadratic preferences. Konüs and Byushgens [1926; 172] noted that one price-quantity observation would suffice to determine the parameters of Cobb-Douglas preferences while $(N+1)/2$ observations would be required to determine all $N(N+1)/2$ parameters for the homogeneous quadratic functional forms, (30) or (31). They also noted that statistical determination of the homogeneous quadratic preferences would be difficult.

Another early contributor to the econometric approach to index numbers was Wald [1937] [1939; 325], who assumed quadratic preferences; recall (24). In addition, he assumed that the demand functions regarded as functions of income (or expenditure) were known functions for the two periods under consideration. (Alternatively, just a knowledge of the income elasticities of demand at the two observed price and quantity points would suffice.) With these assumptions, Wald was able to derive an exact index number formula, so his approach is actually a blend of the exact and econometric approaches.

Wald's *blended* approach seems worthy of further study. However, the *pure* econometric approach has severe limitations. The problem with the latter approach is that in order to provide a second order approximation to general preferences, we require approximately $N^2/2$ parameters in the N good case. Since N is perhaps equal to 50,000 for a typical consumer (a supermarket alone has 15,000 to 20,000 separate items), the required number of parameters to be

[34]Bennet [1920; 462] recognized that not all of the parameters of a quadratic utility function could be identified (due to the unobservable nature of utility) since he wrote of determining the parameters only up to a ratio.

estimated is approximately 1.25 billion, which would require price and quantity observations for about 25,000 periods.[35]

This concludes our survey of the ancient history of index numbers. We turn now to a discussion of some more controversial issues in index number theory.

7. On the Test Approach to Index Number Theory

Although the economic approach to index number theory is perhaps the most compelling approach, it should be mentioned that the test approach has some advantages. In particular, the test approach does not suffer from the following limitations of the economic approach: (i) the economic approach is based on optimizing behavior, an assumption which may not be warranted in general; (ii) the economic approach generally relies on separability assumptions[36] about the underlying aggregator functions, assumptions which are unlikely to be true in general and (iii) in deriving capital rental prices, the economic approach is usually based on ex ante expectations about future prices, expectations which cannot be observed, whereas the test approach can be based on ex post accounting data, which can be observed.[37]

8. On the Chain Principle and Multilateral Indexes

The *chain principle* can be contrasted with the *fixed base principle* for constructing a series of index numbers which extends over three or more periods. Given price and quantity data, p^i, q^i, $i = 1, 2, 3$ for three periods and a bilateral price index function $P(p^1, p^2, q^1, q^2)$ that depends only on the data for two periods, the *fixed base sequence* of aggregate price levels for the three periods would be

$$(35) \qquad 1,\ P(p^1, p^2, q^1, q^2),\ P(p^1, p^3, q^1, q^3)$$

---

[35]These perhaps overly pessimistic calculations lead to the following impossibility theorem: economists will never know the truth, even to the second order. However, recently Diewert and Wales [1988] have devised techniques for obtaining good approximations to completely flexible preferences using only a minimal number of parameters.

[36]For a definitive treatment of separability concepts and duality theory, see Blackorby, Primont and Russell [1978].

[37]See the theoretical discussion in Diewert [1980; 475–476] and the empirical results in Harper, Berndt and Wood [1989] on alternative rental price formulae.

while the *chain sequence* of price levels would be

$$(36) \qquad 1,\ P(p^1,p^2,q^1,q^2),\ P(p^1,p^2,q^1,q^2)P(p^2,p^3,q^2,q^3).$$

Historically, the fixed base principle was the first to be used empirically. In the English language literature, the chain principle was first proposed by Alfred Marshall [1887; 373],[38] basically as a method for overcoming the difficulties in comparing prices over two distant periods, due to the invention of new commodities.[39]

Irving Fisher [1911; 203], who gave the chain system its name, noted that the chain system was invariant to changes in the base period and he also saw the advantage of the method in dealing with the new good problem as the following quotation indicates:[40]

> It may be said that the cardinal virtue of the successive base or chain system is the facility it affords for the introduction of new commodities, the dropping out of obsolete commodities, and the continued readjustment of the system of weighting to new commodities.
>
> Fisher [1911; 204]

While it is true that the use of the chain principle has an advantage in dealing with the introduction of new commodities, it has the following severe disadvantage: it does not satisfy Walsh's multiperiod identity test, (13) above.[41] Thus, as Szulc [1983] and Hill [1988] show, if prices and quantities systematically oscillate around constant values, the use of the chain method will give biased results.

The above difficulty with the chain method was not adequately appreciated by Diewert [1978b; 895] who argued for the use of the chain principle on the

[38]Walsh [1901; 207] attributed the chain principle to Julius Lehr [1885; 45–46], who was motivated to introduce the principle in order to deal with new goods.
[39]Divisia [1926; 44–47] saw his method as being a variant of the chain method, with the basic discrete period being one year (which would minimize seasonal fluctuations). Divisia [1926; 45] also thought that the chain method was the only logical way to make price comparisons over long periods due to the introduction of new goods and the discovery of new inventions. As new goods in his time, he mentioned machine guns (to replace the bow and arrow), submarines, aircraft, the potato and the steam engine.
[40]However, later Fisher [1922; 308–309] preferred the fixed base system. Pigou [1932; 71] criticized Fisher's later position and endorsed the chain principle for the usual reason: with the introduction of new commodities, the chain principle is the only way to make comparisons between distant periods.
[41]Walsh [1901; 401] [1924; 506] made this criticism of the chain system. Szulc [1983; 540] also uses Walsh's test in his evaluation of fixed base versus chain index numbers.

grounds that it would reduce the spread between the Laspeyres and Paasche indexes, (2) and (3) above, and between all known superlative indexes, since price and quantity changes will generally be smaller between adjacent periods than between distant periods. He argued that the spread between the Paasche and Laspeyres indexes would be greater than between the superlative indexes because P_L and P_P only approximate each other to the first order, while superlative indexes approximate each other to the second order (recall our discussion of the Edgeworth second order approximation technique in Section 6.3 above). Diewert [1978b; 894] also presented the results of some numerical experiments using Canadian per capital consumption data for 13 commodity classes over the years 1947–1971. These results showed that the chain method did in fact lead to a smaller spread between P_L defined by (2), P_P defined by (3), P_F defined by (4) and P_T defined by (19) than when a fixed base year, 1947, was used.

Although the chain method will give poor results with oscillating data, Szulc [1983] and Hill [1988] show theoretically that chaining will tend to reduce the spread between the Laspeyres and Paasche price indexes, provided that prices and quantities trend monotonically over the time periods in question. Thus Diewert's [1978b; 894] empirical results could be rationalized by the hypothesis that monotonic trends in the data outweighed oscillatory movements.

While the chain system fails to satisfy Walsh's identity test (13), Hill [1988] showed that the fixed base system fails to satisfy an analogue to the strong identity test (8). Consider the base period to be period 0 and suppose that the bilateral price and quantity indexes, P and Q respectively, are given and they satisfy the product test (14). Then under the fixed base system, price P^* and quantity Q^* comparisons between periods t and $t+1$ are made as follows:

$$(37) \qquad P^*(p^0,p^t,p^{t+1},q^0,q^t,q^{t+1}) \equiv P(p^0,p^{t+1},q^0,q^{t+1})/P(p^0,p^t,q^0,q^t);$$

$$(38) \qquad Q^*(p^0,p^t,p^{t+1},q^0,q^t,q^{t+1}) \equiv Q(p^0,p^{t+1},q^0,q^{t+1})/Q(p^0,p^t,q^0,q^t).$$

The problem is that P^* and Q^* need not satisfy counterparts of (8) even if the underlying bilateral indexes P and Q do satisfy (8); i.e., it will *not* generally be the case that

$$(39) \qquad P^*(p^0,p^t,p^{t+1},q^0,q^t,q^{t+1}) = 1 \quad \text{if } p^t = p^{t+1}.$$

Hill [1988; 6–7] cited the fixed base Paasche quantity index, $Q(p^0,p^t,q^0,q^t) \equiv p^0 \cdot q^t/p^0 \cdot q^0$, normally used in the national accounts, as an example of a fixed base index which generates a P^* which fails to satisfy the Hill identity test (39).[42] Thus both the chain and the fixed base systems fail to satisfy theoretically appropriate identity tests.

We turn now to a discussion of possible alternatives to the use to either the fixed base or chain systems. The first person to propose alternatives was Walsh [1901; 431], but it is convenient to start our discussion by reviewing some proposals due to Irving Fisher [1922; 297–320].

Fisher's [1922; 298] first alternative was to list each and every possible binary comparison and thus the index number user could simply pick out the binary comparisons of interest. Let us call this the *all binary comparisons method.* Fisher actually implemented this method for his data on 36 primary commodities for 6 years using the Fisher ideal price index P_F defined by (4). Thus there were $6(6-1) = 30$ bilateral comparisons. Fisher [1922; 301] then used these bilateral comparisons to determine whether the base invariance test (7) was satisfied to a high degree of approximation:[43] in Fisher's [1922; 302] words, "the differences due to differences of base are trifling."

However, Fisher [1922; 299–305] recognized that it was not practical or worthwhile to list every possible binary comparison: for twenty periods, there would be 380 separate index numbers. Thus Fisher was led to consider other classes of alternatives to the use of either the fixed base or the chain systems. These other methods we call *multiperiod systems*, or in the context of regional comparisons, *multilateral systems.* If there are data for T periods, these multilateral methods make use of the data for all T periods simultaneously to construct a series of T price levels and T quantity levels.

Fisher's [1922; 305] first multiperiod or multilateral method was the *blend system*, which works as follows. Suppose we have price and quantity data, $p^t \equiv (p_1^t, \ldots, p_N^t)$ and $q^t \equiv (q_1^t, \ldots, q_N^t)$ for $t = 1, 2, \ldots, T$ and a bilateral index number formula $P(p^1, p^2, q^1, q^2)$. Then use each period as the base to construct a series of T aggregate price levels using the bilateral function P. Normalize the resulting series so that the first price is unity.[44] The resulting T series or normalized price levels may be written as follows:

$$
(40)\quad
\begin{array}{lll}
1, P(p^1,p^2,q^1,q^2)/P(p^1,p^1,q^1,q^1), & \ldots, P(p^1,p^T,q^1,q^T)/P(p^1,p^1,q^1,q^1); \\
1, P(p^2,p^2,q^2,q^2)/P(p^2,p^1,q^2,q^1), & \ldots, P(p^2,p^T,q^2,q^T)/P(p^2,p^1,q^2,q^1); \\
\vdots \quad \vdots & \vdots \\
1, P(p^T,p^2,q^T,q^2)/P(p^T,p^1,q^T,q^1), & \ldots, P(p^T,p^T,q^T,q^T)/P(p^T,p^1,q^T,q^1).
\end{array}
$$

[42]Actually, Hill [1988; 7] uses the more general proportionality test, $P^*(p^0, p^t, \lambda p^t, q^0, q^t, q^{t+1}) = \lambda$ for $\lambda > 0$, in place of (39).

[43]Fisher [1922; 280–283] also found that P_F satisfied Walsh's multiperiod identity test (13) to a high degree of approximation.

[44]Actually, Fisher [1922; 301] used a slightly different normalization.

Fisher then suggested that the price level for period t, P^t say, should be an arithmetic average of the period t price level in each of the T series in (40); i.e., we have $P^1 = 1$ and for $t = 2, 3, \ldots, T$:

$$(41) \qquad P^t \equiv (1/T) \sum_{k=1}^{T} P(p^k, p^t, q^k, q^t)/P(p^k, p^1, q^k, q^1).$$

Note that the *multiperiod* or *multilateral period t aggregate price level* P^t defined by (41) is a function of the data for all T periods; i.e., $P^t = P^t(p^1, \ldots, p^T, q^1, \ldots, q^T)$. The corresponding *multiperiod* or *multilateral period t aggregate quantity level* Q^t may be defined as

$$(42) \qquad Q^t(p^1, \ldots, p^T, q^1, \ldots, q^T) \equiv p^t \cdot q^t / P^t(p^1, \ldots, p^T, q^1, \ldots, q^T).$$

Fisher's [1922; 307] second multilateral method was the *broadened base system.* In this method, the period t price level P^t was defined as follows:

$$(43) \qquad P^t(p^1, \ldots, p^T, q^1, \ldots, q^T) \equiv (1/T)\Big(\sum_{k=1}^{T} q^k\Big) \cdot p^t, \quad t = 1, \ldots, T.$$

The corresponding quantity levels Q^t can be defined numerically using the multilateral product test relations (42). This method was used by the Economic Commission for Latin America in the early 1960's. It is described in Ruggles [1967; 185] and is referred to as the *market basket method.* Note that (43) is a T period generalization of the two period Edgeworth-Marshall bilateral formula, P_{EM}, defined by (32) above.

Fisher was not the first to suggest the market basket method: Walsh [1901; 431] suggested the same method but called it *Scrope's method with arithmetic weights.* Walsh [1901; 399] also suggested the following system of multilateral price levels P^t, which he called *Scrope's method with geometric weights*:

$$(44) \qquad P^t \equiv \sum_{i=1}^{N} \Big(\prod_{k=1}^{T} q_i^k\Big)^{1/T} p_i^t, \quad t = 1, \ldots, T.$$

The corresponding Q^t can be defined by (42) as usual. It is obvious that (44) is the multilateral generalization of Walsh's bilateral index P_W, defined by (5) above.

The above multilateral methods construct price and quantity levels for each period. It is also possible to devise multilateral methods where the multilateral price index P^{st} gives the level of prices in period t relative to the level of prices in period s and is a function of all of the price and quantity information for the T periods; i.e., $P^{st} = P^{st}(p^1, \ldots, p^T, q^1, \ldots, q^T)$. Thus Walsh [1924; 509] defined the following multiperiod generalization of the Fisher ideal index (4):

$$(45) \quad P^{st} \equiv (p^t \cdot q^1 p^t \cdot q^2 \ldots p^t \cdot q^T / p^s \cdot q^1 p^s \cdot q^2 \ldots p^s \cdot q^T)^{1/T}, \quad s, t = 1, \ldots, T.$$

The corresponding quantity indexes can be defined as follows:

$$(46)\quad Q^{st}(p^1,\ldots,p^T,q^1,\ldots,q^T) \equiv p^t\cdot q^t/p^s\cdot q^s P^{st}(p^1,\ldots,p^T,q^1,\ldots,q^T).$$

Gini [1931; 10] also defined the multilateral price indexes (45) and called the method the *successive weights system*. Both Walsh and Gini noted that (45) collapsed to P_F defined by (4) if $T=2$ and $s=1$ and $t=2$.

Finally, Gini [1931; 12] did propose a new multilateral method which he called the *circular weight system*, because the resulting system of price indexes P^{st} satisfied a multilateral analogue to the circular test (9).[45] Given any bilateral price index function $P(p^1,p^2,q^1,q^2)$, define Gini's multilateral level of prices in period t relative to the period s level as follows:

$$(47)\qquad P^{st} \equiv \left[\frac{P(p^1,p^t,q^1,q^t)}{P(p^1,p^s,q^1,q^s)}\,\frac{P(p^2,p^t,q^2,q^t)}{P(p^2,p^s,q^2,q^s)}\cdots\frac{P(p^t,p^t,q^T,q^t)}{P(p^T,p^s,q^T,q^s)}\right]^{1/T}$$

for $s,t=1,\ldots,T$. The corresponding Q^{st} can be defined by (46). This method was later proposed (using P_F as the P) by Eltetö and Köves [1964] and Szulc [1964] in the multiregional context and is known as the *EKS system*.

Gini [1931; 13–24] tested out his circular weight system (as well as some other alternatives) using the Fisher ideal P_F defined by (4) as his bilateral P, for eight time period observations on five Italian cities. Thus Gini's computations were both multiperiod (between time periods) and multilateral (between locations).

The *ratio* type price indexes, (45) proposed by Walsh and (47) proposed by Gini, can be converted into price *levels* $P^t(p^1,\ldots,p^T,q^1,\ldots,q^T)$ as follows: corresponding to (45), define the period t price level P^t as

$$(48)\qquad P^t \equiv (p^t\cdot q^1 p^t\cdot q^2\ldots p^t\cdot q^T)^{1/T},\quad t=1,\ldots,T,$$

and corresponding to (47), define P^t as

$$(49)\qquad P^t \equiv [P(p^1,p^t,q^1,q^t)P(p^2,p^t,q^2,q^t)\ldots P(p^T,p^t,q^T,q^t)]^{1/T}.$$

In each case, it can be verified that $P^{st}=P^t/P^s$.

If we take the Gini-EKS price levels defined by (49) and divide each of them through by P^1, it can be seen that the resulting normalized price levels P^t/P^1 are closely related to Fisher's (normalized) blended price levels defined by (41): for the Fisher price levels, we take the arithmetic means of the numbers $P(p^k,p^t,q^k,q^t)/P(p^k,p^1,q^k,q^1)$, $k=1,\ldots,T$, while for the Gini-EKS price levels, we take the geometric mean of the same T numbers.

[45]In fact, all of the multilateral indexes defined in this section satisfy the multilateral circularity test.

Walsh [1901; 399] [1924; 509] noted the primary disadvantage of using the multiperiod full information price level functions $P^t(p^1, \ldots, p^T, q^1, \ldots, q^T)$: if the number of periods increases, all of the indexes have to be recomputed.[46] This is not necessarily a fatal objection since it is normal practice for statistical agencies to periodically issue historical revisions and there is no reason why the revisions could not be accomplished using multiperiod indexes.

However, at present, it does not seem prudent to enthusiastically endorse a multiperiod system of index numbers since not enough research has been done on the axiomatic properties of the various multilateral or multiperiod alternatives.[47] Moreover, it would be desirable to develop multiperiod exact and superlative index number formulae and then examine the axiomatic properties of the resulting indexes. In the bilateral case, the Fisher ideal price index P_F emerges as the natural choice of a functional form since it seems to satisfy more reasonable tests than any other known formula and it is superlative as well. We need a multilateral counterpart to this "ideal" bilateral functional form.

To sum up: a comparison of the fixed base, chain, all binary comparisons and multiperiod systems leads to no clear choice at this stage. However, if a definite choice has to be made, I would vote for the chain system used with the bilateral Fisher ideal index P_F.

9. Is the Substitution Bias Small?

The substitution bias in the consumer price index is the discrepancy between the Laspeyres or Paasche price indexes, P_L and P_P defined by (2) and (3) above, and the consumer's true cost of living index, defined by the right hand side of (23). There is an analogous substitution bias in the output price index.

Some researchers argue that these substitution biases can be ignored. For instance, Triplett writes: "Though it has long been a staple of economists' educations, the substitution bias in a fixed-weight price index for consumption is just not very large" (Triplett [1988; 26]). To support the above opinion, Triplett cites the relatively close agreement between the Laspeyres and Paasche price indexes for U.S. aggregate consumption data; see Manser and McDonald [1988]. Triplett's judgment would be correct if in fact these indexes were true microeconomic Paasche and Laspeyres indexes, but they are not: microeconomic samples of price ratios p_i^{t+1}/p_i^t for various goods i are combined with

[46] Walsh [1901; 399] also made the following theoretical objection: "Besides, how is a past variation between two years several years ago to be affected by present variations?"

[47] For a start on this topic, see Diewert [1987] [1988].

base period expenditure shares that are obtained from periodic consumer expenditure surveys. The resulting aggregate indexes are not quite P_L and P_P defined by (2) and (3) above.

It may well be that Triplett is correct in his judgment, but the evidence to support his position has not yet been presented.

Due to the computer revolution, it is now possible to undertake some experiments which could help to determine the extent of the substitution bias. Retail outlets that have computerized price and quantity information on their sales could be sampled. Detailed microeconomic price and quantity vectors p^t and q^t could be constructed and the Laspeyres, Paasche and Fisher indexes defined by (2)–(4) above could be calculated and compared with corresponding official consumer or producer price indexes that covered the same range of goods. Such firm oriented experiments could provide useful information on the size of the substitution bias.[48]

10. Is the New Good Bias Small?

Changes in quality and the introduction of new goods are the source of another bias problem. We first briefly review the ancient literature on methods for quality adjustment.[49]

Some of the early researchers on price measurement were aware of the problem of quality change but the pace and direction of the change did not seem large enough to warrant an explicit treatment.[50]

However, by the latter part of the nineteenth century, Sidgwick realized that not only were improvements in the quality of goods leading to a bias in price comparisons, but also the growth of international and interregional trade (due primarily to transportation improvements) led to the systematic introduction of "entirely new kinds of things" and this too led to a bias in

[48]Such firm based experiments would yield information on the substitution bias in output price indexes. Perhaps some day in the future when consumers use credit or banking cards to pay for all of their purchases, we could obtain an accurate paper trail that could be used to construct true microeconomic Laspeyres, Paasche and Fisher price indexes.

[49]For reviews of the modern literature on quality adjustment using hedonic regression techniques, see Griliches [1990] and Triplett [1990b].

[50]Thus Lowe [1823; Appendix 87] states: "In regard to the quality of our manufactures, we must speak with more hesitation, and can hardly decide whether the balance be in favour of the present or of a former age; for if our fabrics are now much more neat and convenient, they are in a considerable degree less durable."

price comparisons. As the following quotation indicates, Sidgwick thought that utility theory would play a role in eliminating these biases:

> Here again there seems to be no means of attaining more than a rough and approximate solution of the problem proposed; and to reach even this we have to abandon the prima facie exact method of comparing prices, and to substitute the essentially looser procedure of comparing amounts of utility or satisfaction.
>
> Sidgwick [1883;68]

Unfortunately, the mathematical apparatus of consumer theory was not sufficiently developed at that time to enable Sidgwick to make any specific progress on the new good problem.

In a brilliant paper, Marshall not only proposed the tabular standard, the chain system and the Edgeworth-Marshall index number formula (32), he also made the first real progress on the appropriate treatment of new goods, as the following quotation indicates:

> This brings us to consider the great problem of how to modify our unit so as to allow for the invention of new commodities. The difficulty is insuperable, if we compare two distant periods without access to the detailed statistics of intermediate times, but it can be got over fairly well by systematic statistics. A new commodity almost always appears at first at something like a scarcity price, and its gradual fall in price can be made to enter year by year into readjustments of the unit of purchasing power, and to represent fairly well the increased power of satisfying our wants which we derive from the new commodity.
>
> Marshall [1887; 373]

As the above quotation indicates, Marshall was well aware of the product cycle and he felt that the early introduction of new commodities into the consumer price index in the context of the chain system would capture *most* of the benefits due to the introduction of new commodities. As we shall see later, not quite *all* of the benefits are captured using Marshall's suggested method, since his method incorrectly ignores the new good in the first period that it makes its appearance.

Marshall [1887; 373–374] also realized that improvements in transportation led to the general availability of location specific goods, such as fish at the seaside or strawberries at a farm. Marshall correctly felt that these "old" goods that suddenly became available at many locations should be regarded as "new" goods and treated in the same way as a genuinely new good. His words on this important observation are worth quoting:

> This class of consideration is of much more importance than at first sight appears; for a great part of modern agriculture and transport industries are devoted to increasing the periods of time during which

> different kinds of food are available. Neglect of this has, in my opinion, vitiated the statistics of the purchasing power of many in medieval times with regard to nearly all kinds of foods except corn; even the well-to-do would hardly get so simple a thing as fresh meat in winter.
>
> Marshall [1887; 374]

Marshall's suggested treatment of the new good problem (i.e., use the chain system) was acknowledged and adopted by many authors including Irving Fisher [1911; 204] (temporarily) and Pigou [1912; 47]. As we saw earlier in Section 8, Divisia [1926; 45] working from his independent perspective also suggested the use of the chain method as a means of dealing with the new good problem.

The next important contributor to the discussion of new goods in price measurement was Keynes. Keynes [1930; 94] described in some detail one of the most common methods for dealing with the new good problem: simply ignore any new or disappearing goods in the two time periods under consideration and calculate the price index on the basis of the goods that are common to the two situations. The corresponding quantity index was to be obtained residually by deflating the relevant value ratio by this narrowly based price index. Keynes called this method the ***highest common factor method***. This method would be identical to Marshall's chain method if the two time periods were chosen to be adjacent ones. However Keynes [1930; 105–106] advocated his method in the context of a fixed base system of index numbers and he specifically rejected the chain method for three reasons: (i) each time a new product is introduced, a chain index does not take into account the benefits of the expanded choice set, and thus over long periods of time, the chain price index will be biased upwards and the corresponding quantity index will be biased downwards; (ii) the chain index fails Walsh's multiperiod identity test (13) above, and (iii) the chain method was statistically laborious.

Keynes' last objection to the chain method is no longer relevant in this age of computers. Moreover, Keynes was unable to offer any positive alternative to the chain method for comparing situations separated by long periods of time as the following quotation indicates:

> We cannot hope to find a ratio of equivalent substitution for gladiators against cinemas, or for the conveniences of being able to buy motor cars against the conveniences of being able to buy slaves.
>
> Keynes [1930; 96]

However, Keynes' first objection to the chain method (which was later echoed by Pigou [1932; 72][51]) was certainly valid (as was his second objection).

---

[51]Pigou [1932; 71] also had a nice criticism of Keynes' highest common factor method which was later repeated by Hofsten [1952; 59]. Pigou also criticized Fisher's [1922; 308–312] later preference for the fixed base method.

A satisfactory theoretical solution to Keynes' first objection did not occur until Hicks adapted the analytical apparatus of consumer theory to the problem.

When new consumer goods make their appearance for the first time, say in period 2, their prices and quantities can be observed. In period 1, the quantities of the new goods are all obviously zero but what are the corresponding prices? Hicks [1940; 114] provided a theoretical solution:

> They are those prices which, in the 1 situation, would *just* make the demands for these commodities (from the whole community) equal to zero. These prices cannot be estimated, but we can observe that between the two situations the demands for these commodities will have increased from zero to certain positive quantities; and hence it is reasonable to suppose that the 'prices' of these commodities will usually have fallen relatively to other prices. This principle is sufficient to give us a fairly good way of dealing with the case of new goods.
>
> Hicks [1940; 114]

Of course, in the context of the producer price index, the appropriate period 1 shadow prices for the new goods are those prices which just induce each period 2 producer of the new goods to produce zero quantities in period 1.

Hicks' basic idea was used extensively by Hofsten [1952; 95–97] who dealt not only with new goods, but also adapted the Hicksian methodology to deal with disappearing goods as well. Hofsten [1952; 47–50] also presents a nice discussion of various methods that have been used to adjust for quality change.

Frank Fisher and Karl Shell [1972b; 22–26] laid out the formal algebra for constructing the period 1 Hicksian "demand reservation prices" defined in the above quotation by Hicks. Diewert [1980; 498–501] used the Hicksian framework to look at the bias in the Fisher price index P_F defined by (4) when the reservation prices were incorrectly set equal to zero and compared this index to the Fisher price index that simply ignored the existence of the new goods in the two periods under consideration (which is Marshall's method).[52] Diewert [1980; 501–503] also made some suggestions for estimating the appropriate Hicksian reservation prices in an econometric framework.

Is the new good bias large or small? One can only answer this question in the context of the price measurement procedures used by individual statistical agencies. In Diewert [1987; 779], some simple hypothetical examples were given which showed that traditional fixed base procedures could generate much higher measures of price increase than would be generated using the chain method.[53] However, what is needed is empirical evidence.

---

[52]The second index has a smaller bias than the first index.

[53]Since 1978, the U.S. Bureau of Labor Statistics has used a probability sampling approach in the consumer price index which probably reduces some of this fixed weight bias, but the bias is not eliminated.

Numerical computation of alternative methods based on detailed firm data on individual prices and quantities where new goods are carefully distinguished would cast light on the size of the new good bias. Thus the firm oriented experiments suggested at the end of the previous section to cast light on the size of the substitution bias could also be used to study the size of the new good bias.

Another line of empirical work which would be of interest would be to collect industry price and quantity data on various major new goods (e.g., microwave ovens, video recorders, home computers, satellite dishes, etc.) and then attempt to rework the relevant price indexes in the light of this extra data.

11. Has the Theory of the Cost of Living Index Been Exhausted?

Triplett appears to answer the above question in the affirmative as the following quotation indicates:

> The COL index has been subjected to far more research, both theoretical and empirical, than any other price index topic in the history of index numbers. It seems to me that much of the fruit has been picked from this tree.
>
> Triplett [1988; 25]

It seems to me that the harvest is not yet over.

A large gap in our current statistical system is in the area of the consumer's allocation of time. Many years ago, Becker [1965] showed how the consumer's time constraint could be integrated into traditional consumer theory and he applied his new framework to cast light on a wide variety of applied economic problems. Additional applications can be found in a more recent book edited by Juster and Stafford [1985]. In order to implement Becker's theory, information on the consumer's allocation of time is required, broken up into: (i) time at work,[54] (ii) time commuting to work, (iii) time spent shopping, (iv) time spent at housework, and (v) time spent at various leisure activities. Since many productivity improvements involve efficiencies in the consumer's use of time (e.g., a new subway line, an automated banking machine, electronic scanning of prices at the supermarket, etc.), it seems appropriate for statistical agencies to consider the implementation of a version of Becker's framework.

Another area of household statistics which requires further theoretical development and empirical implementation is the area of income statistics: labor

[54]Riddell [1983] discusses the issues involved in incorporating leisure time (defined to be time spent not at work) in measures of the standard of living and he calculates some of his new measures of economic welfare using Canadian data.

income should be decomposed into price and quantity components, income taxes should be taken into account in an appropriate manner and capital gains should be recognized as components of income. The point here is that most of the household measurement theory has concentrated on the commodity demand side and there has not been enough emphasis on the household factor supply and income sides.[55]

12. Conclusion

In Sections 2 to 6 above, we provided an overview of the early literature on price measurement and index number theory in general. In Sections 7 to 11, we discussed various topics that are more controversial. In some cases, we also provided a historical survey of these topics. In Section 8, we reviewed the early literature on the chain principle and various alternatives to it, and in Section 10 we reviewed the early literature on the new good problem.

It seems appropriate to conclude by listing seven recommendations for statistical agencies (the first four agree with those made by Triplett [1988]):

(1) Statistical agencies should be encouraged to provide users with adequate printed documentation.

(2) The decomposition of labor income (on the household side) and labor payments (on the firm side) into price and quantity components needs improvement: weighted index numbers should be used for quantities rather than unweighted personhours, and personhours should be disaggregated into various occupational, educational and demographically homogeneous categories.

(3) In the context of the cost of living index, the flow of services concept should be extended to other classes of consumer durables in addition to housing.

(4) My preferred method for decomposing a value ratio into price and quantity components is the use of a superlative index number formula in the context of the chain method.

(5) Since it is usually impossible to collect complete price and quantity information for each value cell in the relevant accounting framework, it will be necessary to resort to some sort of sampling principle. The appropriate objects to sample are *values* within the relevant cell in the first period. These sampled values would then be broken up into detailed prices and quantities which would then be observed in the following period as well. Finally, Fisher ideal price indexes should be constructed using these sampled values for the two periods and the corresponding quantity indexes should be constructed by deflating the relevant population value ratios by these (sample) price indexes. The entire procedure is explained in some detail by Pigou [1932; 75–77].

[55]See Diewert and Bossens [1992] for an introduction to these topics.

(6) The sizes of the substitution bias and the new good bias are still in doubt. The empirical experiments described in Sections 9 and 10 above would be useful in determining the size of these biases.

(7) More empirical and theoretical work needs to be done on the household supply side; see the suggestions on incorporating Becker's theory of the allocation of time and on the construction of household real income indexes made in Section 11 above.

Essays in Index Number Theory, Volume I
W.E. Diewert and A.O. Nakamura (Editors)

Chapter 3
KONÜS, ALEXANDER ALEXANDROVICH*

W.E. Diewert

Konüs, Alexander Alexandrovich (born 1895). Konüs was born in Moscow on 2 October 1895 and died recently in 1991. During the war years 1914–17, he served in the army and was wounded twice. In 1920, he graduated from the Cooperative Institute in Moscow as a statistician. During the years 1923–30, he worked in the price index section of the Business Cycle (Conjuncture) Institute directed by N.D. Kondratiev and had E.E. Slutsky as a colleague. During the years 1931–45, he taught statistics and undertook analyses of the quality of metal production. In 1945–60, Konüs worked at various research institutes including the Research Institute of Communication. From 1960 to 1982, he was at the Institute of Economic Research in Moscow. He has written over 100 research papers in index number theory, consumer demand, the labor theory of value, statistics and probability theory.

Konüs has made a number of fundamental contributions. In his 1924 paper, he established a revealed preference theory result, he provided the definition of a consumer's true cost of living index as a ratio of cost functions evaluated at different prices but at the same utility level, and he showed that there exists a utility level between the base and current period utility levels such that the true cost of living index evaluated at this intermediate utility level lies between the Paasche and Laspeyres price indexes.

In his 1926 paper written jointly with Byushgens, the idea that preferences can be represented dually in terms of tangential coordinates (i.e. in terms of prices and income instead of quantities) was made, the indirect utility function was introduced and a homogeneous quadratic utility function was shown to be exact for Irving Fisher's ideal index number formula. This is a truly remarkable paper which was well ahead of its time.

The best English language discussion of his later work on consumer demand and index number theory is Konüs [1968].

*First published in *The New Palgrave: A Dictionary of Economics*, Vol. 3, J. Eatwell, M. Milgate and P. Newman (eds.), The Macmillan Press, 1987, p. 62.

Essays in Index Number Theory, Volume I
W.E. Diewert and A.O. Nakamura (Editors)

Chapter 4
LASPEYRES, ERNST LOUIS ETIENNE*

W.E. Diewert

Laspeyres, Ernst Louis Etienne (1834–1913). Laspeyres was born at Halle, Germany, on 28 November 1834 and died on 4 August 1913 at Giessen, Germany.

From 1853 to 1857, he studied at the universities of Tübingen, Berlin, Göttingen and Halle. He received a law degree from the University of Halle in 1857. He studied at the University of Heidelberg from 1857 to 1859, and in 1860 he obtained his PhD from Heidelberg for the thesis, 'The Correlation between Population Growth and Wages'.

From 1860 until 1864 he worked as a lecturer at Heidelberg, where he wrote a history of the economic views of the Dutch (Laspeyres [1863]). In the following ten years, he taught at four different universities: 1864 – Basel; 1866 – the Polytechnic at Riga; 1869 – Dorpat; 1873 – Karlsruhe. Finally, from 1874 to 1900, he taught at the Justus-Liebig University at Giessen.

Laspeyres' main contribution to economics was his development of the index number formula that bears his name. Let the price and quantity of commodity n in period t be p_n^t and q_n^t respectively for $n = 1, \ldots, N$ and $t = 0, 1, \ldots, T$. Then the Laspeyres price index of the N commodities for period t (relative to the base period 0) is defined as

$$P_L \equiv \sum_{n=1}^{N} p_n^t q_n^0 \Big/ \sum_{n=1}^{N} p_n^0 q_n^0.$$

Laspeyres wrote his classic paper [1871] which suggested the above formula partly as an outgrowth of his empirical work on measuring price movements in Germany and partly to criticize the index number formula of Drobisch [1871]. Using the notation defined above, the Drobisch price index for period t is defined as

$$P_D \equiv \left(\sum_{n=1}^{N} p_n^t q_n^t \Big/ \sum_{n=1}^{n} q_n^t\right) \Big/ \left(\sum_{n=1}^{N} p_n^0 q_n^0 \Big/ \sum_{n=1}^{N} q_n^0\right).$$

*First published in *The New Palgrave: A Dictionary of Economics*, Vol. 3, J. Eatwell, M. Milgate and P. Newman (eds.), The Macmillan Press, 1987, pp. 133–134.

Laspeyres criticized this formula by showing that the index generally changed even if all prices remained constant (i.e. P_D does not satisfy an identity test to use modern terminology). An even more effective criticism of P_D is that it is not invariant to changes in the units of measurement (whereas P_L is invariant).

Laspeyres did not write any further papers on index number theory. He wrote papers on economic history, the history of economic thought and on topical economic issues of his time; see Rinne [1981].

Essays in Index Number Theory, Volume I
W.E. Diewert and A.O. Nakamura (Editors)

Chapter 5
INDEX NUMBERS*

W.E. Diewert

The index number problem may be phrased as follows. Suppose we have price data $p^i \equiv (p_1^i, \ldots, p_N^i)$ and quantity data $x^i \equiv (x_1^i, \ldots, x_N^i)$ on N commodities that pertain to economic unit i or that pertain to the same economic unit at time period i for $i = 1, 2, \ldots, I$. The *index number problem* is to find I numbers P^i and I numbers X^i such that

$$(1) \qquad P^i X^i = p^i \cdot x^i \equiv \sum_{n=1}^{N} p_n^i x_n^i \quad \text{for } i = 1, \ldots, I.$$

P^i is the *price index* for period i (or unit i) and X^i is the corresponding *quantity index*. P^i is supposed to be representative of all of the prices p_n^i, $n = 1, \ldots, N$ in some sense, while X^i is to be similarly representative of the quantities x_n^i, $n = 1, \ldots, N$. In what precise sense P^i and X^i represent the individual prices and quantities is not immediately evident and it is this ambiguity which leads to different approaches to index number theory. Note that we require that the product of the price and quantity indexes, $P^i X^i$, equals the actual period (or unit) i net expenditures on the N commodities, $p^i \cdot x^i$. Thus if the P^i are determined, then the X^i may be implicitly determined using equations (1), or vice versa.

Each individual consumes the services of thousands of commodities over a year and most producers utilize and/or produce thousands of individual products and services. Index numbers are used to reduce and summarize this overwhelming abundance of microeconomic information. Hence index numbers intrude themselves on virtually every empirical investigation in economics.

Index number theory splits naturally into two divisions, depending on the size of I. If $I = 2$, so that there are data for only two time periods or two economic units, then we are in the realm of *bilateral index number theory* while if $I > 2$, then we are in the realm of *multilateral indexes*. Bilateral approaches

*First published in *The New Palgrave: A Dictionary of Economics*, Vol. 2, J. Eatwell, M. Milgate and P. Newman (eds.), The Macmillan Press, 1987, pp. 767–780.

are considered in Sections 1–5 below and multilateral approaches are considered in Sections 6–10.

The four main approaches to index number theory are: (i) *statistical* (Section 1), (ii) *test or axiomatic* (Sections 2 and 9), (iii) *microeconomic* which relies on the assumption of maximizing or minimizing behavior (Sections 3, 4 and 5), and (iv) *neostatistical* (Section 10).

1. Statistical Approaches

Let $I = 2$ and consider the following formula for P^2/P^1 due originally to Dutot [1738]:

$$P^2/P^1 = \left(\sum_{n=1}^{N} p_n^2/N\right) \Big/ \left(\sum_{n=1}^{N} p_n^1/N\right). \tag{2}$$

Thus the average level of prices in say period 2 relative to period 1 is set equal to the arithmetic average of the period 2 prices divided by the arithmetic average of the period 1 prices. The right hand side of (2) is called an *index number formula.*

Given an index number formula, we may solve the aggregation problem (1) as follows: set P^2 equal to the index number formula and determine P^1, X^1 and X^2 by:

$$P^1 = 1, \quad X^1 = p^1 \cdot x^1 \quad \text{and} \quad X^2 = p^2 \cdot x^2/P^2. \tag{3}$$

Setting $P^1 = 1$ is regarded as an arbitrary normalization; any other convenient normalization such as $P^1 = 100$ could be chosen, in which case $X^1 \equiv p^1 \cdot x^1/P^1$, $P^2 \equiv (P^2/P^1)P^1$ and $X^2 \equiv p^2 \cdot x^2/P^2$, where P^2/P^1 is the index number formula.

Rather than taking P^2/P^1 to be a ratio of average prices, Carli [1764] suggested taking an average of the price ratios as follows:

$$P^2/P^1 = \sum_{n=1}^{N} (p_n^2/p_n^1)/N. \tag{4}$$

The average of the price ratios in (4) is an arithmetic average. Jevons [1865] suggested using a geometric average:

$$P^2/P^1 = \prod_{n=1}^{N} (p_n^2/p_n^1)^{1/N}. \tag{5}$$

Once the ratio P^2/P^1 has been determined by (4) or (5), P^1, X^1 and X^2 may be determined using (3). But which of the three alternative formulae for P^2/P^1 should we use?

Walsh [1901] criticized the use of the Dutot formula (2) on the grounds that the index was not invariant to changes in the units of measurement. This criticism was a telling one, and virtually nobody uses formula (2) at present. However, formulae (4) and (5) are invariant to changes in the units of measurement, so we must still discriminate between them.

Jevons argued that changes in the quantity of money between the two periods would lead to proportional changes in all prices except for random errors. In particular, Jevons argued that the price ratios, p_n^2/p_n^1, would be independently and symmetrically distributed around a common mean. If this distribution happened to be the normal distribution, then the maximum likelihood estimator for the common mean leads to the index number formula (4). If the ratios p_n^2/p_n^1 happened to be log normally distributed, then statistical considerations would lead us to the index number formula (5).

Bowley [1928] attacked the use of both (4) and (5) on two grounds. First, from an empirical point of view, he showed that price ratios were not symmetrically distributed about a common mean and their logarithms also failed to be symmetrically distributed. Secondly, from a theoretical point of view, he argued that it was unlikely that prices or price ratios were independently distributed. Keynes [1930] developed Bowley's second objection in more detail; he argued that changes in the money supply would not affect all prices at the same time. Moreover, real disturbances in the economy could cause one set of prices to differ in a systematic way from other prices, depending on various elasticities of substitution and complementarity. In other words, prices are not randomly distributed, but are systematically related to each other through the general equilibrium of the economy.

The above criticisms led to a movement away from the use of unweighted averages of price ratios to represent price movements independently of quantity movements. Walsh [1901] and others suggested that the quantity observations x_n^i that were associated with the individual price observations p_n^i should be used as weights in the price index formula.

Scrope [1833] suggested the following formula:

$$P^2/P^1 = p^2 \cdot x/p^1 \cdot x \tag{6}$$

where $x \equiv (x_1, \ldots, x_N)$ was a somewhat vaguely specified quantity vector which was used to weight the price vectors p^1 and p^2 as in (6).

Laspeyres [1871] recommended that x be set equal to x^1, the period 1 quantity vector, while Paasche [1874] suggested that x be set equal to x^2, the period 2 quantity vector. This led to the following formulae:

$$P^2/P^1 = p^2 \cdot x^1/p^1 \cdot x^1 \equiv P_L(p^1, p^2, x^1, x^2); \tag{7}$$

$$P^2/P^1 = p^2 \cdot x^2/p^1 \cdot x^2 \equiv P_P(p^1, p^2, x^1, x^2). \tag{8}$$

Note that we are now following custom in the index number literature by defining an index number formula to be a function $P(p^1, p^2, x^1, x^2)$ of the price and quantity vectors that pertain to the two observations or time periods under consideration: P_L defines the Laspeyres price index while P_P defines the Paasche price index.

Pigou [1912] and Irving Fisher [1922] advocated taking a geometric mean of the Paasche and Laspeyres indexes and the resulting formula has come to be known as the Fisher ideal price index P_F:

$$P^2/P^1 = [P_L P_P]^{1/2} \equiv P_F(p^1, p^2, x^1, x^2). \tag{9}$$

Rather than taking a geometric average of (7) and (8), Walsh [1901] [1921] advocated using formula (6) where the weight vector x was chosen to be the vector of geometric means of the two quantity vectors:

$$\begin{aligned} P^2/P^1 &= \sum\nolimits_{n=1}^{N} p_n^2 (x_n^1 x_n^2)^{1/2} \Big/ \sum\nolimits_{n=1}^{N} p_n^1 (x_n^1 x_n^2)^{1/2} \\ &\equiv P_W(p^1, p^2, x^1, x^2). \end{aligned} \tag{10}$$

Törnqvist [1936] advocated a weighted geometric mean of the price ratios of the following form:

$$P^2/P^1 = \prod\nolimits_{n=1}^{N} (p_n^2/p_n^1)^{s_n} \equiv P_T(p^1, p^2, x^1, x^2) \tag{11}$$

where $s_n \equiv (1/2)(p_n^1 x_n^1 / p^1 \cdot x^1) + (1/2)(p_n^2 x_n^2 / p^2 \cdot x^2)$ is the average expenditure share on good n for $n = 1, \ldots, N$.

It turns out that formulae (7), (8), (9) and (11) are the most widely used formulae for a price index. However, at this point, we have no way of justifying their popularity. Walsh [1901] and Fisher [1922] present scores of functional forms for price indexes — on what basis are we to choose one as being better than the other?

This question leads us to discuss the test or axiomatic approach to index number theory.

2. The Test Approach to Bilateral Indexes

Consider $P(p^1, p^2, x^1, x^2)$, a function of the N period i prices, $p^i \equiv (p_1^i, \ldots, p_N^i)$, and the N period i quantities, $x^i \equiv (x_1^i, \ldots, x_N^i)$ for $i = 1, 2$. The price index P is supposed to represent the level of prices in period 2 relative to period 1. What properties or *tests* should such an index number formula satisfy? The following nine tests (or closely related variants) have been considered in the literature.

BT1: *Identity Test:* $P(p^1, p^2, \alpha x^1, \beta x^2) = 1$ for all numbers $\alpha > 0$, $\beta > 0$ if $p^1 = p^2$ and $x^1 = x^2$.

BT2: *Proportionality Test:* $P(p^1, \alpha p^2, x^1, x^2) = \alpha P(p^1, p^2, x^1, x^2)$ for $\alpha > 0$.

BT3: *Invariance to Changes in Scale Test:* $P(\alpha p^1, \alpha p^2, \beta x^1, \gamma x^2) = P(p^1, p^2, x^1, x^2)$ for all $\alpha > 0$, $\beta > 0$ and $\gamma > 0$.

BT4: *Invariance to Changes in Units (Commensurability) Test:*

$$P(\alpha_1 p_1^1, \ldots, \alpha_N p_N^1;\ \alpha_1 p_1^2, \ldots, \alpha_N p_N^2;\ \alpha_1^{-1} x_1^1, \cdots, \alpha_N^{-1} x_N^1;\ \alpha_1^{-1} x_1^2, \ldots, \alpha_N^{-1} x_N^2) = P(p^1, p^2, x^1, x^2) \quad \text{for } \alpha_1 > 0, \ldots, \alpha_N > 0.$$

BT5: *Symmetric Treatment of Countries or Time (Country or Time Reversal) Test:* $P(p^2, p^1, x^2, x^1) = 1/P(p^1, p^2, x^1, x^2)$.

BT6: *Symmetric Treatment of Commodities (Commodity Reversal) Test:* $P(\tilde{p}^1, \tilde{p}^2, \tilde{x}^1, \tilde{x}^2) = P(p^1, p^2, x^1, x^2)$ where $\tilde{p}^i$ denotes a permutation of the elements of the vector p^i and $\tilde{x}^i$ denotes the same permutation of the elements of x^i, $i = 1, 2$.

BT7: *Monotonicity Test:* $P(p^1, p^2, x^1, x^2) \leq P(p^1, p^3, x^1, x^2)$ if $p^2 \leq p^3$; i.e., if $p_n^2 \leq p_n^3$ for $n = 1, \ldots, N$.

BT8: *Mean Value Test:* $\min_n \{p_n^2/p_n^1\} \leq P(p^1, p^2, x^1, x^2) \leq \max_n \{p_n^2/p_n^1\}$.

BT9: *Circularity Test:* $P(p^1, p^2, x^1, x^2) P(p^2, p^3, x^2, x^3) = P(p^1, p^3, x^1, x^3)$.

Tests BT1 and BT3 may be found in Vartia [1985] who calls BT3 the *strong monetary unit test.* Test BT2 may be found in Walsh [1901], tests BT2 and BT4 are in Fisher [1911], tests BT5 and BT6 are in Fisher [1922], tests BT8 and a stronger version of BT7 are in Eichhorn and Voeller [1976] and test BT9 may be traced back to Westergaard [1890].

BT1 may be interpreted as follows: if prices and quantities are all equal in the two periods (or for the two regions under consideration), then the price index should be unity. This equality should still hold even if all quantities in period 1 are multiplied by the same number α and all quantities in period 2 are multiplied by the same β.

BT2 means if all period 2 prices are multiplied by α, then the new price index should equal α times the old price index.

Tests BT3–BT6 are invariance or symmetry tests. BT3 says that the price index should remain unchanged when each price in both periods is multiplied by the same number α and when quantities in period 1(2) are all multiplied by $\alpha(\beta)$. BT4 says that the index should remain unchanged if each good is measured in different units. BT5 says that if we interchange the role of periods 1 and 2 in our price index, then the new price index should equal the reciprocal of the original index. BT6 says that the index number formula should treat all commodities in an evenhanded way: no commodity can be singled out to play an asymmetric role. For example, suppose $P(p^1, p^2, x^1, x^2) \equiv p_1^2/p_1^1$. Then for $N \geq 2$, this formula, which equals the price ratio for commodity 1 only, fails BT6.

BT7 says that if period 2 prices increase in any manner, then the price index cannot decrease.

BT8 says that the price index should lie between the smallest and largest price ratios over all commodities.

BT9 is a transitivity test which looks beyond the case of only two periods or countries. BT9 says that if we have price and quantity data for three time periods, then the product of the price index going from period 1 to period 2 times the price index going from period 2 to 3 should equal the price index going from period 1 to 3 directly.

All of the above tests seem to be reasonable and desirable.

If $N = 1$ so that there is only one commodity, then BT1 and BT2 imply that $P(p^1, p^2, x^1, x^2)$ must equal p_1^2/p_1^1 and, of course, this index formula will satisfy all of the remaining tests.

In the general N commodity case (assuming that all prices and quantities are positive), which tests are satisfied by the index number formulae defined in the previous section?

It can be shown that the Dutot index (2) satisfies all tests except BT4 (but this is a fatal flaw), the Carli index (4) fails only BT5 and BT9, the Jevons geometric index (5) satisfies all tests, the Laspeyres and Paasche indexes defined by (7) and (8) fail BT5 and BT9, the Fisher and Walsh indexes (9) and (10) fail only BT9, and the Törnqvist index fails BT7 and BT9. Thus from the viewpoint of the test approach, it would appear that the geometric index is best.

The above conclusion is not warranted since our list of desirable tests is incomplete. We shall consider an additional two tests where the geometric index receives a failing grade.

Consider an index number formula that utilizes positive price and quantity information for N commodities, $P^N(p^1, p^2, x^1, x^2)$ say. Now consider the same functional form that uses information on only the first $N-1$ commodities, P^{N-1} say. Then we may want the index number formula to satisfy the following property:

$$(12) \quad \lim_{x_N^1 \to 0, x_N^2 \to 0} P^N(p^1, p^2, x^1, x^2)$$
$$= P^{N-1}(p_1^1, \ldots, p_{N-1}^1, p_1^2, \ldots, p_{N-1}^2, x_1^1, \ldots, x_{N-1}^1, x_1^2, \ldots, x_{N-1}^2).$$

Thus as the quantity of commodity N tends to 0 in both periods (and thus commodity N becomes irrelevant), the N commodity price index tends to the $N-1$ commodity price index, where the prices and quantities of good N have been deleted from the formula. This might be called the *irrelevance of tiny commodities test*, test BT10. Obviously, this test only makes sense if the basic index number formula P satisfies BT6, so that all commodities are treated symmetrically. The geometric price index (5) fails test BT10 as do the other unweighted formulae (2) and (4). However, the quantity weighted price indexes, (7)–(11), all pass this test. Thus from the viewpoint of passing tests, the Fisher and Walsh indexes, (9) and (10), now look just as good as the geometric index (5).

There is another reason for not preferring the geometric price index. Recall our basic aggregation problem (1). It is clear that we can interchange the role of prices and quantities in the two periods and define a *quantity index* $Q(p^1, p^2, x^1, x^2)$ in much the same way that we defined the price index $P(p^1, p^2, x^1, x^2)$. We set $P^2/P^1 = P(p^1, p^2, x^1, x^2)$ and we may set $X^2/X^1 = Q(p^1, p^2, x^1, x^2)$. From (1), we deduce that the product of the price and quantity indexes should equal the value ratio for the two periods; i.e. P and Q should satisfy the following *product test* due to Fisher [1911] but named by Frisch [1930]:

$$P(p^1, p^2, x^1, x^2)Q(p^1, p^2, x^1, x^2) = p^2 \cdot x^2 / p^1 \cdot x^1. \tag{13}$$

However, rather than defining Q independently of P, (13) may be rearranged to yield a *definition* of Q in terms of P. The resulting quantity index is called the *implicit quantity index* that corresponds to P. If we define the implicit quantity index, Q_G say, which corresponds to the geometric price index (5) and consider the quantity counterparts to tests BT1–BT9 above (BT10 does not have a sensible quantity counterpart), we find that Q_G fails the quantity counterpart to BT8, the mean value test. Hence we have another reason for failing the geometric price index and its corresponding implicit quantity index. On the other hand, the implicit Fisher quantity index satisfies tests BT1–BT8 adapted for the quantity context.

The implicit Fisher quantity index Q_F is:

$$\begin{aligned} & Q_F(p^1, p^2, x^1, x^2) \\ & \quad \equiv p^2 \cdot x^2 / p^1 \cdot x^1 P_F(p^1, p^2, x^1, x^2) \\ & \quad = (p^1 \cdot x^2 p^2 \cdot x^2 / p^1 \cdot x^1 p^2 \cdot x^1)^{1/2} \quad \text{using (9)} \\ & \quad = P_F(x^1, x^2, p^1, p^2). \end{aligned} \tag{14}$$

Thus Q_F has the same functional form as P_F except that the role of prices and quantities has been interchanged. We have shown that P_F and Q_F satisfy Fisher's [1922] *factor reversal test* which may be stated as follows:

$$P(p^1, p^2, x^1, x^2)P(x^1, x^2, p^1, p^2) = p^2 \cdot x^2 / p^1 \cdot x^1. \tag{15}$$

Of the price indexes defined in the previous section, only the Fisher price index satisfies (15). However, the Fisher price index is by no means the only index number formula that satisfies the factor reversal test. Walsh [1921] showed how to generate hundreds of formulae that would satisfy the test: take a price index $P(p^1, p^2, x^1, x^2)$, and define its factor antithesis by $p^2 \cdot x^2 / p^1 \cdot x^1 P(x^1, x^2, p^1, p^2)$. Define a new price index by taking the square root of the product of the original index and its factor antithesis. This new index will automatically satisfy the factor reversal test.

The consistency and independence of various bilateral index number tests was studied in some detail by Eichhorn and Voeller [1976]. Our conclusion at this point echoes that of Frisch [1936]: the test approach to index number theory, while extremely useful, does not lead to a single unique index number formula. Thus we turn to economic approaches to index number theory to see if we are led to a more definite conclusion.

3. Microeconomic Approaches to Price Indexes

Before a definition of a microeconomic price index is presented, it is necessary to make a few preliminary definitions.

Let $F(x)$ be a function of N variables, $x \equiv (x_1, \ldots, x_N)$. In the consumer context, F represents a consumer's preferences; i.e. if $F(x^2) > F(x^1)$, then the consumer prefers the commodity vector x^2 over x^1. In this context, F is called a *utility function*. In the producer context, $F(x)$ might represent the output that could be produced using the input vector x. In this context, F is called a *production function*. In order to cover both contexts, we follow the example of Diewert [1976a] and call F an *aggregator function*.

Suppose the consumer or producer faces prices $p \equiv (p_1, \ldots, p_N)$ for the N commodities. Then the economic agent will generally find it is useful to minimize the cost of achieving at least a given utility or output level u; we define the *cost function* or *expenditure function* C as the solution to this minimization problem:

$$C(u, p) \equiv \underset{x}{\text{minimum}}\{p \cdot x : F(x) \geq u\} \tag{16}$$

where $p \cdot x \equiv \sum_{n=1}^{N} p_n x_n$ is the inner product of the price vector p and quantity vector x.

Note that the cost function depends on $1+N$ variables; the utility or output level u and the N commodity prices in the vector p. Moreover, the functional form for the aggregator function F completely determines the functional form for C.

We say that an aggregator function is *neoclassical* if F is: (i) continuous, (ii) positive; i.e. $F(x) > 0$ if $x \gg 0_N$ (which means each component of x is positive), and (iii) linearly homogeneous; i.e. $F(\lambda x) = \lambda F(x)$ if $\lambda > 0$. If F is neoclassical, then the corresponding cost function $C(u,p)$ equals u times the unit cost function, $c(p) \equiv C(1,p)$, where $c(p)$ is the minimum cost of producing one unit of utility or output; i.e.,

$$C(u,p) = uC(1,p) \equiv uc(p). \tag{17}$$

Shephard [1953] formally defined an aggregator function F to be *homothetic* if there exists an increasing continuous function of one variable g such that $g[F(x)]$ is neoclassical. However, the concept of homotheticity was well known to Frisch [1936] who termed it expenditure proportionality. If F is homothetic, then its cost function C has the following decomposition:

$$\begin{aligned} C(u,p) &\equiv \min_x \{p \cdot x : F(x) \geq u\} \\ &= \min_x \{p \cdot x : g[F(x)] \geq g(u)\} \\ &= g(u)c(p) \end{aligned} \tag{18}$$

where $c(p)$ is the unit cost function that corresponds to $g[F(x)]$.

Let $p^1 \gg 0_N$ and $p^2 \gg 0_N$ be positive price vectors pertaining to periods or observations 1 and 2. Let $x > 0_N$ be a nonnegative, nonzero reference quantity vector. Then the Konüs [1924] *price index* or *cost of living index* is defined as:

$$P_K(p^1,p^2,x) \equiv C[F(x),p^2]/C[F(x),p^1]. \tag{19}$$

In the consumer (producer) context, P_K may be interpreted as follows. Pick a reference utility (output) level $u \equiv F(x)$. Then $P_K(p^1,p^2,x)$ is the minimum cost of achieving the utility (output) level u when the economic agent faces prices p^2 relative to the minimum cost of achieving the same u when the agent faces prices p^1. If $N = 1$ so that there is only one consumer good (or input), then it is easy to show that $P_K(p_1^1,p_1^2,x_1) = p_1^2 x_1/p_1^1 x_1 = p_1^2/p_1^1$.

Using the fact that a cost function is linearly homogeneous in its price arguments, it can be shown that P_K has the following homogeneity property: $P_K(p^1,\lambda p^2,x) = \lambda P_K(p^1,p^2,x)$ for $\lambda > 0$ which is analogous to the proportionality test BT2 in the previous section. P_K also satisfies $P_K(p^2,p^1,x) = 1/P_K(p^1,p^2,x)$ which is analogous to the time reversal test, BT5.

Note that the functional form for P_K is completely determined by the functional form for the aggregator function F which determines the functional form for the cost function C.

In general, P_K depends not only on the two price vectors p^1 and p^2, but also on the reference vector x. Malmquist [1953], Pollak [1971a] and Samuelson and Swamy [1974] have shown that P_K is independent of x and is equal to a ratio of unit cost functions, $c(p^2)/c(p^1)$, if and only if the aggregator function F is homothetic.

If we knew the consumer's preferences or the producer's technology, then we would know F and we could construct the cost function C and the Konüs price index P_K. However, we generally do not know F or C and thus it is useful to develop *bounds* that depend on observable price and quantity data but do not depend on the specific functional form for F or C.

Samuelson [1947] and Pollak [1971a] established the following bounds on P_K. Let $p^1 \gg 0_N$, and $p^2 \gg 0_N$. Then for every reference quantity vector $x > 0_N$, we have

$$(20) \qquad \min_n\{p_n^2/p_n^1\} \leq P_K(p^1,p^2,x) \leq \max_n\{p_n^2/p_n^1\};$$

i.e., P_K lies between the smallest and largest price ratios. Unfortunately, these bounds are usually too wide to be of much practical use.

To obtain closer bounds, we now assume that the observed quantity vectors for the two periods, $x^i \equiv (x_1^i,\ldots,x_N^i)$, $i = 1,2$, are solutions to the producer's or consumer's cost minimization problems; i.e., we assume:

$$(21) \qquad p^i \cdot x^i = C(F(x^i),p^i), \qquad p^i \gg 0_N, \quad x^i > 0_N, \qquad i = 1,2.$$

Given the above assumptions, we now have two natural choices for the reference quantity vector x that occurs in the definition of $P_K(p^1,p^2,x)$: x^1 or x^2. The *Laspeyres–Konüs price index* is defined as $P_K(p^1,p^2,x^1)$ and the *Paasche–Konüs price index* is defined as $P_K(p^1,p^2,x^2)$.

Under the assumption of cost minimizing behavior (21), Konüs [1924] established the following bounds:

$$(22) \qquad P_K(p^1,p^2,x^1) \leq p^2 \cdot x^1/p^1 \cdot x^1 \equiv P_L(p^1,p^2,x^1,x^2);$$

$$(23) \qquad P_K(p^1,p^2,x^2) \geq p^2 \cdot x^2/p^1 \cdot x^2 \equiv P_P(p^1,p^2,x^1,x^2),$$

where P_L and P_P are the Laspeyres and Paasche price indexes defined earlier by (7) and (8). If in addition, the aggregator function is homothetic, then Frisch [1936] showed that for any reference vector $x > 0_N$,

$$(24) \qquad P_P \equiv p^2 \cdot x^2/p^1 \cdot x^2 \leq P_K(p^1,p^2,x) \leq p^2 \cdot x^1/p^1 \cdot x^1 \equiv P_L.$$

In the consumer context, it is unlikely that preferences will be homothetic; hence the bounds (24) cannot be justified in general. However, Konüs [1924] showed that bounds similar to (24) would hold even in the general nonhomothetic case, provided that we choose a reference vector $x \equiv \lambda x^1 + (1-\lambda)x^2$ which is a $\lambda, (1-\lambda)$ weighted average of the two observed quantity points. Specifically, Konüs showed that there exists a λ between 0 and 1 such that if $P_P \leq P_L$, then

$$P_P \leq P_K[p^1, p^2, \lambda x^1 + (1-\lambda)x^2] \leq P_L \tag{25}$$

or if $P_P > P_L$, then

$$P_L \leq P_K[p^1, p^2, \lambda x^1 + (1-\lambda)x^2] \leq P_P. \tag{26}$$

The bounds on the microeconomic price index P_K given by (20) and (22)–(26) are the best bounds that we can obtain without making further assumptions on F. In the time series context, the bounds given by (25) or (26) are quite satisfactory: the Paasche and Laspeyres price indexes for consecutive time periods will usually differ by less than 1 percent. However, in the cross section context where the observations represent, for example, production data for two producers in the same industry but in different regions, the bounds are often not very useful since P_L and P_P can differ by 50 percent or more in the cross sectional context; see Ruggles [1967].

In Section 5 below, we will make additional assumptions on the aggregator function F or its cost function dual C that will enable us to determine P_K exactly. Before we do this, in the next section, we will define various quantity indexes that have their origins in microeconomic theory.

4. Microeconomic Approaches to Quantity Indexes

In the one commodity case, a natural definition for a quantity index is x_1^2/x_1^1, the ratio of the single quantity in period 2 to the corresponding quantity in period 1. This ratio is also equal to the expenditure ratio, $p_1^2 x_1^2/p_1^1 x_1^1$, divided by the price ratio, p_1^2/p_1^1. This suggests that in the N commodity case, a reasonable definition for a quantity index would be the expenditure ratio divided by the Konüs price index, P_K. This course of action was suggested by Pollak [1971a]. Thus we define the *Konüs–Pollak quantity index*, Q_K, by:

$$\begin{aligned} & Q_K(p^1, p^2, x^1, x^2, x) \\ & \equiv p^2 \cdot x^2 / p^1 \cdot x^1 P_K(p^1, p^2, x) \\ & = \Big[C[F(x^2), p^2]/C[F(x), p^2]\Big] \Big/ \Big[C[F(x^1), p^1]/C[F(x), p^1]\Big] \end{aligned} \tag{27}$$

where the second line follows from the definition of P_K, (19), and the assumption of cost minimizing behavior in the two periods, (21).

The definition of Q_K depends on the reference vector x which appears in the definition of P_K. The general definition of Q_K simplifies considerably if we choose x to be x^1 or x^2. Thus define the *Laspeyres–Konüs quantity index* as

$$Q_K(p^1, p^2, x^1, x^2, x^1) \equiv C[F(x^2), p^2]/C[F(x^1), p^2] \tag{28}$$

and the *Paasche–Konüs quantity index* as

$$Q_K(p^1, p^2, x^1, x^2, x^2) \equiv C[F(x^2), p^1]/C[F(x^1), p^1]. \tag{29}$$

It turns out that the indexes defined by (28) and (29) are special cases of another class of quantity indexes. For any reference price vector $p \gg 0_N$, define the *Allen* [1949] *quantity index* by

$$Q_A(x^1, x^2, p) \equiv C[F(x^2), p]/C[F(x^1), p]. \tag{30}$$

If p is chosen to be p^1, (30) becomes (29) and if $p = p^2$, then (30) becomes (28).

Using the properties of cost functions, it can be shown that if $F(x^2) \geq F(x^1)$, then $Q_A(x^1, x^2, p) \geq 1$ while if $F(x^2) \leq F(x^1)$, then $Q_A(x^1, x^2, p) \leq 1$. Thus the Allen quantity index correctly indicates whether the commodity vector x^2 is larger or smaller than x^1. It can also be seen that Q_A satisfies a counterpart to the time reversal test; i.e., $Q_A(x^2, x^1, p) = 1/Q_A(x^1, x^2, p)$.

Just as the price index P_K depended on the unobservable aggregator function, so also do the quantity indexes Q_K and Q_A. Thus it is useful to develop bounds for the quantity indexes that do not depend on the particular functional form for F.

Samuelson [1947] and Allen [1949] established the following bounds for (28) and (29):

$$Q_A(x^1, x^2, p^1) = Q_K(p^1, p^2, x^1, x^2, x^2) \leq p^1 \cdot x^2/p^1 \cdot x^1 \equiv Q_L; \tag{31}$$

$$Q_A(x^1, x^2, p^2) = Q_K(p^1, p^2, x^1, x^2, x^1) \geq p^2 \cdot x^2/p^2 \cdot x^1 \equiv Q_P. \tag{32}$$

Note that the observable *Laspeyres* and *Paasche quantity indexes*, Q_L and Q_P, appear on the right hand sides of (31) and (32).

Diewert [1981a], utilizing some results of Pollak [1971a] and Samuelson and Swamy [1974], established the following results: if the underlying aggregator function F is neoclassical and (21) holds, then for all $p \gg 0_N$ and $x \gg 0_N$,

$$Q_P \leq Q_A(x^1, x^2, p) = Q_K(p^1, p^2, x^1, x^2, x) = F(x^2)/F(x^1) \leq Q_L. \tag{33}$$

Thus if the aggregator function F is neoclassical, then the Allen quantity index for all reference vectors p equals the Konüs quantity index for all reference quantity vectors x which in turn equals the ratio of aggregates $(F(x^2)/F(x^1))$. Moreover, Q_A and Q_K are bounded from below by the Paasche quantity index Q_P, and bounded from above by the Laspeyres quantity index Q_L in the neoclassical case.

In the general nonhomothetic case, Diewert [1981a] showed that there exists a λ between 0 and 1 such that $Q_K[p^1, p^2, x^1, x^2, \lambda x^1 + (1-\lambda)x^2]$ lies between Q_P and Q_L and there exists a λ^* between 0 and 1 such that $Q_A[x^1, x^2, \lambda^* p^1 + (1-\lambda^*)p^2]$ also lies between Q_P and Q_L. Thus the observable Paasche and Laspeyres quantity indexes bound both the Konüs quantity index and the Allen quantity index, provided that we choose appropriate reference vectors between x^1 and x^2 and p^1 and p^2 respectively.

Using the linear homogeneity property of the cost function in its price arguments, we can show that the Konüs price index has the desirable homogeneity property, $P_K(p^1, \lambda p^1, x) = \lambda$ for all $\lambda > 0$; i.e., if period 2 prices are proportional to period 1 prices, then P_K equals this common proportionality factor. It would be desirable for an analogous homogeneity property to hold for quantity indexes. Unfortunately, it is not in general true that $Q_K(x^1, \lambda x^1, p^1, p^2, x) = \lambda$ or that $Q_A(x^1, \lambda x^1, p) = \lambda$. Thus we turn to a third microeconomic approach to defining a quantity index which does have the desirable quantity proportionality property.

Let x^1 and x^2 be the observable quantity vectors in the two situations as usual, let $F(x)$ be an increasing, continuous aggregator function, and let $x \gg 0$ be a reference quantity vector. Then the *Malmquist* [1953] *quantity index* Q_M is defined as:

$$Q_M(x^1, x^2, x) \equiv D[F(x), x^2]/D[F(x), x^1] \tag{34}$$

where $D(u, x^i) \equiv \max_k\{k : F(x^i/k) \geq u,\ k > 0\}$ is the *deflation* or *distance function* which corresponds to F. Thus $D[F(x), x^2]$ is the biggest number which will just deflate the quantity vector x^2 onto the boundary of the utility (or production) possibilities set $\{z : F(z) \geq F(x)\}$ indexed by the reference quantity vector x while $D[F(x), x^1]$ is the biggest number which will just deflate the quantity vector x^1 onto the set $\{z : F(z) \geq F(x)\}$ and Q_M is the ratio of these two deflation factors.

Q_M depends on the unobservable aggregator function F and as usual, we are interested in bounds for Q_M.

Diewert [1981a] showed that Q_M satisfied bounds analogous to (20); i.e.,

$$\min_n\{x_n^2/x_n^1\} \leq Q_M(x^1, x^2, x) \leq \max_n\{x_n^2/x_n^1\}. \tag{35}$$

It should be noted that we do not require the assumption of cost minimizing behavior in order to define the Malmquist quantity index or to establish the bounds (35). However, in order to establish the following bounds due to Malmquist [1953] for Q_M, we do need the assumption of cost minimizing behavior (21) and we require the reference vector x to be x^1 or x^2:

(36) $$Q_M(x^1, x^2, x^1) \leq p^1 \cdot x^2 / p^1 \cdot x^1 \equiv Q_L;$$

(37) $$Q_M(x^1, x^2, x^2) \geq p^2 \cdot x^2 / p^2 \cdot x^1 \equiv Q_P.$$

Diewert [1981a] showed that under the hypothesis of cost minimizing behavior, there exists a λ between 0 and 1 such that $Q_M[x^1, x^2, \lambda x^1 + (1-\lambda)x^2]$ lies between Q_P and Q_L. Thus the Paasche and Laspeyres quantity indexes provide bounds for a Malmquist quantity index for some reference indifference or product surface indexed by a quantity vector which is a $\lambda, (1-\lambda)$ weighted average of the two observable quantity vectors, x^1 and x^2.

Pollak [1971a] showed that if F is neoclassical, then we can extend the string of equalities in (33) to include the Malmquist quantity index $Q_M(x^1, x^2, x)$, for any reference quantity vector x. Thus in the case of a linearly homogeneous aggregator function, all three theoretical quantity indexes coincide and this common theoretical index is bounded from below by the Paasche quantity index Q_P and bounded from above by the Laspeyres quantity index Q_L.

In the general case of a nonhomothetic aggregator function, our best theoretical quantity index, the Malmquist index, is also bounded by the Paasche and Laspeyres indexes, provided that we choose a suitable reference quantity vector.

We noted in the price index context that the Paasche and Laspeyres price indexes were usually quite close in the time series context. A similar remark also applies to the Paasche and Laspeyres quantity indexes. Thus taking an average of the Paasche and Laspeyres indexes, such as the Fisher price and quantity indexes, will generally approximate underlying microeconomic price and quantity indexes sufficiently accurately for most practical purposes. However, this observation does not apply to the cross sectional context, where the Paasche and Laspeyres indexes can differ widely. In the following section, we offer another microeconomic justification for using the Fisher indexes that also applies in the context of making interregional and cross country comparisons.

5. Exact and Superlative Indexes

Assume that the producer or consumer is maximizing a neoclassical aggregator function f subject to a budget constraint during the two periods. Under these conditions, it can be shown that the economic agent is also minimizing cost subject to a utility or output constraint. Moreover, the cost function C that corresponds to f can be written as $C[f(x), p] = f(x)c(p)$ where c is the unit cost function (recall (17) above).

Suppose a price index $P(p^1, p^2, x^1, x^2)$ and a quantity index $Q(p^1, p^2, x^1, x^2)$ of the type considered in Sections 1 and 2 are given. The quantity index Q is defined to be *exact* for a neoclassical aggregator function f with unit cost dual c if for every $p^1 \gg 0_N$, $p^2 \gg 0_N$ and $x^i \gg 0_N$ which is a solution to the aggregator maximization problem $\max_x\{f(x) : p^i \cdot x \leq p^i \cdot x^i\} = f(x^i) > 0$ for $i = 1, 2$, we have

$$(38) \qquad Q(p^1, p^2, x^1, x^2) = f(x^2)/f(x^1).$$

Under the same hypothesis, the price index P is *exact* for f and c if we have

$$(39) \qquad P(p^1, p^2, x^1, x^2) = c(p^2)/c(p^1).$$

In (38) and (39), the price and quantity vectors are not regarded as being independent. The p^i can be independent, but the x^i are solutions to the corresponding aggregator maximization problem involving p^i, for $i = 1, 2$. Note that if Q is exact for a neoclassical f, then Q can be interpreted as a Konüs, Allen or Malmquist quantity index and the corresponding P defined implicitly by (13) can be interpreted as a Konüs price index.

The concept of exactness is due to Konüs and Byushgens [1926]. Below, we shall give some examples of exact index number formulae. Additional examples may be found in Afriat [1972b], Pollak [1971a] and Samuelson and Swamy [1974].

Konüs and Byushgens [1926] showed that Irving Fisher's ideal quantity index Q_F defined by (14) and the corresponding price index P_F defined by (9) are exact for the homogeneous quadratic aggregator function f defined by

$$(40) \qquad f(x_1, \ldots, x_N) \equiv \left(\sum_{n=1}^{N} \sum_{m=1}^{N} a_{nm} x_n x_m\right)^{1/2} \equiv (x^T A x)^{1/2}$$

where $A \equiv [a_{nm}]$ is a symmetric $N \times N$ matrix of constants. Thus under the assumption of maximizing behavior, we can calculate $f(x^2)/f(x^1) = Q_F$ and $c(p^2)/c(p^1) = P_F$ where f is defined by (40) and c is the unit cost function that corresponds to f. The important thing to note is that f depends on $N(N+1)/2$ unknown a_{nm} parameters but we do not need to know these parameters in order to evaluate $f(x^2)/f(x^1)$ and $c(p^2)/c(p^1)$.

Diewert [1976a] showed that the Törnqvist price index P_T defined by (11) is exact for the unit cost function $c(p)$ defined by:

$$(41) \qquad \ln c(p) \equiv \alpha_0 + \sum_{n=1}^{N} \alpha_n \ln p_n + (1/2) \sum_{m=1}^{N} \sum_{n=1}^{N} \alpha_{mn} \ln p_m \ln p_n$$

where the parameters α_n and α_{mn} satisfy the following restrictions:

$$\sum_{n=1}^{N} \alpha_n = 1, \quad \sum_{n=1}^{N} \alpha_{mn} = 0 \quad \text{for } m = 1, \ldots, N \quad \text{and}$$

$$\alpha_{mn} = \alpha_{nm} \quad \text{for all } m, n. \tag{42}$$

Thus we may calculate $c(p^2)/c(p^1) = P_T$ and $f(x^2)/f(x^1) = p^2 \cdot x^2/p^1 \cdot x^1 P_T \equiv \widetilde{Q}_T$ where c is the unit cost function defined by (41), f is the aggregator function which corresponds to this c, and $\widetilde{Q}_T$ is the implicit Törnqvist quantity index. Note that we do not have to know the parameters α_n and α_{mn} in order to evaluate $c(p^2)/c(p^1)$ and $f(x^2)/f(x^1)$.

The unit cost function defined by (41) is the *translog* unit cost function defined by Christensen, Jorgenson and Lau [1971]. Since P_T is exact for this translog functional form, P_T is sometimes called the *translog price index*.

Before we present our final example of an exact index number formula, we need to define a family of quantity indexes Q_r that depend on a number, $r \neq 0$:

$$Q_r(p^1, p^2, x^1, x^2) \equiv \left[\sum_{n=1}^{N} s_n^1 (x_n^2/x_n^1)^{r/2}\right]^{1/r} \times \left[\sum_{m=1}^{N} s_m^2 (x_m^2/x_m^1)^{-r/2}\right]^{-1/r} \tag{43}$$

where $s_n^i \equiv p_n^i x_n^i / p^i \cdot x^i$ is the period i expenditure share for good n. For each $r \neq 0$, define the corresponding implicit price index by:

$$\widetilde{P}_r(p^1, p^2, x^1, x^2) \equiv p^2 \cdot x^2/p^1 \cdot x^1 Q_r(p^1, p^2, x^1, x^2). \tag{44}$$

A bit of algebra will show that when $r = 2$, $\widetilde{P}_2 = P_F$, the Fisher price index defined by (9) and when $r = 1$, $\widetilde{P}_1 = P_W$, the Walsh price index defined by (10).

Diewert [1976a] showed that Q_r and $\widetilde{P}_r$ are exact for the *quadratic mean of order r aggregator function* f_r defined by:

$$f_r(x_1, \ldots, x_N) \equiv \left(\sum_{m=1}^{N} \sum_{n=1}^{N} a_{mn} x_m^{r/2} x_n^{r/2}\right)^{1/r} \tag{45}$$

where $A \equiv [a_{mn}]$ is a symmetric matrix of constants. Thus the Walsh price index P_W is exact for $f_1(x)$ defined by (45) when $r = 1$.

Diewert [1974a] defined a linearly homogeneous function f of N variables to be *flexible* if it could provide a second order approximation to an arbitrary twice continuously differentiable linearly homogeneous function. It can be shown that f defined by (40), c defined by (41) and (42) and f_r defined by (45) for each $r \neq 0$ are all examples of flexible functional forms.

Let the price and quantity indexes P and Q satisfy the product test equality, (13). Then Diewert [1976a] defined P and Q to be ***superlative indexes*** if either P is exact for a flexible unit cost function c or Q is exact for a flexible aggregator function f. Thus P_F, P_W, P_T and $\widetilde{P}_r$ are all superlative price indexes.

At this point, it may seem that we are in the same position that we were at the end of Section 2 where we could not find an index number formula that satisfied all reasonable tests; hence we could not single out any formula as being *the* best. In the present context, we have a similar problem: how are we to discriminate between P_F, P_W and P_T? Fortunately, it does not matter very much which of these formulae we choose to use in applications; they will all give the same answer to a reasonably high degree of approximation. Diewert [1978b] showed that all known superlative index number formulae approximate each other to the second order when each index is evaluated at an equal price and quantity point. This means the P_F, P_W, P_T and each $\widetilde{P}_r$ have the same first and second order partial derivatives with respect to all $4N$ arguments when the derivatives are evaluated at a point where $p^1 = p^2$ and $x^1 = x^2$. A similar string of equalities also holds for the corresponding implicit quantity indexes defined using the product test (13). Empirically, it has been found that superlative indexes typically approximate each other to something less than 0.2 percent in the time series context and to about 2 percent in the cross section context; see Fisher [1922] and Ruggles [1967].

Diewert [1978b] also showed that the Paasche and Laspeyres indexes approximate the superlative indexes to the first order at an equal price and quantity point. In the time series context, for adjacent periods, the Paasche and Laspeyres price indexes typically differ by less than 0.5 percent; hence these indexes may also provide acceptable approximations to a superlative index.

Having considered the case of two observations at great length, we now turn our attention to the I observation case. In the next section, we consider the case of I consecutive time series observations of prices and quantities on the same producer or consumer. We shall consider the general case of I observations separated by space (and possibly by time) in Section 9 below.

6. The Fixed Base Versus the Chain Principle

Consider the case of I consecutive observations through time on the prices and quantities of N goods utilized by an economic unit, $p^i \equiv (p_1^i, \ldots, p_N^i)$ and $x^i \equiv (x_1^i, \ldots, x_N^i)$, $i = 1, \ldots, I$.

Suppose that we have decided on a price index P and a quantity index Q where P and Q satisfy (13). This allows us to compare prices and quantities for any two observations. Based on the test approach, we would probably choose P and Q to be the Fisher indexes, P_F and Q_F, since they satisfy the most tests. Since the Fisher indexes also lie between the Paasche and Laspeyres bounds derived in Sections 3 and 4 and moreover are superlative indexes, the choice of the Fisher indexes also fits in well with the microeconomic approach to index number theory.

However, if we choose any superlative price index P in order to make bilateral comparisons of prices, then we are faced with a problem when $I \geq 3$ because the circular test BT9 will not be satisfied. Thus choose observation 1 as the base and make all price comparisons relative to period 1, so that the relative price level in period i is $P(p^1, p^i, x^1, x^i)$. Now choose another observation, say I, as the base so that the price level in period i relative to I is $P(p^I, p^i, x^I, x^i)$. In order to make the price level in period 1 equal to unity, divide each $P(p^I, p^i, x^I, x^i)$ by $P(p^I, p^1, x^I, x^1)$. If the circular test held, then the two price series would coincide; i.e., we would have

$$(46) \qquad P(p^1, p^i, x^1, x^i) = P(p^I, p^i, x^I, x^i)/P(p^I, p^1, x^I, x^1), \quad i = 1, \ldots, I.$$

However, since the circular test does not hold in general, we are faced with a problem: which period should we choose as the base?

There are a number of alternative strategies that could be followed, including: (i) choose the first observation as the base, (ii) take an average over all possible choices of the base period, (iii) abandon the use of a bilateral formula and develop an entirely new multilateral approach, or (iv) use the *chain principle*, which will be explained below.

Alternative (i) seems rather arbitrary but it has simplicity to recommend it. Virtually all official time series indexes are constructed using fixed base Paasche or Laspeyres indexes with the base year being changed every 5 to 15 years.

Alternative (ii) seems attractive at first glance since it treats each period in a symmetric fashion. The problem with the method is that economic history has to be rewritten every time a new observation is added to the initial I observations.

Alternative (iii) may also seem attractive. For example, in making a price comparison between i and j, we may want to utilize the quantity information for all periods, so that the bilateral index number formula $P(p^i, p^j, x^i, x^j)$ could be replaced by $P^*(p^i, p^j, x^1, \ldots, x^I)$. For example, we could use the Scrope index (6) where x could be taken to be the average quantities over all periods, $\sum_{i=1}^{I} x^i/I$, or we could use the Törnqvist index (11) where s_n could be set equal to the average over all periods for the commodity n expenditure share, $\sum_{i=1}^{I}(1/I)p_n^i x_n^i/p^i \cdot x^i$. Both of these examples lead to indexes that satisfy the circularity property for the original I observations. However, as was the case with alternative (ii), these new multilateral indexes would have to be recomputed as new time series observations become available.

Alternative (iv) is to use the ***chain principle***, originally suggested by Marshall [1887], although the term is due to Fisher [1911]. This principle makes use of the natural order provided by the march of time. One first chooses a bilateral index number formula P. The period 1 price level is set equal to unity and the period 2 price level is set equal to $P(p^1, p^2, x^1, x^2)$. The period 3 price level is set equal to $P(p^1, p^2, x^1, x^2)P(p^2, p^3, x^2, x^3)$. The period 4 price level is set equal to the period 3 price level times $P(p^3, p^4, x^3, x^4)$, and so on. Thus the period i price level is not obtained by the direct comparison of period i prices with period 1 prices, $P(p^1, p^i, x^1, x^i)$, but rather as the product of the period by period relative price levels; i.e., by travelling along the links of a chain.

The chain principle has one substantial disadvantage: if $p^i = p^j$ and $x^i = x^j$ and periods i and j are not adjacent, then it is not necessarily the case that the measured price level in period i will coincide with the measured price level in period j.

However, the chain principle has a number of advantages: (i) no single period is singled out to play an asymmetric role, (ii) the price levels for I periods are not changed as additional periods are added to the data set, (iii) if a good disappears or a new good is introduced so that N, the number of commodities, changes, then the chain price indexes (and the corresponding implicit quantity indexes) will still be comparable for all periods before and after the change in N and (iv) all superlative indexes will closely approximate each other if the chain principle is used, since changes in prices and quantities tend to be small for adjacent time periods.

The above mentioned disadvantage of the chain method is due to the lack of circularity in the bilateral formula $P(p^1, p^2, x^1, x^2)$. However, experience has shown (e.g., see Fisher [1922]) that deviations from circularity for superlative index number formulae are small in the time series context. In fact, one can show that if the bilateral index P is superlative (or equal to the Paasche or Laspeyres index), then P satisfies the circular test to the first order; i.e., the first order derivatives of $P(p^1, p^3, x^1, x^3)$ and of $P(p^1, p^2, x^1, x^2)P(p^2, p^3, x^2, x^3)$ with respect to the components of p^1, p^2, p^3, x^1, x^2 and x^3 coincide when evaluated at an equal price and quantity point where $p^1 = p^2 = p^3$ and $x^1 = x^2 = x^3$. (This is a new result.)

Our conclusion at this point is that alternatives (ii) and (iii) do not look very attractive in the time series context: the use of a fixed base or the chain principle seems preferable. However, in the context of cross section data where there is no natural way of ordering the data sequentially, alternatives (ii) and (iii) become much more attractive. We consider genuine multilateral index number formulae in Sections 9 and 10 below.

7. Aggregation Over Consumers

Thus far, our discussion of the microeconomic approach to indexes has been limited to the one consumer or producer case (or the case of two consumers or producers who have identical preferences or technologies). In this section, we relax these restrictions and discuss aggregate consumer or household indexes. In the following section, we discuss aggregate output price and quantity indexes.

Consider the case of two countries or regions (or two time periods for the same country). We suppose that there are H_i households in country i and the preferences of household h in country i are represented by a continuous utility function, $F^{ih}(x, z^{ih})$, where $x \equiv (x_1, \ldots, x_N) \geq 0_N$ is a nonnegative vector of market goods, $p^i \equiv (p_1^i, \ldots, p_N^i) \gg 0_N$ is the corresponding vector of positive prices that each household in region i faces and z^{ih} is a vector of demographic variables or consumption of public goods by household h in country i.

The *restricted cost* or *expenditure function* of household h in country i is defined as:

$$(47) \quad C^{ih}(u^{ih}, p^i, z^{ih}) \equiv \min_x \{p^i \cdot x : F^{ih}(x, z^{ih}) \geq u^{ih}\}, \qquad i = 1, 2; \; h = 1, \ldots, H_i,$$

where u^{ih} is a utility or welfare level. We assume that $x^{ih} > 0_N$, the observable consumption vector for household h in the country i, solves (47) with $u^{ih} = F^{ih}(x^{ih}, z^{ih})$ for $i = 1, 2$ and $h = 1, \ldots, H_i$.

We may use the preferences and observed choices of household h in country i to define a *Konüs price index* P^{ih} for the level of prices in country 2 relative to country 1:

$$(48) \quad P^{ih}(p^1, p^2) \equiv C^{ih}(u^{ih}, p^2, z^{ih}) / C^{ih}(u^{ih}, p^1, z^{ih}), \qquad i = 1, 2; \; h = 1, \ldots, H_i.$$

Suppose that each expenditure function C^{ih} has a translog functional form; i.e., $\ln C^{ih}(u, p, z)$ is a quadratic function in the logarithms of its variables, similar to the right hand side of (41) except that now (u, p, z) replaces p. We note that we can approximate arbitrary preferences to the second order using this functional form; i.e., the translog restricted expenditure function is a flexible functional form. Caves, Christensen and Diewert [1982b] establish the following result: if household h in country 1 and household k in country 2 have translog expenditure functions with identical coefficients on the second order terms in commodity prices (this forces some similarity in preferences), then the geometric mean of the two theoretical Konüs price indexes, P^{1h} and P^{2k}, equals the observable Törnqvist or translog price index $P_T(p^1, p^2, x^{1h}, x^{2k})$; i.e., we have

$$(49) \quad [P^{1h}(p^1, p^2) P^{2k}(p^1, p^2)]^{1/2} = P_T(p^1, p^2, x^{1h}, x^{2k}), \qquad h = 1, \ldots, H_1; \; k = 1, \ldots, H_2.$$

Rather than deal with the individual household indexes p^{ih} defined by (48), one can define an average index, where the average is taken over all households in the country. Thus for a nonnegative weights vector $\alpha^i \equiv (\alpha_1^i, \ldots, \alpha_{H_i}^i)$ such that $\sum_{h=1}^{H_i} \alpha_h^i = 1$, define the *country i average price index* P^i as an α^i weighted geometric mean of the individual indexes:

$$(50) \qquad P^i(p^1, p^2, \alpha^i) \equiv \prod_{h=1}^{H_i} P^{ih}(p^1, p^2)^{\alpha_h^i}, \quad i = 1, 2.$$

The theoretical index P^i defined in (50) utilizes the price vectors p^1 and p^2 in both countries but utilizes only the preferences of households in country i. As a final bit of averaging, we take the geometric mean of P^1 and P^2 to obtain a final theoretical price index that treats each household in each country in a symmetric fashion:

$$(51) \qquad P(p^1, p^2, \alpha^1, \alpha^2) \equiv [P^1(p^1, p^2, \alpha^1) P^2(p^1, p^2, \alpha^2)]^{1/2}.$$

The natural choices for the household weighting vectors α^1 and α^2 are: (i) *democratic weights* and (ii) *plutocratic weights*. In case (i), each household in each country is given an equal weight; i.e., $\alpha_h^i \equiv 1/H_i$, $i = 1, 2$, and $h = 1, \ldots, H_i$. In case (ii), each household gets a weight that is proportional to its share of consumption in its own country; i.e., $\alpha_h^i \equiv p^i \cdot x^{ih}/p^i \cdot x^i \equiv s_h^i$ for $i = 1, 2$ and $h = 1, \ldots, H_i$, where $x_i \equiv \sum_{h=1}^{H_i} x^{ih}$ is country i's aggregate consumption vector. If each household has a translog restricted expenditure function with identical coefficients on the second order terms in commodity prices, then making repeated use of (49), we can deduce that the aggregate price index (51) in case (i) is:

$$(52) \quad P(p^1, p^2, 1/H_1, \ldots, 1/H_1,\ 1/H_2, \ldots, 1/H_2) = \prod_{h=1}^{H_1} \prod_{k=1}^{H_2} P_T(p^1, p^2, x^{1h}, x^{2k})^{1/H_1 H_2}$$

while in case (ii), the aggregate price index is

$$(53) \qquad P(p^1, p^2, s^1, s^2) = P_T(p^1, p^2, x^1, x^2)$$

where $s^i \equiv (s_1^i, \ldots, s_{H_i}^i)$ is the country i expenditure share vector for households in country i and $x^i \equiv \sum_{h=1}^{H_i} x^{ih}$ is the aggregate country i consumption vector. Thus if individual household consumption data x^{ih} are available, then the *democratic aggregate price index* defined by (52) can be evaluated as a geometric mean of individual household translog price indexes. If only aggregate country consumption data x^i are available, then the *plutocratic aggregate price index* defined by (53) can be evaluated as a translog price index (recall (11)), using the aggregate consumption vectors x^1 and x^2 as quantity weights.

The terms democratic and plutocratic are due to Prais [1959]. For other approaches to aggregate consumer price indexes, see Prais [1959], Pollak [1981] and Diewert [1983a].

We turn now to the construction of an aggregate quantity index for the two periods. Let the preferences of household h in country i be represented by the continuous, increasing utility function $F^{ih}(x)$ where we have absorbed the vector of demographic variables into the function F^{ih}. For $i = 1, 2$, and $h = 1, \ldots, H_i$, define the household h in country i *deflation function* $D^{ih}(u, x)$ for $x > 0_N$ and u in the range of F by

$$(54) \qquad D^{ih}(u, x) \equiv \max_{\delta}\{\delta : F^{ih}(x/\delta) \geq u, \ \delta \geq 0\}.$$

As in Section 5, define the Malmquist quantity index for $x^* > 0_N$ relative to $x > 0_N$ using the preferences of household h in country i by:

$$(55) \qquad Q^{ih}(x, x^*, u) \equiv D^{ih}(u, x^*)/D^{ih}(u, x), \quad i = 1, 2; \ h = 1, \ldots, H_i.$$

Let the observed consumption vector of household h in country i be $x^{ih} > 0_N$ and define the corresponding utility level by $u^{ih} \equiv F^{ih}(x^{ih})$.

Define the *index of average household consumption* in country j relative to i by:

$$(56) \quad Q^{ij} \equiv \sum_{k=1}^{H_j} Q^{jk}\left(\sum_{h=1}^{H_i} x^{ih}/H_i, x^{jk}, u^{jk}\right) \Big/ H_i; \quad i, j = 1, 2, \ i \neq j.$$

To explain the meaning of (56), define the country i average or per capita consumption vector by

$$(57) \qquad \bar{x}^i \equiv \sum_{h=1}^{H_i} x^{ih}/H_i, \qquad i = 1, 2.$$

Then Q^{12} is an average of the individual country 2 Malmquist indexes $Q^{2k}(\bar{x}^1, x^{2k}, u^{2k})$ which in turn compares the observed consumption vector of household k in country 2 with the average consumption vector $\bar{x}^1$ for country 1, using the indifference surface through x^{2k} of household k in country 2 as the reference indifference surface. Similarly, Q^{21} is an average of the individual country 1 Malmquist indexes $Q^{1k}(\bar{x}^2, x^{1k}, u^{1k})$.

Diewert [1986] showed that under the assumption of expenditure minimizing behavior on the part of consumers in both countries, Q^{12} has the lower bound $p^2 \cdot \bar{x}^2/p^2 \cdot \bar{x}^1$, a Paasche quantity index in per capita quantities, and $[Q^{21}]^{-1}$ has the upper bound $p^1 \cdot \bar{x}^2/p^1 \cdot \bar{x}^1$, a Laspeyres quantity index in per capita quantities. Diewert also shows that there exists a $0 \leq \lambda \leq 1$ such that

$\lambda Q^{12} + (1-\lambda)(Q^{21})^{-1}$ lies between these per capita Paasche and Laspeyres quantity indexes. This suggests that we can approximate the theoretical index $\lambda Q^{12} + (1-\lambda)(Q^{21})^{-1}$ by an average of these Paasche and Laspeyres indexes such as the Fisher index $Q_F(p^1, p^2, \bar{x}^1, \bar{x}^2) \equiv (p^2 \cdot \bar{x}^2 / p^2 \cdot \bar{x}^1)^{1/2} (p^1 \cdot \bar{x}^2 / p^1 \cdot \bar{x}^1)^{1/2}$ where the per capita quantity vectors $\bar{x}^i$ are defined by (57).

To summarize this section: we have shown that the translog price index P_T and the Fisher quantity index Q_F, which had very satisfactory economic interpretations in the case of one consumer, also have reasonable economic interpretations in the many consumer case.

8. Aggregation Over Producers

We assume that there are two regions or countries that are to be compared. The two countries could represent the same country at different time periods. The private production sector in country i uses a vector of primary inputs $v^i \equiv (v_1^i, \ldots, v_M^i)$, where v_m^i is the amount of input m used in country i. These inputs are different types of labor, capital, land and other natural resources. There are N net outputs that can be produced by the private production sector in each country. These goods are different types of consumer and investment goods, exports and imports. The net output vector for country i is $y^i \equiv (y_1^i, \ldots, y_N^i)$ and the corresponding price vector is the positive vector $w^i \equiv (w_1^i, \ldots, w_N^i) \gg 0_N$. We assume $w^i \cdot y^i \equiv \sum_{n=1}^N w_n^i y_n^i > 0$ for $i = 1, 2$. If $y_n^i < 0$, then good n is utilized as an input in country i; this good could be an imported good or it could be an intermediate input that is produced by the government sector in country i. Note that in contrast to the consumer case, we no longer assume that quantity vectors are nonnegative.

The technology set for the private production sector in country i is the set $S^i \equiv \{(y^i, v^i)\}$, a feasible set of net output vectors y^i and primary input vectors v^i. If knowledge is freely transferable across countries, then $S^i = S$ for $i = 1, 2$ so that there is a common technology set across countries. However, we do not require this assumption in what follows.

Define country i's *private national product function* g^i by

(58) $$g^i(w, v) \equiv \max_y \{w \cdot y : (y, v) \text{ belongs to } S^i\}, \quad i = 1, 2.$$

The number $g^i(w, v)$ is the maximum value of outputs (less the value of imports) that the private production sector of country i can produce, given that each producer in the country faces the price vector w and the aggregate private economy has at its disposal the primary input vector v. If each producer faces the same price vector w and behaves competitively, then we do not have to concern ourselves with individual producer output vectors: all that matters is the aggregate net output vector. The national product function was introduced into the economics literature by Samuelson [1953–54]; it is sometimes called a variable or restricted profit function or a net revenue function.

The Fisher–Shell [1972a] *output price index* of country 2 relative to 1 using the country i technology set S^i and primary input vector v^i is defined as

$$(59) \qquad P^i(w^1, w^2) \equiv g^i(w^2, v^i)/g^i(w^1, v^i), \qquad i = 1, 2.$$

We assume optimizing behavior on the part of producers in each country so that the observed country i price and net output vectors, w^i and y^i, satisfy $w^i \cdot y^i = g^i(w^i, v^i) > 0$ for $i = 1, 2$.

Assume that the private national product function $g^i(w, v)$ has a translog functional form for each i (recall (41) except that now $\ln g^i(w, v)$ is a quadratic form in the logarithms of w_n and v_m) and further assume that the coefficients for the quadratic terms in the logarithms of output prices are the same across the two countries. Then Diewert [1986] showed that the geometric mean of the two theoretical output price indexes defined by (59) is exactly equal to the observable translog price index $P_T(w^1, w^2, y^1, y^2)$, where P_T is defined by (11); i.e.,

$$(60) \qquad [P^1(w^1, w^2)P^2(w^1, w^2)]^{1/2} = P_T(w^1, w^2, y^1, y^2).$$

Thus the translog price index P_T again turns out to have a strong microeconomic justification.

The Malmquist quantity index has been applied to the problem of constructing output indexes by Caves, Christensen and Diewert [1982b], and Diewert [1986]. It is first necessary to define the country i output deflation function d^i: for a primary input vector v and a net output vector y, define

$$(61) \qquad d^i(y, v) \equiv \min_\delta\{\delta : (y/\delta, v) \text{ belongs to } S^i,\ \delta^i \geq 0\}, \quad i = 1, 2$$

where S^i is the technology set for country i. Thus $d^i(y, v)$ denotes the amount the net output vector y must be deflated so that the deflated output vector and the reference input vector v are just on the frontier of the country i production possibilities set S^i.

The Malmquist output index of country 2 relative to country 1 using the country i technology and primary input vector v^i is:

$$(62) \qquad Q^i(y^1, y^2) \equiv d^i(y^2, v^i)/d^i(y^1, v^i), \qquad i = 1, 2.$$

Assume the observed country i net output vector y^i solves the country i private product maximization problem, $\max_y\{w^i \cdot y : (y, v^i) \text{ belongs to } S^i\}$

for $i = 1, 2$. Then Diewert [1986] established the following bounds for the Malmquist indexes Q^i defined by (62):

$$(63)\qquad Q^1(y^1, y^2) \geq w^1 \cdot y^2/w^1 \cdot y^1 \equiv Q_L(w^1, w^2, y^1, y^2);$$

$$(64)\qquad Q^2(y^1, y^2) \leq w^2 \cdot y^2/w^2 \cdot y^1 \equiv Q_P(w^1, w^2, y^1, y^2)$$

where Q_L and Q_P are the Laspeyres and Paasche quantity indexes.

It is also possible to define a deflation function d that uses a convex combination of the input vectors, $\lambda v^1 + (1 - \lambda)v^2$, and a convex combination of the technology sets for the two countries, $\lambda S^1 + (1 - \lambda)S^2$:

$$(65)\ d(y, \lambda) \equiv \min_{\delta > 0} \left\{ \delta : \left[y/\delta, [\lambda v^1 + (1 - \lambda)v^2] \right] \text{ belongs to } \lambda S^1 + (1 - \lambda)S^2 \right\}.$$

The *Malmquist λ weighted average output index* for country 2 relative to 1 may be defined as:

$$(66)\qquad Q^\lambda(y^1, y^2) \equiv d(y^2, \lambda)/d(y^1, \lambda).$$

Assuming maximizing behavior, Diewert [1986] shows that there exists a λ such that $0 \leq \lambda \leq 1$ and $Q^\lambda(y^2, y^2)$ lies between the Laspeyres and Paasche quantity indexes Q_L and Q_P defined in (63) and (64). Thus the Fisher quantity index, $Q_F \equiv (Q_L Q_P)^{1/2}$, should provide an adequate approximation to the theoretical output index $Q^\lambda(y^1, y^2)$.

Our conclusion is that the translog price index P_T and the Fisher quantity index Q_F have reasonably strong justifications in the bilateral aggregate private production context.

We now turn our attention to the multilateral case.

9. Multilateral Test Approaches

We are finally in a position to study the multilateral index number problem which was set out in the introduction. To review the notation, there are I positive price vectors $p^i \equiv (p_1^i, \ldots, p_N^i)$ and I quantity vectors $x^i \equiv (x_1^i, \ldots, x_N^i)$ with $p^i \cdot x^i > 0$ for $i = 1, \ldots, I$. We wish to find $2I$ positive numbers P^i (price indexes) and X^i (quantity indexes) such that $P^i X^i = p^i \cdot x^i$ for $i = 1, \ldots, I$. The I data points (p^i, x^i) will typically be observations on production or consumption units that are separated spatially but yet are still comparable. For the sake of definiteness, we shall refer to the I data points as countries. Each commodity n is supposed to be the same across all countries. This can always be done by a suitable extension of the list of commodities.

Our first approach to the construction of a system of multilateral price and quantity indexes is based on the use of a bilateral quantity index Q. In this method, the first step is to pick the 'best' bilateral index number formula: e.g., the Fisher index Q_F defined by (14) or the implicit translog quantity index defined by $\widetilde{Q}_T(p^1, p^2, x^1, x^2) \equiv p^2 \cdot x^2 / p^1 \cdot x^1 P_T(p^1, p^2, x^1, x^2)$ where P_T is defined by (11). Secondly, pick a numeraire country, say country 1, and then calculate the aggregate quantity for each country i relative to country 1 by evaluating the quantity index $Q(p^1, p^i, x^1, x^i)$. In order to put these relative quantity measures on a symmetric footing, we convert each relative to country 1 quantity measure into a share of world quantity by dividing through by $\sum_{k=1}^{I} Q(p^1, p^k, x^1, x^k)$. For a general numeraire country j, define the *share of world quantity for country i, using country j as the numeraire country*, by:

$$(67) \qquad \sigma_i^j(p, x) \equiv Q(p^j, p^i, x^j, x^i) / \sum_{k=1}^{I} Q(p^j, p^k, x^j, x^k), \quad i = 1, \dots, I,$$

where $p \equiv (p^1, \dots, p^I)$ is the $N \times I$ matrix of price data and $x \equiv (x^1, \dots, x^I)$ is the $N \times I$ matrix of quantity data. Once the numeraire country j has been chosen and the country i shares σ_i^j calculated, we may set $X^i \equiv \sigma_i^j$ and $P^i \equiv p^i \cdot x^i / X^i$ for $i = 1, \dots, I$. Thus we have provided a solution to the multilateral index number problem (1). Of course, one is free to renormalize the resulting P^i and X^i if desired; i.e., all X^i can be multiplied by a number provided all P^i are divided by this same number. Kravis [1984] calls this method the *star system*, since the numeraire country plays a starring role: all countries are compared with it and it alone.

We shall assume throughout this section that the index number formula Q satisfies the quantity counterpart to the bilateral price tests BT1 through BT6. (The quantity counterpart to BT1 is $Q(\alpha p^1, \beta p^2, x^1, x^2) = 1$ if $p^1 = p^2$ and $x^1 = x^2$, the quantity counterpart to BT2 is $Q(p^1, p^2, x^1, \alpha x^2) = \alpha Q(p^1, p^2, x^1, x^2)$ for $\alpha > 0$, and so on.) This assumption is not restrictive since our best bilateral formulae, Q_F and $\widetilde{Q}_T$, both satisfy these tests.

Of course, the problem with the star system for making multilateral comparisons is its lack of invariance to the choice of the numeraire or star country. Different choices for the base country will in general give rise to different indexes P^i and X^i. This problem can be traced to the lack of circularity of the bilateral formula Q: if Q satisfies the time reversal test BT5 and the circular test BT9 for quantity indexes, then $\sigma_i^j = \sigma_i^k$ for all i, j and k; i.e., the shares σ_i^j defined by (67) do not depend on the choice of the numeraire country j. However, given that the bilateral formula Q does not satisfy the circularity test (as is the case with Q_F and $\widetilde{Q}_T$), how can we generate multilateral indexes that treat each country symmetrically?

Fisher [1922] recognized that the simplest way of achieving symmetry was to average base specific index numbers over all possible bases. Thus define country i's share of world output $S_i(p, x)$ by

$$S_i(p,x) \equiv \sum_{j=1}^{I} \sigma_i^j(p,x)/I, \quad I = 1, \ldots, I \tag{68}$$

where the σ_i^j are defined by (67). We can now define country i quantities and prices by

$$X^i \equiv S_i(p,x), \; P^i \equiv p^i \cdot x^i / X^i, \quad i = 1, \ldots, I. \tag{69}$$

Fisher [1922] called this method of constructing multilateral indexes the *blend method* while Diewert [1986] called it the *democratic weights method*, since each share of world output using each country as the base is given an equal weight in the formation of the average.

Of course, there is no need to use an arithmetic average of the σ_i^j as in (68); one can use a geometric average:

$$\sigma_i(p,x) \equiv \left[\prod_{j=1}^{I} \sigma_i^j(p,x)\right]^{1/I}, \quad i = 1, \ldots, I. \tag{70}$$

Using (70), the resulting shares no longer sum to one in general, so country i's share of world output is now defined as:

$$S_i(p,x) \equiv \sigma_i(p,x) / \sum_{k=1}^{I} \sigma_k(p,x), \quad i = 1, \ldots, I. \tag{71}$$

If the Fisher index Q_F is used in the definition of the σ_i^j, then

(72)

$$S_i(p,x)/S_j(p,x) = \left[\prod_{k=1}^{I} Q_F(p^k, p^i, x^k, x^i) / \prod_{m=1}^{I} Q_F(p^m, p^j, x^m, x^j)\right]^{1/I}$$

and in this case, the multilateral method defined by (70) reduces to a method recommended by Eltetö and Köves [1964] and Szulc [1964], the *EKS method*. Instead of using the Fisher formula in (72), Caves, Christensen and Diewert [1982a] advocated the use of the translog quantity index Q_T while Diewert [1986] suggested the use of the implicit translog quantity index $\tilde{Q}_T$, since $\tilde{Q}_T$ is well defined even in the case where some quantities x_n^i are negative (whereas Q_T is not). We call the indexes generated by (69) and (71) for a general bilateral index Q, *generalized EKS* indexes.

When forming averages of the σ_i^j as in (68) or (70), there is no necessity to use equal weights: one can define country j's value share of world output as $\beta_j \equiv p^j \cdot x^j / \sum_{k=1}^{I} p^k \cdot x^k$ (this requires all prices to be measured in units of a common currency) and then we may define a plutocratic share weighted average of the σ_i^j:

$$S_i(p,y) \equiv \sum_{j=1}^{I} \beta_j(p,x)\sigma_i^j(p,x). \tag{73}$$

Diewert [1986] called this method of constructing multilateral indexes the *plutocratic weights method.*

Another multilateral method that is based on a bilateral index Q may be described as follows. Define

$$\alpha_i(p,x) \equiv \sum_{j=1}^{I} [Q(p^j,p^i,x^j,x^i)^{-1}]^{-1}, \quad i=1,\ldots,I. \tag{74}$$

If there is only one commodity so that $N=1$ and the bilateral index Q satisfies BT1, BT2 and BT3, then $\alpha_i = \left[\sum_{j=1}^{I}(x^i/x^j)^{-1}\right]^{-1} = \left[\sum_{j=1}^{I} x^j/x^i\right]^{-1} = x^i/\sum_{j=1}^{I} x^j$ which is country i's share of world product. In the general case where $N>1$, the 'shares' α_i do not necessarily sum up to unity, so it is necessary to normalize them:

$$S_i(p,x) \equiv \alpha_i(p,x) \Big/ \sum_{k=1}^{I} \alpha_k(p,x), \quad i=1,\ldots,I. \tag{75}$$

Diewert [1986] called this the *own share method* for making multilateral comparisons.

The above methods for achieving consistency and symmetry rely on averaging over various bilateral index number comparisons. Fisher [1922] realized that symmetry could be achieved by making comparisons with an average; he called this broadening the base. Thus the *basket method* (which corresponds to Fisher's [1922] formula 6053 and to method 8 described in Ruggles [1967]) may be described as follows. The price level of country i relative to country j is set equal to $p^i \cdot (\sum_{k=1}^{J} x^k/I)/p^j \cdot (\sum_{k=1}^{J} x^k/I)$. This index number formula is a Scrope index (6), where the reference quantity vector is chosen to be the average market basket, $\sum_k x^k/I$. The same result could be achieved if we chose x to be the total market basket, $\sum_k x^k$. Now define $Q^{ji} \equiv p^i \cdot x^i/p^j \cdot x^j [p^i \cdot (\sum_k x^k)/p^j \cdot (\sum_k x^k)]$ to be the implicit output of country i relative to j. Choose a j as a numeraire country and calculate country i's share of world output as:

$$\begin{aligned} S_i(p,x) &\equiv Q^{ji} \Big/ \sum_{k=1}^{I} Q^{jk} \\ &= \left(p^i \cdot x^i/p^i \cdot \sum_k x^k\right) \Big/ \sum_{m=1}^{I} \left(p^m \cdot x^m/p^m \sum_k x^k\right), \quad i=1,\ldots,I. \end{aligned} \tag{76}$$

Note that the final expression for S_i does not depend on the choice of the numeraire country j. As usual, once the share functions, S_i, have been defined, the aggregate X^i and P^i may be defined by (69).

A variation on the basket method due to Geary [1958] and Khamis [1972] is defined by (77)–(79) below:

$$\text{(77)} \qquad \pi_n \equiv \sum_{i=1}^{I} p_n^i x_n^i / P^i \sum_{k=1}^{I} x_n^k, \qquad n = 1, \ldots, N;$$

$$\text{(78)} \qquad P^i \equiv \sum_{n=1}^{N} p_n^i x_n^i \Big/ \sum_{m=1}^{N} \pi_m x_m^i, \quad i = 1, \ldots, I;$$

$$\text{(79)} \qquad X^i \equiv p^i \cdot x^i / P^i, \qquad i = 1, \ldots, I.$$

π_n is interpreted as an average international price for good n. From (78), it can be seen that P^i, the price level or purchasing power parity for country i, is a Paasche-like price index for country i except that the base prices are chosen to be the international prices π_n. The π_n and $(P^i)^{-1}$ can be solved for as a system of simultaneous linear equations (up to a scalar normalization) or the $(P^i)^{-1}$ may be determined as the components of the eigenvector that corresponds to the maximal positive eigenvalue of a certain matrix. The P^i can be normalized so that the quantities X^i defined by (79) sum up to unity. This *GK method* for making multilateral comparisons has been widely used in empirical applications: see Kravis *et al.* [1975].

We have defined seven methods for making multilateral comparisons: the star method (67), the democratic (68) and plutocratic (73) weights methods, the generalized EKS method (71), the own share method (74), the basket method (76) and the GK method (79). How can we discriminate among them?

One helpful approach would be to define a system of *multilateral tests* and then evaluate how the above methods satisfy these tests.

In the bilateral situation, it was natural to phrase the tests in terms of the price index P or the quantity index Q, since if either of these functions were given (along with a single normalization such as $P^1 = 1$), then the aggregates P^1, P^2, X^1 and X^2 were all determined.

In the multilateral situation, it seems natural to phrase the tests in terms of the properties of the system of world output share functions, $S(p, x) \equiv [S_1(p, x), \ldots, S_I(p, x)]$, since given these share functions, we may set $X^i \equiv S_i(p, x)$ and $P^i = p^i \cdot x^i / X^i$ for $i = 1, \ldots, I$.

The tests MT1 to MT6 listed below are multilateral counterparts to the bilateral tests BT1 to BT6 applied to quantity indexes rather than price indexes. MT0 is a preliminary test that does not have a bilateral counterpart. Recall that $p \equiv (p^1, \ldots, p^I)$ and $x \equiv (x^1, \ldots, x^I)$.

MT0: *Share Test:* $\sum_{i=1}^{I} S_i(p, x) = 1$.

MT1: *Multilateral Identity Test:* $S_i(\alpha_1 p^1, \ldots, \alpha_1 p^I, \beta_1 x^1, \ldots, \beta_I x^I) = \beta_i$ for $i = 1, \ldots, I$ for all $\alpha_i > 0$, $\beta_i > 0$ if $p^1 = \cdots = p^I$, $x^1 = \cdots = x^I$ and $\sum_j \beta_j = 1$.

MT2: *Proportionality Test:* For $i = 1, \ldots, I$ and $\lambda_i > 0$, $S_i(p, x^1, \ldots, x^{i-1}, \lambda_i x^i, x^{i+1}, \ldots, x^I)/S_j(p, x^1, \ldots, x^{i-1}, \lambda_i x^i, x^{i+1}, \ldots, x^I) = \lambda_i S_i(p,x)/S_j(p,x)$ for $j = 1, \ldots, i-1, i+1, \ldots, I$.

MT3: *Invariance to Changes in Scale Test:* $S_i(\alpha_1 p^1, \ldots, \alpha_I p^I, \beta x^1, \ldots, \beta x^I) = S_i(p,x)$ for all $\alpha_i > 0$, $\beta > 0$, $i = 1, \ldots, I$.

MT4: *Invariance to Changes in Units Test:* $S_i(\alpha_1 p_1^1, \ldots, \alpha_N p_N^1; \ldots; \alpha_1 p_1^I, \ldots, \alpha_N p_N^I; \alpha_1^{-1} x_1^1, \ldots, \alpha_N^{-1} x_N^1; \ldots; \alpha_1^{-1} x_1^I, \ldots, \alpha_N^{-1} x_N^I) = S_i(p,x)$ for $i = 1, \ldots, I$ and $\alpha_1 > 0, \ldots, \alpha_N > 0$.

MT5: *Symmetric Treatment of Countries Test:* Let $\hat{p}$ denote a permutation of the I columns of the $N \times I$ matrix p, let $\hat{x}$ denote the same permutation of the I columns of x, and let $\widehat{S}(p,x)$ denote the same permutation of the I columns of the row vector $S(p,x)$. Then $\widehat{S}(p,x) = S(\hat{p}, \hat{x})$.

MT6: *Symmetric Treatment of Commodities Test:* Let $\tilde{p}$ denote a permutation of the N rows of p and let $\tilde{x}$ denote the same permutation of the N rows of x. Then $S_i(\tilde{p}, \tilde{x}) = S_i(p,x)$ for $i = 1, \ldots, I$.

MT7: *Country Partitioning Test:* Let $S_j^I \equiv S_j(p^1, \ldots, p^I; x^1, \ldots, x^I)$ for $j = 1, \ldots, I$ and let $0 < \lambda_i < 1$. Define $S_j^{I+1} \equiv S_j[p^1, \ldots, p^I, p^i; x^1, \ldots, x^{i-1}, \lambda_i x^i, x^{i+1}, \ldots, x^I, (1-\lambda_i)x^i]$ for $j = 1, \ldots, I, I+1$. Then $S_j^I = S_j^{I+1}$ for $j = 1, \ldots, i-1, i+1, \ldots, I$, $S_i^{I+1} = \lambda_i S_i^I$ and $S_{I+1}^{I+1} = (1-\lambda_i) S_i^I$. This property is to hold no matter which country i is partitioned.

The functions S_j^I are the share functions for the initial world economy that consists of I countries. The functions S_j^{I+1} are the share functions for a new world economy, where the original country i with price vector p^i and quantity vector x^i, has been partitioned into two countries with price vectors p^i and p^i and quantity vectors $\lambda_i x^i$ and $(1-\lambda_i)x^i$. MT7 says that under these conditions, the original country i share S_i^I splits into $\lambda_i S_i^I$ and $(1-\lambda_i)S_i^I$ and the remaining shares are unaffected.

MT8: *Irrelevance of Tiny Countries Test:* Let $\lambda_i > 0$ and define $S_j^I(\lambda_i) \equiv S_j(p, x^1, \ldots, x^{i-1}, \lambda_i x^i, x^{i+1}, \ldots, x^I)$ for $j = 1, \ldots, I$. Define $S_j^{I-1} \equiv S_j(p^1, \ldots, p^{i-1}, p^{i+1}, \ldots, p^I; x^1, \ldots, x^{i-1}, x^{i+1}, \ldots, x^I)$ for $j = 1, \ldots, I-1$. Then $\lim_{\lambda_i \to 0} S_j^I(\lambda_i) = S_j^{I-1}$ for $j = 1, \ldots, i-1$ and $\lim_{\lambda_i \to 0} S_j^I(\lambda_i) = S_{j-1}^{I-1}$ for $j = i+1, i+2, \ldots, I$. This property is to hold for all choices of the disappearing country i.

In the above test, the quantity vector for country i is deflated down to a zero vector. Consider the resulting system of limiting share functions. For all countries except i, the limiting share is equal to the share we would get if we simply deleted the data for country i and defined a system of share functions for a world economy consisting of only $I-1$ countries. MT8 is a country counterpart to the irrelevance of tiny commodities test, (12).

The above multilateral tests were proposed by Diewert [1986]. Our final multilateral test is a new one.

MT9: *Bilateral Property Test:* When $I = 2$, $S_2(p^1, p^2, x^1, x^2)/S_1(p^1, p^2, x^1, x^2)$ satisfies the bilateral tests BT1–BT6 for quantity indexes.

When $I = 2$, the multilateral system collapses down to a bilateral system. Hence it seems perfectly sensible to demand that S^2/S^1 should satisfy tests BT1–BT6 at least, since our best bilateral indexes, Q_F and $\widetilde{Q}_T$, satisfied these tests.

The seven multilateral methods mentioned above satisfy most of the multilateral tests, assuming that the bilateral index Q satisfies BT1–BT6 (recall that all of the multilateral methods utilize a bilateral index Q except the basket and GK methods).

The star system fails MT5 (it is obviously not symmetric) and it fails MT8 when the tiny country is chosen to be the numeraire country.

The plutocratic weights method fails MT2 and MT3; thus the resulting quantity indexes are not invariant to country inflation rates, a very severe defect.

The democratic weights method fails the multilateral proportionality test MT2 and the two consistency in aggregation tests, MT7 and MT8. This method is dominated by the generalized EKS method which fails only MT7 and MT8.

The basket and GK methods fail MT2 and MT9: both methods fail BT2 when $I = 2$. When $I = 2$, $S_2/S_1 = p^2 \cdot x^2 p^1 \cdot (x^1 + x^2)/p^1 \cdot x^1 p^2 \cdot (x^1 + x^2)$ for the basket method, and $S_2/S_1 = p^2 \cdot x^2/p^1 \cdot x^1 P_{GK}(p^1, p^2, x^1, x^2)$ for the GK method where the GK bilateral price index is defined as

$$(80) \quad P_{GK}(p^1, p^2, x^1, x^2) \equiv \left[\sum_{n=1}^{N} p_n^2 x_n^1 x_n^2/(x_n^1 + x_n^2)\right] \Big/ \left[\sum_{k=1}^{N} p_k^1 x_k^1 x_k^2/(x_k^1 + x_k^2)\right].$$

The five multilateral methods that use a bilateral index Q as a building block all have the property that $S_2/S_1 = Q$ when $I = 2$, which explains why these methods pass MT9.

Unfortunately, none of the above multilateral methods satisfies all nine multilateral tests. Our tentative conclusion is that if a symmetric multilateral method is desired, then the choice seems to be between the EKS (which fails MT7 and MT8, the consistency in aggregation properties) and the own share method (which fails the multilateral proportionality test, MT2). However, the systematic study of multilateral methods has only begun, so it may well be that better methods will be discovered in the future.

We turn now to another class of methods for constructing multilateral indexes.

10. Neostatistical Approaches to Multilateral Indexes

In one of the early statistical approaches to the construction of a price index, P^2/P^1, the index was found by minimizing the sum of squared residuals, $\sum_{n=1}^{N}(p_n^2/p_n^1 - P^2/P^1)^2$ with respect to P^2/P^1. The resulting price index turns out to equal (4). Note that this price index was defined independently of quantities.

Theil [1960] initiated a *neostatistical approach* where the price and quantity indexes, P^i and X^i, are simultaneously determined. *Theil's best linear price and quantity indexes* may be found by solving the following constrained minimization problem:

$$\min_{P^1,\ldots,P^I,X^1,\ldots,X^J} \sum_{i=1}^{I} \sum_{j=1}^{I} (p^i \cdot x^j - P^i X^j)^2 \tag{81}$$

subject to a normalization on the P^i or X^i such as

$$\sum_{i=1}^{I} X^i = 1, \tag{82}$$

which means that we can interpret the Theil X^i as shares. We may define $e_{ij} \equiv p^i \cdot x^j - P^i X^j$ as an error and then the interpretation of (81) becomes straightforward: we choose the P^i and X^i to minimize the sum of squared errors subject to (82).

The Theil index numbers have not been widely used, since they do not satisfy the product test equalities:

$$P^i X^i = p^i \cdot x^i, \qquad i = 1, \ldots, I. \tag{83}$$

Kloek and de Wit [1961] suggested a number of variants of the Theil indexes including one where the constraints (83) were imposed. Thus define the Kloek and de Wit multilateral indexes as the P^i and X^i which solve (81) subject to (82) and (83). Unfortunately, the resulting indexes do not have very satisfactory properties: they fail the multilateral tests MT2, MT3, MT7, MT8 and MT9 (when $I = 2$, the resulting bilateral quantity index fails BT2 and BT3 for quantity indexes.)

Another neostatistical approach has been suggested by van Yzeren [1956] which he called the *balanced method*. The price indexes P^i are determined (up to a normalization or factor of proportionality) by solving the following minimization problem:

$$\min_{P^1,\ldots,P^I} \sum_{i=1}^{I} \sum_{j=1}^{I} (P^i)^{-1} (p^i \cdot x^j / p^j \cdot x^j) P^j. \tag{84}$$

The X^i are then determined by (83) and the price normalization may be chosen so that (82) is satisfied. In this case, the errors e_{ij} may be defined by $(P^iX^j)^{-1}(p^i \cdot x^j/p^j \cdot x^j)P^jX^j = 1 + e_{ij}$ and the P^i may be found by minimizing

$$\sum\nolimits_{i=1}^{I} \sum\nolimits_{j=1}^{I} e_{ij}$$

subject to a normalization.

The van Yzeren system of share functions does rather well in the multilateral test examination: the share functions pass all tests except the consistency in aggregation tests MT7 and MT8. When $I = 2$, X^2/X^1 turns out to equal the Fisher quantity index, Q_F. Thus the balanced method satisfies the same tests as the EKS method. However, the EKS method has a more satisfactory economic interpretation and is easier to construct numerically.

Many other neostatistical approaches to the construction of multilateral indexes could be explored. However, the resulting methods seem to be rather arbitrary and, moreover, they lack economic interpretations.

11. Other Aspects

There are many aspects of index number theory that we cannot cover in this brief survey, such as: (i) *sampling problems* (see Fisher [1922] and Allen [1975]), (ii) the treatment of *seasonality* (see Turvey [1979], Balk [1980] and Diewert [1983c]), (iii) consistency in aggregation and the theory of *subindexes* (see Vartia [1976a], Pollak [1975] and Diewert [1983a]), (iv) productivity indexes (see Jorgenson and Griliches [1967], Caves, Christensen and Diewert [1982b] and Denny and Fuss [1983]), and (v) *econometric approaches* to cost of living indexes (see Jorgenson and Slesnick [1983]).

However, one area of concern that must be discussed is the *new goods problem*. Suppose that we are in the time series context and we have price and quantity data for $N-1$ commodities in periods 1 and 2, p_n^t and x_n^t for $t = 1, 2$ and $n = 1, \ldots, N-1$. Suppose in addition, that x_N^2 units of a new good are sold at the price p_N^2 during period 2. How are we to compute the bilateral price index, $P(p^1, p^2, x^1, x^2)$, when we do not know p_N^1, the price of the new good in period 1? Of course, we can assume $x_N^1 = 0$, so determining the quantity of the new good in period 1 is no problem.

From the viewpoint of the microeconomic approach to index number theory, Hicks [1940] provided a formal solution to this new good problem: if we are in the consumer context, p_N^1 should be the price which would just make the consumer's demand for good N in period 1 equal to zero. The practical problem is that this shadow price is not observable: we require a knowledge of the consumer's indifference surfaces to calculate it. Of course, econometric techniques could be used to estimate these shadow prices (see Diewert [1980] for an example of such a technique in the producer context), but most index number practitioners will find it inconvenient to resort to econometrics. In practice, most official indexes ignore the existence of new goods.

In order to illustrate the price index bias that can result from the omission of new goods, we shall present a hypothetical example. Suppose that there are three periods, one 'old' good with constant price and quantity, $p_1^t = x_1^t = 1$ for $t = 1, 2, 3$ and one 'new' good which appears in period 2, so that $x_2^1 = 0$. Typically, new goods follow a product cycle: they are introduced at a relatively high price and then the price declines over time. Thus we assume that $p_2^2 = 2$ and $p_2^3 = 1$, so that the period 2 price for the new good is twice as high as the period 3 price. We assume that the quantity purchased of the new good in period 2 is $f > 0$, where f is a fraction which represents the period 2 proportion of new goods to old goods. We assume that the quantity purchased of the new good in period 3 is $2f$ and that the shadow price p_2^1 in period 1 that would make the demand for the new good equal to zero is 4.

If the new good is ignored, we find that $P^t = 1$ for $t = 1, 2, 3$ for any reasonable index number formula. The true chain Laspeyres price indexes which do not ignore the new good are: $P^1 = 1$, $P^2 = 1$ and $P^3 = (1+f)/(1+2f)$. The reader can verify that the same indexes result no matter what shadow price p_2^1 is chosen. Evaluating P^3 for various reasonable values of f yields the following period 3 price indexes: if $f = 0.01$, then $P^3 = 0.9902$; if $f = 0.02$, then $P^3 = 0.9808$ and if $f = 0.05$, then $P^3 = 0.9545$. Thus the conventional Laspeyres price index which ignores the existence of new goods will have an *upward bias* of about 1 to 4.5 percent compared with the true Laspeyres index.

In order to evaluate the bias in the conventional chained Paasche and Fisher price indexes, we have to use our assumption that $p_2^1 = 4$. Under our assumptions, we obtain the following values for the true chained Paasche price indexes in period 3: if $f = 0.01$, then $P^3 = 0.9619$; if $f = 0.02$, then $P^3 = 0.9273$ and if $f = 0.05$, then $P^3 = 0.8403$. We also obtain the following values for the true chained Fisher price indexes in period 3: if $f = 0.01$, then $P^3 = 0.9759$; if $f = 0.02$, then $P^3 = 0.9537$ and if $f = 0.05$, then $P^3 = 0.8956$. Thus the conventional Fisher ideal price index will have an upward bias of about 2.5 to 10.5 percent in period 3, depending on the fraction f of new goods introduced in period 2.

The above analysis of bias is only illustrative but it does indicate that ignoring new goods could lead to a substantial overestimation of price inflation and a corresponding underestimation of real growth rates, especially in advanced market economies where millions of new goods are introduced each year.

Essays in Index Number Theory, Volume I
W.E. Diewert and A.O. Nakamura (Editors)

Chapter 6
DUALITY APPROACHES TO MICROECONOMIC THEORY*

W.E. Diewert

1. Introduction

What do we mean when we say that there is a "duality" between cost and production functions? Suppose that a *production function* F is given and that $u = F(x)$, where u is the maximum amount of output that can be produced by the technology during a certain period if the vector of input quantities $x \equiv (x_1, x_2, \ldots, x_N)$ is utilized during the period. Thus, the production function F describes the technology of the given firm. On the other hand, the firm's *minimum total cost* of producing at least the output level u given the input prices $(p_1, p_2, \ldots, p_N) \equiv p$ is defined as $C(u,p)$ and it is obviously a function of u, p and the given production function F. What is not so obvious is that (under certain regularity conditions) the cost function $C(u,p)$ also completely describes the technology of the given firm; i.e., given the firm's cost function C, it can be used in order to define the firm's production function F. Thus, there is a *duality* between cost and production functions in the sense that either of these functions can describe the technology of the firm equally well in certain circumstances.

In the first part of this chapter, we develop this duality between cost and production functions in more detail. In Section 2, we derive the regularity conditions that a cost function C *must* have (irrespective of the functional form or specific regularity properties for the production function F), and we show how a production function can be constructed from a given cost function. In Section 3, we develop this duality between cost and production functions in a more formal manner.

---

*This paper is an abridged version of IMSSS Technical Report No. 281, Stanford University, 1978. An even more abridged version was published in K.J. Arrow and M.D. Intriligator (eds.), *Handbook of Mathematical Economics*, Amsterdam: North-Holland Publishing Co., 1982, pp. 535–599. The author would like to thank C. Blackorby and R.C. Allen for helpful comments and the Canada Council for financial support. My thanks also to May McKee and Shelley Hey for typing a difficult manuscript. A preliminary version of this paper was presented at the New York meetings of the Econometric Society, December 1977.

In Section 4, we consider the duality between a (direct) production function F and the corresponding indirect production function G. Given a production function F, input prices $p \equiv (p_1, p_2, \ldots, p_N)$ and an input budget of y dollars, the *indirect production function* $G(y,p)$ is defined as the maximum output $u = F(x)$ that can be produced, given the budget constraint on input expenditures $p^T x \equiv \sum_{i=1}^N p_i x_i \leq y$. Thus, the indirect production function $G(y,p)$ is a function of the maximum allowable budget y, the input prices which the producer faces p, and the producer's production function F. Under certain regularity conditions, it turns out that G can also completely describe the technology and thus there is a duality between direct and indirect production functions.

The above dualities between cost, production and indirect production functions also have an interpretation in the context of consumer theory: simply let F be a consumer's *utility function*, x be a vector of commodity purchases (or rentals), u be the consumer's utility level, and y be the consumer's "income" or expenditure on the N commodities. Then $C(u,p)$ is the minimum cost of achieving utility level u given that the consumer faces the commodity prices p and there is a duality between the consumer's utility function F and the function C, which is often called the *expenditure function* in the context of consumer theory. Similarly, $G(y,p)$ can now be defined as the maximum utility that the consumer can attain given that he faces prices p and has income y to spend on the N commodities. In the consumer context, G is called the consumer's *indirect utility function*.

Thus, each of our duality theorems has two interpretations: one in the producer context and one in the consumer context. In Section 2, we will use the producer theory terminology for the sake of concreteness. However, in subsequent sections, we use a more neutral terminology which will cover both the producer and the consumer interpretations: we call a production or utility function F an *aggregator function*, a cost or expenditure function C a *cost function*, and an indirect production or utility function G an *indirect aggregator function*.

In Section 5, the distance function $D(u,x)$ is introduced. The *distance function* provides yet another way in which tastes or technology can be characterized. The main use of the distance function is in constructing the Malmquist [1953] quantity index.

In Sections 2–5, we will provide proofs of theorems so that the reader will be able to appreciate the techniques involved. In the remainder of the paper, results will often only be stated (with some exceptions where new results are presented).

In Section 6, we discuss a variety of other duality theorems: i.e., we discuss other methods for equivalently describing tastes or technology, either locally or globally, in the one output, N inputs context. The reader who is primarily interested in applications can skip Sections 3–6.

The mathematical theorems presented in Sections 2–6 may appear to be only theoretical results (of modest mathematical interest perhaps) devoid of practical applications. However, this is not the case. In Sections 7–10, we survey some of the applications of the duality theorems developed earlier. These applications fall into two main categories: (i) the measurement of technology or preferences (Sections 10 and 11), and (ii) the derivation of comparative statics results (Sections 7–9 inclusive).

In Section 11, we consider firms that can produce *many* outputs while utilizing many inputs (whereas earlier, we dealt only with the *one* output case). We state some useful duality theorems and then note some applications of these theorems.

Finally, in Section 12, we show how duality theory can be modified to deal with noncompetitive situations and, in Section 13, we briefly note some of the other areas of economics where duality theory has been applied.

2. Duality between Cost (Expenditure) and Production (Utility) Functions: A Simplified Approach

Suppose we are given an N input production function $F : u = F(x)$, where u is the amount of output produced during a period and $x \equiv (x_1, x_2, \ldots, x_N) \geq 0_N$[1] is a nonnegative vector of input quantities utilized during the period. Suppose, further, that the producer can purchase amounts of the inputs at the fixed positive prices $(p_1, p_2, \ldots, p_N) \equiv p \gg 0_N$ and that the producer does not attempt to exert any monopsony power on input markets.[2]

The producer's *cost function* C is defined as the solution to the problem of minimizing the cost of producing at least output level u, given that the producer faces the input price vector p:

$$C(u,p) \equiv \min_x \{p^T x : F(x) \geq u\} \tag{2.1}$$

In this section, it is shown that the cost function C satisfies a surprising number of regularity conditions, *irrespective* of the functional form for the production function F, provided only that solutions to the cost minimization problem (2.1) exist. In a subsequent section, it is shown how these regularity conditions on the cost function may be used in order to prove comparative

[1]Notation: $x \geq 0_N$ means each component of the N dimensional vector x is nonnegative, $x \gg 0_N$ means that each component is positive, and $x > 0_N$ means that $x \geq 0_N$ but $x \neq 0_N$.

[2]In Section 12 of this chapter, the assumption of competitive behavior is relaxed.

statics theorems about derived demand functions for inputs (cf. Samuelson [1947; ch. 4]).

Before establishing the properties of the cost function C, it is convenient to place the following minimal regularity condition on the production function F:

ASSUMPTION 1 ON F: F is continuous from above; i.e., for every $u \in$ range F,[3] $L(u) \equiv \{x : x \geq 0_N, F(x) \geq u\}$ is a closed set.

If F is a continuous function, then of course F will also be continuous from above. Assumption 1 is sufficient to imply that solutions to the cost minimization problem (2.1) exist, as the following lemma indicates.

LEMMA 1. *If F satisfies assumption 1 above and $p \gg 0_N$, then for every $u \in$ range F, $\min_x\{p^T x : x \geq 0_N, F(x) \geq u\}$ exists.*

Proof: Let $u \in$ range F. Then there exists $x^* \geq 0_N$ such that $F(x^*) \geq u$. Define the set $S^* \equiv \{x : p^T x \leq p^T x^*, x \geq 0_N\}$. Since $p \gg 0_N$, S^* is a closed and bounded set. Thus

$$\begin{aligned} C(u,p) &\equiv \min_x\{p^T x : x \geq 0_N, F(x) \geq u\} \\ &= \min_x\{p^T x : x \in L(u)\} \\ &= \min_x\{p^T x : x \in L(u) \cap S^*\} \end{aligned}$$

since if $x \geq 0_N$ and $x \notin S^*$, then $p^T x^* < p^T x$ and x could not be a solution to the cost minimization problem. Thus we can restrict attention to the closed and bounded set of feasible x's, $L(u) \cap S^*$, where the minimum of $p^T x$ will be attained. Q.E.D.

The following seven properties for the cost function C can now be derived, assuming *only* that the production function F satisfies assumption 1.

PROPERTY 1 FOR C: For every $u \in$ range F and $p \gg 0_N$, $C(u,p) \geq 0$; i.e., C is a *nonnegative* function.

Proof:

$$\begin{aligned} C(u,p) &\equiv \min_x\{p^T x : x \geq 0_N, F(x) \geq u\} \\ &= p^T x^* \text{ say, where } x^* \geq 0_N \text{ and } F(x^*) \geq u \\ &\geq 0 \text{ since } p \gg 0_N \text{ and } x^* \geq 0_N. \end{aligned}$$

Q.E.D.

[3]This simply means that the output u can be produced by the technology. Throughout this section, range F can be replaced with the smallest convex set containing range F.

PROPERTY 2 FOR C: For every $u \in$ range F, if $p \gg 0_N$ and $k > 0$, then $C(u, kp) = kC(u,p)$; i.e., the cost function is (positively) *linearly homogeneous* in input prices for any fixed output level.

Proof: Let $p \gg 0_N$, $k > 0$ and $u \in$ range F. Then

$$\begin{aligned} C(u,kp) &\equiv \min_x \{(kp)^T x : F(x) \geq u\} \\ &= k \min_x \{p^T x : F(x) \geq u\} \\ &\equiv kC(u,p). \end{aligned}$$

Q.E.D.

PROPERTY 3 FOR C: If any combination of input prices increases, then the minimum cost of producing any feasible output level u will not decrease; i.e. if $u \in$ range F and $p^1 > p^0$, then $C(u,p^1) \geq C(u,p^0)$.

Proof:

$$\begin{aligned} C(u,p^1) &\equiv \min_x \{p^{1T} x : F(x) \geq u\} \\ &= p^{1T} x^1 \text{ say, where } x^1 \geq 0_N \text{ and } F(x^1) \geq u \\ &\geq p^{0T} x^1 \text{ since } p^1 > p^0 \text{ and } x^1 \geq 0_N \\ &\geq \min_x \{p^{0T} x : F(x) \geq u\} \text{ since } x^1 \text{ is feasible for the cost} \\ &\qquad \text{minimization problem but not necessarily optimal} \\ &\equiv C(u,p^0). \end{aligned}$$

Q.E.D.

Thus far, the properties of the cost function have been intuitively obvious from an economic point of view. However, the following important property is not an intuitively obvious one.

PROPERTY 4 FOR C: For every $u \in$ range F, $C(u,p)$ is a concave function[4] of p.

Proof: Let $u \in$ range F, if $p^0 \gg 0_N$, $p^1 \gg 0_N$ and $0 \leq \lambda \leq 1$. Then

$$C(u,p^0) \equiv \min_x \{p^{0T} x : F(x) \geq u\} = p^{0T} x^0$$

say, and

$$C(u,p^1) \equiv \min_x \{p^{1T} x : F(x) \geq u\} = p^{1T} x^1$$

[4]A function $f(z)$ of n variables defined over a convex set S is concave iff $z^1, z^2 \in S$, $0 \leq \lambda \leq 1$ implies $f[\lambda z^1 + (1-\lambda)z^2] \geq [\lambda f(z^1) + (1-\lambda) f(z^2)]$. A set S is a convex set iff $z^1, z^2 \in S$, $0 \leq \lambda \leq 1$ implies $[\lambda z^1 + (1-\lambda)z^2] \in S$.

say. Now

$$\begin{aligned}C[u,\lambda p^0+(1-\lambda)p^1] &\equiv \min_x\{[\lambda p^0+(1-\lambda)p^1]^T x : F(x)\geq u\}\\ &= [\lambda p^0+(1-\lambda)p^1]^T x^\lambda, \quad \text{say,}\\ &= \lambda p^{0T}x^\lambda+(1-\lambda)p^{1T}x^\lambda\\ &\geq \lambda p^{0T}x^0+(1-\lambda)p^{1T}x^1,\end{aligned}$$

since x^λ is feasible for the cost minimization problems associated with the input price vectors p^0 and p^1 but is not necessarily optimal for those problems

$$= \lambda C(u,p^0)+(1-\lambda)C(u,p^1).$$

Q.E.D.

The basic idea in the above proof is used repeatedly in duality theory. Owing to the nonintuitive nature of property 4, it is perhaps useful to provide a geometric interpretation of it in the two input case (i.e., $N=2$).

Suppose that the producer must produce the output level u. The u isoquant is drawn in Figure 2.1. Define the set S^0 as the set of nonnegative input combinations which are either on or below the optimal isocost line when the producer faces prices p^0; i.e., $S^0 \equiv \{x : p^{0T}x \leq C(u,p^0), x\geq 0_N\}$, where $C^0\equiv C(u,p^0)=p^{0T}x^0$ is the minimum cost of producing output u given that the producer faces input prices $p^0 \gg 0_N$. Note that the vector of inputs x^0 solves the cost minimization problem in this case. Now suppose that the producer faces the input prices $p^1\gg 0_N$ and define S^1, C^1 and x^1 analogously; i.e., $S^1\equiv\{x : p^{1T}x\leq C(u,p^1), x\geq 0_N\}$, $C^1\equiv C(u,p^1)=p^{1T}x^1$, where the vector of inputs x^1 solves the cost minimization problem when the producer faces prices p^1.

Let $0<\lambda<1$ and suppose now that the producer faces the "average" input prices $\lambda p^0+(1-\lambda)p^1$. Define S^λ, C^λ and x^λ as before:

$$\begin{aligned}S^\lambda &\equiv \{x : [\lambda p^0+(1-\lambda)p^1]^T x\leq C[u,\lambda p^0+(1-\lambda)p^1],\ x\geq 0_N\},\\ C^\lambda &\equiv C[u,\lambda p^0+(1-\lambda)p^1]=[\lambda p^0+(1-\lambda)p^1]^T x^\lambda\end{aligned}$$

where x^λ solves the cost minimization problem when the producer faces the average prices $\lambda p^0+(1-\lambda)p^1$. Finally, consider the isocost line which would result if the producer spends an "average" of the two initial costs, $\lambda C^0+(1-\lambda)C^1$, facing the "average" input prices, $\lambda p^0+(1-\lambda)p^1$. The set of nonnegative input combinations which are either on or below this isocost line is defined as the set

$$S^*\equiv\{x : [\lambda p^0+(1-\lambda)p^1]^T x\leq[\lambda C^0+(1-\lambda)C^1],\ x\geq 0_N\}.$$

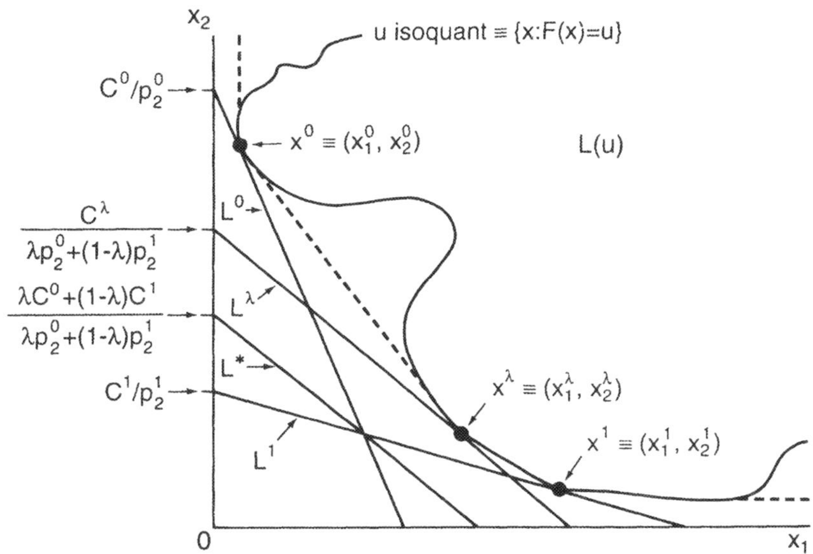

Figure 2.1

In order to show the concavity property for C, we need to show that $C^\lambda \geq \lambda C^0 + (1-\lambda)C^1$ or, equivalently, we need to show that S^λ contains the set S^*. It can be shown that the isocost line associated with the set S^*,

$$L^* \equiv \{x : [\lambda p^0 + (1-\lambda)p^1]^T x = [\lambda C^0 + (1-\lambda)C^1]\},$$

passes through the intersection of the isocost line associated with the sets S^0 and S^1.[5] On the other hand, the isocost line associated with the set S^λ,

$$L^\lambda \equiv \{x : [\lambda p^0 + (1-\lambda)p^1]^T x = C^\lambda\}$$

is obviously parallel to L^*. Finally, L^λ must be either coincident with or lie above L^*, since if L^λ were below L^*, then there would exist a point on

[5]Let $x^* \in L^0 \cap L^1$. Then $p^{0T}x^* = C^0$ and $p^{1T}x^* = C^1$. Thus $[\lambda p^0 + (1-\lambda)p^1]^T x^* = \lambda C^0 + (1-\lambda)C^1$ and $x^* \in S^*$. This also follows from the readily proven proposition that $S^0 \cap S^1 \subset S^* \subset S^0 \cup S^1$ (cf. Diewert [1974a; 157–158]).

the u isoquant which would lie below at least one of the isocost lines $L^0 \equiv \{x : p^{0T}x = C^0\}$ or $L^1 \equiv \{x : p^{1T}x = C^1\}$ which would contradict the cost minimizing nature of x^0 or x^1.[6]

PROPERTY 5 FOR C: For every $u \in$ range F, $C(u,p)$ is *continuous* in p for $p \gg 0_N$.

Proof: This property is a mathematical consequence of property 4 for C, concavity in p for fixed u. For proofs, see Fenchel [1953; 75] or Rockafellar [1970; 82]. Q.E.D.

PROPERTY 6 FOR C: $C(u,p)$ is nondecreasing in u for fixed p; i.e., if $p \gg 0_N$, $u^0, u^1 \in$ range F and $u^0 \leq u^1$, then $C(u^0,p) \leq C(u^1,p)$.

Proof: Let $p \gg 0_N$, $u^0, u^1 \in$ range F and $u^0 \leq u^1$. Thus

$$\begin{aligned} C(u^1,p) &\equiv \min_x\{p^Tx : F(x) \geq u^1\} \\ &\geq \min_x\{p^Tx : F(x) \geq u^0\}, \text{ since if } u^0 \leq u^1 \text{ then} \\ &\quad \{x : F(x) \geq u^1\} \subset \{x : F(x) \geq u^0\}, \\ &\quad \text{and the minimum of } p^Tx \text{ over a larger set cannot increase} \\ &\equiv C(u^0,p). \end{aligned}$$

Q.E.D.

In contrast to the previous properties for the cost function, the following property requires some heavy mathematical artillery. Since these mathematical results are useful not only in the present section, but also in subsequent sections, we momentarily digress and state these results.

In the following definitions, let S denote a subset of R^M, T a subset of R^K, $\{x^n\}$ a sequence of points of S and $\{y^n\}$ a sequence of points of T. For a more complete discussion of the following definitions and theorems, see Green and Heller [1981].

DEFINITION: ϕ is a *correspondence* from S into T if, for every $x \in S$, there exists a nonempty image set $\phi(x)$ which is a subset of T.

DEFINITIONS: A correspondence ϕ is *upper semicontinuous* (or alternatively, *upper hemicontinuous*) at the point $x^0 \in S$ if $\lim_n x^n = x^0$, $y^n \in \phi(x^n)$, $\lim_n y^n = y^0$ implies $y^0 \in \phi(x^0)$. A correspondence ϕ is *lower semicontinuous* at $x^0 \in S$ if $\lim_n x^n = x^0$, $y^0 \in \phi(x^0)$ implies that there exists a sequence $\{y^n\}$ such that $y^n \in \phi(x^n)$ and $\lim_n y^n = y^0$. A correspondence ϕ is *continuous* at $x^0 \in S$ if it is both upper and lower semicontinuous at x^0.

[6]It can be seen that the approximating set $S^a \equiv \{x : p^{0T}x \geq C^0,\ x \geq 0_N\} \cap \{x : p^{1T}x \geq C^1,\ x \geq 0_N\}$ *contains* the true technological set $L(u) \equiv \{x : F(x) \geq u\}$ and thus the minimum cost associated with S^a will generally be *lower* than the cost associated with $L(u)$.

LEMMA 2. (Berge [1963; 111-112]): *ϕ is an upper semicontinuous correspondence over S iff graph $\phi \equiv \{(x,y) : x \in S,\ y \in \phi(x)\}$ is a closed set in $S \times T$.*[7]

UPPER SEMICONTINUITY MAXIMUM THEOREM. (Berge [1963; 116]): *Let f be a continuous from above function*[8] *defined over $S \times T$ where T is a compact (i.e., closed and bounded) subset of R^K. Suppose that ϕ is a correspondence from S into T and that ϕ is upper semicontinuous over S. Then the function g defined by $g(x) \equiv \max_y\{f(x,y) : y \in \phi(x)\}$ is well defined and is continuous from above over S.*

MAXIMUM THEOREM. (Debreu [1952; 889–890], [1959, 19], Berge [1963, 116]): *Let f be a continuous real valued function defined over $S \times T$, where T is a compact subset of R^K. Let ϕ be a correspondence from S into T and let ϕ be continuous at $x^0 \in S$. Define the (maximum) function g by $g(x) \equiv \max_y\{f(x,y) : y \in \phi(x)\}$ and the (set of maximizers) correspondence ξ by $\xi(x) \equiv \{y : y \in \phi(x)$ and $f(x,y) = g(x)\}$. Then the function g is continuous at x^0 and the correspondence ξ is upper semicontinuous at x^0.*

PROPERTY 7 FOR C: For every $p \gg 0_N$, $C(u,p)$ is *continuous from below*[9] in u; i.e., if $p^* \gg 0_N$, $u^* \in$ range F, $u^n \in$ range F for all n, $u^1 \leq u^2 \leq \ldots$ and $\lim u^n = u^*$, then $\lim_n C(u^n,p^*) = C(u^*,p^*)$.

Proof: Define the correspondence L for $u \in$ range F by $L(u) \equiv \{x : F(x) \geq u,\ x \geq 0_N\}$. Since F is continuous from above (recall assumption 1), it can be shown that (see Rockafellar [1970; 51] that the graph of L, graph $L \equiv \{(u,x) : x \geq 0_N,\ u \in$ range $F,\ u \leq F(x)\}$ is a closed set and hence by Lemma 2 above, L is an upper semicontinuous correspondence over range F. Let $p^* \gg 0_N$, $u^* \in$ range F and let x^* be a solution to the cost minimization problem

$$C(u^*,p^*) \equiv \min_x\{p^{*T}x : x \geq 0_N,\ F(x) \geq u^*\} = p^{*T}x^*.$$

Define $S^* \equiv \{x : p^{*T}x \leq p^{*T}x^*,\ x \geq 0_N\}$. For $u \in$ range F and $u \leq u^*$, it can be seen that

$$\begin{aligned} C(u,p^*) &\equiv \min_x\{p^{*T}x : x \geq 0_N,\ F(x) \geq u\} \\ &= \min_x\{p^{*T}x : x \in L(u) \cap S^*\} \\ &= -\max_x\{-p^{*T}x : x \in \phi(u)\}, \end{aligned}$$

---

[7] $S \times T$ is the set of (x,y) such that $x \in S$ and $y \in T$.

[8] A real valued function f defined over $S \times T$ is continuous from above (or alternatively, is upper semicontinuous) at $z^0 \in S \times T$ iff either of the following conditions is satisfied: (i) for every $\varepsilon > 0$, there exists a neighborhood of z^0, $N(z^0)$, such that $z \in N(z^0)$ implies $f(z) < f(z^0) + \varepsilon$, or (ii) if $z^n \in S \times T$, $\lim_n z^n = z^0$, $f(z^n) \geq f(z^0)$, then $\lim_n f(z^n) = f(z^0)$. f is continuous from above over $S \times T$ if it is continuous from above at each point of $S \times T$. See Green and Heller [1981].

[9] A function f is continuous from below iff $-f$ is continuous from above.

where $\phi(u) \equiv L(u) \cap S^*$, a compact set for $u \in$ range F and $u \leq u^*$. It can be verified that ϕ is an upper semicontinuous correspondence at u^* and that $-p^{*T}x$ is continuous in x and u and hence continuous from above. Thus by the Upper Semicontinuity Maximum Theorem $-C(u,p^*) = \max_x\{-p^{*T}x : x \in \phi(u)\}$ is continuous from above in u at u^* so that $C(u,p^*)$ is continuous from below at u^*. Q.E.D.

In order to illustrate the last property of C, the reader may find it useful to let $N = 1$ and let the production function $F(x)$ be the following (continuous from above) step function (cf. Shephard [1970; 89]):

$$F(x) \equiv \{0, \text{ if } 0 \leq x < 1;\ 1, \text{ if } 1 \leq x < 2;\ 2, \text{ if } 2 \leq x < 3; \ldots\}.$$

For $p > 0$, the corresponding cost function $C(u,p)$ is the following (continuous from below) step function:

$$C(u,p) \equiv \{0, \text{ if } 0 = u;\ p, \text{ if } 0 < u \leq 1;\ 2p, \text{ if } 1 < u \leq 2; \ldots\}.$$

The above properties of the cost function have some empirical implications, as we shall see later. However, one application can be mentioned at this point. Suppose that we can observe cost, input prices and output for a firm and suppose further that we have econometrically estimated the following linear cost function:[10]

$$C(u,p) = \alpha + \beta^T p + \gamma u \tag{2.2}$$

where α and γ are constants and β is a vector of constants. Could (2.2) be the firm's true cost function? The answer is *no* if the firm is competitively minimizing costs and if either one of the constants α and γ is nonzero, for in this case, C does not satisfy property 2 (linear homogeneity in input prices).

Suppose now that we have somehow determined the firm's true cost function C, but that we do not know the firm's production function F (except that it satisfies assumption 1). How can we use the given cost function $C(u,p)$ (satisfying properties 1–7 above) in order to construct the firm's underlying production function $F(x)$?

Equivalent to the production function $u = F(x)$ are the family of isoproduct surfaces $\{x : F(x) = u\}$ or the family of level sets $L(u) \equiv \{x : F(x) \geq u\}$. For any $u \in$ range F, the cost function can be used in order to construct an outer approximation to the set $L(u)$ in the following manner. Pick input prices $p^1 \gg 0_N$ and graph the isocost surface $\{x : p^{1T}x = C(u,p^1)\}$. The set $L(u)$ must lie above (and intersect) this set, because $C(u,p^1) \equiv \min_x\{p^{1T}x : x \in L(u)\}$; i.e., $L(u) \subset \{x : p^{1T}x \geq C(u,p^1)\}$. Pick additional input price vectors

[10]This type of cost function is often estimated by economists; e.g., see Walters [1961] survey article on cost and production functions.

$p^2 \gg 0_N$, $p^3 \gg 0_N$, ... and graph the isocost surfaces $\{x : p^{iT}x = C(u, p^i)\}$. It is easy to see that $L(u)$ must be a subset of each of the sets $\{x : p^{iT}x \geq C(u, p^i)\}$. Thus

$$(2.3) \qquad L(u) \subset \bigcap\nolimits_{p \gg 0_N} \{x : p^T x \geq C(u, p)\} \equiv L^*(u);$$

i.e., the true production possibilities set $L(u)$ must be contained in the outer approximation production possibilities set $L^*(u)$ which is obtained as the intersection of all of the supporting total cost half spaces to the true technology set $L(u)$. In Figure 2.1, $L^*(u)$ is indicated by dashed lines. Note that the boundary of this set forms an approximation to the true u isoquant and that this approximating isoquant coincides with the true isoquant in part, but it does not have the backward bending and nonconvex portions of the true isoquant.

Once the family of approximating production possibilities sets $L^*(u)$ has been constructed, the approximating production function F^* can be defined as

$$(2.4) \qquad \begin{aligned} F^*(x) &\equiv \max_u \{u : x \in L^*(u)\} \\ &= \max_u \{u : p^T x \geq C(u, p) \text{ for every } p \gg 0_N\} \end{aligned}$$

for $x \geq 0_N$. Note that the maximization problem defined by (2.4) has an infinite number of constraints (one constraint for each $p \gg 0_N$). However, (2.4) can be used in order to define the approximating production function F^* given only the cost function C.

It is clear (recall Figure 2.1) that the approximating production function F^* will not in general coincide with the true function F. However, it is also clear that from the viewpoint of observed market behavior, if the producer is competitively cost minimizing, then it does not matter whether the producer is minimizing cost subject to the production function constraint given by F or F^*: observable market data will never allow us to determine whether the producer has the production function F or the approximating function F^*.

It is also clear that if we want the approximating production function F^* to coincide with the true production function F, then it is necessary that F satisfy the following two assumptions:

ASSUMPTION 2 ON F: F is *nondecreasing*; i.e., if $x^2 \geq x^1 \geq 0_N$, then $F(x^2) \geq F(x^1)$.

ASSUMPTION 3 ON F: F is a *quasiconcave* function; i.e., for every $u \in$ range F, $L(u) \equiv \{x : F(x) \geq u\}$ is a convex set.

If F satisfies assumption 2, then backward bending isoquants cannot occur, while if F satisfies assumption 3, then nonconvex isoquants of the type drawn in Figure 2.1 cannot occur.

It is not too difficult to show that if F satisfies assumptions 1–3 and the cost function C is computed via (2.1), then the approximating production function

F^* computed via (2.4) will coincide with the original production function F; i.e., there is a *duality* between cost functions satisfying properties 1–7 and production functions satisfying assumptions 1–3. The first person to prove a formal duality theorem of this type was Shephard [1953].

In the following section, we will prove a similar duality theorem after placing somewhat stronger conditions on the underlying production function F.

The following result is the basis for most theoretical and empirical applications of duality theory.

LEMMA 3. (Hicks [1946; 331], Samuelson [1947; 68], Karlin [1959; 272] and Gorman [1976]): *Suppose that the production function F satisfies assumption 1 and that the cost function C is defined by (2.1). Let $u^* \in$ range F, $p^* \gg 0_N$ and suppose that x^* is a solution to the problem of minimizing the cost of producing output level u^* when input prices p^* prevail; i.e.,*

$$C(u^*, p^*) \equiv \min_x \{p^{*T}x : F(x) \geq u^*\} = p^{*T}x^*. \tag{2.5}$$

If in addition, C is differentiable with respect to input prices at the point (u^*, p^*), then

$$x^* = \nabla_p C(u^*, p^*) \tag{2.6}$$

where

$$\nabla_p C(u^*, p^*) \equiv [\partial C(u^*, p_1^*, \ldots, p_N^*)/\partial p_1, \ldots, \partial C(u^*, p_1^*, \ldots, p_N^*)/\partial p_N]^T$$

is the vector of first order partial derivatives of C with respect to the components of the input price vector p.

Proof: Given any vector of positive input prices $p \gg 0_N$, x^* is feasible for the cost minimization problem defined by $C(u^*, p)$ but it is not necessarily optimal; i.e., for every $p \gg 0_N$, we have the following inequality:

$$p^T x^* \geq C(u^*, p). \tag{2.7}$$

For $p \gg 0_N$, define the function $g(p) \equiv p^T x^* - C(u^*, p)$. From (2.7), $g(p) \geq 0$ for $p \gg 0_N$ and from (2.5), $g(p^*) = 0$. Thus, $g(p)$ attains a global minimum at $p = p^*$. Since g is differentiable at p^*, the first order necessary conditions for a local minimum must be satisfied:

$$\nabla_p g(p^*) = x^* - \nabla_p C(u^*, p^*) = 0_N$$

which implies (2.6). Q.E.D.

Thus differentiation of the producer's cost function $C(u, p)$ with respect to input prices p yields the producer's system of cost minimizing input demand functions, $x(u, p) = \nabla_p C(u, p)$.

The above lemma should be carefully compared with the following result.

LEMMA 4. (Shephard [1953; 11]): *If the cost function $C(u,p)$ satisfies properties 1–7 and, in addition, is differentiable with respect to input prices at the point (u^*,p^*), then*

$$x(u^*,p^*) = \nabla_p C(u^*,p^*) \tag{2.8}$$

where $x(u^*,p^*) \equiv [x_1(u^*,p^*),\ldots,x_N(u^*,p^*)]^T$ is the vector of cost minimizing input quantities needed to produce u^* units of output given input prices p^*, where the underlying production function F^* is defined by (2.4), $u^* \in$ range F^* and $p^* \gg 0_N$.

The difference between Lemma 3 and Lemma 4 is that Lemma 3 assumes the existence of the production function F and does not specify the properties of the cost function other than differentiability, while Lemma 4 assumes only the existence of a cost function satisfying the appropriate regularity conditions and the corresponding production function F^* is defined using the given cost function. Thus, from an econometric point of view, Lemma 4 is more useful than Lemma 3: in order to obtain a valid system of input demand functions, all we have to do is postulate a functional form for C which satisfies the appropriate regularity conditions and differentiate C with respect to the components of the input price vector p. It is not necessary to compute the corresponding production function F^* nor is it necessary to endure the sometimes painful algebra involved in deriving the input demand functions from the production function via Lagrangian techniques.[11]

For formal proofs of Lemma 4, see the following section and the accompanying references.

Historical Notes:

The proposition that there are two or more equivalent ways of representing preferences or technology forms the core of duality theory. The mathematical basis for the economic theory of duality is Minkowski's [1911] Theorem:[12] every closed convex set can be represented as the intersection of its supporting halfspaces. Thus, under certain conditions, the closed convex set $L(u) \equiv \{x : F(x) \geq u, x \geq 0_N\}$ can be represented as the intersection of the halfspaces generated by the isocost surfaces tangent to the production possibilities set $L(u), \bigcap_p \{x : p^T x \geq C(u,p)\}$.

If the consumer (or producer) has a budget of $y > 0$ to spend on the N commodities (or inputs), then the maximum utility (or output) that he can

[11]For an exposition of the Lagrangian method for deriving demand functions and comparative statics theorems, see Intriligator [1981].

[12]See Fenchel [1953; 48–50] or Rockafellar [1970; 95–99].

obtain given that he faces prices $p \gg 0_N$ can generally be obtained by solving the equation $y = C(u,p)$ or by solving

$$1 = C(u, p/y) \tag{2.9}$$

(where we have used the linear homogeneity of C in p) for u as a function of the normalized prices, p/y. Call the resulting function G so that $u = G(p/y)$. Alternatively, G can be defined directly from the utility (or production) function F in the following manner for $p \gg 0_N$, $y > 0$:

$$G^*(p,y) \equiv \max_x \{F(x) : p^T x \leq y, \ x \geq 0_N\} \tag{2.10}$$

or

$$G(p/y) \equiv \max_x \{F(x) : (p/y)^T x \leq 1, \ x \geq 0_N\}.$$

Houthakker [1951–52; 157] called the function G the *indirect utility function*, and like the cost function C, it also can characterize preferences or technology uniquely under certain conditions (cf. Section 4 below). Our reason for introducing it at this point is that historically, it was introduced into the economics literature before the cost function by Antonelli [1971; 349] in 1886 and then by Konüs [1924]. However, the first paper which recognized that preferences could be equivalently described by a direct or indirect utility function appears to be by Konüs and Byushgens [1926; 157] who note that the equations $u = F(x)$ and $u = G(p/y)$ are equations for the same surface, but in different coordinate systems: the first equation is in pointwise coordinates while the second is in planar or tangential coordinates. Konüs and Byushgens [1926; 159] also set up the minimization problem that allows one to derive the direct utility function from the indirect function and, finally, they graphed various preferences in price space for the case of two goods.

The English language literature on duality theory seems to have started with two papers by Hotelling [1932][1935], who was perhaps the first economist to use the word "duality":

> Just as we have a utility (or profit) function u of the quantities consumed whose derivatives are the prices, there is, dually, a function of the prices whose derivatives are the quantities consumed.
>
> Hotelling [1932; 594]

Hotelling [1932; 597] also recognized that "the cost function may be represented by surfaces which will be concave upward"; i.e., he recognized that the cost function $C(u,p)$ would satisfy a curvature condition in p.

Hotelling [1932; 590] [1935; 68] also introduced the *profit function* Π which provides yet another way by which a decreasing returns to scale technology can be described. Using our notation, Π is defined as

$$\Pi(p) \equiv \max_x \{F(x) - p^T x\}. \tag{2.11}$$

Hotelling indicated that the profit maximizing demand functions, $x(p) \equiv [x_1(p), \ldots, x_N(p)]^T$, could be obtained by differentiating the profit function Π; i.e., $x(p) = -\nabla_p \Pi(p)$. Thus, if Π is twice continuously differentiable, one can readily deduce *Hotelling's* [1935; 69] *symmetry conditions*:

$$-\frac{\partial x_i}{\partial p_j}(p) = \frac{\partial^2 \Pi}{\partial p_i \partial p_j}(p) = \frac{\partial^2 \Pi}{\partial p_j \partial p_i}(p) = -\frac{\partial x_j}{\partial p_i}(p). \tag{2.12}$$

The next important contribution to duality theory was made by Roy who independently recognized that preferences could be represented by pointwise or tangential coordinates:

> Il vient alors tout naturellement à l'esprit d'invoquer le principle de dualité qui permet d'utiliser les équations tangentielles au lieu des équations ponctuelles; ainsi apparaît-il possible de présenter les équations d'équilibre sous une forme nouvelle et susceptible d'interprétation fécondes.
>
> Roy [1942; 18–19]

Roy [1942; 20] defined the indirect utility function G^* as in equation (2.10) above and then he derived the counterpart to Lemma 3 above, which is called *Roy's Identity* [1942; 18–19]).

$$x(p/y) = -\frac{\nabla_p G^*(p,y)}{\nabla_y G^*(p,y)}, \tag{2.13}$$

where $x(p/y) \equiv [x_1(p/y), \ldots, x_N(p/y)]^T$ is the vector of utility (or output) maximizing demand functions given that the consumer (or producer) faces input prices $p \gg 0_N$ and has a budget $y > 0$ to spend. Roy [1942; 24–27] showed that G^* was decreasing in the prices p, increasing in income y and homogeneous of degree 0 in (p, y); i.e., $G^*(\lambda p, \lambda y) = G^*(p, y)$ for $\lambda > 0$. Thus, $G^*(p, y) = G^*(p/y, 1) \equiv G(p/y) = G(v)$, where $v \equiv p/y$ is a vector of *normalized prices*. In his 1947 paper, Roy derived the following version of Roy's Identity [1947; 219] where the indirect utility function G is used in place of G^*:

$$x_i(v) = \frac{\partial G(v)}{\partial v_i} \Big/ \sum\nolimits_{j=1}^{N} v_j \frac{\partial G(v)}{\partial v_j}; \qquad i = 1, 2, \ldots, N. \tag{2.14}$$

The French mathematician Ville [1951–52; 125] also derived the useful relations (2.14) in 1946, so perhaps (2.14) should be called *Ville's Identity*. Ville [1951–52; 126] also noted that if the direct utility function $F(x)$ is linearly homogeneous, then the indirect function $G(v) \equiv \max_x\{F(x) : v^T x \le 1,\ x \ge 0_N\}$ is homogeneous of degree -1; i.e., $G(\lambda v) = \lambda^{-1}G(v)$ for $\lambda > 0$, $v \gg 0_N$, and thus $-G(v) = \sum_{j=1}^N v_j(\partial G(v)/\partial v_j)$. Substitution of the last identity into (2.14) yields the simpler equations (see also Samuelson [1972]) if $G(v)$ is positive:

$$x_i(v) = -\partial \ln G(v)/\partial v_i, \qquad i = 1, 2, \ldots, N. \tag{2.15}$$

At this point, it should be mentioned that Antonelli [1971; 349] obtained a version of Roy's Identity in 1886 and Konüs and Byushgens [1926; 159] *almost* derived it in 1926 in the following manner: they considered the problem of *minimizing* indirect utility $G(v)$ with respect to the normalized prices v subject to the constraint $v^T x = 1$. As Houthakker [1951–52; 157–158] later observed, it turns out that this constrained minimization problem generates the direct utility function; i.e., we have for $x \gg 0_N$:

$$F(x) = \min_v\{G(v) : v^T x \le 1,\ v \ge 0_N\}. \tag{2.16}$$

Konüs and Byushgens obtained the first order conditions for the problem (2.16): $\nabla_v G(v) = \mu x$. If the Lagrange multiplier μ is eliminated from this last system of equations using the constraint $v^T x = 1$, we obtain $x = \nabla_v G(v) \ / \ v^T \nabla_v G(v)$, which is (2.14) written in vector notation. However, Konüs and Byushgens did not explicitly carry out this last step.

Another notable early paper was written by Wold [1943, 1944] who defined the indirect utility function $G(v)$ (he called it a "price preference function") and showed that the indifference surfaces of price space were either convex to the origin or possibly linear; i.e., he showed that $G(v)$ was a quasiconvex function[13] in the normalized prices v. Wold's early work is summarized in Wold [1953; 145–148].

Malmquist [1953; 212] also defined the indirect utility function $G(v)$ and indicated that it was a quasiconvex function in v.

If the production function F is subject to constant returns to scale (i.e., $F(\lambda x) = \lambda F(x)$ for every $\lambda \ge 0$, $x \ge 0_N$) in addition to being continuous from above, then the corresponding cost function decomposes in the following manner: let $u > 0$, $p \gg 0_N$; then

$$\begin{aligned} C(u,p) &\equiv \min_x\{p^T x : F(x) \ge u\} \\ &= \min_x\{u p^T (x/u) : F(x/u) \ge 1\} \\ &= u \min_z\{p^T z : F(z) \ge 1\} \\ &\equiv uC(1,p). \end{aligned} \tag{2.17}$$

[13] A function G is quasiconvex if and only if $-G$ is quasiconcave.

(The above proof assumes that there exists at least one $x^* > 0_N$ such that $F(x^*) > 0$ so that the set $\{z : F(z) \geq 1\}$ is not empty). Samuelson [1953–54] assumed that the production function F was linearly homogeneous and subject to a "generalized law of diminishing returns", $F(x'+x'') \geq F(x')+F(x'')$, which is equivalent to concavity of F when F is linearly homogeneous. Samuelson [1953–54; 15] then defined the unit cost function $C(1,p)$ and indicated that $C(1,p)$ satisfied the same properties in p that F satisfied in x. Samuelson [1953–54; 15] also noted that a flat on the unit output production surface (a region of infinite substitutability) would correspond to a corner on the unit cost surface, a point which was also made by Shephard [1953; 27–28].

Shephard's 1953 monograph appears to be the first modern, rigorous treatment of duality theory. Shephard [1953; 13–14] notes that the cost function $C(u,p)$ can be interpreted as the support function for the convex set $\{x : F(x) \geq u\}$, and he uses this fact to establish the properties of $C(u,p)$ with respect to p. Shephard [1953; 13] also explicitly mentions Minkowski's [1911] Theorem on convex sets and Bonnesen and Fenchel's [1934] monograph on convex sets. It should be mentioned that Shephard did not develop a direct duality between production and cost functions; he developed a duality between production and distance functions (which we will define in a later section) and then between distance and cost functions.

Shephard [1953; 41] defined a production function F to be *homothetic* if it could be written as

$$F(x) = \phi[f(x)]$$

where f is a homogeneous function of degree one and ϕ is a continuous, positive monotone increasing function of f. Let us formally introduce the following additional conditions on F (or f):

ASSUMPTION 4 ON F: F is (nonnegatively) *linearly homogeneous*; i.e., if $x \geq 0_N$, $\lambda \geq 0$, then $F(\lambda x) = \lambda F(x)$.

ASSUMPTION 5 ON F: F is *weakly positive*; i.e., for every $x \geq 0_N$, $F(x) \geq 0$ but $F(x^*) > 0$ for at least one $x^* > 0_N$.

Now let us assume that $\phi(f)$ is a continuous, monotonically increasing function of one variable for $f \geq 0$ with $\phi(0) = 0$. Under these conditions the inverse function ϕ^{-1} exists and has the same properties as ϕ, with $\phi^{-1}[\phi(f)] = f$ for all $f \geq 0$. If $f(x)$ satisfies assumptions 1, 4 and 5 above, then the cost function which corresponds to $F(x) \equiv \phi[f(x)]$ decomposes as follows: let $u > 0$, $p \gg 0_N$; then

$$\begin{aligned} C(u,p) &\equiv \min_x\{p^T x : \phi[f(x)] \geq u\} \\ &= \min_x\{p^T x : f(x) \geq \phi^{-1}(u)\} \\ &= \phi^{-1}(u) \min_x\{p^T(x/\phi^{-1}(u)) : f(x/\phi^{-1}(u)) \geq 1\}, \\ &\qquad \text{where } \phi^{-1}(u) > 0 \text{ since } u > 0, \\ &= \phi^{-1}(u)c(p), \end{aligned} \tag{2.18}$$

where $c(p) \equiv \min_z\{p^T z : f(z) \geq 1\}$ is the *unit cost function* which corresponds to the linearly homogenous function f, a nonnegative, (positively) linearly homogenous, nondecreasing, concave and continuous function of p (recall properties 1–5 above). As usual, we will not be able to derive the original production function $\phi[f(x)]$ from the cost function (2.18) unless f also satisfies assumptions 2 and 3 above. Shephard [1953; 43] obtained the factorization (2.18) for the cost function corresponding to a homothetic production function.

Finally, Shephard [1953; 28–29] noted several practical uses for duality theory: (i) as an aid in aggregating variables, (ii) in econometric studies of production when input data are not available but cost, input price and output data are available, and (iii) as an aid in deriving certain comparative statics results. Thus, Shephard either derived or anticipated many of the theoretical results and practical applications of duality theory.

Turning now to the specific results obtained in this section, McFadden [1966] showed that the minimum in definition (2.1) exists if F satisfies assumption 1. Property 1 was obtained by Shephard [1953; 14], property 2 by Shephard [1953; 14] and Samuelson [1953–54; 15], property 3 by Shephard [1953; 14], property 4 by Shephard [1953; 15] (our method of proof is due to McKenzie [1956–57; 185]), properties 5 and 6 by Uzawa [1964; 217], and finally property 7 was obtained by Shephard [1970; 83].

The method for constructing the approximating production possibilities sets $L^*(u)$ in terms of the cost function is due to Uzawa [1964].

The very important point that the approximating isoquants do not have any of the backward bending or nonconvex parts of the true isoquants was made in the context of consumer theory by Hotelling [1935; 74], Wold [1943; 231] [1953; 164] and Samuelson [1950b; 359–360] and in the context of producer theory by McFadden [1966] [1978a]. It is worth quoting Hotelling and Samuelson at some length in order to emphasize this point:

> If indifference curves for purchases be thought of as possessing a wavy character, convex to the origin in some regions and concave in others, we are forced to the conclusion that it is only the portions convex to the origin that can be regarded as possessing any importance, since the others are essentially unobservable. They can be detected only by the discontinuities that may occur in demand with variation in price-ratios, leading to an abrupt jumping of a point of tangency across a chasm when the straight line is rotated. But, while such discontinuities may reveal the existence of chasms, they can never measure their depth. The concave portions of the indifference curves and their many-dimensional generalizations, if they exist, must forever remain in unmeasurable obscurity.
>
> Hotelling [1935; 74]

> It will be noted that any point where the indifference curves are convex rather than concave *cannot be observed in a competitive market.*

> Such points are shrouded in eternal darkness — unless we make our consumer a *monopsonist* and let him choose between goods lying on a very convex 'budget curve' (along which he is affecting the prices of what he buys). In this monopsony case, we could still deduce the slope of the man's indifference curve from the slope of the observed constraint at the equilibrium point.
>
> Samuelson [1950b; 359–360]

Our proof of Lemma 3 follows a proof attributed by Diamond and McFadden [1974; 4] to W.M. Gorman; however the same method of proof was also used by Karlin [1959; 272]. Hicks' and Samuelson's proof of Lemma 3 assumed differentiability of the production function and utilized the first order conditions for the cost minimization problem along with the properties of determinants. Our earlier quotation by Hotelling [1932; 594] indicates that he also obtained the Hicks [1946; 331], Samuelson [1947; 68] [1953–54; 15–16] results in a slightly different context.

References to some of the more recent literature on duality will be given in subsequent sections.

3. Duality between Cost and Aggregator (Production or Utility) Functions

In this section, we assume that the aggregator function F satisfies the following properties:

CONDITIONS I ON F:

(i) F is a real valued function of N variables defined over the nonnegative orthant $\Omega \equiv \{x : x \geq 0_N\}$ and is *continuous* on this domain.

(ii) F is *increasing*; i.e., $x'' \gg x' \geq 0_N$ implies $F(x'') > F(x')$.

(iii) F is a *quasiconcave* function.

Note that properties (i) and (ii) above are stronger than assumptions 1 and 2 on F made in the previous section, so that we should be able to deduce somewhat stronger conditions on the cost function $C(u,p)$ which corresponds to an $F(x)$ satisfying conditions I above.

Let U be the range of F. From I(i) and (ii), it can be seen that $U \equiv \{u : \bar{u} \leq u < \bar{\bar{u}}\}$, where $\bar{u} \equiv F(0_N) < \bar{\bar{u}}$. Note that the least upper bound $\bar{\bar{u}}$ could be a finite number or $+\infty$. In the context of production theory, typically $\bar{u} = 0$ and $\bar{\bar{u}} = +\infty$, but for consumer theory applications, there is no reason to restrict the range of the utility function F in this manner.

Define the set of positive prices $P \equiv \{p : p \gg 0_N\}$.

THEOREM 1. *If F satisfies conditions I, then $C(u,p) \equiv \min_x\{p^T x : F(x) \geq u\}$ defined for all $u \in U$ and $p \in P$ satisfies conditions II below.*

CONDITIONS II ON C:

(i) $C(u,p)$ is a real valued function of $N+1$ variables defined over $U \times P$ and is jointly *continuous* in (u,p) over this domain.

(ii) $C(\bar{u},p) = 0$ for every $p \in P$.

(iii) $C(u,p)$ is *increasing in u* for every $p \in P$; i.e., if $p \in P$, $u', u'' \in U$, with $u' < u''$, then $C(u',p) < C(u'',p)$.

(iv) $C(\bar{\bar{u}},p) = +\infty$ for every $p \in P$; i.e., if $p \in P$, $u^n \in U$, $\lim_n u^n = \bar{\bar{u}}$, then $\lim_n C(u^n,p) = +\infty$.

(v) $C(u,p)$ is (positively) *linearly homogeneous in p* for every $u \in U$; i.e., $u \in U$, $\lambda > 0$, $p \in P$ implies $C(u,\lambda p) = \lambda C(u,p)$.

(vi) $C(u,p)$ is *concave in p* for every $u \in U$.

(vii) $C(u,p)$ is *increasing in p* for $u > \bar{u}$ and $u \in U$.

(viii) C is such that the function $F^*(x) \equiv \max_u\{u : p^T x \geq C(u,p)$ for every $p \in P$, $u \in U\}$ is continuous for $x \geq 0_N$.

Proof: (i) By I(i), F is continuous and hence continuous from above. Thus, by Lemma 1 in the previous section, $C(u,p)$ is well defined as a minimum for $(u,p) \in U \times P$. In order to prove the continuity of C, we will use the Maximum Theorem, so it is first necessary to show that the correspondence

$$L(u) \equiv \{x : x \geq 0_N,\ F(x) \geq u\} \tag{3.1}$$

is continuous for $u \in U$. Since F is continuous from above, it can be seen that graph $L \equiv \{(x,u) : x \geq 0_N, F(x) \geq u\}$ is a closed set in R^{N+1}, and thus by Lemma 2, L is an upper semicontinuous correspondence over U. To show that L is lower semicontinuous over U, let

$$u^0 \in U, \quad x^0 \in L(x^0), \quad u^n \in U, \quad \lim_n u^n = u^0. \tag{3.2}$$

Since $x^0 \in L(u^0)$, by (19), $F(x^0) \geq u^0$. We must consider two cases.

CASE 1: $F(x^0) = u^0 + \lambda$ where $\lambda > 0$. By (3.2), there exists n^* such that for $n \geq n^*$, $u^n \leq u^0 + \lambda$. For $n < n^*$, let x^n be any point such that $x^n \in L(u^n)$ while for $n \geq n^*$, define $x^n \equiv x^0$ so that $F(x^n) = F(x^0) = u^0 + \lambda \geq u^n$ and thus $x^n \in L(u^n)$ and $\lim_n x^n = x^0$.

CASE 2: $F(x^0) = u^0$. If $u^n \leq u^0$, then define $x^n = x^0$ so that $F(x^n) = F(x^0) = u^0 \geq u^n$ and $x^n \in L(u^n)$. If $u^n > u^0$, then define the scalar k^n by $f(k^n) \equiv F(x^0 + k^n 1_N) = u^n$ where 1_N is an N dimensional vector of ones. Since $f(0) = u^0 < u^n$ and $f(k)$ is a continuous monotonically increasing function of k by I(i) and (ii), it can be seen that k^n is well defined. Note that as n becomes large k^n tends to 0 since u^n tends to u^0. Now define $x^n = x^0 + k^n 1_N$. Thus $x^n \in L(u^n)$ and $\lim_n x^n = x^0$ in this case also. Thus $L(u)$ is both lower and upper semicontinuous over U.

We cannot immediately apply the Maximum Theorem at this point since $L(u)$ is not a compact set.

Let $u^0 \in U$, $p^0 \in P$. Define the following sets:

$$(3.3) \qquad \begin{aligned} U_\delta(u^0) &\equiv \{u : \bar{u} \leq u \leq u^0 + \delta\}, \\ P_\delta(p^0) &\equiv \{p : (p - p^0)^T(p - p^0) \leq \delta^2\}. \end{aligned}$$

Choose $\delta > 0$ small enough so that $P_\delta(p^0) \subset P$ and $U_\delta(u^0) \subset U$. Now let $x^* > 0_N$ be any point such that

$$(3.4) \qquad F(x^*) \geq u^0 + \delta.$$

Now for every $p \in P_\delta(p^0)$, define the compact set $B_p \equiv \{x : p^T x \leq p^T x^*,\ x \geq 0_N\}$. For $i = 1, 2, \ldots, N$, define $m_i \equiv \max_p\{p^T x^*/p_i : p \equiv (p_1, p_2, \ldots, p_N)^T,\ p \in P_\delta(p^0)\}$. Since $P_\delta(p^0)$ is compact and each component of the vector p is positive if $p \in P_\delta(p^0)$, m_i is well defined as a maximum. Define $m = \max\{m_i : i = 1, 2, \ldots, N\}$. Define the compact set B as $B \equiv \{x : x \geq 0_N,\ x \leq m 1_N\}$ where 1_N is a vector of ones. It is obvious that

$$(3.5) \qquad B_p \equiv \{x : p^T x \leq p^T x^*,\ x \geq 0_N\} \subset B \text{ for } p \in P_\delta(p^0).$$

For $(u, p) \in U_\delta(u^0) \times P_\delta(p^0)$, we have

$$\begin{aligned} C(u,p) &\equiv \min_x\{p^T x : x \in L(u),\ x \geq 0_N\} \\ &= \min_x\{p^T x : x \in L(u),\ x \geq 0_N,\ p^T x \leq p^T x^*\} \\ &\qquad \text{since by (3.3) and (3.4), } x^* \text{ is feasible when } u \in U_\delta(u^0) \\ &= \min_x\{p^T x : x \in L(u) \cap B\} \\ &\qquad \text{for } p \in P_\delta(p^0) \text{ using (3.5).} \end{aligned}$$

Since $L(u)$ is a continuous correspondence and since B is a (constant) compact set, the correspondence $\phi(u,p) \equiv L(u) \cap B$ for $(u,p) \in U_\delta(u^0) \times P_\delta(p^0)$ is continuous with compact image sets and thus continuity of C follows via the Maximum Theorem.

(ii) Let $p \in P$. From property 1 in the previous section, $C(\bar{u}, p) \geq 0$. Since $F(0_N) = \bar{u}$, $C(\bar{u}, p) \equiv \min_x\{p^T x : F(x) \geq \bar{u}\} \leq p^T 0_N = 0$. Thus $C(\bar{u}, p) = 0$.

(iii) Let $p \in P$ and $\bar{u} \leq u' < u'' < \bar{\bar{u}}$. Then

$$\begin{aligned} C(u'',p) &\equiv \min_x\{p^T x : F(x) \geq u''\} \\ &= p^T x'' \text{ where } F(x'') = u'' \\ &> p^T k' x'' \text{ where } F(k'x'') = u' < u'' \text{ and } 0 \leq k' < 1, \\ &\qquad \text{using I(i) and (ii)} \\ &\geq \min_x\{p^T x : F(x) \geq u'\} \text{ since } k'x'' \text{ is feasible but not} \\ &\qquad \text{necessarily optimal for the cost minimization problem} \\ &\equiv C(u'p). \end{aligned}$$

(iv) Let $u^n \in U$, $\lim_n u^n = \bar{\bar{u}}$ and $p \gg 0_N$. Then $C(u^n, p) = p^T x^n$ where $x^n \geq 0_N$ and $F(x^n) = u^n$. *Suppose* the components of x^n remain bounded from above for all n; i.e., $x^n \leq k^* 1_N$ for all n. Then each $x^n \in S \equiv \{x : 0_N \leq x \leq k^* 1_N\}$, a compact set, and thus $\{x^n\}$ contains at least one convergent subsequence, $\{x^{n_k}\}$ say, with $\lim x^{n_k} = x^*$. Thus $\bar{\bar{u}} = \lim u^{n_k} = \lim F(x^{n_k}) = F(\lim x^{n_k}) = F(x^*)$ using the continuity of F. But then using I(ii), $F(x^* + 1_N) > F(x^*) = \bar{\bar{u}}$, which is impossible since $\bar{\bar{u}}$ is the least upper bound for the range of F. Thus our supposition is false, and at least one component of x^n tends to $+\infty$. Since $p \gg 0_N$, $p^T x^n$ also tends to $+\infty$.

(v) Since F is continuous, it is continuous from above and thus linear homogeneity of C in p follows from property 2 of the previous section.

(vi) Concavity of C in p follows from property 4 of the previous section.

(vii) Let $u \in U$, $u > \bar{u}$, $p', p'' \in P$. Then

$$\begin{aligned} C(u, p' + p'') &\equiv 2C(u, \tfrac{1}{2}p' + \tfrac{1}{2}p'') && \text{using II(v)} \\ &\geq 2[\tfrac{1}{2}C(u, p') + \tfrac{1}{2}C(u, p'')] && \text{using II(vi)} \\ &= C(u, p') + C(u, p'') \\ &> C(u, p'), \end{aligned}$$

since for $u > \bar{u}$, II(ii) and (iii) imply that $C(u, p'') > 0$.

(viii) It is first necessary to show that $F^*(x)$ is well defined as a maximum. Let $x \geq 0_N$ and $p \gg 0_N$. Then the set $I_p(x) \equiv \{u : u \in U, C(u, p) \leq p^T x\}$ is a compact interval containing the point $\bar{u}$, using II(i), (ii) and (iii). Thus

$$\begin{aligned} F^*(x) &\equiv \max_u \{u : C(u, p) \leq p^T x \quad \text{for every } p \in P, u \in U\} \\ &= \max_u \{u : u \in I_p(x) \quad \text{for every } p \in P\} \\ &= \max_u \{u : u \in I(x)\}, \end{aligned} \tag{3.6}$$

where $I(x) \equiv \bigcap_{p \in P}\{I_p(x)\}$ is a compact interval containing $\bar{u}$. Thus $F^*(x)$ is well defined as a maximum.

At this point, it is useful to extend the domain of definition of C from $p \gg 0_N$ to $p \geq 0_N$. This can be done by utilizing the *Fenchel closure operation*: for each $u \in U$, define the hypograph of $C(u, p)$ as the (convex) set $G(u) \equiv \{(k, p) : p \gg 0_N,\ k \leq C(u, p)\}$, let $\overline{G}(u)$ denote the closure of $G(u)$ in R^{N+1}, and now define $C(u, p)$ for $p \geq 0_N$ as $C(u, p) \equiv \max_k \{k : (k, p) \in \overline{G}(u)\}$. It can be seen (cf. Fenchel [1953; 78] or Rockafellar [1970, 85]) that for each $u \in U$, the extended C is continuous in p for $p \in \Omega \equiv \{p : p \geq 0_N\}$.[14]

[14]It can also be shown that the extended function C is jointly continuous over $U \times \Omega$ (see Rockafellar [1970, 89]. However, $C(u, p)$ need not be strictly increasing in u when p is on the boundary of Ω; e.g., consider the function $F(x_1, x_2) \equiv x_1$ which has the dual cost function $C(u, p_1, p_2) \equiv p_1 u$ which is not increasing in u when $p_1 = 0$.

Once the domain of definition of C has been extended in the above continuous manner, F^* can now be defined as

$$(3.7) \qquad F^*(x) \equiv \max_u\{u : C(u,p) \le p^T x \quad \text{for every } p \in \Omega,\ u \in U\}.$$

We now show that F^* is continuous over Ω by showing that $F^* = F$. Let $x' \ge 0_N$ and $u' \equiv F(x')$. Then for any $p \in P$,

$$(3.8) \qquad C(u',p) \equiv \min_x\{p^T x : x \in L(u')\} \le p^T x'$$

since x' is feasible but not necessarily optimal for the minimization problem. By continuity, (3.8) is also valid for all $p \in \Omega$. Thus $F^*(x') \equiv \max_u\{u : u \in U, C(u,p) \le p^T x'$ for every $p \in \Omega\} \ge u'$ since by (3.8), u' is feasible for all of the constraints in the maximization problem.

Suppose $F^*(x') = u'' > u'$. Then u'' satisfies the inequalities

$$(3.9) \qquad C(u'',p) \le p^T x' \text{ for every } p \in \Omega.$$

Since $L(u'') \equiv \{x : F(x) \ge u''\}$ is a closed, convex set by I(i) and (iii), it is equal to the intersection of its supporting halfspaces by Minkowski's [1911] Theorem. By I(ii), the surface $\{x : F(x) = u''\}$ never bends backwards. Hence $L(u'')$ is unbounded from above, and it can be seen that[15] $L(u'') = \{x : C(u'',p) \le p^T x$ for every $p \in \Omega\}$. Thus by (3.9), $x' \in L(u'')$ which implies that $F(x') \ge u'' > u'$, which is a contradiction since $F(x') = u'$. Thus our supposition is false and $F^*(x') = u' = F(x')$. Q.E.D.

Note that we have proven the following corollaries to Theorem 1.

COROLLARY 1.1. *If $C(u,p)$ satisfies conditions II above, then the domain of definition of C can be extended from $U \times P$ to $U \times \Omega$. The extended function C is continuous in p for $p \in \Omega \equiv \{p : p \ge 0_N\}$ for each $u \in U$.*

COROLLARY 1.2. *For every $x \ge 0_N$, $F^*(x) = F(x)$, where F^* is the function defined by the cost function C in part (viii) of conditions II.*

Corollary 1.2 shows that the cost function can completely describe a production function which satisfies conditions I; i.e., to use McFadden's [1966] terminology, the cost function is a *sufficient statistic* for the production function.

The proof of Theorem 1 is straightforward, with the exception of parts (i) and (viii), the parts that involve the *continuity* properties of the cost or production function. These continuity complexities appear to be the only difficult concepts associated with duality theory: this is why we tried to avoid them in the previous section as much as possible. For further discussion on continuity

[15]Recall our discussion of equation (2.3) in the previous section.

problems, see Shephard [1970], Friedman [1972], Diewert [1974a], Blackorby, Primont and Russell [1978] and Blackorby and Diewert [1979].

Property I(ii), increasingness of F, is required in order to prove the correspondence $L(u)$ continuous and thus that $C(u,p)$ is continuous over $U \times P$.[16] If property I(ii) is replaced by a weak monotonicity assumption (such as our old assumption 2 on F of the previous section), then plateaus on the graph of F ("thick" indifference surfaces to use the language of utility theory) will imply discontinuities in C with respect to u (cf. Friedman [1972; 169]).

Note that II(ii) and (iii) imply that $C(u,p) > 0$ for $u > \bar{u}$ and $p \gg 0_N$ and that II(vii) is not an independent property of C since it follows from II(ii), (iii), (v) and (vi). Note also that we have not assumed that F be strictly quasiconcave; i.e., that the production possibility sets $L(u) \equiv \{x : F(x) \geq u\}$ be strictly convex.

Finally, it is evident that given only a firm's total cost function C, we can use the function F^* defined in terms of the cost function by (3.7) in order to generate the firm's production function. This is formalized in the following theorem.

THEOREM 2. *If C satisfies conditions II above, then F^* defined by (3.7) satisfies conditions I. Moreover, if $C^*(u,p) \equiv \min_x\{p^T x : F^*(x) \geq u\}$ is the cost function which is defined by F^*, then $C^* = C$.*

Proof: (i) Extend the domain of definition of C from $U \times P$ to $U \times \Omega$ via the Fenchel closure operation. The extended C is then continuous over $U \times \Omega$ by Corollary 1.1 above. In the proof of Theorem 1 above, we have seen that $F^*(x)$ defined by (3.7) is well defined for $x \geq 0_N$. Property II(viii) implies that F^* is continuous over Ω.

(ii) It is first necessary to define F^* in yet another way: for $x \geq 0_N$,

$$\begin{aligned} F^*(x) &\equiv \max_u\{u : C(u,p) \leq p^T x \text{ for every } p \geq 0_N,\ u \in U\} \\ &= \max_u\{u : C(u,p) - p^T x \leq 0 \\ &\quad \text{for } p \geq 0_N \text{ and } 1_N^T p = 1,\ u \in U\} \\ &\quad \text{using II(i) and (v)} \\ (3.10) \qquad &= \max_u\{u : H(u,x) \leq 0,\ u \in U\} \end{aligned}$$

[16]Friedman [1972] shows that I(ii) plus continuity from above (assumption 1 on F in the previous section) is sufficient to imply the joint continuity of C over $U \times P$ (and indeed over $U \times \Omega$ if we make use of Rockafellar's [1970; 89] result). However, unless we assume the additional property on F of continuity from below, we cannot conclude that $C(u,p)$ is increasing in u for $p \in P$, a property which follows from I(i) and I(ii).

where

$$H(u,x) \equiv \max_p \{C(u,p) - p^T x : p \geq 0_N,\ 1_N^T p = 1\}. \tag{3.11}$$

Since $C(u,p) - p^T x$ is continuous in p over the compact set $S \equiv \{p : p \geq 0_N,\ 1_N^T p = 1\}$, $H(u,x)$ is well defined as a maximum.[17] Moreover, since $C(u,p) - p^T x$ is continuous in u, x and p, the Maximum Theorem implies that $H(u,x)$ will be continuous over $U \times \Omega$. We can also show that $H(u,x)$ is nondecreasing in u. Let $x \geq 0_N$, $u', u'' \in U$ with $u' < u''$. Then

$$\begin{aligned} H(u',x) &\equiv \max_p \{C(u',p) - p^T x : p \in S\} \\ &= C(u',p') - p'^T x \text{ where } p' \in S \\ &\leq C(u'',p') - p'^T x \text{ using property II(iii)}^{18} \\ &\leq \max_p \{C(u'',p) - p^T x : p \in S\} \text{ since } p' \text{ is feasible but} \\ &\quad \text{not necessarily optimal for the maximization problem} \\ &\equiv H(u'',x). \end{aligned}$$

Also properties II(ii) and II(iv) imply that $H(\bar{u},x) \leq 0$ and $H(u,x)$ tends to $+\infty$ as u tends to $\bar{\bar{u}}$. Thus if u^* solves the maximization problem (3.10), then $H(u^*,x) = 0$.

Now let $0_N \leq x' \ll x''$. Then

$$F^*(x'') \equiv \max_u \{u : H(u,x'') \leq 0, u \in U\} = u'' \text{ say,}$$

where

$$\begin{aligned} 0 &= H(u'',x'') \\ &= C(u'',p'') - p''^T x'' \quad \text{for some } p'' \geq 0_N \text{ such that } 1_N^T p'' = 1 \\ &< C(u'',p'') - p''^T x' \quad \text{since } x' \ll x'' \text{ and } p'' > 0_N \\ &\leq H(u'',x') \quad \text{using definition (3.11).} \end{aligned}$$

---

[17]The function $H(u,x)$ is called the difference function by Blackorby and Diewert [1979]. It is equal to the negative of the conjugate function to the concave function of p, $C(u,p)$, for each u. For material on conjugate concave (or convex) functions, see Fenchel [1953; 88–92], Karlin [1959; 226], Rockafellar [1970; 104] or Jorgenson and Lau [1974b].

[18]The continuity of C and property II(iii), $C(u',p) < C(u'',p)$ if $u' < u''$ and $p \in P$ imply only that $C(u',p) \leq C(u'',p)$ when p belongs to the boundary of P.

Thus u'' is not a feasible solution for the maximization problem

$$\max_u \{u : H(u, x') \leq 0, u \in U\} = F^*(x')$$

since $H(u'', x') > 0$. Since H is nondecreasing in u, if $u \geq u''$, $H(u, x') > 0$ also. Thus $F^*(x') < u'' = F^*(x'')$.

(iii) Let $x' \geq 0_N$, $x'' \geq 0_N$, $0 \leq \lambda \leq 1$, $F^*(x') \geq u^*$ and $F^*(x'') \geq u^*$. Then by the definition of F^*, (3.7), and property II(iii) of C:

$$C(u^*, p) \leq C[F^*(x'), p] \leq p^T x' \text{ for every } p \in P \text{ and}$$
$$C(u^*, p) \leq C[F^*(x''), p] \leq p^T x'' \text{ for every } p \in P.$$

Thus $C(u^*, p) \leq \lambda p^T x' + (1 - \lambda) p^T x'' = p^T [\lambda x' + (1 - \lambda) x'']$ for every $p \in P$. Hence

$$\begin{aligned} F^*[\lambda x' + (1-\lambda)x''] &\equiv \max_u \{u : C(u,p) \leq p^T[\lambda x' + (1-\lambda)x''] \text{ for every } p \in P\} \\ &\geq u^* \text{ since } u^* \text{ is feasible for the maximization problem.} \end{aligned}$$

Thus F^* is a quasiconcave function.

It remains to show that C^*, the cost function of F^*, equals C. Let $u^* \in U$ and $p^* \in P$. Then

$$\begin{aligned} C^*(u^*, p^*) &\equiv \min_x \{p^{*T} x : F^*(x) \geq u^*\} \\ &= \min_x \{p^{*T} x : F^*(x) = u^*\} \text{ using properties I(i) and (ii)} \\ &= \min_x \{p^{*T} x : \max_u \{u : C(u,p) \leq p^T x \text{ for every } p \in S\} = u^*, \\ &\qquad x \geq 0_N\} \text{ using definition (3.10) for } F^* \\ &= \min_x \{p^{*T} x : C(u^*, p) \leq p^T x \text{ for every } p \in S \text{ with equality} \\ &\qquad \text{holding for at least one } p \in S,\ x \geq 0_N\} \\ &= p^{*T} x^* \end{aligned} \tag{3.12}$$

where x^* is any supergradient[19] of the concave function of p, $g(p) \equiv C(u^*, p)$, at the point p^*. The last equality in (3.12) follows since by the definition of x^* being a supergradient,[20] we have $C(u^*, p) \leq p^T x^*$ for every $p \in P$ (and hence also for every $p \in S$ using the continuity of C) and $C(u^*, p^*) = p^{*T} x^*$. This last equality in conjunction with (3.12) establishes that $C^*(u^*, p^*) = C(u^*, p^*)$.

Q.E.D.

---

[19]In general, x^* is a supergradient to a function g at the point p^* iff $g(p) \leq g(p^*) + x^{*T}(p - p^*)$ for all $p \in P$. If g is a concave function over the set $P \equiv \{p : p \gg 0_N\}$, then Rockafellar [1970; 214–215] shows that for every $p^* \in P$, the set of supergradients to g at the point p^*, $\partial g(p^*)$, is a nonempty, closed convex set. If g is differentiable at p^*, then $g(p^*)$ reduces to the single point $\nabla g(p^*)$, the gradient vector of g. Finally, if g is positively linearly homogeneous over P, then it can be seen that x^* is a supergradient to g at p^* iff $g(p) \leq x^{*T} p$ for every $p \in P$ and $g(p^*) = x^{*T} p^*$.

[20]Since $C(u, p)$ is increasing in p for $p \in P$, $x^* \geq 0_N$ also.

COROLLARY 2.1. *The set of supergradients to C with respect to p at the point $(u^*,p^*) \in U \times P$, $\partial C(u^*,p^*)$, is the solution set to the cost minimization problem $\min_x\{p^{*T}x : F^*(x) \geq u^*\}$ where F^* is the aggregator function which corresponds to the given cost function satisfying conditions II via definitions (3.6), (3.7), or (3.10). (The supergradients satisfy $x^* \in \partial C(u^*,p^*)$ iff $C(u^*,p) \leq C(u^*,p^*) + x^{*T}(p-p^*)$ for every $p \gg 0_N$.)*

Proof: If $x^* \in \partial C(u^*,p^*)$, we have already shown that x^* is a solution to the cost minimization problem (3.12). On the other hand, if $x' \geq 0_N$ is *not* a supergradient to the linearly homogeneous function $g(p) \equiv C(u^*,p)$ at the point $p^* \gg 0_N$, then we must have *either* $C(u^*,p') > p'^T x'$ for some $p' \in P$ in which case x' is not feasible for the minimization problem above (3.12), *or* $C(u^*,p^*) < p^{*T}x'$ but, in this case, x' cannot be optimal for the minimization problem above (3.12), since at least one supergradient $x^* \geq 0_N$ exists which satisfies the constraints of (3.12) and $C(u^*,p^*) = p^{*T}x^* < p^{*T}x'$. Q.E.D.

COROLLARY 2.2. (*Shephard's* [1953; 11] *Lemma*): *If C satisfies conditions II and, in addition, is differentiable with respect to input prices at the point $(u^*,p^*) \in U \times P$, then the solution x^* to the cost minimization problem $\min_x\{p^{*T}x : F^*(x) \geq u^*\}$ is unique and is equal to the vector of partial derivatives of $C(u^*,p^*)$ with respect to the components of the input price vector p; i.e.,*

$$x^* = \nabla_p C(u^*,p^*). \tag{3.13}$$

Proof: Apply the above corollary, noting that $\partial C(u^*,p^*)$ reduces to the single point $\nabla_p C(u^*,p^*)$ when C is differentiable with respect to p at the point (u^*,p^*). Q.E.D.

The above two theorems provide a version of the Shephard [1953] [1970] Duality Theorem between cost and aggregator functions. The conditions on C which correspond to our conditions I on F seem to be straightforward with the exception of II(viii), which essentially guarantees the continuity of the aggregator function F^* corresponding to the given cost function C. Condition II(viii) can be dropped if we strengthen II(iii) to $C(u,p)$ increasing in u for every p in $S \equiv \{p : p \geq 0_N, 1_N^T p = 1\}$. The resulting F^* can be shown to be continuous (cf. Blackorby, Primont and Russell [1978]); however, many useful functional forms do not satisfy the strengthened condition II(iii).[21] An alternative method of dropping II(viii), which preserves continuity of the direct aggregator function F^* corresponding to a given cost function C, is to develop local duality theorems; i.e., assume that C satisfies conditions II(i)–(vii) for $(u,p) \in U \times P$, where P is now restricted to be a *compact*, convex subset of the positive orthant. A (locally) valid continuous F^* can then be defined from C which in turn has C as its cost function over $U \times P$. This approach is pursued in Blackorby and Diewert [1979].

[21] E.g., consider $C(u,p) \equiv b^T pu$ where $b > 0_N$ but b is not $\gg 0_N$. This corresponds to a Leontief or fixed coefficients aggregator function.

Historical Notes

Duality theorems between F and C have been proven under various regularity conditions by Shephard [1953] [1970], McFadden [1962], Chipman [1970], Hanoch [1978b], Diewert [1971a] [1974a], Afriat [1973a] and Blackorby, Primont and Russell [1978].

Duality theorems between C and the level sets of F, $L(u) \equiv \{x : F(x) \geq u\}$, have been proven by Uzawa [1964], McFadden [1966] [1978a], Shephard [1970], Jacobsen [1970] [1972], Diewert [1971a], Friedman [1972], and Sakai [1973].

4. Duality Between Direct and Indirect Aggregator Functions

We assume that the direct aggregator (utility or production) function F satisfies conditions I listed in the previous section. The basic optimization problem that we wish to consider in this section is the problem of maximizing utility (or output) $F(x)$ subject to the budget constraint $p^T x \leq y$ where $p \gg 0_N$ is a vector of given commodity (or input) prices and $y > 0$ is the amount of money the consumer (or producer) is allowed to spend. Since $y > 0$, the constraint $p^T x \leq y$ can be replaced with $v^T x \leq 1$ where $v \equiv p/y$ is the vector of *normalized* prices. The *indirect aggregator function* $G(v)$ is defined for $v \gg 0_N$ as

$$G(v) \equiv \max_x \{F(x) : v^T x \leq 1, x \geq 0_N\}. \tag{4.1}$$

THEOREM 3. *If the direct aggregator function F satisfies conditions I, then the indirect aggregator function G defined by (4.1) satisfies the following conditions:*

CONDITIONS III ON G: (i) $G(v)$ is a real valued function of N variables defined over the set of positive normalized prices $V \equiv \{v : v \gg 0_N\}$ and is a *continuous* function over this domain.

(ii) G is *decreasing*; i.e., if $v'' \gg v' \gg 0_N$, then $G(v'') < G(v')$.

(iii) G is *quasiconvex* over V.

(iv) G[22] is such that the function $\widehat{F}(x) \equiv \min_v\{G(v) : v^T x \le 1, v \ge 0_N\}$ defined for $x \gg 0_N$ is continuous over this domain and has a continuous extension[23] to the nonnegative orthant $\Omega \equiv \{x : x \ge 0_N\}$.

Proof: (i) For $v \gg 0_N$, the constraint set $\rho(v) \equiv \{v : x \ge 0_N, v^T x \le 1\}$ for the maximization problem (4.1) is compact so that $G(v)$ is well defined as a maximum. It can be verified that $\rho(v)$ is a continuous correspondence for $v \gg 0_N$ and that there exists a compact set B such that $\rho(v) \subset B$ if $v^0 \gg 0_N$, $v \in N_\delta(v^0)$ where $N_\delta(v^0) \equiv \{v : (v - v^0)^T(v - v^0) \le \delta^2\}$ and $\delta > 0$ is chosen small enough so that $N_\delta(v^0) \subset V \equiv \{v : v \gg 0_N\}$. Thus for $v \in N_\delta(v^0)$, $G(v) \equiv \max_v\{F(x) : x \in \rho(v)\} = \max_v\{F(x) : x \in \rho(v) \cap B\}$ and continuity of G follows from the continuity of F and the Maximum Theorem.

(ii) Let $0_N \ll v' \ll v''$. Then

$$\begin{aligned} G(v'') &\equiv \max_x\{F(x) : v''^T x \le 1, x \ge 0_N\} \\ &= \max_x\{F(x) : v''^T x = 1, x \ge 0_N\} \text{ using I(ii)} \\ &= F(x'') \text{ say where } v''^T x'' = 1 \text{ and } x'' \ge 0_N. \end{aligned}$$

Since $v' \ll v''$, $v'^T x'' < 1$ and thus $\varepsilon^* \equiv (1 - v'^T x'')/v'^T 1_N > 0$. Thus

$$\begin{aligned} G(v') &\equiv \max_x\{F(x) : v'^T x \le 1, x \ge 0_N\} \\ &\ge F(x'' + \varepsilon^* 1_N) \text{ since } x'' + \varepsilon^* 1_N \ge 0_N \text{ is feasible for the} \\ &\qquad \text{maximization problem as } v'^T(x'' + \varepsilon^* 1_N) = 1 \\ &> F(x'') \text{ using condition I(ii) on } F \\ &= G(v''). \end{aligned}$$

---

[22] G here is the extension of G to the nonnegative orthant that is defined by the Fenchel closure operation; i.e., define the epigraph of the original G as $\Gamma \equiv \{(u, v) : v \gg 0_N, u \ge G(v)\}$, define the closure of Γ as $\overline{\Gamma}$ and define the extended G as $G(v) \equiv \inf_u\{u : (u, v) \in \overline{\Gamma}\}$ for $v \ge 0_N$. The resulting extended G is continuous from below (the sets $\{v : G(v) \le u, v \ge 0_N\}$ are closed for all u). If the range of F is $U \equiv \{u : \bar{u} \le u < \bar{\bar{u}}\}$ where $\bar{u} < \bar{\bar{u}}$, then the range of the unextended G is $\{u : \bar{u} < u < \bar{\bar{u}}\}$ and the range of the extended G is $\{u : \bar{u} < u \le \bar{\bar{u}}\}$ so that if $\bar{\bar{u}} = +\infty$, then $G(v) = +\infty$ for $v = 0_N$ and possibly for other points v on the boundary of the nonnegative orthant.

[23] Again $\widehat{F}$ is extended to the nonnegative orthant by the Fenchel closure operation: define the hypograph of the original $\widehat{F}$ as $\Delta \equiv \{(u, x) : x \gg 0_N, u \le \widehat{F}(x)\}$, define the closure of Δ as $\overline{\Delta}$ and define the extended $\widehat{F}$ as $F(x) \equiv \sup_u\{u : (u, x) \in \overline{\Delta}\}$ for $x \ge 0_N$. Since the unextended $\widehat{F}$ is continuous for $x \gg 0_N$, the extended $\widehat{F}$ can easily be shown to be continuous from above for $x \ge 0_N$. Condition III(iv) implies that the extended F is continuous from below for $x \ge 0_N$ as well.

(iii) Let $v' \gg 0_N$, $v'' \gg 0_N$, $0 \leq \lambda \leq 1$, $G(v') \leq u^*$ and $G(v'') \leq u^*$. Define the sets $H' \equiv \{x : v'^T x \leq 1, x \geq 0_N\}$, $H'' \equiv \{x : v''^T x \leq 1, x \geq 0_N\}$ and $H^\lambda \equiv \{x : [\lambda v' + (1-\lambda)v'']^T x \leq 1, x \geq 0_N\}$. Then, as in Section 1, it can be seen that $H^\lambda \subset H' \cup H''$. Thus

$$\begin{aligned} G[\lambda v' + (1-\lambda)v''] &\equiv \max_x \{F(x) : x \in H^\lambda\} \\ &\leq \max_x \{F(x) : x \in H' \cup H''\} \text{ since } H^\lambda \subset H' \cup H'' \\ &\leq u^* \text{ since } F(x) \leq u^* \text{ if } x \in H' \text{ or if } x \in H''. \end{aligned}$$

(iv) Extend G to $v \geq 0_N$ using the Fenchel closure operation. The resulting extended G is continuous from below and thus $\min_v\{G(v) : v^T x \leq 1, v \geq 0_N\}$ will exist and be finite for $x \gg 0_N$ using a result due to Berge [1963; 76]. Thus $\widehat{F}(x)$ is well defined for $x \gg 0_N$. We show that $\widehat{F}$ has a continuous extension to $x \geq 0_N$ by showing that $\widehat{F}(x) = F(x)$ for $x \gg 0_N$.

Let $x^* \gg 0_N$ and $u^* \equiv F(x^*)$. Since x^* is on the boundary of the closed convex set $L(u^*) \equiv \{x : F(x) \geq u^*, x \geq 0_N\}$ (where we have used I(i), (ii) and (iii)), there exists at least one supporting hyperplane $v^* \neq 0_N$ to $L(u^*)$ at the point x^*; i.e., v^* is such that $x \in L(u^*)$ implies $v^{*T}x \geq v^{*T}x^*$. By property I(ii) on F, $v^* > 0_N$ and we can normalize v^* so that

$$v^{*T}x^* = 1. \tag{4.2}$$

By property I(ii) on F, v^* also has the property that

$$x \in \text{interior } L(u^*) \text{ implies } v^{*T}x > v^{*T}x^* = 1. \tag{4.3}$$

Now

$$\begin{aligned} G(v^*) &\equiv \sup_x \{F(x) : v^{*T}x \leq 1, x \geq 0_N\} \\ &\geq F(x^*) \equiv u^* \text{ since by (4.2), } x^* \text{ is feasible.} \end{aligned} \tag{4.4}$$

If $F(x) > u^*$, then $x \in$ interior $L(u^*)$ and (4.3) implies that $v^{*T}x > 1$ so that x is not feasible for the maximization problem in (4.4). Thus

$$G(v^*) = F(x^*) = u^*. \tag{4.5}$$

Now

$$\begin{aligned} \widehat{F}(x^*) &\equiv \min_v \{G(v) : v^T x^* \leq 1, v \geq 0_N\} \\ &\leq G(v^*) \text{ since by (4.2), } v^* \text{ is feasible.} \end{aligned} \tag{4.6}$$

Also

$$\begin{aligned} \widehat{F}(x^*) &\equiv \min_v \{G(v) : v^T x^* \leq 1, v \geq 0_N\} \\ &= G(v') \text{ say where } v'^T x^* = 1, v' \geq 0_N \qquad (4.7) \\ &\equiv \sup_x \{F(x) : v'^T x \leq 1, x \geq 0_N\} \\ &\geq F(x^*) \text{ since by (4.7), } x^* \text{ is feasible} \\ &= G(v^*) \quad \text{by (4.5).} \qquad (4.8) \end{aligned}$$

(4.5), (4.6) and (4.8) imply that $F(x^*) = G(v^*) = \widehat{F}(x^*)$. Q.E.D.

COROLLARY 3.1. *The direct aggregator function F can be recovered from the indirect aggregator function G; i.e., for $x \gg 0$, $F(x) = \min_v\{G(v) : v^T x \leq 1, v \geq 0_N\}$.*

COROLLARY 3.2. *Let F satisfy conditions I and let $x^* \gg 0_N$. Define the closed convex set of normalized supporting hyperplanes at the point x^* to the closed convex set $\{x : F(x) \geq F(x^*), x \geq 0_N\}$ by $H(x^*)$.*[24] *Then: (i) $H(x^*)$ is the solution set to the indirect utility (or production) minimization problem $\min_v\{G(v) : v^T x^* \leq 1, v \geq 0_N\}$, where G is the indirect function which corresponds to F via definition (4.1), and (ii) if $v^* \in H(x^*)$, then x^* is the solution to the direct utility (or production) maximization problem $\max_x\{F(x) : v^{*T} x \leq 1, x \geq 0_N\}$.*

Proof: (i) If $v^* \in H(x^*)$, we have shown in the proof of Theorem 3(iv) that v^* is a solution to

$$\min_v\{G(v) : v^T x^* \leq 1, v \geq 0_N\} = \min_v\{G(v) : v^T x^* = 1,\ v \geq 0_N\} \tag{4.9}$$

where the equality in (4.9) follows from III(ii) on G. Now assume that v' is feasible for the second minimization problem in (4.9); i.e., $v' \geq 0_N$ and $v'^T x^* = 1$. Then

$$G(v') \equiv \max_x\{F(x) : v'^T x \leq 1, x \geq 0_N\} > F(x^*)$$

where the above inequality follows if v' is *not* a supporting hyperplane to the set $\{x : F(x) \geq F(x^*)\}$. Thus v' will *not* be a solution to (4.9) since $G(v^*) = F(x^*) < G(v')$ where $v^* \in H(x^*)$. Thus the solution set to (4.9) is precisely $H(x^*)$.

(ii) This part follows directly from (4.4) and (4.5). Q.E.D.

COROLLARY 3.3. (*Hotelling* [1935; 71], *Wold* [1944; 69–71] [1953; 145] *Identity*): *If F satisfies conditions I and in addition is differentiable at $x^* \gg 0_N$ with a nonzero gradient vector $\nabla F(x^*) > 0_N$, then x^* is a solution to the direct (utility or production) maximization problem $\max_x\{F(x) : v^{*T} x \leq 1, x \geq 0_N\}$ where*

$$v^* \equiv \frac{\nabla F(x^*)}{x^{*T} \nabla F(x^*)}. \tag{4.10}$$

Proof: Under the stated conditions, the set of normalized supporting hyperplanes $H(x^*)$ reduces to the single point v^* defined by (4.10) (note that

---

[24] If $v^* \in H(x^*)$, then $v^{*T} x^* = 1$, $v^* \geq 0_N$ and $F(x) \geq F(x^*)$ implies $v^{*T} x \geq v^{*T} x^* = 1$. The closedness and convexity of $H(x^*)$ is shown in Rockafellar [1970; 215].

$x^{*T}v^* = x^{*T}\nabla F(x^*)/x^{*T}\nabla F(x^*) = 1$). The present corollary now follows from part (ii) of the previous corollary. Q.E.D.

The system of equations (4.10) is known as the system of *inverse demand functions*; the ith equation

$$p_i/y \equiv v_i^* = [\partial F(x^*)/\partial x_i] \Big/ \left[\sum\nolimits_{j=1}^{N} x_j^* \partial F(x^*)/\partial x_j\right]$$

gives the ith commodity price p_i divided by expenditure y as a function of the quantity vector x^* which the producer or consumer would choose if he were maximizing $F(x)$ subject to the budget constraint $v^{*T}x = 1$.

We now assume that a well behaved indirect aggregator function G is given and we show that it can be used in order to define a well behaved direct aggregator function F which has G as its indirect function.

THEOREM 4. *Suppose G satisfies conditions III. Then $\widehat{F}(x)$ defined for $x \gg 0_N$ by*

$$\widehat{F}(x) \equiv \min_v\{G(v) : v^T x \leq 1, v \geq 0_N\} \tag{4.11}$$

has an extension to $x \geq 0_N$ which satisfies conditions I. Moreover, if we define $G^*(v) \equiv \max_x\{\widehat{F}(x) : v^T x \leq 1, x \geq 0_N\}$ for $v \gg 0_N$, then $G^*(v) = G(v)$ for all $v \gg 0_N$.

Proof: For $x \gg 0_N$, $\widehat{F}(x)$ is well defined as a minimum (see the proof of Theorem 3). Now extend $\widehat{F}$ to $\Omega \equiv \{x : x \geq 0_N\}$ via the Fenchel closure operation. Continuity of the extended $\widehat{F}$ follows directly from III(iv). To show that F is increasing and quasiconcave over $x \gg 0_N$, repeat the proofs of parts (ii) and (iii) of Theorem 3 with the obvious changes due to the fact that we are now dealing with the minimization problem (4.11) instead of the maximization problem (4.1). The extended $\widehat{F}$ will also have the properties of increasingness and quasiconcavity over Ω. Finally, the proof that $G^*(v) = G(v)$ for $v \gg 0_N$ proceeds analogously to the proof in Theorem 3 that $\widehat{F}(x) = F(x)$ for $x \gg 0_N$. Q.E.D.

COROLLARY 4.1. *Let G satisfy conditions III and let $v^* \gg 0_N$. Define the closed convex set of normalized supporting hyperplanes at the point v^* to the closed convex set $\{v : G(v) \leq G(v^*), v \geq 0_N\}$ by $H^*(v^*)$. Then: (i) $H^*(v^*)$ is the solution set to the direct maximization problem $\max_x\{\widehat{F}(x) : v^{*T}x \leq 1, x \geq 0_N\}$, where $\widehat{F}$ is the direct function which corresponds to the given indirect function G via definition (4.11), and (ii) if $x^* \in H(v^*)$, then v^* is a solution to the indirect minimization problem $\min_v\{G(v) : v^T x^* \leq 1, v \geq 0_N\}$.*

The proof of Corollary 4.1 follows in an analogous manner to the proof of Corollary 3.2.

COROLLARY 4.2. (*Ville* [1946; 35], *Roy* [1947; 222] *Identity*): *If G satisfies conditions III and, in addition, is differentiable at* $v^* \gg 0_N$ *with a nonzero gradient vector* $\nabla G(v^*) < 0_N$, *then* x^* *is the unique solution to the direct maximization problem* $\max_x\{\widehat{F}(x) : v^{*T}x \leq 1, x \geq 0_N\}$, *where*

$$x^* \equiv \nabla G(v^*)/v^{*T}\nabla G(v^*). \tag{4.12}$$

Proof: Under the stated conditions, the set of normalized supporting hyperplanes $H^*(v^*)$ reduces to the single point $x^* > 0_N$ defined by (4.12). Thus from part (i) of the previous corollary, x^* is the unique solution to the direct maximization problem. Q.E.D.

It can be seen that (4.12) provides the counterpart to Shephard's Lemma in the previous section. Shephard's Lemma and Roy's Identity are the basis for a great number of theoretical and empirical applications as we shall see later.

Finally, we note that although condition III(iv) appears to be a bit odd, it enables us to derive a continuous direct aggregator function from a given indirect function satisfying conditions III.[25]

Historical Notes

Duality theorems between direct and indirect aggregator functions have been proven by Samuelson [1965] [1969b] [1972], Newman [1965; 138–165], Lau [1969], Shephard [1970; 105–113], Hanoch [1978b], Weddepohl [1970; ch. 5], Katzner [1970; 59–62], Afriat [1972c] [1973c] and Diewert [1974a].

For closely related work relating assumptions on systems of consumer demand functions to the direct aggregator function F (the integrability problem), see Samuelson [1950b], Hurwicz and Uzawa [1971], Hurwicz [1971] and Afriat [1973a] [1973b].

For a geometric interpretation of Roy's Identity, see Darrough and Southey [1977], and for some extensions, see Weymark [1980].

---

[25]Without condition III(iv), we could still deduce continuity of $\widehat{F}(x)$ over $x \gg 0_N$ but the resulting $\widehat{F}$ would not necessarily have a continuous extension to $x \geq 0_N$ (since $\widehat{F}$ is not necessarily concave, but is only quasiconcave over $x \gg 0_N$, its extension is not necessarily continuous). For discussion and examples of these continuity problems, see Diewert [1974a; 121–123].

5. Duality between Direct Aggregator Functions and Distance or Deflation Functions

In this section, we consider a *fourth* alternative method of characterizing tastes or technology, a method which proves to be extremely useful for defining a certain class of index number formulae due to Malmquist [1953; 232].

As usual, let $F(x)$ be an aggregator function satisfying conditions I listed in Section 3 above. For u belonging to the interior of the range of F (i.e., $u \in \text{Int } U$, where $U \equiv \{u : \bar{u} \leq u < \bar{\bar{u}}\}$) and $x \gg 0_N$, define the *distance* or *deflation function*[26] D as

$$D(u, x) \equiv \max_k \{k : F(x/k) \geq u, \; k > 0\}. \tag{5.1}$$

Thus $D(u^*, x^*)$ is the biggest number which will just deflate (inflate if $F(x^*) < u^*$) the given point $x^* \gg 0_N$ onto the boundary of the utility (or production) possibility set $L(u^*) \equiv \{x : F(x) \geq u^*\}$. If $D(u^*, x^*) > 1$, then $x^* \gg 0_N$ produces a higher level of utility or output than the level indexed by u^*.

In turns out that the mathematical properties of $D(u, x)$ with respect to x are the same as the properties of $C(u, p)$ with respect to p, but the properties of D with respect to u are reciprocal to the properties of C with respect to u, as the following theorem shows.

THEOREM 5. *If F satisfies conditions I, then D defined by (5.1) satisfies conditions IV below.*

CONDITIONS IV ON D: (i) $D(u, x)$ is a real valued function of $N + 1$ variables defined over $\text{Int } U \times \text{Int } \Omega = \{u : \bar{u} < u < \bar{\bar{u}}\} \times \{x : x \gg 0_N\}$ and is *continuous* over this domain.

(ii) $D(\bar{u}, x) = +\infty$ for every $x \in \text{Int } \Omega$; i.e., $u^n \in \text{Int } U$, $\lim u^n = \bar{u}$, $x \in \text{Int } \Omega$ implies $\lim_n D(u^n, x) = +\infty$.

(iii) $D(u, x)$ is *decreasing in u* for every $x \in \text{Int } \Omega$; i.e., if $x \in \text{Int } \Omega$, $u', u'' \in \text{Int } U$ with $u' < u''$, then $D(u', x) > D(u'', x)$.

(iv) $D(\bar{\bar{u}}, x) = 0$ for every $x \in \text{Int } \Omega$; i.e., $u^n \in \text{Int } U$, $\lim u^n = \bar{\bar{u}}$, $x \in \text{Int } \Omega$ implies $\lim_n D(u^n, x) = 0$.

(v) $D(u, x)$ is (positively) *linearly homogeneous in x* for every $u \in \text{Int } U$; i.e., $u \in \text{Int } U$, $\lambda > 0$, $x \in \text{Int } \Omega$ implies $D(u, \lambda x) = \lambda D(u, x)$.

---

[26]Shephard [1953; 6][1970; 65] introduced the distance function into the economics literature, using the slightly different but equivalent definition: $D(u, x) \equiv 1/\min_\lambda\{\lambda : F(\lambda x) \geq u, \lambda > 0\}$. McFadden [1978a] and Blackorby, Primont and Russell [1978] call D the *transformation* function, while in the mathematics literature (e.g., Rockafellar [1970; 28]), D is termed a *guage* function. The term *deflation* function for D would seem to be more descriptive from an economic point of view.

(vi) $D(u,x)$ is *concave in* x for every $u \in \text{Int } U$.

(vii) $D(u,x)$ is *increasing in* x for every $u \in \text{Int } U$; i.e., $u \in \text{Int } U$, $x', x'' \in \text{Int } \Omega$ implies $D(u, x' + x'') > D(u, x')$.

(viii) D is such that the function

$$\widetilde{F}(x) \equiv \{u : u \in \text{Int } U,\ D(u,x) = 1\} \tag{5.2}$$

defined for $x \gg 0_N$ has a continuous extension to $x \geq 0_N$.

Proof: (i) We first show that D defined by (5.1) is well defined. Let $x^* \gg 0_N$ and define the function $g_{x^*}(k) \equiv F(x^*/k)$ for $k > 0$. Note that $\lim_{k\to\infty} g_{x^*}(k) = \lim_{k\to\infty} F(x^*/k) = F(0_N) = \bar{u}$ using I(i) and $\lim_{k\to 0} g_{x^*}(k) = \lim_{k\to 0} F(x^*/k) = \bar{\bar{u}}$ using $x^* \gg 0_N$, I(i), I(ii) and the definition of $\bar{\bar{u}}$. Using I(ii), it is easy to show that $g_{x^*}(k)$ is a monotonically decreasing function of k. Finally, from I(i), $g_{x^*}(k)$ is a continuous function of k. Thus range $g_{x^*} = \text{Int } U$ and for every $u^* \in \text{Int } U$, there exists a *unique* $k^* > 0$ such that $g_{x^*}(k^*) \equiv F(x^*/k^*) = u^*$. By I(ii), if $k > k^*$, then $F(x^*/k) < F(x^*/k^*) = u^*$. Thus

$$\begin{aligned} D(u^*, x^*) &\equiv \max_k\{k : F(x^*/k) \geq u^*,\ k > 0\} \\ &= \{k^* : F(x^*/k^*) = u^*,\ k^* > 0\}. \end{aligned} \tag{5.3}$$

In what follows, we will use (5.3) in order to define D instead of (5.1).[27] We show that $D(u,x)$ is continuous in (u,x) over $\text{Int } U \times \text{Int } \Omega$ by showing that the upper and lower level sets are closed in $\text{Int } U \times \text{Int } \Omega$.

Let $(u^n, x^n) \in \text{Int } U \times \text{Int } \Omega$ with $\lim_n (u^n, x^n) \equiv (u^0, x^0) \in \text{Int } U \times \text{Int } \Omega$ and $D(u^n, x^n) \leq k^* > 0$ for every n. Define $k^0 > 0$ by $F(x^0/k^0) = u^0$, and $k^n > 0$ by $F(x^n/k^n) = u^n$. Thus $D(u^n, x^n) \equiv k^n \leq k^*$ implies, using I(ii), that $F(x^n/k^*) \leq F(x^n/k^n) = u^n$. Thus $u^0 \equiv \lim_n u^n \geq \lim_n F(x^n/k^*) = F(x^0/k^*)$ using I(i). But $u^0 = F(x^0/k^0) \geq F(x^0/k^*)$ implies $k^* \geq k^0 \equiv D(u^0, x^0)$ using I(ii).

Now let $(u^n, x^n) \in \text{Int } U \times \text{Int } \Omega$ with $\lim_n (u^n, x^n) \equiv (u^0, x^0) \in \text{Int } U \times \text{Int } \Omega$ and $D(u^n, x^n) \geq k^* > 0$ for every n. Define $k^0 > 0$ by $F(x^0/k^0) = u^0$ and $k^n > 0$ by $F(x^n/k^n) = u^n$. Thus $D(u^n, x^n) \equiv k^n \geq k^*$ implies that $F(x^n/k^*) \geq F(x^n/k^n) = u^n$. Thus $u^0 \equiv \lim_n u^n \leq \lim_n F(x^n/k^*) = F(x^0/k^*)$ using I(i) again. But $u^0 = F(x^0/k^*) \leq F(x^0/k^*)$ implies $k^* \leq k^0 \equiv D(u^0, x^0)$.

(ii) Let $x^* \gg 0_N$, $\bar{u} < u^n < \bar{\bar{u}}$ and $\lim_n u^n = \bar{u}$. Define $k^n \equiv D(u^n, x^*)$ so that $F(x^*/k^n) = u^n$. Since $\lim_n u^n = \bar{u}$ and since $x = 0_N$ is the unique solution for the equation $F(x) = \bar{u}$, continuity of F implies that $\lim_n x^*/k^n = 0$ so that $\lim_n k^n = +\infty$.

[27] The reason why we did not define D directly by (5.3) is that definition (5.1) provides a valid definition for D when F satisfies weaker regularity conditions (such as our assumptions 1, 2 and 3 in Section 1).

(iii) This follows directly from (5.3) and property I(ii) on F.

(iv) Let $x^* \gg 0_N$, $\bar{u} < u^n < \bar{\bar{u}}$ and $\lim_n u^n = \bar{\bar{u}}$. Define $k^n \equiv D(u^n, x^*)$ so that $F(x^*/k^n) = u^n$. If $x \geq 0_N$ satisfies the equation $F(x) = \bar{\bar{u}}$, then at least one component of x must be $+\infty$. Since $\bar{\bar{u}} = \lim_n = \lim_n F(x^*/k^n) = F[\lim_n (x^*/k^n)]$ using I(i), it follows that we must have $\lim_n k^n = 0$.

(v) Let $u \in \text{Int } U$, $x \gg 0_N$, $\lambda > 0$. Then using (5.3), $D(u, \lambda x) \equiv \{k : k > 0, F(\lambda x/k) = u\} = \lambda\{\lambda^{-1}k : \lambda^{-1}k > 0, F(x/\lambda^{-1}k) = u\} = \lambda\{k' : k' > 0, F(x/k') = u\} \equiv \lambda D(u, x)$.

(vi) Let $u \in \text{Int } U$, $x' \gg 0_N$, $x'' \gg 0_N$ and $0 \leq \lambda \leq 1$. Define $k' \equiv D(u, x')$, $k'' \equiv D(u, x'')$ and $k^\lambda \equiv D[u, \lambda x' + (1-\lambda)x'']$. Then $F(x', k') = u$, $F(x''/k'') = u$ and $F[[\lambda x' + (1-\lambda)x'']/k^\lambda] = u$. If we define $\lambda^* \equiv k'\lambda/[(1-\lambda)k'' + \lambda k']$ and $k^* \equiv (1-\lambda)k'' + \lambda k'$, it can be verified that $0 \leq \lambda^* \leq 1$, $k^* > 0$, and $[\lambda x' + (1-\lambda)x'']/k^* = \lambda^*(x'/k') + (1-\lambda^*)(x''/k'')$. Thus by I(iii), $F[[\lambda x' + (1-\lambda)x'']/k^*] \geq u$. Using I(ii), the last inequality implies $D[u, \lambda x' + (1-\lambda)x''] \equiv k^\lambda \geq k^* \equiv \lambda k' + (1-\lambda)k'' \equiv \lambda D(u, x') + (1-\lambda)D(u, x'')$.

(vii) Let $u \in \text{Int } U$, $x' \gg 0_N$, $x'' \gg 0_N$. Properties IV(ii) to (iv) imply that $D(u, x'') > 0$. Thus

$$\begin{aligned} D(u, x' + x'') &= 2D[u, (1/2)x' + (1/2)x''] && \text{using (v) above} \\ &\geq 2[(1/2)D(u, x') + (1/2)D(u, x'')] && \text{using (vi) above} \\ &> D(u, x') && \text{using } D(u, x'') > 0. \end{aligned}$$

(viii) Let $x^* \gg 0_N$ and define $u^* \equiv F(x^*)$. Thus using I(ii) and definition (5.3) it can be seen that $D(u^*, x^*) \equiv \{k : F(x^*/k) = u^*, k > 0\} = 1$. Using IV(iii), $\tilde{F}(x^*) \equiv \{u : D(u, x^*) = 1, u \in \text{Int } U\} = u^*$. Thus $F(x^*) = \tilde{F}(x^*)$ for every $x^* \gg 0_N$ and since F is continuous over Ω, $\tilde{F}$ has F as its continuous extension. Q.E.D.

COROLLARY 5.1. *$\tilde{F}(x) \equiv \{u : u \in \text{Int } U, D(u, x) = 1\} = F(x)$ for every $x \gg 0_N$ and thus $\tilde{F} = F$; i.e., the original aggregator function F is recovered from the distance function D via definition (5.2) if F satisfies conditions I.*

As was the case with the cost function $C(u, p)$ studied in Section 3 above, D satisfying conditions IV over Int $U \times$ Int Ω can be uniquely extended to Int $U \times \Omega$ using the Fenchel closure operation. It can be verified that the extended D satisfies conditions (v), (vi) and (vii) over Int $U \times \Omega$, but the joint continuity condition IV(i) and the monotonicity conditions in u are no longer necessarily satisfied.[28] It should also be noted that if condition I(iii) (quasiconcavity of F) were dropped, then Theorem 5 would still be valid except that condition IV(vi) (concavity of D in x) would have to be dropped.

The following theorem shows that the deflation function D can also be used in order to define a continuous aggregator function $\tilde{F}$.

[28]If conditions IV(i) to (vii) were satisfied by D over Int $U \times \Omega$, then we could derive the corresponding continuous F from D without using the somewhat unusual condition IV(viii).

THEOREM 6. *If D satisfies conditions IV above, then $\widetilde{F}$ defined by (5.2) for $x \in \text{Int } \Omega$ has an extension to Ω which satisfies conditions I. Moreover, if we define the deflation function D^* which corresponds to $\widetilde{F}$ by*

$$D^*(u,x) \equiv \{k : \widetilde{F}(x/k) = u,\ k > 0\}, \tag{5.4}$$

then $D^*(u,x) = D(u,x)$ for $(u,x) \in \text{Int } U \times \text{Int } \Omega$.

Proof: (i) Since for every $x \in \text{Int } \Omega$, range $D(u,x)$ as a function of $u \in \text{Int } U$ is $(0,+\infty)$ and since D is a continuous, monotonically decreasing function of u over this domain (we have used IV(i)–(iv) here), we see that $\widetilde{F}(x)$ is well defined by (5.2) for $x \gg 0_N$. Property IV(viii) implies that $\widetilde{F}$ has a continuous extension to $x \geq 0_N$. Since $\widetilde{F}$ is continuous over Ω, we need only prove properties I(ii) and (iii) for $x \in \text{Int } \Omega$.

(ii) Let $0_N \ll x' \ll x''$ and define $u', u'' \in \text{Int } U$ by the equations $D(u',x'') = 1$ and $D(u'',x'') = 1$. Thus by (5.2), $\widetilde{F}(x') = u'$ and $\widetilde{F}(x'') = u''$. Now

$$\begin{aligned} 1 &= D(u'',x'') \\ &= D(u',x') \\ &< D(u',x'') \text{ since } x' \ll x'' \text{ using IV(vii)}. \end{aligned}$$

But $D(u'',x'') < D(u',x'')$ implies $\widetilde{F}(x'') = u'' > u' = \widetilde{F}(x')$ using IV(iii).

(iii) Let $x', x'' \in \text{Int } \Omega$, $0 \leq \lambda \leq 1$, $u* \in \text{Int } U$ with $\widetilde{F}(x') \geq u^*$ and $\widetilde{F}(x'') \geq u^*$. Then $D(u^*,x') \geq 1$ and $D(u^*,x'') \geq 1$, since $D[F(x'),x'] = 1$ and $D[\widetilde{F}(x''),x''] = 1$ using (5.2) and IV(iii). Note that $D\Big[\widetilde{F}[\lambda x' + (1-\lambda)x''],$ $\lambda x' + (1-\lambda)x''\Big] = 1$ also using definition (5.2). Thus $D[u^*, \lambda x' + (1-\lambda)x''] \geq \lambda D(u^*,x') + (1-\lambda)D(u^*,x'')$ using IV(vi) $\geq \lambda 1 + (1-\lambda)1 = 1$. Again using IV(iii), we conclude that $\widetilde{F}[\lambda x' + (1-\lambda)x''] \geq u^*$.

To prove the moreover part of the theorem, let $u \in \text{Int } U$, $x \in \text{Int } \Omega$ and define $k \equiv D^*(u,x) > 0$. Then by definition (5.4), $\widetilde{F}(x/k) = u$. By definition (5.2), the last equality implies

$$\begin{aligned} 1 &= D(u,x/k) \\ &= (1/k)D(u,x) \quad \text{using IV(v)} \end{aligned}$$

or

$$k \equiv D^*(u,x) = D(u,x).$$

Q.E.D.

COROLLARY 6.1. (Shephard[29] [1953; 10–13], Hanoch [1978; 7]: *If D satisfies conditions IV and, in addition, is continuously differentiable at $(u^*, x^*) \in$ Int $U \times$ Int Ω with $D(u^*, x^*) = 1$ and $\partial D(u^*, x^*)/\partial u < 0$, then x^* is a solution to the direct maximization problem* $\max_x\{\widetilde{F}(x) : v^{*T}x \leq 1, x \geq 0_N\}$, *where $\widetilde{F}$ is defined by (5.2) and $v^* > 0_N$ is defined by*

$$v^* \equiv \nabla_x D(u^*, x^*). \tag{5.5}$$

Moreover, $\widetilde{F}$ is continuously differentiable at x^* with

$$\nabla_x \widetilde{F}(x^*) = -\frac{\nabla_x D(u^*, x^*)}{\partial D(u^*, x^*)/\partial u}. \tag{5.6}$$

Proof: Since $\widetilde{F}(x)$ is implicitly defined by the equation $D[\widetilde{F}(x), x] = 1$ for x in a neighborhood of $x^* \gg 0_N$, the Implicit Function Theorem (see Rudin [1953; 177–182]) implies that $\widetilde{F}$ is continuously differentiable at x^* with partial derivatives given by (5.6), since the Jacobian of the transformation, $\partial D(u^*, x^*)/\partial u$, is nonzero. Since $D(u^*, x)$ is linearly homogenous in x, multiplying both sides of (5.6) by x^{*T} yields $x^{*T}\nabla\widetilde{F}(x^*) = -x^{*T}\nabla_x D(u^*, x^*)$ / $\nabla_u D(u^*, x^*) = -D(u^*, x^*)/\nabla_u D(u^*, x^*) = -1/\nabla_u D(u^*, x^*) > 0$ using Euler's Theorem on homogeneous functions. Therefore,

$$\nabla_x D(u^*, x^*) = -\nabla_x \widetilde{F}(x^*)\nabla_u D(u^*, x^*) = \nabla_x \widetilde{F}(x^*)/x^{*T}\nabla_x F(x^*). \tag{5.7}$$

Since $\nabla_x D(u^*, x^*) > 0_N$, $\nabla_x \widetilde{F}(x^*) > 0_N$ also. Now apply Corollary 3.3. Equations (4.10) and (5.7) imply (5.5). Q.E.D.

Thus, the consumer's system of inverse demand functions can be obtained by differentiating the deflation function D satisfying conditions IV (plus differentiability) with respect to the components of the vector x.

Historical Notes

Duality theorems between distance or deflation functions D and aggregator functions $\widetilde{F}$ have been proven by Shephard [1953] [1970], Hanoch [1978b], McFadden [1978a] and Blackorby, Primont and Russell [1978].

There are a number of interesting relationships (and further duality theorems) between direct and indirect aggregator, cost and deflation functions. For example, Malmquist [1953; 214] and Shephard [1953; 18] showed that the deflation function for the indirect aggregator function, $\max_k\{k : G(v/k) \leq u, k > 0\}$, equals the cost function, $C(u, v)$. A complete description of these interrelationships and further duality theorems under various regularity conditions may be found in Hanoch [1978b] and Blackorby, Primont and Russell [1978]. For some applications see Deaton [1979].

For a local duality theorem between deflation and aggregator functions, see Blackorby and Diewert [1979].

---

[29]The result can readily be deduced from several separate equations in Shephard but it is explicit in Hanoch's paper.

6. Other Duality Theorems

Concave functions can be characterized by their *conjugate* functions. Furthermore, it turns out that closed convex sets can also be described by a conjugate function under certain conditions.[30] Thus, a direct aggregator function F, having convex level sets $L(u) \equiv \{x : F(x) \geq u\}$, can also be characterized by its conjugate function as well as by its cost, deflation or indirect aggregator function. This conjugacy approach was initiated by Hotelling [1932; 590–592] [1935; 68–70] and extended by Samuelson [1947; 36–39] [1960] [1972], Lau [1969] [1976] [1978a], Jorgenson and Lau [1974a] [1974b], and Blackorby, Primont and Russell [1978]. We will not review this approach in detail, although in a later section we will review the closely related duality theorems between profit and transformation functions.

Another class of duality theorems (which also has its origins in the work of Hotelling [1935; 75] and Samuelson [1960]) is obtained by partitioning the commodity vector $x \geq 0_N$ into two vectors, x^1 and x^2 say, and then defining the consumer's *variable indirect aggregator*[31] function g as

$$(6.1) \qquad g(x^1, p^2, y^2) \equiv \max_{x^2}\{F(x^1, x^2) : p^{2T}x^2 \leq y^2, x^2 \geq 0_{N_2}\}$$

where $p^2 \gg 0_{N_2}$ is a positive vector of prices the consumer faces for the goods x^2 and $y^2 > 0$ is consumer's budget which he has allocated to spend on the x^2 goods. The solution set to (6.1), $x^2(x^1, p^2, y^2)$, is the consumer's *conditional* (on x^1) *demand correspondence*. If g satisfies appropriate regularity conditions, conditional demand functions can be generated by applying Roy's Identity (4.12) to the function $G(v^2) \equiv g(x^1, v^2, 1)$, where $v^2 \equiv p^2/y^2$. For formal duality theorems between direct and variable indirect aggregator functions, see Epstein [1975], Diewert [1978a] and Blackorby, Primont and Russell [1977a]. For various applications of this duality, see Epstein [1975] (for applications to consumer choice under uncertainty) and Pollak [1969] and Diewert [1978a] (estimation of preferences for public goods using market demand functions). Finally, the variable indirect utility function can be used to prove versions of Hicks' [1946; 312–313] composite good theorem — that a group of goods behaves just as if it were a single commodity if the prices of the group of goods change in the same proportion — under less restrictive conditions than were employed by Hicks; see Pollak [1969], Diewert [1978a] and Blackorby, Primont and Russell [1977a].

We turn now to a brief discussion of a vast literature; i.e., the implications of various special structures on *one* of our many equivalent representations

---

[30]See Rockafellar [1970; 102–105] and Karlin [1959; 226–227].

[31]Pollak [1969] uses the alternative terminology, "conditional indirect utility function."

of tastes or technology (such as the direct or indirect aggregator function or the cost function) on the other representations. For example, Shephard [1953] showed that *homotheticity* of the direct function implied that the cost function factored into $\phi^{-1}(u)c(p)$ (recall equation (2.18) above). Another example of a special structure is *separability*.[32] References which deal with the implications of separability and/or homotheticity include Shephard [1953] [1970], Samuelson [1953–54], [1965] [1969b] [1972], Gorman [1959] [1976], Lau [1969] [1978a], McFadden [1978a], Hanoch [1975] [1978b], Pollak [1972], Diewert [1974a], Jorgenson and Lau [1975], and Blackorby, Primont and Russell [1975a] [1975b] [1977a] [1977b] [1977d] [1978]. For the implications of separability and/or homotheticity on Slutsky coefficients or on partial elasticities of substitution,[33] see Sono [1945], Pearce [1961], Goldman and Uzawa [1964], Geary and Morishma [1973], Berndt and Christensen [1973a], Russell [1975], Diewert [1974a; 150–153] and Blackorby and Russell [1976]. For the implications of Hicks' [1946; 312–313] Aggregation Theorem on aggregate elasticities of substitution, see Diewert [1974c].

For empirical tests of the separability assumption, see Berndt and Christensen [1973b] [1974], Burgess [1974] and Jorgenson and Lau [1975]; for theoretical discussions of these testing procedures, see Blackorby, Primont and Russell [1977c], Jorgenson and Lau [1975], Lau [1977c], Woodland [1978] and Denny and Fuss [1977].

For the implications of assuming *concavity* of the direct aggregator function or of assuming *convexity* of the indirect aggregator function, see Diewert [1978a].

The duality theorems referred to above have been "global" in nature. A "local" approach has been initiated by Blackorby and Diewert [1979], where it

---

[32] Loosely speaking, $F(x) = F(x^1, x^2, \ldots, x^M)$ is separable in the partition $(x^1, x^2, \ldots, x^M)$ if there exist functions $\widehat{F}$, $F^1, \ldots, F^M$ such that $F(x) = \widehat{F}[F^1(x^1), F^2(x^2), \ldots, F^M(x^M)]$ and F is additively separable if there exist functions F^*, $F^1, F^2, \ldots, F^M$ such that $F(x) = F^*[F^1(x^1) + F^2(x^2) + \cdots + F^M(x^M)]$. For historical references and more precise definitions, see Blackorby, Primont and Russell [1977d][1978].

[33] Uzawa [1962] observed that the Allen [1938; 504] partial elasticity of substitution between inputs i and j, $\sigma_{ij}(u,p) = C(u,p)C_{ij}(u,p)/C_i(u,p)C_j(u,p)$ where $C(u,p) \equiv \min_x\{p^T x : F(x) \geq u\}$ is the cost function dual to the aggregator function F and C_i denotes the partial derivative with respect to the ith price in p, p_i, and C_{ij} denotes the second order partial derivative of C with respect to p_i and p_j. Shephard's Lemma implies that $C_{ij}(u,p) = \partial x_i(u,p)/\partial p_j = \partial x_j(u,p)/\partial p_i = C_{ji}(u,p)$ assuming continuous differentiability of C, where $x_i(u,p)$ and $x_j(u,p)$ are the cost minimizing demand functions. Thus σ_{ij} can be regarded as a normalization of the response of the cost minimizing x_i to a change in p_j.

is assumed that a given cost function $C(u,p)$ satisfies conditions II above over $U \times P$, where U is a *finite* interval and P is a closed, convex, and *bounded* subset of positive prices. They then construct the corresponding direct aggregator, indirect aggregator and deflation functions which are dual to the given "locally" valid cost function C. The proofs of these "local" duality theorems turn out to be much simpler than the corresponding "global" duality theorems presented in this paper (and elsewhere), since troublesome continuity problems do not arise due to the assumption that $U \times P$ is compact. These "local" duality theorems are useful in empirical applications, since econometrically estimated cost functions frequently do not satisfy the appropriate regularity conditions for all prices, but the conditions may be satisfied over a smaller subset of prices which is the empirically relevant set of prices.

Epstein has extended duality theory to cover more general maximization problems. In an unpublished working paper of his, the following utility maximization problem which arises in the context of choice under uncertainty is considered:

$$(6.2) \quad \max_{x,x^1,x^2} \Big\{F(x,x^1,x^2) : x \geq 0_N,\ x^1 \geq 0_{N_1},\ x^2 \geq 0_{N_2}, \\ p^T x + p^{1T} x^1 \leq y^1,\ p^T x + p^{2T} x^2 \leq y^2\Big\}$$

where x represents current consumption, there are two future uncertain states of nature, x^i represents consumption in state i $(i = 1,2)$, p is the current price vector, p^i is the discounted future price vector which will prevail if state i occurs, and $y^i > 0$ is the consumer's discounted income if state i occurs. In Epstein [1981a], the following maximization problem is considered:

$$(6.3) \quad \max_x \{F(x) : x \geq 0_N, c(x,\alpha) \leq 0\}$$

where c is a given constraint function which depends on a vector of parameters α.

We will not attempt to provide a detailed analysis of Epstein's results but rather we will present a more abstract version of his basic technique which will hopefully capture the essence of duality theory.

The basic maximization problem we study is $\max_x\{F(x) : x \in B(v)\}$ where F is a function of N real variables x defined over some set S and $B(v)$ is a constraint set which depends on a vector of M parameters v, which in turn can vary over a set V. Our assumptions on the sets S and V and the constraint set correspondence B are:

(6.4) (i) S and V are nonempty compact sets in R^N and R^M.

(ii) For every $v \in V$, $B(v)$ is nonempty and $B(v) \subset S$.

(iii) For every $x \in S$, the inverse correspondence[34] $B^{-1}(x)$ is nonempty and $B^{-1}(x) \subset V$.

(iv) The correspondence B is continuous over V.

(v) The correspondence B^{-1} is continuous over S.

Our assumptions on the primal function F are:

(6.5) (i) F is a real valued function of N variables defined over S and is *continuous* over S.

(ii) For every $x^* \in S$, there exists $v^* \in V$ such that

$$F(x^*) = \max_x\{F(x) : x \in B(v^*)\}.$$

The function G dual to F is defined for $v \in V$ by

$$G(v) \equiv \max_x\{F(x) : x \in B(v)\}. \tag{6.6}$$

THEOREM 7. *If S, V and B satisfy (6.4) and F satisfies (6.5), then G defined by (6.6) satisfies the following conditions:*

(6.7) *(i) G is a real valued function of M variables defined over V and is continuous over V.*

(ii) For every $v^* \in V$, there exists $x^* \in S$ such that

$$G(v^*) = \min_v\{G(v) : v \in B^{-1}(x^*)\}.$$

Moreover, if we define the function F^* dual to G for $x \in S$ by

$$F^*(x) \equiv \min_v\{G(v) : v \in B^{-1}(x)\}, \tag{6.8}$$

then $F^*(x) = F(x)$ for every $x \in S$.

Proof: (6.7)(i) follows from (6.4)(i), (ii), (iv); (6.5)(i) and the Maximum Theorem.

(ii) Let $v^* \in V$ and let $x^* \in S$ be any solution to the following maximization problem:

$$\max_x\{F(x) : x \in B(v^*)\} = F(x^*) = G(v^*) \tag{6.9}$$

where (6.9) follows using definition (6.6). By definition (6.8)

$$F^*(x^*) \equiv \min_v\{G(v) : v \in B^{-1}(x^*)\} \leq G(v^*) \tag{6.10}$$

[34] $B^{-1}(x) \equiv \{v : v \in V \text{ and } x \in B(v)\}$.

since by (6.9), $x^* \in B(v^*)$ and thus $v^* \in B^{-1}(x^*)$ and thus v^* is feasible for the minimization problem in (6.10). For every $v \in B^{-1}(x^*)$, $x^* \in B(v)$ and thus using (6.9), we have

$$(6.11) \qquad G(v) \equiv \max_x\{F(x) : x \in B(v)\} \geq F(x^*) = G(v^*)$$

since x^* is feasible for the maximization problem in (6.11) and the last equality in (6.11) follows from (6.9). Since (6.11) is satisfied for every $v \in B^{-1}(x^*)$,

$$(6.12) \qquad F^*(x^*) \equiv \min_v\{G(v) : v \in B^{-1}(x^*)\} \geq G(v^*).$$

Thus (6.9), (6.10) and (6.12) imply that $F(x^*) = G(v^*) = F^*(x^*)$ and thus (6.10) becomes an equality, which establishes property (6.7)(ii) for G.

Notice that we have not yet used property (6.5)(ii) for F. However, it will be used in order to prove the moreover part of the theorem. Let $x^* \in S$. Then using (6.5)(ii), there exists $v^* \in V$ such that $F(x^*) = \max\{F(x) : x \in B(v^*)\} \equiv G(v^*)$. Thus we have an $x^* \in S$ and $v^* \in V$ such that (6.9) is satisfied and now we can repeat the proof of part (ii) above, showing that $F(x^*) = F^*(x^*)$. Q.E.D.

COROLLARY 7.1. *Let $x^* \in S$ and define $H(x^*)$ to be the set of $v^* \in V$ such that $F(x^*) = \max_x\{F(x) : x \in B(v^*)\}$. If $v^* \in H(x^*)$, then x^* is a solution to $\max_x\{F(x) : x \in B(v^*)\}$, and v^* is a solution to $\min_v\{G(v) : v \in B^{-1}(x^*)\}$.*

Notice that property (6.5)(ii) of F is the replacement for our old quasiconcavity assumption in Section 4, and the set $H(x^*)$ defined in Corollary 7.1 replaces the set of normalized supporting hyperplanes which occurred in Corollary 3.2.

Owing to the symmetric nature of our assumptions, it can be seen that the proof of the following theorem is the same as the proof of Theorem 7, except that the inequalities are reversed.

THEOREM 8. *If S, V and B satisfy (6.4) and G satisfies (6.7), then F^* defined by (6.8) satisfies (6.5). Moreover, if we define the function G^* dual to F^* for $v \in V$ by*

$$(6.13) \qquad G^*(v) \equiv \max_x\{F^*(x) : x \in B(v)\},$$

then $G^*(v) = G(v)$ for every $v \in V$.

COROLLARY 8.1. *Let $v^* \in V$ and define $H^*(v^*)$ to be the set of $x^* \in S$ such that $G(v^*) = \min_v\{G(v) : v \in B^{-1}(x^*)\}$. If $x^* \in H^*(v^*)$, then v^* is a solution to $\min_v\{G(v) : v \in B^{-1}(x^*)\}$, and x^* is a solution to $\max_x\{F^*(x) : x \in B(v^*)\}$.*

Note that condition (6.7)(ii) on G replaces our old quasiconvexity condition on G in Section 4, and the set $H^*(v^*)$ defined in Corollary 8.1 replaces the set of normalized supporting hyperplanes which occurred in Corollary 4.1.

We cannot establish counterparts to Corollary 3.3 (Hotelling–Wold Identity) and Corollary 4.2 (Ville–Roy Identity) since these corollaries made use of the differentiable nature of F or G and the relevant constraint function. Thus in order to derive counterparts to Corollaries 3.3 and 4.2 in the present context, we need to make additional assumptions on F (or G) and the constraint correspondence B.[35] However, the above theorems (due essentially to Epstein) do illustrate the underlying structure of duality theory. They can also be interpreted as examples of local duality theorems.

7. Cost Minimization and the Derived Demand for Inputs

Assume that the technology of a firm can be described by the production function F where $u = F(x)$ is the maximum output that can be produced using the nonnegative vector of inputs $x \geq 0_N$. Assume that F satisfies assumption 1 of Section 2 (i.e., the production function is continuous from above). If the firm takes the prices of inputs $p \gg 0_N$ as given (i.e., the firm does not behave monopsonistically with respect to inputs), then we saw in Section 2 that the firm's total cost function $C(u;p) \equiv \min_x\{p^T x : F(x) \geq u\}$ was well defined for all $p \gg 0_N$ and $u \in$ range F. Moreover, $C(u,p)$ was linearly homogeneous and concave in prices p for every u and was nondecreasing in u for each fixed p.

Now suppose that C is twice continuously differentiable[36] with respect to its arguments at a point (u^*, p^*) where $u^* \in$ range F and $p^* \equiv (p_1^*, \ldots, p_N^*) \gg 0_N$. From Lemma 3 in Section 2, the cost minimizing input demand functions $x_1(u,p), \ldots, x_N(u,p)$ exist at (u^*, p^*) and they are in fact equal to the partial derivatives of the cost function with respect to the N input prices:

$$(7.1) \qquad x_i(u^*, p^*) = \partial C(u^*, p^*)/\partial p_i; \quad i = 1, \ldots, N.$$

Thus, the assumption that C be twice continuously differentiable at (u^*, p^*) ensures that the cost minimizing input demand functions $x_i(u,p)$ exist and are once continuously differentiable at (u^*, p^*).

Define $(\partial x_i/\partial p_j) \equiv [\partial x_i(u^*, p^*)/\partial p_j]$ to be the $N \times N$ matrix of derivatives of the N input demand functions $x_i(u^*, p^*)$ with respect to the N prices p_j^*, $i, j = 1, 2, \ldots, N$. From (7.1), it follows that

$$(7.2) \qquad (\partial x_i/\partial p_j) = \nabla_{pp}^2 C(u^*, p^*)$$

[35] Epstein [1981a] derives counterparts to 4.2 in the context of his specific models.

[36] By this assumption, we mean that the second order partial derivatives of C exist and are continuous functions for a neighborhood around (u^*, p^*).

where $\nabla^2_{pp}C(u^*,p^*) \equiv [\partial^2 C(u^*,p^*)/\partial p_i \partial p_j]$ is the Hessian matrix of the cost function with respect to the input prices evaluated at (u^*,p^*). Twice continuous differentiability of C with respect to p at (u^*,p^*) implies (via Young's Theorem) that $\nabla^2_{pp}C(u^*,p^*)$ is a symmetric matrix, so that using (7.2),

$$(7.3) \qquad (\partial x_i/\partial p_j) = (\partial x_i/\partial p_j)^T = (\partial x_j/\partial p_i),$$

i.e., $\partial x_i(u^*,p^*)/\partial p_j = \partial x_j(u^*,p^*)/\partial p_i$ for all i and j.

Since C is *concave* in p and is twice continuously differentiable with respect to p around the point (u^*,p^*), it follows[37] that $\nabla^2 C(u^*,p^*)$ is a negative semidefinite matrix. Thus by (7.2),

$$(7.4) \qquad z^T(\partial x_i/\partial p_j)z \leq 0 \quad \text{for all vectors } z.$$

Thus, in particular, letting $z = e_i$ (the ith unit vector), (7.4) implies

$$(7.5) \qquad \partial x_i(u^*,p^*)/\partial p_i \leq 0, \quad i = 1,2,\ldots,N;$$

i.e., the ith cost minimizing input demand function cannot slope upwards with respect to the ith input price for $i = 1,2,\ldots,N$.

Since C is linearly homogeneous in p, we have $C(u^*,\lambda p^*) = \lambda C(u^*,p^*)$ for all $\lambda > 0$. Partially differentiating this last equation with respect to p_i for λ close to 1 yields the equation $C_i(u^*,\lambda p^*)\lambda = \lambda C_i(u^*,p^*)$, where $C_i(u^*,p^*) \equiv \partial C(u^*,p^*)/\partial p_i$. Thus, $C_i(u^*,\lambda p^*) = C_i(u^*,p^*)$ and differentiation of this last equation with respect to λ yields (when $\lambda = 1$)

$$\sum_{j=1}^{N} p_j^* \partial^2 C(u^*,p^*)/\partial p_i \partial p_j = 0 \quad \text{for } i = 1,2,\ldots,N.$$

Thus, using (7.2), we find that the input demand functions $x_i(u^*,p^*)$ satisfy the following N restrictions:

$$(7.6) \qquad (\partial x_i/\partial p_j)p^* = \nabla^2_{pp}C(u^*,p^*)p^* = 0_N$$

where $p^* \equiv (p_1^*,p_2^*,\ldots,p_N^*)^T$.

The final general restriction that we can obtain on the derivatives of the input demand functions is obtained as follows: for λ near 1, differentiate both sides of $C(u^*,\lambda p^*) = \lambda C(u^*,p^*)$ with respect to u and then differentiate the resulting equation with respect to λ. When $\lambda = 1$, the last equation becomes

$$\sum_{j=1}^{N} p_j^* \partial^2 C(u^*,p^*)/\partial u \partial p_j = \partial C(u^*,p^*)/\partial u.$$

[37]See Fenchel [1953; 87–88] or Rockafellar [1970; 27].

Note that the twice continuous differentiability of C and (7.1) implies that

$$\partial^2 C(u^*,p^*)/\partial u \partial p_j = \partial^2 C(u^*,p^*)/\partial p_j \partial u$$
$$= \partial[\partial C(u^*,p^*)/\partial p_j]/\partial u = \partial x_j(u^*,p^*)/\partial u.$$

Thus

$$\sum_{j=1}^N p_j^* \frac{\partial^2 C(u^*,p^*)}{\partial u \partial p_j} = \sum_{j=1}^N p_j^* \frac{\partial x_j(u^*,p^*)}{\partial u}$$

$$(7.7) \qquad = \frac{\partial C(u^*,p^*)}{\partial u} \geq 0.$$

The inequality $\partial C(u^*,p^*)/\partial u \geq 0$ follows from the nondecreasing in u property of C. The inequality (7.7) tells us that the changes in cost minimizing input demands induced by an increase in output cannot all be negative; i.e., not all inputs can be inferior.

With the additional assumption that F be linearly homogeneous (and there exists $x > 0_N$ such that $F(x) > 0$), we can deduce (cf. Section 2) that $C(u,p) = uc(p)$, where $c(p) \equiv C(1,p)$. Thus, when F is linearly homogeneous,

$$(7.8) \qquad x_i(u^*,p^*) = u^* \partial c(p^*)/\partial p_i, \quad i = 1,\ldots,N,$$

and $\partial x_i(u^*,p^*)/\partial u = \partial c(p^*)/\partial p_i$. Thus if $x_i^* \equiv x_i(u^*,p^*) > 0$ for $i = 1,2,\ldots,N$, using (69) we can deduce the additional restrictions

$$(7.9) \qquad \frac{\partial x_i(u^*,p^*)}{\partial u}\frac{u^*}{x_i^*} = \frac{u^* \partial c(p^*)/\partial p_i}{x_i^*} = 1$$

if F is linearly homogeneous; i.e., all of the input elasticities with respect to output are unity.

For the general two input case, the general restrictions (7.3)–(7.7) enable us to deduce the following restrictions on the partial derivatives of the two input demand functions, $x_1(u^*,p_1^*,p_2^*)$ and $x_2(u^*,p_1^*,p_2^*)$: $\partial x_1/\partial p_1 \leq 0$, $\partial x_2/\partial p_2 \leq 0$, $\partial x_1/\partial p_2 \geq 0$, $\partial x_2/\partial p_1 \geq 0$ (and if any one of the above inequalities holds strictly, then they all do, since $p_1^* \partial x_1/\partial p_1 = -p_2^* \partial x_1/\partial p_2 = -p_2^* \partial x_2/\partial p_1 = (p_2^*)^2(p_1^*)^{-1}\partial x_2/\partial p_2)$ and $p_1^* \partial x_1/\partial u + p_2^* \partial x_2/\partial u \geq 0$. Thus, the signs of $\partial x_1/\partial u$ and $\partial x_2/\partial u$ are ambiguous, but if one is negative, then the other must be positive. For the constant returns to scale two input case, the ambiguity disappears: we have $\partial x_1(u^*,p^*)/\partial u \geq 0$, $\partial x_2(u^*,p^*)/\partial u \geq 0$ and at least one of the inequalities must hold strictly if $u^* > F(0_2)$.

An advantage in deriving these well known comparative statics results using duality theory is that the restrictions (7.2)–(7.7) are valid in cases where the direct production function F is not even differentiable. For example, a Leontief production function has a linear cost function $C(u,p) = ua^T p$, where $a^T \equiv (a_1,a_2,\ldots,a_N) > 0_N^T$ is a vector of constants. It can be verified that the restrictions (7.2)–(7.7) are valid for this nondifferentiable production function.

Historical Notes

Analogues to (7.3) and (7.4) in the context of profit functions were obtained by Hotelling [1932; 594] [1935; 69–70]. Hicks [1946; 311 and 331] and Samuelson [1947; 69] obtained all of the relations (7.2)–(7.6) and Samuelson [1947; 66] also obtained (7.7). All of these authors assumed that the primal function F was differentiable and their proofs used the first order conditions for the cost minimization (or utility maximization) problem plus the properties of determinants in order to prove their results.

Our proofs of (7.3)–(7.6), using only differentiability of the cost function plus Lemma 3 in Section 1, are due to McKenzie [1956–57; 188–189] and Karlin [1959; 273]. McFadden [1978a] also provides alternative proofs.

If F is only homothetic rather than being linearly homogeneous, then the relations (7.9) are no longer true. If F is homothetic, then by (2.18), $C(u,p) = \phi^{-1}(u)c(p)$ where ϕ^{-1} is a monotonically increasing function of one variable. Thus, under our differentiability assumptions, $x_i(u^*,p^*) = \phi^{-1}(u^*)\partial c(p^*)/\partial p_i$ and $\partial x_i(u^*,p^*)/\partial u = [d\phi^{-1}(u^*)/du][\partial c(p^*)/\partial p_i]$, so that if $x_i^* \equiv x_i(u^*,p^*) > 0$,

$$\text{(7.10)} \qquad \frac{\partial x_i(u^*,p^*)}{\partial u}\frac{u^*}{x_i^*} = \frac{u^*[d\phi^{-1}(u^*)]/du}{\phi^{-1}(u^*)} \equiv \eta(u^*) \geq 0 \text{ for } i = 1,2,\ldots,N.$$

Thus, in the case of a homothetic production function, the input elasticities with respect to output are all equal to the same nonnegative number independent of the input prices, but dependent in general on the output level u^*. Furthermore, assuming homotheticity of F, we can solve the equation $C(u,p) = \phi^{-1}(u)c(p) = y$ for $u = \phi[y/c(p)] = \phi[1/c(p/y)] \equiv G(p/y)$, where $y > 0$ is the producer's allowable expenditure on inputs. If we replace u^* by $\phi(y^*/c(p^*))$ in the system of input demand functions $x_i(u^*,p^*)$, we obtain the system of "market" demand functions

$$\begin{aligned} x_i\left[\phi[y^*/c(p^*)],p^*\right] &= \phi^{-1}[\phi(y^*/c(p^*)]\partial c(p^*)/\partial p_i \\ &= [y^*/c(p^*)]\partial c(p^*)/\partial p_i \quad \text{for } i = 1,2,\ldots,N. \end{aligned}$$

Thus if $x_i^* \equiv x_i(u^*,p^*) > 0$,

$$\text{(7.11)} \qquad \frac{\partial x_i}{\partial y}\left[\phi[\frac{y^*}{c(p^*)}],p^*\right]\frac{y^*}{x_i^*} = 1, \quad i = 1,2,\ldots,N;$$

i.e., all inputs have unitary "income" (or expenditure) elasticity of demand if the underlying aggregator function F is homothetic. Note the close resemblance of (7.11) to (7.9). That homotheticity of F implies the relations (7.11) dates back to Frisch [1936; 25] at least. For further references, see Chipman [1974a; 27].

8. The Slutsky Conditions for Consumer Demand Functions

Assume that a consumer has a utility function $F(x)$ defined over $x \geq 0_N$ which is continuous from above. Then we have seen in Section 2 that $C(u,p) \equiv \min_x\{p^T x : F(x) \geq u\}$ is well defined for $u \in$ range F and $p \gg 0_N$. Moreover, the cost function C has a number of properties including nondecreasingness in u for each $p \gg 0_N$ and linear homogeneity and concavity in p for each $u \in$ range F.

Assume that the consumer faces prices $p^* \gg 0_N$ and has income $y^* > 0$ to spend on commodities. Then the consumer will wish to choose the largest u such that his cost minimizing expenditure on the goods is less than or equal to his available income. Thus, the consumer's equilibrium utility level will be u^* defined by

$$u^* \equiv \max_u\{u : C(u,p^*) \leq y^*, u \in \text{range } F\}.$$

Now assume that C is twice continuously differentiable with respect to its arguments at the point (u^*,p^*) with

$$\partial C(u^*,p^*)/\partial u > 0. \tag{8.1}$$

The fact that C is nondecreasing in u implies that $\partial C(u^*,p^*)/\partial u \geq 0$; however, the slightly stronger assumption (8.1) enables us to deduce that the consumer will actually spend all of his income on purchasing (or renting) commodities; i.e., (8.1) implies that

$$C(u^*,p^*) = y^*. \tag{8.2}$$

Furthermore, since C is linearly homogeneous in p, (8.2) implies

$$C(u^*,p^*/y^*) = 1. \tag{8.3}$$

Our differentiability assumptions plus (8.1) and (8.3) imply (using the Implicit Function Theorem) that (8.3) can be solved for u as a function of p/y in a neighborhood of p^*/y^*. The resulting function $G(p/y)$ is the consumer's ***indirect utility function***, which gives the maximum utility level the consumer can attain, given that he faces commodity prices p and has income y to spend on commodities. The Implicit Function Theorem also implies that G will be twice continuously differentiable with respect to its arguments at p^*/y^*. Note that

$$u^* = G(p^*/y^*). \tag{8.4}$$

The consumer's system of *Hicksian* [1946; 331] *or constant real income demand functions*[38] $f_1(u,p),\ldots,f_N(u,p)$ is defined as the solution to the expenditure minimization problem $\min_x\{p^T x : F(x) \geq u\}$. Since we have assumed that C is differentiable with respect to p at (u^*,p^*), by Lemma 3 in Section 2,

$$(8.5) \qquad f_i(u^*,p^*) = \partial C(u^*,p^*)/\partial p_i, \quad i = 1,\ldots,N;$$

i.e., the Hicksian demand functions can be obtained by differentiating the cost function with respect to the commodity prices. On the other hand, the consumer's system of *ordinary market demand functions*, $x_1(y,p),\ldots,x_N(y,p)$, can be obtained from the Hicksian system (8.5), if we replace u by $G(p/y)$, the maximum utility the consumer can obtain when he has income y and faces prices p. Thus,

$$(8.6) \qquad x_i(y^*,p^*) \equiv f_i[G(p^*/y^*),p^*], \quad i = 1,\ldots,N.$$

Thus, the consumer's system of market demand functions can be obtained from the cost function as well as by using the Ville–Roy Identity (4.12). Finally, it can be seen that if we replace y in the consumer's system of market demand functions by $C(u,p)$, then we should obtain precisely the system of Hicksian demand functions (8.5); i.e., we have

$$(8.7) \qquad x_i[C(u^*,p^*),p^*] = \partial C(u^*,p^*)/\partial p_i, \quad i = 1,\ldots,N.$$

Differentiating both sides of (8.7) yields:

$$\begin{aligned}
\frac{\partial^2 C(u^*,p^*)}{\partial p_i \partial p_j} &= \frac{\partial x_i(y^*,p^*)}{\partial p_j} + \frac{\partial x_i(y^*,p^*)}{\partial y}\frac{\partial C(u^*,p^*)}{\partial p_j} && \text{using (8.2)} \\
&= \frac{\partial x_i(y^*,p^*)}{\partial p_j} + f_j(u^*,p^*)\frac{\partial x_i(y^*,p^*)}{\partial y} && \text{using (8.5)} \\
&= \frac{\partial x_i(y^*,p^*)}{\partial p_j} + x_j(y^*,p^*)\frac{\partial x_i(y^*,p^*)}{\partial y} && \text{using (8.4) and (8.6)}
\end{aligned}$$

$$(8.8) \qquad \equiv k_{ij}^*, \quad i,j = 1,2,\ldots,N,$$

where k_{ij}^* is known as the *ij*th *Slutsky coefficient*. Note that the $N \times N$ matrix of these Slutsky coefficients, $K^* \equiv [k_{ij}^*]$, can be calculated from a knowledge of the market demand functions, $x_i(y,p)$, and their first order derivatives at the point (y^*,p^*). (8.8) shows that $K^* \equiv \nabla_{pp}^2 C(u^*,p^*)$ and thus (recall equations (7.3), (7.4) and (7.6) of the previous section) K^* satisfies the following *Slutsky–Samuelson–Hicks Conditions*:

$$(8.9) \qquad \text{(i)} \quad K^* = K^{*T}.$$

$$\text{(ii)} \quad z^T K^* z \leq 0 \text{ for every } z.$$

$$\text{(iii)} \quad K^* p^* = 0_N.$$

[38]In the previous section, these functions are denoted as $x_1(u,p),\ldots,x_N(u,p)$.

Historical Notes

Slutsky [1915] deduced (8.9)(i) and part of (8.9)(ii); i.e., that $k^*_{ii} \leq 0$. Samuelson [1938; 348] and Hicks [1946; 311] deduced the entire set of restrictions (8.9) under the assumption that F was twice continuously differentiable at an equilibrium point $x^* > 0_N$ and F satisfied the additional property that $v^T \nabla^2_{xx} F(x^*) v < 0$ for all $v \neq 0_N$ such that $v \neq k\nabla F(x^*)$ for any scalar k. In fact, under these hypotheses, Samuelson and Hicks were able to deduce the following strengthened version of (8.9)(ii): $z^T K^* z < 0$ for every $z \neq 0_N$ such that $z \neq kp^*$ for any scalar k.

Our proof of conditions (8.9) is due to McKenzie [1956–57] and Karlin [1959; 267–273]. See also Arrow and Hahn [1971; 105]. This method of proof again has the advantage that differentiability of F does not have to be assumed; essentially, all that is required is differentiability of the demand functions. Afriat [1972c] makes this point.

For a derivation of conditions (8.9) which utilizes only the properties of the indirect utility function G, see Diewert [1977; 356].

For a "traditional" derivation of (8.9), see Intriligator [1981].

9. Consumption Theorems in Terms of Over and Under Compensation Revisited

The task of this section is to cast some light on the following somewhat enigmatic footnote in a paper by Samuelson:

> This can be seen by writing utility as a function of the overcompensating changes in prices or $U = U(q_1, \ldots) = U[F^1(p_1, \ldots, p_n), \ldots] = V(p_1, \ldots, p_n)$ with $[\partial V(p_1, \ldots, p_n)/\partial p_i]$ proportional to $(q_i - q_i^0)$ and vanishing at $p_i = p_i^0$. Hence, at p_i^0, $[\partial(q_i - q_i^0)/\partial p_j]$ is proportional to $[\partial^2 V(p_1^0, \ldots, p_n^0) \;/\; \partial p_i \partial p_j]$, which is symmetric; this last matrix is also negative semi-definite[39] because the price ratios at (p^0) give the lowest utility possible … .
>
> To handle the case of undercompensation, note that around any initial point, $q_i^0 = D^i(p_i^0, \ldots, p_n^0, I^0)$, we can solve the implicit set of equations
>
> $$q_i = D^i(p_1, \ldots, p_n, X)$$
> $$\Sigma p_i^0 D^i(p_1, \ldots, p_n, X) = \Sigma p_i^0 q_i^0$$

[39]This is an obvious slip; Samuelson means positive rather than negative semi-definiteness.

for $q_i = f^i(p_1, \ldots, p_n)$ and $X = X(p_1, \ldots, p_n)$; then it can be shown that for $U = U(q_1, \ldots) = U(f^1, \ldots) = W(p_1, \ldots, p_n)$, $[\partial f^i(p_1^0, \ldots, p_n^0) / \partial p_j] = (\partial^2 W/\partial p_i \partial p_j)$ is symmetric, and negative semi-definite by virtue of the fact that the price ratios at (p^0) maximise U or W.

It can be shown that taking a mean of overcompensated and undercompensated changes — as e.g. $\frac{1}{2}[f(p_1, \ldots, p_n) + F(p_1, \ldots, p_n)]$ — gives a change that agrees locally around (p^0) with an indifference change up to derivatives of still higher order: such a locus osculates the indifference surface so as to have not only the same slope but also the same curvature.[40]

Samuelson [1953; 8]

Suppose that the consumer's utility function F is defined and continuous from above for nonnegative commodity vectors $x \geq 0_N$. Then, as we have seen in Section 2, the consumer's cost or expenditure function $C(u,p) \equiv \min_x\{p^T x : F(x) \geq u\}$ is well defined for all positive commodity price vectors $p \gg 0_N$ and all utility levels $u \in U$, where U is the smallest convex set containing the range of F. Moreover, C satisfies properties 1–7 of Section 2. Suppose that C is twice continuously differentiable in some neighborhood around the point (u^0, p^0) where $u^0 \in U$ and $p^0 \gg 0_N$, and in addition:

(9.1) $$\partial C(u^0, p^0)/\partial u \equiv \nabla_u C(u^0, p^0) > 0 \text{ and}$$
(9.2) $$\nabla_p C(u^0, p^0) \equiv x^0 > 0_N.$$

By Lemma 3, the consumer's system of constant utility (or Hicksian) demand functions, $x(u,p) \equiv [x_1(u,p), \ldots, x_N(u,p)]^T$, can be obtained by differentiating C with respect to the commodity prices; i.e., for (u,p) close to (u^0, p^0), we have $x(u,p) = \nabla_p C(u,p)$. Thus $x^0 \equiv (x_1^0, \ldots, x_N^0)^T$ in (9.2) can be interpreted as the consumer's initial demand vector.

Samuelson's *overcompensated indirect utility function*, $u = V(p)$, can be defined as the solution to the following equation involving u and p:

(9.3) $$C(u,p) = p^T x^0.$$

---

[40]Samuelson and Swamy [1974; 582] add the following explanatory note on the above footnote: "The truth of this finding, that the Ideal index gives a second-order or osculating approximation to the true homothetic index, could have been vaguely suspected from the finding in Samuelson [1953; p. 8, n. 1] that the symmetric mean of overcompensated and undercompensated demand functions provides a high-order osculating approximation to the Slutsky-Hicks just-compensated demand along the indifference contours." A symmetric mean $m(x,y)$ of two nonnegative numbers x and y is usually defined to be any function which satisfies (i) $m(x,y) = m(y,x)$, (ii) $m(x,x) = x$, and (iii) $\text{Min}\{x,y\} \leq m(x,y) \leq \text{Max}\{x,y\}$.

Thus the consumer is given a new budget constraint indexed by the commodity price vector p and given just enough income, $y \equiv C(u,p)$, so that he can purchase his initial consumption vector x^0 at the new prices: this is the economic interpretation of equation (9.3) which implicitly defines $u = V(p)$. Our differentiability assumptions on C plus assumption (9.1) are sufficient to imply the existence of $V(p)$ for $p \in B_\delta(p^0)$ where $B_\delta(p^0) \equiv \{p : (p-p^0)^T(p-p^0) < \delta^2\}$ is the open ball of radius $\delta > 0$ around the point p^0. We choose $\delta > 0$ small enough so that $B_\delta(p^0)$ is a subset of the positive orthant and so that $C(V(p),p)$ is twice continuously differentiable with respect to the components of p with $\nabla_u C(V(p),p) > 0$ for $p \in B_\delta(p^0)$. This will imply that for $p \in B_\delta(p^0)$,[41]

$$(9.4) \qquad V(p) = \max_u \{u : C(u,p) \leq p^T x^0,\ u \in U\}.$$

The following inequality is valid for every $p \gg 0_N$:

$$(9.5) \qquad p^T x^0 \geq \min_x \{p^T x : F(x) \geq u^0 \equiv F(x^0)\} = C(u^0,p),$$

since x^0 is feasible for the minimization problem in (9.5), but is not necessarily optimal. Thus, since C is nondecreasing in u, u^0 is feasible for the maximization problem in (9.4), and thus, for every $p \gg 0_N$,

$$V(p) \geq u^0$$

with $V(P^0) = u^0$. Thus V does in fact attain a global minimum with respect to prices p when $p = p^0$.

The partial derivatives of $V(p)$ for $p \in B_\delta(p^0)$ can be obtained by replacing u in (9.3) by $V(p)$ and differentiating the resulting equation. For $p \in B_\delta(p^0)$, we find that

$$(9.6) \qquad \nabla_p V(p) = \left[x^0 - \nabla_p C[V(p),p]\right] / \nabla_u C[V(p),p]$$

so that when $p = p^0$, (using (9.2) as well):

$$(9.7) \qquad \nabla_p V(p^0) = 0_N.$$

Now differentiate the system of equations (9.6) with respect to p. When $p = p^0$, using (9.7) we find that:

$$(9.8) \qquad \nabla^2_{pp} V(p^0) = -\nabla^2_{pp} C(u^0,p^0)/\nabla_u C(u^0,p^0).$$

Note that $\nabla^2_{pp} V(p^0)$ is a positive semidefinite symmetric matrix, since $C(u^0,p)$ is concave and twice continuously differentiable with respect to p at $p = p^0$, and thus $\nabla^2_{pp} C(u^0,p^0)$ is a negative semidefinite symmetric matrix.

---

[41]In fact, (9.4) can be used to define $V(p)$ for all $p \gg 0_N$.

Now define the consumer's system of *overcompensated demand functions*, $d(p) \equiv [d_1(p), \ldots, d_N(p)]^T$, for $p \in B_\delta(p^0)$, by replacing u in the consumer's system of Hicksian demand functions, $x(u,p) \equiv \nabla_p C(u,p)$, by $u = V(p)$; i.e., $d(p) \equiv \nabla_p C(V(p),p)$. Now differentiate this last system of equations with respect to p in order to form the $N \times N$ matrix of overcompensated demand derivatives $[\partial d_i(p)/\partial p_j] \equiv \nabla_p d(p)$. Using (9.7) when evaluating the derivatives at $p = p^0$, we find that

$$(9.9) \qquad \nabla_p d(p^0) = \nabla_{pp}^2 C(u^0, p^0) = k^0 \nabla_{pp}^2 V(p^0)$$

where the last equality follows from (9.8) with $k^0 \equiv -1/\nabla_u C(u^0, p^0) < 0$. Thus, the matrix of derivatives of the overcompensated demand functions is precisely *equal* to the matrix of derivatives of the Hicksian demand functions (which in turn is equal to the matrix of Slutsky coefficients[42]), when both matrices are evaluated at $p = p^0$.

We turn now to the system of undercompensated demand functions. The *undercompensated indirect utility function* $u = W(p)$ is defined for $p \in B_\delta(p^0)$ to be the solution to the following equation involving u and p (if the solution exists):

$$(9.10) \qquad p^{0T} \nabla_p C(u,p) = p^{0T} x^0.$$

An economic interpretation of $W(p)$ can be obtained as follows: given a price vector $p \in B_\delta(p^0)$ and a utility level u near u^0, calculate the consumer's Hicksian demand vector $x(u,p) \equiv \nabla_p C(u,p)$. Then choose $u \equiv W(p)$ so that the resulting demand vector $x[W(p),p]$ will just be on the consumer's original budget constraint $p^{0T} \nabla_p C(u,p^0) = p^{0T} x^0$.

When $p = p^0$, (9.10) becomes $p^{0T} \nabla_p C(u,p^0) = p^{0T} x^0$, and this last equation has the unique solution $u = u^0$. (Using $p^{0T} \nabla_p C(u,p^0) = C(u,p^0)$, by Euler's Theorem on homogeneous functions, $p^{0T} x^0 = p^{0T} \nabla_p C(u^0,p^0) = C(u^0,p^0)$, $\nabla_u C(u^0,p^0) > 0$, and $C(u,p^0)$ is nondecreasing in u). Thus since $\partial[p^{0T} \nabla_p C(u^0,p^0)]/\partial u = \nabla_u C(u^0,p^0) > 0$ by (9.1), our differentiability assumptions on C plus the Implicit Function Theorem imply the existence of $u = W(p)$ satisfying (9.10) for $p \in B_\delta(p^0)$ for some $\delta > 0$.

Since the maximum utility the consumer could attain in the original budget constraint was $u^0 \equiv F(x^0) = \max_x\{F(x) : p^{0T} x \leq p^{0T} x^0, x \geq 0_N\}$, it is easy to see that $W(p) \leq u^0 = W(p^0)$ for all $p \in B_\delta(p^0)$. Thus $\max_p\{W(p) : p \in B_\delta(p^0)\} = W(p^0)$ and thus W attains at least a local maximum at $p = p^0$.[43]

The partial derivatives of $W(p)$ can be obtained by replacing u in (9.10) by $W(p)$ and differentiating the resulting equation with respect to p for $p \in B_\delta(p^0)$:

$$(9.11) \qquad p^{0T} \nabla_{pp}^2 C[W(p),p] + p^{0T} \nabla_{pu}^2 C[W(p),p] \nabla_p^T W(p) = 0_N^T.$$

[42]Recall equation (8.8) in Section 8.

[43]W will not in general be defined for all $p \gg 0_N$ whereas V will be.

When $p = p^0$, $W(p^0) = u^0$, $p^{0T}\nabla^2_{pu}C(u^0,p^0) = \nabla_u C(u^0,p^0) > 0$ and $p^{0T}\nabla^2_{pp}C(u^0,p^0) = 0_N^T$, so that (9.11) yields

$$\nabla_p W(p^0) = 0_N. \tag{9.12}$$

Now differentiate (9.11) with respect to the components of p and evaluate the resulting system of equations when $p = p^0$. Using (9.12) and the identities $p^{0T}\nabla^2_{pu}C(u^0,p^0) = \nabla_u C(u^0,p^0)$ and $\nabla_p\left[p^{0T}\nabla^2_{pp}C[W(p),p]\right] = -\nabla^2_{pp}C(u^0,p^0)$, when $p = p^0$,[44] we find that

$$\nabla^2_{pp}W(p^0) = \nabla^2_{pp}C(u^0,p^0)/\nabla_u C(u^0,p^0) = -\nabla^2_{pp}V(p) \tag{9.13}$$

where the last equality follows from (9.8). Thus the Hessian matrix of W evaluated at p^0 is a negative semidefinite symmetric matrix which is proportional to the matrix of Slutsky coefficients $\nabla^2_{pp}C(u^0,p^0)$ and is equal to minus the Hessian matrix of V evaluated at p^0.

Define the consumer's system of *undercompensated demand functions*, $D(p) \equiv [D_1(p),\ldots,D_N(p)]^T$ for $p \in B_\delta(p^0)$, by replacing u in the consumer's system of Hicksian demand functions, $x(u,p) \equiv \nabla_p C(u,p)$, by $u = W(p)$; i.e. $D(p) \equiv \nabla_p C[W(p),p]$. Now differentiate this last system of equations with respect to p in order to form the $N \times N$ matrix of undercompensated demand derivatives $[\partial D_i(p)/\partial p_j] \equiv \nabla_p D(p)$. Using (9.12), at $p = p^0$

$$\nabla_p D(p^0) = \nabla^2_{pp}C(u^0,p^0). \tag{9.14}$$

The above results establish counterparts to the Samuelson results using duality theory. We have obtained a strengthening of Samuelson's results in the sense that it is not necessary to take a symmetric mean of the over and under compensated demand systems: both systems when differentiated with respect to p yield *precisely* the consumer's matrix of Hicksian demand derivatives when $p = p^0$.

10. Empirical Applications using Cost or Indirect Utility Functions

Suppose that the technology of an industry can be characterized by a constant returns to scale production function f which has the following properties:[45]

(10.1) f is a (i) positive, (ii) linearly homogeneous, and (iii) concave function defined over the positive orthant in R^N.

[44] Since for all $p \in B_\delta(p^0)$, $p^T\nabla^2_{pp}C[W(p),p] = 0_N^T$, then $\nabla_p\left[p^T\nabla^2_{pp}C[W(p),p]\right] = 0_{N\times N} = \nabla^2_{pp}C[W(p),p] + \nabla_p\left[p^{0T}\nabla^2_{pp}C[W(p),p]\right]$ where $p^0 \equiv p$ is treated as a constant vector when differentiating the last term.

[45] f can be uniquely extended to the nonnegative orthant by using the Fenchel closure operation.

It can be shown[46] that the cost function which corresponds to f has the following form: for $u \geq 0$, $p \gg 0_N$,

$$C(u,p) \equiv \min_{x}\{p^T x : f(x) \geq u,\ x \geq 0_N\} = uc(p) \tag{10.2}$$

where $c(p) \equiv C(1,p)$ is the unit cost function and it also satisfies the three properties listed in (10.1).

The producer's system of input demand functions, $x(u,p) \equiv [x_1(u,p), \ldots, x_N(u,p)]^T$, can be obtained as the set of solutions to the programming problem (10.2) if we are given a functional form for the production function f. Thus, one method for obtaining a system of derived input demand functions that are consistent with the hypothesis of cost minimization is to postulate a (differentiable) functional form for f and then use the usual Lagrangian techniques in order to solve (10.2).

The problem with this first method for obtaining the system of input demand functions $x(u,p)$ is that it is usually very difficult to obtain an algebraic expression for $x(u,p)$ in terms of the (unknown) parameters which characterize the production function f, particularly if we assume that f is a *flexible*[47] linearly homogeneous functional form.

A second method for obtaining a system of input demand functions $x(u,p)$ makes use of Lemma 4 (Shephard's Lemma): simply postulate a functional form for the cost function $C(u,p)$ which satisfies the appropriate regularity conditions and, in addition, is differentiable with respect to input prices. Then, $x(u,p) = \nabla_p C(u,p)$ and the system of derived demand functions can be obtained by differentiating the cost function with respect to input prices.

For example, suppose that the unit cost function is defined by

$$c(p) \equiv \sum_{i=1}^{N}\sum_{j=1}^{N} b_{ij} p_i^{\frac{1}{2}} p_j^{\frac{1}{2}} \text{ with } b_{ij} = b_{ji} \geq 0 \text{ for all } i,j.$$

---

[46]See Samuelson [1953–54] and Diewert [1974a; 110–112].

[47]f is a flexible functional form if it can provide a second order (differential) approximation to an arbitrary twice continuously differentiable function f^* at a point x^*. f differentially approximates f^* at x^* iff (i) $f(x^*) = f^*(x^*)$, (ii) $\nabla f(x^*) = \nabla f^*(x^*)$, and (iii) $\nabla^2 f(x^*) = \nabla^2 f^*(x^*)$, where both f and f^* are assumed to be twice continuously differentiable at x^* (and thus the two Hessian matrices in (iii) will be symmetric). Thus a general flexible functional form f must have at least $1 + N + N(N+1)/2$ free parameters. If f and f^* are both linearly homogeneous, then $f^*(x^*) = x^{*T} f^*(x^*)$ and $\nabla^2 f^*(x^*)x^* = 0_N$, and thus a flexible linearly homogeneous functional form f need have only $N + N(N-1)/2 = N(N+1)/2$ free parameters. The term "flexible" is due to Diewert [1974a; 113] while the term "differential approximation" is due to Lau [1974; 183].

Then, if at least one $b_{ij} > 0$, the resulting function c satisfies (10.1), and the input demand functions are

$$x_i(u,p) = \sum_{j=1}^{N} b_{ij}(p_j/p_i)^{\frac{1}{2}}u; \; i = 1,2,\ldots,N. \tag{10.3}$$

Note that the system of input demand equations (10.3) is linear in the unknown parameters, and thus linear regression techniques can be used in order to estimate the b_{ij}, if we are given data on output, inputs and input prices. Note also that b_{ij} in the ith input demand equation should equal b_{ji} in the jth equation for $j \neq i$. These are the Hotelling [1932; 594], Hicks [1946; 311 and 331], Samuelson [1947; 64] *symmetry restrictions* (7.3) and we can statistically test for their validity. If some of the b_{ij} are negative, then the system of input demand equations can still be locally valid.[48] Finally, note that if $b_{ij} = 0$ for $i \neq j$, then (10.3) becomes $x_i(u,p) = b_{ii}u$, $i = 1,2,\ldots,N$, which is the system of input demand functions that corresponds to the Leontief [1941] production function, $f(x_1, x_2, \ldots, x_N) \equiv \min\{x_i/b_{ii} : i = 1,2,\ldots,N\}$. In the general case, the production function which corresponds to (10.3) is called the *generalized Leontief production* function.[49] It can also be shown that the corresponding unit cost function, $\sum_i \sum_j b_{ij} p_i^{\frac{1}{2}} p_j^{\frac{1}{2}}$, is a flexible linearly homogeneous functional form.[50]

As another example of the second method for obtaining input demand functions, consider the following *translog cost function*:

(10.4)

$$\ln C(u,p) \equiv \alpha_0 + \sum_{i=1}^{N} \alpha_i \ln p_i + \frac{1}{2}\sum_{i=1}^{N}\sum_{j=1}^{N} \gamma_{ij} \ln p_i \ln p_j + \delta_0 \ln u + \sum_{i=1}^{N} \delta_i \ln p_i \ln u + \frac{1}{2}\varepsilon_0 (\ln u)^2$$

where the parameters satisfy the following restrictions:

$$\sum_{i=1}^{N} \alpha_i = 1; \quad \gamma_{ij} = \gamma_{ji} \text{ for all } i,j;$$

$$\sum_{j=1}^{N} \gamma_{ij} = 0 \text{ for } i = 1,2,\ldots,N; \quad \text{and} \sum_{i=1}^{N} \delta_i = 0. \tag{10.5}$$

The restrictions (10.5) ensure that C defined by (10.4) is linearly homogeneous in p. The additional restrictions

$$\delta_0 = 1; \quad \delta_i = 0, \text{ for } i = 1,2,\ldots,N; \quad \text{and} \quad \varepsilon_0 = 0 \tag{10.6}$$

[48] See Blackorby and Diewert [1979] and Diewert [1974a; 113-114].
[49] See Diewert [1971a].
[50] See Diewert [1974a; 115].

ensure that $C(u,p) = uC(1,p)$ so that the corresponding production function is linearly homogeneous. Finally, with the additional restrictions $\gamma_{ij} = 0$ for all i, j and $\alpha_i \geq 0$ for $i = 1, 2, \ldots, N$, C defined by (10.4) reduces to a Cobb-Douglas cost function.

The "translog" functional form defined by (10.4) is due to Christensen, Jorgenson and Lau [1971], Griliches and Ringstad [1971] (for two inputs) and Sargan [1971; 154–146] (who calls it the *log quadratic production function*). In general, C defined by (10.4) will not satisfy the appropriate regularity conditions (e.g., conditions II in Section 3) *globally*, but it can provide a good *local* approximation to an arbitrary twice differentiable, linearly homogeneous in p, cost function;[51] i.e., the translog function form (10.4) is flexible.

The cost minimizing input demand functions $x_i(u,p)$ which (10.4) generates via Shephard's Lemma are not linear in the unknown parameters. However, it is easy to verify that the factor share functions

$$s_i(u,p) \equiv p_i x_i(u,p) / \sum_{k=1}^{N} p_k x_k(u,p) = p_i x_i(u,p)/C(u,p) = \partial \ln C(u,p)/\partial \ln p_i$$

are linear in the unknown parameters:

$$s_i(u,p) = \alpha_i + \sum_{j=1}^{N} \gamma_{ij} \ln p_j + \delta_i \ln u, \quad i = 1, \ldots, N. \tag{10.7}$$

However, since the shares sum to unity, only $N - 1$ of the N equations defined by (10.7) can be statistically independent. Moreover, notice that the parameters α_0, δ_0 and ε_0 do not appear in (10.7). However, all of the parameters can be statistically determined given data on output, inputs and input prices if we append equation (10.4) (which is also linear in the unknown parameters) to $N - 1$ of the N equations in (10.7).

The above two examples illustrate how simple it is to use the second method for generating systems of input demand functions which are consistent with the hypothesis of cost minimization.

Just as Shephard's Lemma (3.13) can be used to derive systems of cost minimizing input demand functions, Roy's Identity (4.12) can be used to derive systems of utility maximizing commodity demand functions in the context of consumer theory. For example, consider the following *translog indirect utility function*:[52] for $v \equiv p/y \gg 0_N$, define

$$G(v) \equiv \alpha_0 + \sum_{i=1}^{N} \alpha_i \ln v_i + \frac{1}{2} \sum_{i=1}^{N} \sum_{j=1}^{N} \gamma_{ij} \ln v_i \ln v_j; \ \gamma_{ij} = \gamma_{ji}. \tag{10.8}$$

[51]See Lau [1974; 186].

[52]See Jorgenson and Lau [1970] and Christensen, Jorgenson and Lau [1975]. This translog function can locally approximate any twice continuously differentiable indirect utility function. However, $G(v)$ defined by (10.8) will, in general, not satisfy conditions III globally.

Roy's Identity (4.12) applied to G defined by (10.8) yields the following system of consumer demand functions where $v \equiv (v_1, \ldots, v_N)^T = p^T/y$, $p^T \equiv (p_1, \ldots, p_N)$ is a vector of positive commodity prices, and $y > 0$ is the consumer's expenditure on the N goods:

$$(10.9) \quad x_i(p/y) = \frac{p_i^{-1} y\left(\alpha_i + \sum_{j=1}^{N} \gamma_{ij} \ln p_j - \sum_{j=1}^{N} \gamma_{ij} \ln y\right)}{\sum_{k=1}^{N} \alpha_k + \sum_{k=1}^{N} \sum_{m=1}^{N} \gamma_{km} \ln p_m - \sum_{k=1}^{N} \sum_{m=1}^{N} \gamma_{km} \ln y}, \quad i = 1, 2, \ldots, N.$$

Note that the demand functions are homogeneous of degree 0 in all of the parameters taken together. Thus, in order to identify the parameters, a normalization such as

$$(10.10) \qquad \sum_{i=1}^{N} \alpha_i = -1$$

must be appended to equations (10.9). Note also that the parameter α_0 which occurs in (10.8) cannot be identified if we have data only on consumer purchases (rentals in the case of durable goods) x, prices p, and total expenditure y. Moreover, only $N - 1$ of the N equations in (10.9) are independent and equation (10.8) cannot be added to the independent equations in (10.9) to give N independent estimating equations because the left hand side of (10.8) is the unobservable variable, utility u. Thus the econometric procedures used to estimate consumer preferences are not entirely analogous to the procedures used to estimate production functions, even though from a theoretical point of view, the duality between cost and production functions is entirely isomorphic to the duality between expenditure and utility functions.

The system of commodity demand functions defined by (10.9) is not linear in the unknown parameters and thus nonlinear regression techniques will have to be used in order to estimate econometrically the unknown parameters. We generally obtain nonlinear demand equations using Roy's Identity if we assume that G is defined by a flexible functional form.[53]

---

[53]However, if we assume that the direct utility function F is linearly homogeneous, then the corresponding indirect utility function G will be homogeneous of degree -1. An indirect utility function G which is flexible homogeneous of degree -1 can be obtained by using the translog functional form (10.8) with the additional restrictions $\alpha_0 = 0$, $\sum_{i=1}^{N} \alpha_i = -1$ and $\sum_{j=1}^{N} \gamma_{ij} = 0$ for $i = 1, \ldots, N$. In this case, the consumer's system of commodity share equations becomes $s_i \equiv p_i x_i / y = -\alpha_i - \sum_{j=1}^{N} \gamma_{ij} \ln p_j$, $i = 1, \ldots, N$, which is linear in the unknown parameters. However, the assumption of linear homogeneity (or even homotheticity) for F is highly implausible in the consumer context, since it leads to unitary income elasticities for all goods (cf. Frisch [1936; 25]).

The system of demand equations defined by (10.9) could be utilized given microeconomic data on a single utility maximizing consumer (with constant preferences) or given cross section data on a number of consumers, assuming that each utility maximizing consumer in the sample had the same preferences. However, could we legitimately apply the system (10.9) to market data; i.e., assume that x_i represented total market demand for commodity i divided by the number of independent consuming units, p_i is the price of commodity i, and y is total market expenditure on all goods divided by the number of consumer units? The answer is no in general.[54] However, if we have information on the *distribution* $\phi(y)$ of expenditure y by the different households in the market and we are willing to assume that each household has the same tastes, then the market demand functions X_i can be obtained by integrating over the individual demand functions $x_i(p/y)$:

$$(10.11) \qquad X_i(p) = N\int_0^{\infty} x_i(p/y)\phi(y)dy, \quad i = 1, \ldots, N,$$

where N is the number of households in the market and $\int_0^{\infty} \phi(y)dy = 1$. The integrations in (10.11) can readily be performed using the $x_i(p/y)$ defined by (10.9) if we impose the following normalizations on the parameters of the translog indirect utility function defined by (10.8): (i) $\alpha_0 = 0$; (ii) $\sum_{i=1}^{N}\alpha_i = -1$; and (iii) $\sum_{i=1}^{N}\sum_{j=1}^{N}\gamma_{ij} = 0$. The effect of these three normalizations is to make G homogeneous of degree -1 along the ray of equal prices; i.e., $G(\lambda 1_N) = \lambda^{-1}G(1_N)$ for all $\lambda > 0$, and this in turn is simply a harmless (from a theoretical point of view but not necessarily from an econometric point of view) cardinalization of utility so that utility is proportional to income when the prices the consumer faces are all equal. This approach for obtaining systems of market demand functions consistent with microeconomic theory has been pursued by Diewert [1974a; 127–130] and Berndt, Darrough and Diewert [1977].

There is a simpler method for obtaining systems of market demand functions consistent with individual utility maximizing behavior which is due to Gorman [1953]: assume that each household's preferences can be represented by a cost function of the form

$$(10.12) \qquad C(u,p) = b(p) + uc(p)$$

where b and c are unit cost functions which satisfy conditions (10.1), $p \gg 0_N$ and $c(p)u \geq y - b(p) \geq 0$ where y is household expenditure. Blackorby,

[54]The reader is referred to the considerable body of literature on the implications of microeconomic theory for systems of market (excess) demand functions, which is reviewed by Shafer and Sonnenschein [1980] and Diewert [1976c].

Boyce and Russell [1978] call a functional form for C which has the structure (10.12) a *Gorman polar form*. If $y - b(p) \geq 0$, the indirect utility function which corresponds to (10.12), is $G(v) \equiv [1/c(v)] - [b(v)/c(v)] = [y/c(p)] - [b(p)/c(p)]$ where $v \equiv p/y$, then Roy's Identity (4.12) yields the following system of individual household demand functions if the unit cost functions b and c are differentiable:

$$(10.13) \qquad x(p/y) = \nabla_p b(p) + [c(p)]^{-1}[y - b(p)]\nabla_p c(p); \;\; y \geq b(p).$$

The interesting thing about the system of consumer demand functions defined by (10.13) is that they are *linear* in the household's income or expenditure y. Thus, if every household in the market under consideration has the same preferences which are dual to C defined by (10.12) and each household has income $y \geq b(p)$, then the system of market demand functions $X(p)$ defined by (10.11) is independent of the distribution of income; in fact

$$(10.14) \qquad X(p)/N^* = \nabla_p b(p) + [c(p)]^{-1}[y^* - b(p)]\nabla_0 c(p)$$

where $X(p)/N^*$ is the per capita market demand vector and $y^* \equiv \int y\phi(y)dy$ is average or per capital expenditure. Comparing (10.14) with (10.13), we see that the per capita market demand system has the same functional form as the individual demand vector for a single decision making unit. The advantage of this approach over the previous approach is that it does not require information on the distribution of expenditure: all that is required is information on market expenditure by commodity, commodity (rental) prices, and the number of consumers or households.[55]

Several flexible functional forms for cost functions have been estimated empirically, using Shephard's Lemma in order to derive systems of input demand functions: see Parks [1971], Denny [1972][1974], Binswanger [1974], Hudson and Jorgenson [1974], Woodland [1975], Berndt and Wood [1975], Burgess [1974] [1975], and Khaled [1978]. Khaled also develops a very general class of functional forms which contains most of the other commonly used functional forms as special cases.

There are also many applications of the above theory to the problem of estimating consumer preferences. For empirical examples, see Lau and Mitchell [1970], Diewert [1974d], Christensen, Jorgenson and Lau [1975], Jorgenson and Lau [1975], Boyce [1975], Boyce and Primont [1976], Christensen and Manser [1977], Darrough [1977], Blackorby, Boyce and Russell [1978], Howe, Pollak and Wales [1979], Donovan [1977] and Berndt, Darrough and Diewert [1977].[56]

---

[55]For other generalizations of the Gorman polar form (10.12) which have useful aggregation properties, see Gorman [1959; 476], Muellbauer [1975][1976] and Lau [1977a][1977b]. Diewert [1978a] shows that functional forms of the type (10.12) are flexible. See also Lau [1977c].

[56]Lau [1978b] considers the problems of testing for or imposing the various monotonicity and curvature conditions on the cost of indirect utility functions. On the issue of how flexible are flexible functional forms, see Wales [1977] and Byron [1977].

11. Profit Functions

Up to now, we have considered the case of a firm which produces only a single output, using many inputs. However, in the real world most firms produce a variety of outputs, so that it is now necessary to consider the problems of modelling a multiple output, multiple input firm.

For econometric applications, it is convenient to introduce the concept of a firm's *variable profit function* $\Pi(p, x)$: it simply denotes the maximum revenue minus variable input expenditures that the firm can obtain given that it faces prices $p \gg 0_I$ for variable inputs and outputs and given that another vector of inputs $x \geq 0_J$ is held fixed. We denote the variable inputs and outputs by the I dimensional vector $u \equiv (u_1, u_2, \ldots, u_I)$, the fixed inputs by the J dimensional vector $-x \equiv (-x_1, \ldots, -x_J)$, and the set of all feasible combinations of inputs and outputs is denoted by T, the firm's *production possibilities set.* Outputs are denoted by positive numbers and inputs are denoted by negative numbers, so if $u_i > 0$, then the ith variable good is an output produced by the firm. Formally, we define Π for $p \gg 0_I$ and $-x \leq 0_J$ by

$$\Pi(p, x) \equiv \max_u \{p^T u : (u, -x) \in T\}. \tag{11.1}$$

If T is a closed nonempty, convex cone in Euclidean $I + J$ dimensional space with the additional properties: (i) if $(u, -x) \in T$, then $x \geq 0_J$ (the last J goods are always inputs), (ii) if $(u', -x') \in T$, $u'' \leq u'$ and $-x'' \leq -x'$, then $(u'', -x'') \in T$ (free disposal) and (iii) if $(u, -x) \in T$, then the components of u are bounded from above (bounded outputs for bounded fixed inputs), then Π has the following properties: (i) $\Pi(p, x)$ is a nonnegative real valued function defined for every $p \gg 0_I$ and $x \geq 0_J$ such that $\Pi(p, x) \leq p^T b(x)$ for every $p \gg 0_J$; (ii) for every $x \geq 0_J$, $\Pi(p, x)$ is (positively) linearly homogeneous, convex and continuous in p; and (iii) for every $p \gg 0_I$, $\Pi(p, x)$ is (positively) linearly homogeneous, concave, continuous and nondecreasing in x. Moreover, it can be shown[57] that T can be constructed using Π as follows:

$$T = \{(u, -x) : p^T u \leq \Pi(p, x), \text{ for every } p \gg 0_I;\ x \geq 0_J\}. \tag{11.2}$$

Thus, there is a duality between production possibilities sets T and variable profit functions Π satisfying the above regularity conditions. Moreover, in a manner which is analogous to the proof of Shephard's Lemma (3.13) and Roy's Identity (4.12), the following result can be proven:

[57] See Gorman [1968b], McFadden [1966], or Diewert [1973a].

HOTELLING'S LEMMA. [1932; 594]: *If a variable profit function* Π *satisfies the regularity conditions below (11.1) and is in addition differentiable with respect to the variable quantity prices at* $p^* \gg 0_I$ *and* $x^* \geq 0_J$*, then* $\partial\Pi(p^*, x^*)/\partial p_i = u_i(p^*, x^*)$ *for* $i = 1, 2, \ldots, I$*, where* $u_i(p^*, x^*)$ *is the profit maximizing amount of net output* i *(of input* i *if* $\partial\Pi(p^*, x^*)/\partial p_i < 0$*) given that the firm faces the vector of variable prices* p^* *and has the vector* x^* *of fixed inputs at its disposal.*

Hotelling's Lemma can be used in order to derive systems of variable output supply and input demand functions. We need only postulate a functional form for $\Pi(p, x)$ which is consistent with the appropriate regularity conditions for Π and is differentiable with respect to the components of p. For example, consider the translog variable profit function Π defined as:

$$\begin{aligned} \ln \Pi(p, x) \equiv \alpha_0 &+ \sum_{i=1}^{I} \alpha_i \ln p_i + \frac{1}{2} \sum_{i=1}^{I} \sum_{h=1}^{I} \gamma_{ih} \ln p_i \ln p_h \\ &+ \sum_{i=1}^{I} \sum_{j=1}^{J} \delta_{ij} \ln p_i \ln x_j + \sum_{j=1}^{J} \beta_j \ln x_j \\ &+ \frac{1}{2} \sum_{j=1}^{J} \sum_{k=1}^{J} \phi_{jk} \ln x_j \ln x_k \end{aligned} \tag{11.3}$$

where $\gamma_{ih} = \gamma_{hi}$ and $\phi_{jk} = \phi_{kj}$. It is easy to see that Π defined by (11.3) is homogeneous of degree one in p if and only if

$$\begin{aligned} &\sum_{i=1}^{I} \alpha_i = 1; \quad \sum_{i=1}^{I} \delta_{ij} = 0 \text{ for } j = 1, \ldots, J; \\ &\sum_{h=1}^{I} \gamma_{ih} = 0 \text{ for } i = 1, \ldots, I. \end{aligned} \tag{11.4}$$

Similarly, $\Pi(p, x)$ is homogeneous of degree one in x[58] if and only if

$$\begin{aligned} &\sum_{j=1}^{J} \beta_j = 1; \quad \sum_{j=1}^{J} \delta_{ij} = 0 \text{ for } i = 1, \ldots, I; \\ &\sum_{k=1}^{J} \phi_{jk} = 0 \text{ for } j = 1, \ldots, J. \end{aligned} \tag{11.5}$$

If $\Pi(p, x) > 0$, define the ith variable net supply share by $s_i(p, x) \equiv p_i u_i(p, x)/\Pi(p, x)$. Hotelling's Lemma applied to the translog variable profit

[58]If we drop the assumption that the production possibilities set T be a cone (so that we no longer assume constant returns to scale in all inputs and outputs), then $\Pi(p, x)$ does not have to be homogeneous of degree one in x. Thus the restrictions (11.5) can be used to test whether T is a cone or not. If we drop the assumption that T be convex, then $\Pi(p, x)$ need not be concave (or even quasiconcave) in x.

function defined by (11.3) yields the following system of net supply share functions:

$$s_i(p,x) = \alpha_i + \sum_{h=1}^{I} \gamma_{ih} \ln p_h + \sum_{j=1}^{J} \delta_{ij} \ln x_j;\ i = 1,\ldots,I. \tag{11.6}$$

Since the shares sum to unity, only $I-1$ of the equations (11.6) are independent. $I-1$ of equations (11.6) plus equation (11.3) can be used in order to estimate the parameters of the translog variable profit function. Note that these equations are linear in the unknown parameters as are the restrictions (11.4) and (11.5) so that modifications of linear regression techniques can be used.

Alternative functional forms for variable profit functions have been suggested by McFadden [1978b], Diewert [1973a] and Lau [1974]. Empirical applications have been made by Kohli [1978], Woodland [1977c], Harris and Appelbaum [1977], and Epstein [1977].

A concept which is closely related to the variable profit function notion, is the concept of a *joint cost function*, $C(u,w) \equiv \min_x\{w^T x : (u,-x) \in T\}$, where T is the firm's production possibilities set as before, and $w \gg 0_J$ is a vector of positive input prices. As usual, if $C(u,w)$ is differentiable with respect to input prices w (and satisfies the appropriate regularity conditions), then Shephard's Lemma can be used in order to derive the producer's system of cost minimizing input demand functions $x(u,w)$; i.e., we have

$$x(u,w) = \nabla_w C(u,w). \tag{11.7}$$

Joint cost functions have been empirically estimated by Burgess [1976a] (who utilized a functional form suggested by Hall [1973]), Brown, Caves and Christensen [1979], and Christensen and Greene [1976] (who utilized a translog functional form for C(u,w) analogous to the translog variable profit function defined by (11.3)).

Historical Notes

Samuelson [1953–54; 20] introduced the concept of the variable profit function[59] and stated some of its properties. Gorman [1968b] and McFadden [1966] [1978a] established duality theorems between a production possibilities set satisfying various regularity conditions and the corresponding variable profit function.[60] Alternative duality theorems are due to Shephard [1970], Diewert

[59]Samuelson called it the national product function.

[60]Gorman uses the term "gross profit function" and McFadden uses the term "restricted profit function" to describe $\Pi(p,x)$.

[1973a] [1974b], Sakai [1974] and Lau [1976]. For the special case of a single fixed input, see Shephard [1970; 248–250] or Diewert [1974b].

McFadden [1966] [1978a] introduced the joint cost function, stated its properties, and proved formal duality theorems between it and the firm's production possibilities set T, as did Shephard [1970] and Sakai [1974].

There are also very simple duality theorems between production possibilities sets and *transformation functions*, which give the maximum amount of one output that the firm can produce (or the minimum amount of input required) given fixed amounts of the remaining inputs and outputs. For examples of these theorems, see Diewert [1973a], Jorgenson and Lau [1974a] [1974b], and Lau [1976].

As usual, Hotelling's Lemma can be generalized to cover the case of a nondifferentiable variable profit function: the gradient of Π with respect to p is replaced by the set of subgradients. This generalization was first noticed by Gorman [1968b; 150–151] and McFadden [1966; 11] and repeated by Diewert [1973a; 313] and Lau [1976; 142].

If $\Pi(p, x)$ is differentiable with respect to the components of the vector of fixed inputs, then $w_j \equiv \partial\Pi(p, x)/\partial x_j$ can be interpreted as the worth to the firm of a marginal unit of the jth fixed input; i.e., it is the "shadow price" for the jth input (cf. Lau [1976; 142]). Moreover, if the firm faces the vector of rental prices $w \gg 0_J$ for the "fixed" inputs, and during some period the "fixed" inputs can be varied, then if the firm minimizes the cost of producing a given amount of variable profits we will have (cf. Diewert [1974a; 140])

$$(11.8) \qquad w = \nabla_x \Pi(p, x)$$

and these relations can also be used in econometric applications.

The translog variable profit was independently suggested by Russell and Boyce [1974] and Diewert [1974a; 139]. Of course, it is a straightforward modification of the translog functional form due to Christensen, Jorgenson and Lau [1971], and Sargan [1971].

The comparative statics properties of $\Pi(p, x)$ or $C(u, w)$ have been developed by Samuelson [1953–54], McFadden [1966] [1978a], Diewert [1974a; 142–146], and Sakai [1974].

In international trade theory, it is common to assume the existence of sectoral production functions, fixed domestic resources x, and fixed prices of internationally traded goods p. If we now attempt to maximize the net value of internationally traded goods produced by the economy, we obtain the economy's variable profit function, $\Pi(p, x)$, or Samuelson's [1953–54] *national product function*. If the sectoral production functions are subject to constant returns to scale, $\Pi(p, x)$ will have all the usual properties mentioned above. However, the existence of sectoral technologies will imply additional comparative statics restrictions on the national product function π: see Chipman [1966], [1972],

[1974b], Samuelson [1966], Ethier [1974], Woodland [1977a][1977b], Diewert and Woodland [1977], and Jones and Scheinkman [1977] and the many references included in these papers.

Finally, note that the properties of $\Pi(p, x)$ with respect to x are precisely the properties that a neoclassical production function possesses. If x is a vector of primary inputs, then $\Pi(p, x)$ can be interpreted as a value added function. If the prices p vary (approximately) in proportion over time, then $\Pi(p, x)$ can be deflated by the common price trend and the resulting real value added function has all of the properties of a neoclassical production function; see Khang [1971], Bruno [1978] and Diewert [1978a][1980].

12. Duality and Noncompetitive Approaches to Microeconomic Theory

Up to now, we have assumed that producers and consumers take prices as given and optimize with respect to the quantity variables they control. We indicate in this section how duality theory can be utilized even if there is monopsonistic or monopolistic behavior on the part of consumers or producers. We will not attempt to be comprehensive but will illustrate the techniques involved by means of our four approaches to modeling nonprice taking behavior.

Approach 1: The Monopoly Problem

Suppose that a monopolist produces output x_0 by means of the production function $F(x)$, where $x \geq 0_N$ is a vector of variable inputs. Suppose, further, that he faces the (inverse) demand function $p_0 = wD(x_0)$; i.e., $p_0 \geq 0$ is the price at which he can sell $x_0 > 0$ units of output, D is a continuous positive function of x_0, and the variable $w > 0$ represents the influence on demand of "other variables". That is to say, if the monopolist is selling to consumers, w might equal disposal income for the period under consideration; if the monopolist is selling to producers, w might be a linearly homogeneous function of the prices that those other producers face.[61] Finally, suppose that the monopolist behaves competitively on input markets, taking as given the vector $p \gg 0_N$ of input prices. The monopolist's profit maximization problem

[61]If nonquantity variables do not influence the inverse demand function that the monopolist faces for the periods under consideration, then w can be set equal to 1 in each period.

may be written as

$$\max_{p_0,x_0,x}\{p_0x_0 - p^Tx : x_0 = F(x),\ p_0 = wD(x_0),\ x \geq 0_N\}$$
$$= \max_x\{wD[F(x)]F(x) - p^Tx : x \geq 0_N\}$$
$$= \max_x\{wF^*(x) - p^Tx : x \geq 0_N\}$$
$$(12.1)\qquad \equiv \Pi^*(w,p)$$

where $F^*(x) \equiv D[F(x)]F(x) = p_0x_0/w$ is the deflated (by w) revenue function or pseudo production function and Π^* is the corresponding pseudo profit function (recall Section 11) which corresponds to F^*.[62] Notice that w plays the role of a price of $F^*(x)$. If F^* is a concave function, then $\Pi^*(1,p/w)$ will be the conjugate function to F^* (recall the Samuelson [1960], Lau [1969][1978a], and Jorgenson and Lau [1974a][1974b] conjugacy approach to duality theory) and Π^* will be dual to F^* (i.e., F^* can be recovered from Π^*). Even if F^* is not concave, if the maximum in (12.1) exists over the relevant range of (w,p) prices, then Π^* can be used to represent the relevant part of F^* (i.e., the free disposal convex hull of F^* can be recovered from Π^*). Moreover if Π^* is differentiable at (w^*,p^*) and w_0^*, p_0^*, x^* solve (12.1), then Hotelling's Lemma implies

$$(12.2)\qquad u_0^* \equiv p_0^*x_0^*/w^* = \nabla_w\Pi^*(w^*,p^*) \text{ and } -x^* = \nabla_p\Pi^*(w^*,p^*).$$

Moreover, if Π^* is twice continuously differentiable at (w^*,p^*), then we can deduce the usual comparative statics results on the derivatives of the deflated sales function $u_0(w^*,p^*) \equiv \nabla_w\Pi^*(w^*,p^*)$ and the input demand functions $-x(w^*,p^*) \equiv \nabla_p\Pi^*(w^*,p^*)$: namely $\nabla^2\Pi^*(w^*,p^*)$ is a positive semidefinite symmetric matrix and $(w^*,p^{*T})\nabla^2\Pi^*(w^*,p^*) = 0_{N+1}^T$.

Equations (12.2) can be used in order to estimate econometrically the parameters of Π^* and hence indirectly of F^*: simply postulate a functional form for Π^*, differentiate Π^*, and then fit (12.2), given a time series of observations on p_0, p, w, x_0 and x. The drawbacks to this method are: (i) we cannot disentangle D from F; (ii) we cannot test whether the producer is in fact behaving competitively on the output market; and (iii) we cannot use our estimated equations to predict output x_0 or selling price p_0 separately.

Approach 2: The Monopsony Problem

Consider the problem of a consumer maximizing a utility function $F(x)$ satisfying conditions I but now we no longer assume that the consumer faces

[62] Note that we have suppressed mention of any fixed inputs. We assume sufficient regularity on F and D so that the maximum in (12.1) exists.

fixed prices for the commodities he purchases, but rather he is able to monopsonistically exploit one or more of the suppliers that he faces. Then in period r, he faces a nonlinear budget constraint of the form $h_r(x) = 0$ where $x \geq 0_N$ is his vector of purchases (or rentals). Let $x^r > 0_N$ be a solution to the period r constrained utility maximization problem, so that

$$(12.3) \qquad \max_x\{F(x) : h_r(x) = 0,\ x \geq 0_N\} = F(x^r); \quad r = 1, \ldots, T.$$

Suppose, further, that the rth budget constraint function h_r is differentiable at x^r with $\nabla_x h(x^r) \gg 0_N$ for each r. Then we may linearize the rth budget constraint around $x = x^r$ by taking a first order Taylor series expansion. The linearized rth budget constraint is $h_r(x^r) + [\nabla_x h_r(x^r)]^T(x - x^r) = 0$ or $[\nabla h_r(x^r)]^T(x - x^r) = 0$ since $h_r(x^r) = 0$ using (12.3). It is easy to see that the utility surface $\{x : F(x) = F(x^r), x \geq 0_N\}$ is tangent not only to the original nonlinear budget surface $\{x : h_r(x) = 0, x \geq 0_N\}$ at $x = x^r$, but also to the linearized budget constraint surface $\{x : [\nabla h_r(x^r)]^T(x - x^r) = 0, x \geq 0_N\}$ at $x = x^r$. Since we assume F is quasiconcave, the set $\{x : F(x) \geq F(x^r), x \geq 0_N\}$ is convex and the linearized budget constraint is a supporting hyperplane to this set; i.e.,

$$(12.4) \qquad \max_x\{F(x) : p^{rT}x \leq p^{rT}x^r, x \geq 0_N\} = F(x^r), \quad r = 1, \ldots, T$$

where $p^r \equiv \nabla h_r(x^r)$ for $r = 1, 2, \ldots, T$. But now (12.4) is just a series of aggregator maximization problems of the type we have studied in Section 4 (the rth vector of normalized prices is defined as $v^r \equiv p^r/p^{rT}x^r$) and the estimation techniques outlined in Section 10 above (recall equations (10.9) for example) can be used in order to estimate the parameters of the indirect utility function dual to F.

When we were dealing with linear budget constraints in Section 4, it was irrelevant whether F was quasiconcave or not (recall our discussion and diagram in Section 2). However, now we require the additional assumption that F be quasiconcave in order to rigorously justify the replacement of (12.3) by (12.4). Note also that in order to implement the above procedure, it is necessary to know the vector of derivatives $\nabla_x h_r(x^r)$ for each r; i.e., we have to know the derivatives of the supply functions that the consumer is "exploiting" each period — information which was not required in approach 1.

The monopsony model presented here is actually much broader than the classical model of monopsonistic exploitation: prices that the consumer faces can vary with the quantity purchased for a large number of reasons, including search and transactions costs, quantity discounts, and the existence of progressive taxes on labor earnings. Most tax systems lead to budget constraints with "kinks" or nondifferentiable points. This does not cause any problems with the above procedure unless the consumer's observed consumption-leisure choice falls precisely on a kink in his budget constraint.[63]

[63]See Wales [1973] and Wales and Woodland [1976][1977][1979] for econometric treatments of this last problem.

Approach 3: The Monopoly Problem Revisited

Consider again the monopoly problem outlined above. Suppose $x_0^r > 0$, $x^r > 0_N$ is a solution to the period r monopoly profit maximization problem which can be rewritten as

$$(12.5)\quad \max_{x_0,x}\{w^r D(x_0)x_0 - p^{rT}x : x_0 = F(x), x \geq 0_N\}$$
$$= w^r D(x_0^r)x_0^r - p^{rT}x^r; \quad r = 1, 2, \ldots, T,$$

where $p_0^r \equiv w^r D(x_0^r) > 0$ is the observed selling price of the output during period r, $w^r D(x_0)$ is the period r inverse demand function, and $p^r \gg 0_N$ is the period r input price vector. If the production function F is continuous and concave (so that the production possibility set $\{(x_0, x) : x_0 \leq F(x), x \geq 0_N\}$ is closed and convex) and if the inverse demand function D is differentiable at x_0^r for $r = 1, \ldots, T$, then the objective function for the rth maximization problem in (12.5) can be linearized around (x_0^r, x^r) and this linearized objective function will be tangent to the production surface $x_0 = F(x)$ at (x_0^r, x^r). Thus,

$$(12.6)\quad \max_{x_0,x}\{\tilde{p}_0^r x_0 - p^{rT}x : x_0 = F(x), x \geq 0_N\} \equiv \Pi(\tilde{p}_0^r, p^r)$$
$$= \tilde{p}_0^r x_0^r - p^{rT}x^r, \quad r = 1, \ldots, T,$$

where $\tilde{p}_0^r \equiv w^r D(x_0^r) + w^r D'(x_0^r)x_0^r = p_0^r + w^r D'(x_0^r)x_0^r > 0$ is the period r *shadow* or *marginal* price of output ($\tilde{p}_0^r < p_0^r$ if $w^r > 0$ and $D'(x_0^r) < 0$) and Π is the firm's *true* profit function which is dual to the production function F (recall Π^* defined in approach 1 was dual to the convex hull of $D[F(x)]F(x) \equiv F^*(x)$). Thus, the true nonlinear monopolistic profit maximization problems (12.6) which have the usual structure once the appropriate marginal output prices $\tilde{p}_0^r$ have been calculated so that the usual econometric techniques can be applied (recall equations (11.6) in Section 11).[64]

Comparing approach 3 with approach 1, it can be seen that approach 3 requires the extra assumption that the production function be concave (convex technology) and requires additional information; i.e., a knowledge of the slope of the demand curve the monopolist is exploiting is required for each period.

It is easy to see how this approach can be generalized to a multiproduct firm which simultaneously exploits several output and input markets: all that is required is the assumption of a convex technology and a (local) knowledge of the demand and supply curves that the firm is exploiting so that the appropriate shadow prices can be calculated.

Of course, the above techniques can also be used in situations where the firm is not behaving monopolistically or monopsonistically in an exploitive sense, but merely faces prices for its outputs or inputs that depend on the quantity sold or purchased for any number of reasons, including transactions costs or quantity discounts.

---

[64]The notation has been changed and we are now holding fixed inputs fixed for all r, so that we can suppress mention of these fixed inputs in (12.5).

Approach 4: The Monopoly Problem Once Again

Suppose now that the production function satisfies conditions I and, as usual, we suppose that $x_0^r > 0$, $x^r > 0_N$ is the solution to the period r monopolistic profit maximization problem (12.5), which we rewrite as

$$\text{(12.7)} \quad \max_{x_0}\{w^r D(x_0)x_0 - C(x_0, p^r) : x_0 \geq 0\}$$
$$= w^r D(x_0^r)x_0^r - p^{rT}x^r, \qquad r = 1, \ldots, T$$

where C is the cost function dual to F. If the inverse demand function D is differentiable at $x_0^r > 0$ and $\partial C(x_0^r, p^r)/\partial x_0$ exists, then the first order conditions for the rth maximization problem in (12.7) yield the condition $w^r D(x_0^r) + w^r D(x_0^r)x_0^r - \partial C(x_0^r, p^r)/\partial x_0 = 0$ or, recalling that $p_0^r \equiv w^r D(x_0^r)$ is the observed selling price of output in period r,

$$\text{(12.8)} \qquad p_0^r = -w^r D(x_0^r)x_0^r + \partial C(x_0^r, p^r)/\partial x_0, \quad r = 1, \ldots, T.$$

If the cost function C is differentiable with respect to input prices at (x_0^r, p^r) for each r, then Shephard's Lemma implies the additional equations

$$\text{(12.9)} \qquad x^r = \nabla_p C(x_0^r, p^r), \qquad r = 1, \ldots, T.$$

Suppose that the part of the inverse demand function that depends on x_0, $D(x_0)$, can be adequately approximated over the relevant x_0 range by the following function:

$$\text{(12.10)} \qquad D(x_0) \equiv \alpha - \beta \ln x_0$$

where $\alpha > 0$, $\beta \geq 0$ are constants. Substitution of (12.10) into (12.8) yields the equations

$$\text{(12.11)} \qquad p_0^r = w^r \beta + \partial C(x_0^r, p^r)/\partial x_0, \qquad r = 1, \ldots, T.$$

Given the observable price and quantity decisions of the firm, p_0^r, p^r, x_0^r, x^r and data on w^r (we can assume $w^r \equiv 1$ if this is appropriate), the system of equations (12.9) and (12.11) can be jointly econometrically estimated once we assume a differentiable functional form for the cost function $C(x_0, p)$. Note that if $\beta = 0$ in equations (12.11), then the producer is behaving competitively, selling output at a price p_0^r equal to marginal cost, $\partial C(x_0^r, p^r)/\partial x_0$. Equations (12.11) are also consistent with the producer behaving like a "naive" markup monopolist (depending on what w^r is). Thus, we now have the basis for a statistical test of market structure: (i) if $\beta = 0$, then the producer's behavior is consistent with competitive price taking behavior, (ii) if $\beta > 0$ and $\beta w^r/p_0^r < 1$

for $r = 1, 2, \ldots, T$, then we have consistency with classical monopolistic behavior,[65] (iii) if $\beta > 0$ but $\beta w^r/p_0^r \geq 1$ for some r, then we have consistency with markup monopolistic behavior, and (iv) if $\beta < 0$, then we have inconsistency with all three of the above types of behavior.[66]

This approach offers several advantages over the previous approaches: (i) we can now statistically test for competitive behavior, (ii) informational requirements are low — we do not require exogenous information on the elasticity of demand (this information is endogenously generated), (iii) we do not have to assume that the production function F is concave so that the model is consistent with an increasing returns to scale production function, and finally, (iv) the procedure is particularly simple — just insert the term βw^r into the competitive equation, price equals marginal cost.

Historical Notes

Approach 1 is essentially due to Lau [1974; 193–194] [1978][67] but it has its roots in Hotelling [1932; 609]. Approach 2 is in Diewert [1971b] but it has its roots in the work of Frisch [1936; 14–15]. Approach 3 (which is isomorphic to approach 2) is outlined in Diewert [1974a; 155]. Approach 4 is due to Appelbaum [1975], who makes somewhat different assumptions on the functional form of the inverse demand function.[68] Appelbaum [1975][1979] also indicates how his approach can be extended to several monopolistically supplied outputs or monopsonistically demanded inputs and he presents an empirical example based on the U.S. crude petroleum and natural gas industry. Another empirical example of his technique based on Canada–U.S. trade is in Appelbaum and Kohli [1979]. Approach 4 has also been applied by Schworm [1980] in the context of investment theory where the price of investment goods purchased by a firm depends on the quantity purchased.

---

[65]Using (12.10) and $p_0^r \equiv w^r D(x_0^r)$, we find that the first order conditions (12.8) translate into $w^r D(x_0^r)\left[1 + [D(x_0^r)x_0^r/D(x_0^r)]\right] = p_0^r[1 - (\beta w^r/p_0^r)] = \partial C(x_0^r, p^r)/\partial x_0 > 0$ or $[1 - (\beta w^r/p_0^r)] = (p_0^r)^{-1}\partial C/\partial x_0 > 0$ which implies $\beta w^r/p_0^r < 1$. The second order necessary conditions for (12.7) require $-\beta \leq (x_0^r/w^r)\partial^2 C(x_0^r, p^r)/\partial x_0^2$ which will be satisfied if $\beta \geq 0$ and $\partial^2 C(x_0^r, p^r)/\partial x_0^2 \geq 0$ (nondecreasing marginal costs or nonincreasing returns to scale).

[66]Of course, these tests are conditional on the assumed functional form for C, the assumed functional form for the inverse demand function $w_f D(x_0)$ where D is defined by (12.10), and the assumption of price taking behavior on input markets.

[67]Lau uses a normalized profit function and does not assume that $p_0 = wD(x_0)$, but simply assumes that $p_0 = D(x_0)$.

[68]He models more explicitly the demand function that the producer is exploiting.

13. Conclusion

We have attempted to give a fairly comprehensive treatment of the foundations of the duality approach to microeconomic theory in Sections 2–6 of this chapter. In Sections 7 and 8 we showed how duality theory could be used in order to derive the usual comparative statics theorems for producer and consumer theory, while in Section 9 some additional partial equilibrium comparative statics theorems were derived. In Sections 10 and 11, we showed how duality theory has been used as an aid in the econometric estimation of preferences and technology. Finally, in Section 12, we indicated how duality theory could be applied in certain noncompetitive situations.

The number of papers using duality theory during the last decade is so large that, unfortunately, we are unable to review (or even reference) them. Additional topics and references can be found in my earlier survey paper (Diewert [1974a]) and the comments on it (Jacobsen [1974], Lau [1974] and Shephard [1974]) as well as in Fuss and McFadden [1978] which provides a comprehensive treatment of the duality approach to production theory.

We have mentioned the aggregation over consumers problem in Section 10 above but we have not mentioned the corresponding aggregation over producers problem: for results and references to this literature, see Hotelling [1935; 67–70], Gorman [1968b], Sato [1975] and Diewert [1980; Part III].

Although we have used duality theory to derive several partial equilibrium comparative statics theorems, we have not mentioned the corresponding general equilibrium literature: see Jones [1965][1972], Diewert [1974e][1974f][1978d], Epstein [1974], Woodland [1974] and Burgess [1976b] for various applications. The related literature on optimal taxation often makes use of duality theory: for references to this literature, see Mirrlees [1981], and Deaton [1979].

Finally, some recent references that utilize duality theory in the context of continuous time optimization problems are Lau [1974; 190–193], Appelbaum [1975], Cooper and McLaren [1977], Epstein [1978] [1981b], McLaren and Cooper [1980], and Schworm [1980].

Essays in Index Number Theory, Volume I
W.E. Diewert and A.O. Nakamura (Editors)

Chapter 7

THE ECONOMIC THEORY OF INDEX NUMBERS: A SURVEY*

W.E. Diewert

1. Introduction

The literature on index numbers is so vast that we can cover only a small fraction of it in this chapter. Frisch [1936] distinguishes three approaches to index number theory: (i) 'statistical' approaches, (ii) the test approach, and (iii) the functional approach, which Wold [1953; 135] calls the preference field approach and Samuelson and Swamy [1974; 573] call the economic theory of index numbers. We shall mainly cover the essentials of the third approach. In the following two sections, we define the different economic index number concepts that have been suggested in the literature and develop various numerical bounds. Then in Section 4, we briefly survey some of the other approaches to index number theory. In Section 5, we relate various functional forms for utility or production functions to various index number formulae. In Section 6, we develop the link between 'flexible' functional forms and 'superlative' index number formulae. The final section offers a few historical notes and some comments on some related topics such as the measurement of consumer surplus and the Divisia index.

2. Price Indexes and the Konüs Cost of Living Index

We assume that a consumer is maximizing a utility function $F(x)$ subject to the expenditure constraint $p^T x \equiv \sum_{i=1}^N p_i x_i \leq y$ where $x \equiv (x_1, \ldots, x_N)^T \geq 0_N$ is a

*First published in *Essays in the Theory and Measurement of Consumer Behaviour in Honour of Sir Richard Stone*, edited by A. Deaton, London: Cambridge University Press, 1981, pp. 163–208. The financial support of the Canada Council is gratefully acknowledged as are the helpful comments of R.C. Allen, C. Blackorby, and A. Deaton, who are not responsible for the remaining shortcomings of this chapter. It is a pleasure to dedicate this chapter to Professor Stone, since the author first learned of the existence of the index number problem as a graduate student at Berkeley by reading some of Professor Stone's work.

nonnegative vector of commodity rentals, $p \equiv (p_1, \ldots, p_N)^T \gg 0_N$ is a positive vector of commodity prices[1] and $y > 0$ is expenditure on the N commodities. We could also assume that a producer is maximizing a production function $F(x)$ subject to the expenditure constraint $p^T x \leq y$ where $x \geq 0_N$ is now an input vector, $p \gg 0_N$ is an input price vector and $y > 0$ is expenditure on the inputs. In order to cover both the consumer and producer theory applications, we shall call the utility or production function F an *aggregator function* in what follows.

The consumer's (or producer's) aggregator maximization problem can be decomposed into two stages: in the first stage, the consumer (or producer) attempts to minimize the cost of achieving a given utility (or output) level, and, in the second stage, he chooses the maximal utility (or output) level that is just consistent with his budget constraint.

The solution to the first stage problem defines the consumer's (or producer's) cost function C:

(1) $$C(u,p) \equiv \min_x \{p^T x : F(x) \geq u, \ x \geq 0_N\}$$

The cost function C turns out to play a pivotal role in the economic approach to index number theory.

Throughout much of this chapter, we shall assume that the aggregator function F satisfies the following *conditions I:* F is a real valued function of N variables defined over the nonnegative orthant $\Omega \equiv \{x : x \geq 0_N\}$ which has the three properties of (i) continuity, (ii) increasingness[2] and (iii) quasiconcavity.[3]

Let U be the range of F. From I(i) and (ii), it can be seen that $U \equiv \{u : \bar{u} \leq u \leq \bar{\bar{u}}\}$ where $\bar{u} \equiv F(0_N) < \bar{\bar{u}}$. Note that the least upper bound $\bar{\bar{u}}$ could be a finite number or $+\infty$. In the context of production theory, typically $\bar{u} = 0$ and $\bar{\bar{u}} = +\infty$, but, for consumer theory applications, there is no reason to restrict the range of the utility function F in this manner.

Define the set of positive prices $P \equiv \{p : p \gg 0_N\}$. It can be shown that (see Diewert [1978c]) if F satisfies conditions I, then the cost function C defined by (1) satisfies the following *conditions II*:

(i) $C(u,p)$ is a real valued function of $N+1$ variables defined over $U \times P$ and is jointly *continuous* in (u,p) over this domain.

---

[1]Notation: $x \geq 0_N$ means each component of the column vector x is nonnegative, $x \gg 0_N$ means each component is positive, $x > 0_N$ means $x \geq 0_N$ but $x \neq 0_N$ where 0_N is an N dimensional vector of zeros, and x^T denotes the transpose of x.

[2]If $x'' \gg x' \geq 0_N$, then $F(x'') > F(x')$.

[3]For every $u \in$ range F, the upper level set $L(u) \equiv \{x : F(x) \geq u\}$ is a convex set. A set S is convex iff $x' \in S$, $x'' \in S$, $0 \leq \lambda \leq 1$ implies $\lambda x' + (1-\lambda)x'' \in S$: i.e. the line segment joining any two points belonging to S also belongs to S.

(ii) $C(\bar{u}, p) = 0$ for every $p \in P$.

(iii) $C(u, p)$ is *increasing in u* for every $p \in P$; i.e., if $p \in P$, u', $u'' \in U$, with $u' < u''$, then $C(u', p) < C(u'', p)$.

(iv) $C(\bar{\bar{u}}, p) = +\infty$ for every $p \in P$; i.e., if $p \in P$, $u^n \in U$, $\lim_n u^n = \bar{u}$, then $\lim_n C(u^n, p) = +\infty$.

(v) $C(u, p)$ is (positively) *linearly homogenous* in p for every $u \in U$; i.e., $u \in U$, $\lambda > 0$, $p \in P$ implies $C(u, \lambda p) = \lambda C(u, p)$.

(vi) $C(u, p)$ is *concave in p* for every $u \in U$; i.e., if $p' \gg 0_N$, $p'' \gg 0_N$, $0 \leq \lambda \leq 1$, $u \in U$, then $C(u, \lambda p' + (1-\lambda)p'') \geq \lambda C(u, p') + (1-\lambda)C(u, p'')$.

(vii) $C(u, p)$ is increasing in p for $u > \bar{u}$ and $u \in U$.

(viii) C is such that the function $F^*(x) \equiv \max_u\{u : p^T x \geq C(u, p)$ for every $p \in P,\ u \in U\}$ is continuous for $x \geq 0_N$.

For some of the theorems to be presented in this chapter, we can weaken the regularity conditions on the aggregator function F to just *continuity from above*.[4] Under this weakened hypothesis on F, the cost function C defined by (1) will still satisfy many of the properties in conditions II above.[5]

Finally, some of the theorems below make use of the following (stronger) regularity conditions on the aggregator function: we say that F is a *neoclassical aggregator function* if it is defined over the positive orthant $\{x : x \gg 0_N\}$ and is (i) *positive*, i.e. $F(x) > 0$ for $x \gg 0_N$, (ii) (positively) *linearly homogeneous*, and (iii) *concave* over $\{x : x \gg 0_N\}$. Under these conditions (let us call them *conditions III*) F can be extended to the nonnegative orthant Ω, and the extended F will be nonnegative, linearly homogeneous, concave, increasing and continuous over Ω (see Diewert [1978c]). Moreover, if F is neoclassical, then F's cost function C factors into

$$C(u, p) \equiv uC(1, p) \equiv uc(p) \tag{2}$$

for $u \geq 0$ and $p \gg 0_N$ where $c(p) \equiv C(1, p)$ is F's *unit cost function*. It can be shown that c satisfies the same regularity conditions as F; i.e. c is also a

---

[4] F is continuous from above over $x \geq 0_N$ iff for every $u \in$ range F, $L(u) \equiv \{x : F(x) \geq u\}$ is a closed set.

[5] Specifically, Diewert [1978c] shows that C will satisfy the following conditions II″: (i) $C(u, p)$ is a real valued function of $N + 1$ variables defined over $U \times P$ and is continuous in p for fixed u and continuous from below in u for fixed p (the set U is now the convex hull of the range of F), (ii) $C(u, p) \geq 0$ for every $u \in U$ and $p \in P$, (iii) $C(u, p)$ is nondecreasing in u for fixed p, (iv) $C(u, p)$ is nondecreasing in p for fixed u, and properties (v) and (vi) are the same as (v) and (vi) of conditions II.

neoclassical function. Also, if we are given a neoclassical unit cost function c, then the underlying aggregator function F can be defined for $x \gg 0_N$ by

$$F(x) \equiv \max_u \{u : C(u,p) \leq p^T x \text{ for every } p > 0_N\}$$

$$= \max_u \{u : uc(p) \leq p^T x \text{ for every } p \geq 0_N,\ p^T x = 1\}$$

$$= \min_p \{1/c(p) : p \geq 0_N,\ p^T x = 1\} \tag{3}$$

$$= 1/\max_p \{c(p) : p^T x = 1,\ p \geq 0_N\}. \tag{4}$$

Now that we have disposed of the mathematical preliminaries, we can define the Konüs [1924] *cost of living index*[6] P_K: for $p^0 \gg 0_N$, $p^1 \gg 0_N$ and $x > 0_N$

$$P_K(p^0, p^1, x) \equiv C[F(x), p^1]/C[F(x), p^0]. \tag{5}$$

Thus P_K depends on three sets of variables: (i) p^0, a vector of period 0 or base period prices, (ii) p^1, a vector of period 1 or current period prices,[7] and (iii) x, a reference vector of quantities.[8] In the consumer context, P_K can be interpreted as follows. Pick a reference indifference surface indexed by the quantity vector $x > 0_N$. Then $P_K(p^0, p^1, x)$ is the minimum cost of achieving the standard of living indexed by x when the consumer faces period 1 prices p^1 relative to the minimum cost of achieving the same standard of living when the consumer faces period 0 prices p^0. Thus P_K can be interpreted as a level of prices in period 1 relative to a level of prices in period 0. If the number of goods is only one (i.e. $N = 1$), then it is easy to see that $P_K(p_1^0, p_1^1, x_1) = p_1^1/p_1^0$ for all $x_1 > 0$.

Note that the mathematical properties of P_K with respect to p^0, p^1 and x are determined by the mathematical properties of F and C given by conditions I and II above. In particular, for $\lambda > 0$, $p^0 \gg 0_N$, $p^1 \gg 0_N$ and $x \gg 0_N$, we have $P_K(p^0, \lambda p^0, x) = \lambda$ and $P_K(p^0, p^1, x) = 1/P_K(p^1, p^0, x)$. Thus if period 1 prices are proportional to period 0 prices, then P_K is equal to the common factor of proportionality for any reference quantity vector x. However, if prices are not proportional, then in general P_K depends on the reference vector x, except when preferences are homothetic as is shown in the following result.

[6]Or *cost of production index* in the producer context.

[7]In the theory of international comparisons, p^0 and p^1 can be interpreted as price vectors that a given consumer (whose utility level is indexed by the quantity vector x) faces in countries 0 and 1.

[8]The index P_K can also be written as $P_K(p^0, p^1, u) \equiv C(u, p^1)/C(u, p^0)$ where u is the reference output or utility level. Written in this form, the symmetry of the Konüs price index P_K with the Malmquist quantity index to be introduced later becomes apparent. However, our present notation for P_K is more convenient when we set the reference consumption vector x equal to the observed consumption vector x^r in period r.

THEOREM 1. (Malmquist [1953; 215], Pollak [1971a; 31], Samuelson and Swamy [1974; 569–570]): *Let the aggregator function F satisfy conditions I. Then $P_K(p^0, p^1, x)$ is independent of x if and only if F is homothetic.*[9]

Proof: If F is homothetic, then, by definition, there exists a continuous, monotonically increasing function of one variable G, with $G(\bar{u}) = 0$ such that $G[F(x)] \equiv f(x)$ is a neoclassical aggregator function (i.e. f satisfies conditions III above). Under these conditions, F's cost function decomposes as follows: for $u > 0$, $p \gg 0_N$,

$$\begin{aligned} C(u,p) &\equiv \min_x\{p^T x : F(x) \geq u\} \\ &= \min_x\{p^T x : G[F(x)] \geq G(u)\} \\ &= G(u)c(p) \qquad (6) \end{aligned}$$

where c is the unit cost function which corresponds to the neoclassical aggregator function f. Thus for $p^0 \gg 0_N$, $p^1 \gg 0_N$ and $x > 0_N$, we have

$$\begin{aligned} P_K(p^0,p^1,x) &\equiv C[F(x),p^1]/C[F(x),p^0] \\ &= G[F(x)]c(p^1)/G[F(x)]c(p^0) \\ &= c(p^1)/c(p^0) \qquad (7) \end{aligned}$$

which is independent of x.

Conversely, if P_K is independent of x, then we must have the factorization (7); i.e. we must have for every $x \gg 0_N$, $p \gg 0_N$

$$C(F(x),p) = G[F(x)]c(p) \qquad (8)$$

for some functions G and c, whose regularity properties must be such that C satisfies conditions II. It can be verified that the regularity conditions on C and the decomposition (8) imply that the functions c and $G(F)$ both satisfy conditions III,[10] so that, in particular, $G[F(x)]$ is (positively) linearly homogeneous in x. Thus F is homothetic. Q.E.D.

---

[9]It seems clear that earlier researchers such as Frisch [1936; 25] also knew this result, but they had some difficulty in stating it precisely, since the concept of homotheticity was not invented until 1953 (by Shephard [1953] and Malmquist [1953]).

[10]Linear homogeneity of $G(F)$ follows from the following identity which can be derived in a manner analogous to (4): $G[F(x)] = 1/\max_p\{c(p) : p \geq 0_N, p^T x = 1\}$ for every $x \gg 0_N$.

Thus in the case of a homothetic aggregator function, the Konüs cost of living index $P_K(p^0, p^1, x)$ is independent of the reference quantity vector x and is equal to a ratio of unit cost functions, $c(p^1)/c(p^0)$.

If we knew the consumer's preferences (or the producer's production function), then we could construct the cost function $C(u,p)$ and the Konüs price index P_K. However, usually we do not know F or C and thus it is useful to develop ***nonparametric bounds*** on P_K; i.e. bounds that do not depend on the functional form for the aggregator function F (or its cost function dual C).

THEOREM 2. (Lerner [1935–36], Joseph [1935–36; 149], Samuelson [1947; 159], Pollak [1971a; 12]): *If the aggregator function F is continuous from above, then, for every $p^0 \equiv (p_1^0, \ldots, p_N^0)^T \gg 0_N$, $p^1 \equiv (p_1^1, \ldots, p_N^1)^T \gg 0_N$ and $\bar{x} > 0_N$ where $F(\bar{x}) > F(0_N)$,*

$$(9)\quad \min_i\{p_i^1/p_i^0 : i = 1, \ldots, N\} \leq P_K(p^0, p^1, \bar{x}) \leq \max_i\{p_i^1/p_i^0 : i = 1, \ldots, N\};$$

i.e. P_K lies between the smallest and the largest price ratio.

Proof: Let $p^0 \gg 0_N$, $p^1 \gg 0_N$, $\bar{x} > 0_N$ where $F(\bar{x}) > F(0_N)$ and let $x^0 \geq 0_N$ and $x^1 \geq 0$ solve the following cost minimization problems:

$$(10)\quad C[F(\bar{x}), p^0] \equiv \min_x\{p^{0T}x : F(x) \geq F(\bar{x})\} = p^{0T}x^0 > 0$$

$$(11)\quad C[F(\bar{x}), p^1] \equiv \min_x\{p^{1T}x : F(x) \geq F(\bar{x})\} = p^{1T}x^1 > 0.$$

Then

$$C[F(\bar{x}), p^1] \equiv \min_x\{p^{1T}x : F(x) \geq F(\bar{x})\}$$

$$\geq \min_x\{p^{1T}x : p^{0T}x \geq p^{0T}x^0, x \geq 0_N\}$$

$$\text{since } \{x : F(x) \geq F(\bar{x})\} \subset \{x : p^{0T}x \geq p^{0T}x^0,\ x \geq 0_N\}$$

$$(12)\quad = \min_i\{p_i^1(p^{0T}x^0/p_i^0) :\ i = 1, \ldots, N\}$$

since the solution to the linear programming problem $\min_x\{p^{1T}x : p^{0T}x \geq p^{0T}x^0,\ x \geq 0_N\}$ can be taken to be a corner solution. Similarly,

$$C[F(\bar{x}), p^0] \geq \min_i\{p_i^0(p^{1T}x^1/p_i^1) : i = 1, \ldots, N\}$$

or

$$(13)\quad 1/C[F(\bar{x}), p^0] \leq \max_i\{p_i^1/p_i^0 p^{1T}x^1 :\ i = 1, \ldots, N\}.$$

Since $P_K(p^0, p^1, \bar{x}) \equiv C[F(\bar{x}), p^1]/C[F(\bar{x}), p^0]$, (10) and (12) imply the lower limit of (9) while (11) and (13) imply the upper limit. Q.E.D.

The geometric idea behind the above algebraic proof is that the sets $\{x : p^{0T}x \geq p^{0T}x^0,\ x \geq 0_N\}$ and $\{x : p^{1T}x^1 \geq p^{1T}x^1,\ x \geq 0_N\}$ form outer approximations to the true utility (or production) possibility set $\{x : F(x) \geq F(\bar{x})\}$. Moreover, it can be seen that the bounds on P_K given by (9) are the best possible,[11] i.e., if $F(x) \equiv p^{0T}x$, then P_K will attain the lower bound while, if $F(x) \equiv p^{1T}x$, then P_K will attain the upper bound in (9).

It is natural to assume that we can observe the consumer's (or producer's) quantity choices, $x^0 > 0_N$ and $x^1 > 0_N$, made during periods 0 and 1 in addition to the prices which prevailed during those periods, $p^0 \gg 0_N$ and $p^1 \gg 0_N$. In the remainder of this section, we shall also assume that the consumer (or producer) is engaging in cost minimizing behavior during the two periods. Thus we assume:

$$(14)\quad p^{0T}x^0 = C[F(x^0), p^0];\ p^{1T}x^1 = C[F(x^1), p^1];\ p^0, p^1 \gg 0_N;\ x^0, x^1 > 0_N.$$

Given the above assumptions, we now have two natural choices for the quantity vector x which occurs in the definition of the Konüs cost of living index $P_K(p^0, p^1, x)$: x^0 or x^1. The *Laspeyres–Konüs cost of living index* is defined as $P_K(p^0, p^1, x^0)$ and the *Paasche–Konüs cost of living index* is defined as $P_K(p^0, p^1, x^1)$.[12] It turns out that the Laspeyres–Konüs index $P_K(p^0, p^1, x^0)$ is related to the *Laspeyres price index* $P_L(p^0, p^1, x^0, x^1) \equiv p^{1T}x^0/p^{0T}x^0$ while the Paasche–Konüs index $P_K(p^0, p^1, x^1)$ is related to the *Paasche price index* $P_P(p^0, p^1, x^0, x^1) \equiv p^{1T}x^1/p^{0T}x^1$.

THEOREM 3. (Konüs [1924; 17–19]): *Suppose F is continuous from above and (14) holds. Then*

$$(15)\qquad P_K(p^0, p^1, x^0) \leq p^{1T}x^0/p^{0T}x^0 \equiv P_L \text{ and}$$

$$(16)\qquad P_K(p^0, p^1, x^1) \geq p^{1T}x^1/p^{0T}x^1 \equiv P_P.$$

Proof:

$$\begin{aligned} P_K(p^0, p^1, x^0) &\equiv C[F(x^0), p^1]/C[F(x^0), p^0] \\ &= C[F(x^0), p^1]/p^{0T}x^0 \text{ using (14)} \\ &\equiv \min_x\{p^{1T}x : F(x) \geq F(x^0)\}/p^{0T}x^0 \\ &\leq p^{1T}x^0/p^{0T}x^0 \end{aligned}$$

[11]This point is made by Pollak [1971a; 28].

[12]The terminology is due to Wold [1953; 136].

since x^0 is feasible for the cost minimization problem (but is not necessarily optimal), which proves (15). Similarly,

$$\begin{aligned} P_K(p^0, p^1, x^1) &= p^{1T}x^1/C[F(x^1), p^0] \\ &= p^{1T}x^1/\min_x\{p^{0T}x : F(x) \geq F(x^1)\} \\ &\geq p^{1T}x^1/p^{0T}x^1. \end{aligned}$$

Q.E.D.

COROLLARY 3.1. (Pollak [1971a; 17]):

$$\min_i\{p_i^1/p_i^0 : i = 1, \ldots, N\} \leq P_K(p^0, p^1, x^0) \leq p^{1T}x^0/p^{0T}x^0 \equiv P_L. \tag{17}$$

COROLLARY 3.2. (Pollak [1971a; 18]):

$$P_P \equiv p^{1T}x^1/p^{0T}x^1 \leq P_K(p^0, p^1, x^1) \leq \max_i\{p_i^1/p_i^0 : i = 1, \ldots, N\}. \tag{18}$$

COROLLARY 3.3. (Frisch [1936; 25]): *If in addition F is homothetic, then for $x \gg 0_N$,*

$$P_P \equiv p^{1T}x^1/p^{0T}x^1 \leq P_K(p^0, p^1, x) \leq p^{1T}x^0/p^{0T}x^0 \equiv P_L. \tag{19}$$

The first two corollaries follow from Theorems 2 and 3, while the third corollary follows from Theorems 1 and 2. Note that

$$\begin{aligned} P_L \equiv p^{1T}x^0/p^{0T}x^0 &= \sum_{i=1}^N (p_i^1/p_i^0)(p_i^0 x_i^0/p^{0T}x^0) \\ &\equiv \sum_{i=1}^N (p_i^1/p_i^0)s_i^0 \leq \max_i\{p_i^1/p_i^0 :\ i = 1, 2, \ldots, N\} \end{aligned}$$

since a share weighted average of the price ratios p_i^1/p_i^0 will always be equal to or less than the maximum price ratio. Thus the bounds given by (17) will generally be sharper than the Joseph–Pollak bounds given by (9). Similarly,

$$\begin{aligned} P_P \equiv p^{1T}x^1/p^{0T}x^1 &\equiv \sum_{i=1}^N (p_i^1/p_i^0)(p_i^0 x_i^1/p^{0T}x^1) \\ &\geq \min_i\{p_i^1/p_i^0 : i = 1, 2, \ldots, N\}, \end{aligned}$$

so that the bounds (18) are generally sharper than the bounds (9).

The geometric idea behind the proof of Theorem 3 is that the sets $\{x : x = x^0\}$ and $\{x : x = x^1\}$ form inner approximations to the true utility (or production) possibility sets $\{x : F(x) \geq F(x^0)\}$ and $\{x : F(x) \geq F(x^1)\}$ respectively. Moreover, it can be seen that the bounds on P_K given by (15) and (16) are attainable if F is a Leontief aggregator function (so that the corresponding cost function is linear in prices).[13]

[13] Pollak [1971a; 20] makes this well known point. F is a Leontief aggregator function if $F(x_1, x_2, \ldots, x_N) \equiv \min_i\{x_i/a_i : i = 1, 2, \ldots, N\}$ where $a^T \equiv (a_1, a_2, \ldots, a_N) > 0_N$. In this case $C(u, p) = up^T a$.

THEOREM 4. (Konüs [1924; 20–21]): *Let F satisfy conditions I and suppose (14) holds. Then there exists a* λ^* *such that* $0 \leq \lambda^* \leq 1$ *and* $P_K[p^0, p^1, \lambda^* x^1 + (1-\lambda^*)x^0]$ *lies between* P_L *and* P_P *; i.e. either*

(20) $$P_L \equiv p^{1T}x^0/p^{0T}x^0 \leq P_K[p^0, p^1, \lambda^* x^1 + (1-\lambda^*)x^0] \leq p^{1T}x^1/p^{0T}x^1 \equiv P_P$$

or

(21) $$P_P \leq P_K[p^0, p^1, \lambda^* x^1 + (1-\lambda^*)x^0] \leq P_L.$$

Proof: Define $h(\lambda) \equiv P_K(p^0, p^1, \lambda x^1 + (1-\lambda)x^0) \equiv C[F(\lambda x^1 + (1+\lambda)x^0), p^1]/C[F(\lambda x^1 + (1-\lambda)x^0), p^0]$. Since both F and C are continuous with respect to their arguments, h is continuous over the closed interval $[0, 1]$. Note that $h(0) = P_K(p^0, p^1, x^0)$ and $h(1) = P_K(p^0, p^1, x^1)$. There are $4! = 24$ possible inequalities between the four numbers P_L, P_P, $h(0)$ and $h(1)$. However, from Theorem 3, we have the restrictions $h(0) \leq P_L$ and $P_P \leq h(1)$. These restrictions imply that there are only six possible inequalities between the four numbers: (1) $h(0) \leq P_L \leq P_P \leq h(1)$, (2) $h(0) \leq P_P \leq P_L \leq h(1)$, (3) $h(0) \leq P_P \leq h(1) \leq P_L$, (4) $P_P \leq h(0) \leq P_L \leq h(1)$, (5) $P_P \leq h(1) \leq h(0) \leq P_L$ and (6) $P_P \leq h(0) \leq h(1) \leq P_L$. Since $h(\lambda)$ is continuous over $(0, 1)$ and thus assumes all intermediate values between $h(0)$ and $h(1)$, it can be seen that we can choose λ between 0 and 1 so that $P_L \leq h(\lambda^*) \leq P_P$ for case (1) or so that $P_P \leq h(\lambda^*) \leq P_L$ for cases (2) to (6), which establishes (20) or (21). Q.E.D.

It should be noted that λ^* can be chosen so that (20) or (21) is satisfied and in addition $F[\lambda^* x^1 + (1-\lambda^*)x^0]$ lies between $F(x^0)$ and $F(x^1)$. Thus the Paasche and Laspeyres indexes provide bounds for the Konüs cost of living index for some reference indifference surface which lies between the period 0 and period 1 indifference surfaces.

The above theorems provide bounds for the Konüs price index $P_K(p^0, p^1, x)$ under various hypotheses. We cannot improve upon these bounds unless we are willing to make specific assumptions about the functional form for the aggregator function F, a strategy we will pursue in Sections 5 and 6.

3. The Konüs, Allen and Malmquist Quantity Indexes

In the case of only one commodity, a quantity index could be defined as x_1^1/x_1^0, the ratio of the quantity in period 1 to the quantity in period 0. This ratio is also equal to the ratio of expenditures in the two periods, $p_1^1 x_1^1/p_1^0 x_1^0$, divided by the price index p_1^1/p_1^0. This suggests that a reasonable notion of a quantity index in the general N commodity case could be the expenditure ratio deflated by the Konüs cost of living index. Thus we define the *Konüs–Pollak* [1971a; 64] *implicit quantity index* for $p^0 \gg 0_N$, $p^1 \gg 0_N$, $x^0 > 0_N$, $x^1 > 0_N$ and $x > 0_N$ as

$$\widetilde{Q}_K(p^0, p^1, x^0, x^1, x) \equiv p^{1T}x^1/p^{0T}x^0 P_K(p^0, p^1, x) \tag{22}$$

$$= \frac{C[F(x^1), p^1]}{C[F(x^0), p^0]} \Big/ \frac{C[F(x), p^1]}{C[F(x), p^0]} \tag{23}$$

where (23) follows if the consumer or producer is engaging in cost minimizing behavior during the two periods; i.e. (23) follows if (14) is true. Note that $\widetilde{Q}_K$ depends on the period 0 prices and quantities, p^0 and x^0, the period 1 prices and quantities, p^1 and x^1, and the reference indifference surface indexed by the quantity vector x.

The following result shows that $\widetilde{Q}_K$ gives the correct answer (at least ordinally) if the reference quantity vector x is chosen appropriately.

THEOREM 5. *Suppose F satisfies conditions I and (14) holds. (i) If $F(x^1) > F(x^0)$, then for every $x \geq 0_N$ such that $F(x^1) \geq F(x) \geq F(x^0)$, $\widetilde{Q}_K(p^0, p^1, x^0, x^1, x) > 1$. (ii) If $F(x^1) = F(x^0)$, then, for every $x \geq 0_N$ such that $F(x) = F(x^1) = F(x^0)$, $\widetilde{Q}_K(p^0, p^1, x^0, x^1, x) = 1$. (iii) If $F(x^1) < F(x^0)$, then for every $x \geq 0_N$ such that $F(x^1) \leq F(x) \leq F(x^0)$, $\widetilde{Q}_K(p^0, p^1, x^0, x^1, x) < 1$.*

Proof of (i):

$$\widetilde{Q}_K(p^0, p^1, x^0, x^1, x) = \frac{C[F(x^1), p^1]}{C[F(x), p^1]} \frac{C[F(x), p^0]}{C[F(x^0), p^0]} \quad \text{using (23)}$$

$$> 1$$

since $F(x^1) \geq F(x)$ implies $C[F(x^1), p^1] \geq C[F(x), p^1]$ and $F(x) \geq F(x^0)$ implies $C[F(x), p^0] \geq C[F(x^0), p^0]$ with at least one of the inequalities holding strictly, using property (iii) on the cost function C.

Parts (ii) and (iii) follow in an analogous manner. Q.E.D.

It can be verified that if $F(x^1) > F(x^0) > F(x)$, then, if F is not homothetic, it is not necessarily the case that $\widetilde{Q}_K(p^0, p^1, x^0, x^1, x) > 1$. However, if we choose x to be x^0 or x^1, then the resulting $\widetilde{Q}_K$ will have the desirable properties outlined in Theorem 5. Thus define the *Laspeyres–Konüs implicit quantity index* as

$$\widetilde{Q}_K(p^0, p^1, x^0, x^1, x^0) \equiv p^{1T}x^1/p^{0T}x^0 P_K(p^0, p^1, x^0)$$

$$= C[F(x^1), p^1]/C[(F(x^0), p^0] \cdot (C[F(x^0), p^1]/C[F(x^0), p^0])$$
using (5) and (14)

$$= C[F(x^1), p^1]/C[F(x^0), p^1] \tag{24}$$

and the *Paasche–Konüs implicit quantity index* as

$$\widetilde{Q}_K(p^0,p^1,x^0,x^1,x^1) \equiv p^{1T}x^1/p^{0T}x^0 P_K(p^0,p^1,x^1)$$
$$= C[F(x^1),p^0]/C[F(x^0),p^0] \tag{25}$$

where (25) follows using definition (5) for P_K and the assumptions (14) of cost minimizing behavior.

It turns out that the quantity indexes defined by (24) and (25) are special cases of another class of quantity indexes. For $x^0 > 0_N$, $x^1 > 0_N$ and $p \gg 0_N$, define the *Allen* [1949; 199] *quantity index* as

$$Q_A(x^0,x^1,p) \equiv C[F(x^1),p]/C[F(x^0),p]. \tag{26}$$

Note that $\widetilde{Q}_K(p^0,p^1,x^0,x^1,x) = Q_A(x^0,x,p^0)Q_A(x,x^1,p^1)$ and that the *Laspeyres–Allen quantity index* $Q_A(x^0,x^1,p^0)$ equals the Paasche–Konüs implicit quantity index $\widetilde{Q}_K(p^0,p^1,x^0,x^1,x^1)$ while the *Paasche–Allen quantity index* $Q_A(x^0,x^1,p^1)$ equals $\widetilde{Q}_K(p^0,p^1,x^0,x^1,x^0)$, assuming that (14) holds.

THEOREM 6. *Suppose F satisfies conditions I. (i) If $F(x^1) > F(x^0) > \bar{u}$, then $Q_A(x^0,x^1,p) > 1$ for every $p \gg 0_N$. (ii) If $F(x^1) = F(x^0) > \bar{u}$, then $Q_A(x^0,x^1,p) = 1$ for every $p \gg 0_N$. (iii) If $\bar{u} < F(x^1) < F(x^0)$, then $Q_A(x^0,x^1,p) < 1$ for every $p \gg 0_N$.*

The proof of the above lemma follows directly from definition (26) and property (iii) for the cost function $C(u,p)$: increasingness in u.[14]

It turns out that Allen quantity indexes do not satisfy bounds analogous to those given by Theorem 2 for the Konüs price indexes. However, there is a counterpart to Theorem 3.

THEOREM 7. (Samuelson [1947; 162], Allen [1949; 199]): *If the aggregator function F is continuous from above and (14) holds, then*

$$Q_A(x^0,x^1,p^0) \le p^{0T}x^1/p^{0T}x^0 \equiv Q_L(p^0,p^1,x^0,x^1) \quad \text{and} \tag{27}$$

$$Q_A(x^0,x^1,p^1) \ge p^{1T}x^1/p^{1T}x^0 \equiv Q_P(p^0,p^1,x^0,x^1); \tag{28}$$

i.e. the Laspeyres–Allen quantity index is bounded from above by the Laspeyres quantity index Q_L and the Paasche–Allen quantity index is bounded below by the Paasche quantity index Q_P.

Proof:

$$\begin{aligned} Q_A(x^0,x^1,p^0) &= C[F(x^1),p^0]/p^{0T}x^0 \quad \text{using (26) and (14)} \\ &\equiv \min_x\{p^{0T}x : F(x) \ge F(x^1)\}/p^{0T}x^0 \\ &\le p^{0T}x^1/p^{0T}x^0 \end{aligned}$$

[14]We also utilize property (ii) for $C : C(\bar{u},p) = 0$ for every $p \gg 0_N$.

since x^1 is feasible for the minimization problem. Similarly,

$$\begin{aligned}Q_A(x^0,x^1,p^1) &= p^{1T}x^1 / \min_x\{p^{1T}x : F(x) \geq F(x^0)\}\\ &\geq p^{1T}x^1/p^{1T}x^0\end{aligned}$$

since x^0 is feasible for the minimization problem and $p^{1T}x^0 > 0$. Q.E.D.

THEOREM 8. *If F is homothetic (so that there exists a continuous, monotonically increasing function of one variable such that $G[F(x)]$ is neoclassical) and (14) holds, then for every $x \gg 0_N$ and $p \gg 0_N$*

$$\begin{aligned}\widetilde{Q}_K(p^0,p^1,x^0,x^1,x) &= Q_A(x^0,x^1,p)\\ &= G[F(x^1)]/G[F(x^0)].\end{aligned} \tag{29}$$

Proof:

$$\begin{aligned}\widetilde{Q}_K(p^0,p,x^0,x^1,x) &= \frac{C[F(x^1),p^1]}{C[F(x^0),p^0]} \Big/ \frac{C[F(x),p^1]}{C[F(x),p^0]} \quad \text{using (23)}\\ &= \frac{G[F(x^1)]c(p^1)}{G[F(x^0)]c(p^0)} \Big/ \frac{G[F(x)]c(p^1)}{G[F(x)]c(p^0)}\\ &\quad \text{by homotheticity of } F\\ &= G[F(x^1)]/G[F(x^0)]\\ &= G[F(x^1)]c(p)/G[F(x^0)]c(p)\\ &= C[F(x^1),p]/C[F(x^0),p]\\ &\quad \text{by homotheticity again}\\ &\equiv Q_A(x^0,x^1,p).\end{aligned}$$

Q.E.D.

COROLLARY 8.1. (Samuelson and Swamy [1974; 570]): *If $Q_A(x^0,x^1,p)$ is independent of p and F satisfies conditions I, then F must be homothetic.*

Proof: If $Q_A(x^0,x^1,p)$ is independent of p, then $C[F(x^1),p]/C[F(x^1),p]$ is independent of p for all $x^0 \gg 0_N$ and $x^1 \gg 0_N$. Thus we must have $C[F(x),p] = G(F(x)]c(p)$ for some functions G and c which implies that F is homothetic. Q.E.D.

Corollary 8.2. *If F is neoclassical (so that $G(u) \equiv u$) and (14) holds, then for every $x \gg 0_N$, and every $p \gg 0_N$:*

$$\tilde{Q}_K(p^0, p^1, x^0, x^1, x) = Q_A(x^0, x^1, p) = F(x^1)/F(x^0). \tag{30}$$

Corollary 8.3. *If F is homothetic and (14) holds, then for every $x \gg 0_N$ and $p \gg 0_N$:*

$$\begin{aligned} Q_P \equiv p^{1T}x^1/p^{1T}x^0 &\le \tilde{Q}_K(p^0, p^1, x^0, x^1, x) = Q_A(x^0, x^1, p) \\ &\le p^{0T}x^1/p^{0T}x^0 \equiv Q_L. \end{aligned} \tag{31}$$

Proof: From (28),

$$\begin{aligned} Q_P \le Q_A(x^0, x^1, p^1) = \tilde{Q}_K(p^0, p^1, x^0, x^1, x) &= Q_A(x^0, x^1, p) \\ &= Q_A(x^0, x^1, p^0) \quad \text{using (29)} \\ &\le Q_L \quad \text{using (27).} \end{aligned}$$

Q.E.D

Thus if the aggregator function is homothetic, then the Allen and implicit Konüs quantity indexes coincide for all reference vectors p and x, and their common value is bounded from below by the Paasche quantity index Q_P and above by the Laspeyres quantity index Q_L. Note that Q_P and Q_L can be computed from observable data.

In the general case when F is not necessarily homothetic, the following results give bounds for $\tilde{Q}_K$ and Q_A.

Theorem 9. *Let F satisfy conditions I and suppose (14) holds. Then there exists a λ^* such that $0 \le \lambda^* \le 1$ and $Q_K[x^0, x^1, p^0, p^1, \lambda^* x^1 + (1-\lambda^*)x^0]$ lies between Q_P and Q_L.*

Proof: From Theorem 4, either (20) or (21) holds for $P_K[p^0, p^1, \lambda^* x^1 + (1-\lambda^*)x^0]$ for some λ^* between 0 and 1. If (20) holds, then, using definition (22):

$$\begin{aligned} Q_L = (p^{1T}x^1/p^{0T}x^0)/P_P &\le \tilde{Q}_K[x^0, x^1, p^0, p^1, \lambda^* x^1 + (1-\lambda^*)x^0] \\ &\le (p^{1T}x^1/p^{0T}x^0)/P_L = Q_P. \end{aligned}$$

Similarly, if (21) holds then $Q_P \le Q_K[x^0, x^1, p^0, p^1, \lambda^* x^1 + (1-\lambda^*)x^0] \le Q_L$.

Q.E.D.

THEOREM 10. *Let F be continuous from above and suppose (14) holds. Then there exists a λ^* such that $0 \leq \lambda^* \leq 1$ and $Q_A[x^0, x^1, \lambda^* p^1 + (1-\lambda^*)p^0]$ lies between Q_L and Q_P.*

Proof: Define $h(\lambda) \equiv Q_A[x^0, x^1, \lambda p^1 + (1-\lambda)p^0] \equiv C[F(x^1), \lambda p^1 + (1-\lambda)p^0]/C[F(x^0), \lambda p^1 + (1-\lambda)p^0]$. Since F is continuous from above, $C(u,p)$ is continuous in p and thus $h(\lambda)$ is continuous for $0 \leq \lambda \leq 1$. Note that $h(0) = Q_A(x^0, x^1, p)$ and $h(1) = Q_A(x^0, x^1, p^1)$. From Theorem 7, $h(0) \leq Q_L$ and $Q_P \leq h(1)$. Now repeat the proof of Theorem 9 with Q_L and Q_P replacing P_L and P_P. Q.E.D.

Thus the Paasche and Laspeyres quantity indexes (which are observable) bound both the implicit Konüs quantity index $\widetilde{Q}_K$ and the Allen quantity index Q_A, provided that we choose appropriate reference vectors between x^0 and x^1 or p^0 and p^1 respectively. However, it is also necessary to assume cost minimizing behavior on the part of the consumer or producer during the two periods in order to derive the above bounds.

Recall that the Konüs price index P_K had the desirable property that $P_K(p^0, \lambda p^0, x) = \lambda P_K(p^0, p^0, x)$ for all $\lambda > 0$, $p^0 \gg 0_N$, and $x \gg 0_N$; i.e. if the current period prices were proportional to the base period prices, then the price index equalled this common factor of proportionality λ. It would be desirable if an analogous homogeneity property held for the quantity indexes. Unfortunately, it is *not* always the case that $\widetilde{Q}_K(x^0, \lambda x^0, p^0, p^1, x) = \lambda$ or that $Q_A(x^0, \lambda x^0, p) = \lambda$. However, the following quantity index does have this desirable homogeneity property.

For $\bar{x} \gg 0_N$, $x^0 \gg 0_N$, $x^1 \gg 0_N$, define the *Malmquist* [1953; 232] *quantity index* as

$$(32) \qquad Q_M(x^0, x^1, \bar{x}) \equiv D[F(\bar{x}), x^1]/D[F(\bar{x}), x^0]$$

where $D[u, \bar{x}] \equiv \max_k\{k : F(\bar{x}/k) \geq u, k > 0\}$ is the *deflation function*[15] which corresponds to the aggregator function F. Thus $D[F(\bar{x}), x^1]$ is the biggest number which will just deflate the period 1 quantity vector x^1 onto the boundary of

[15]If F satisfies *conditions I*, then it can be shown (e.g., see Diewert, [1978c]), that the deflation function D satisfies *conditions IV*: (i) $D(u,x)$ is a real valued function of $N+1$ variables defined over Int $U \times$ Int $\Omega = \{u : \bar{u} < u < \bar{\bar{u}}\} \times \{x : x \gg 0_N\}$ and is *continuous* over this domain, (ii) $D(\bar{u}, x) = +\infty$ for every $x \in$ Int Ω; i.e., $u^n \in$ Int U, $\lim u^n = \bar{u}$, $x \in$ Int Ω implies $\lim_n D(u^n, x) = +\infty$, (iii) $D(u,x)$ is *decreasing in u* for every $x \in$ Int Ω; i.e., if $x \in$ Int Ω, $u', u'' \in$ Int U with $u' < u''$, then $D(u', x) > D(u'', x)$, (iv) $D(\bar{\bar{u}}, x) = 0$ for every $x \in$ Int Ω; i.e. $u'' \in$ Int U, $\lim u'' = \bar{\bar{u}}$, $x \in$ Int Ω implies $\lim_n D(u^n, x) = 0$, (v) $D(u,x)$ is (positively) *linearly homogeneous in x* for every $u \in$ Int U; i.e., $u \in$ Int U, $\lambda > 0$, $x \in$ Int Ω implies $D(u, \lambda x) = \lambda D(u,x)$, (vi) $D(u,x)$ is *concave in x* for every $u \in$ Int U, (vii) $D(u,x)$ is increasing in x for every $u \in$ Int U; i.e., $u \in$ Int U, x', $x'' \in$ Int Ω implies $D(u, x' + x'') > D(u, x')$, and (viii) D is such that the function $\widetilde{F}(x) \equiv \{u : u \in$ Int U, $D(u,x) = 1\}$ defined for $x \gg 0_N$ has a continuous extension to $x \geq 0_N$.

the utility (or production) possibility set $[x : F(x) \geq F(\bar{x}), x \geq 0_N]$ indexed by the quantity vector $\bar{x}$ while $D[F(\bar{x}), x^0]$ is the biggest number which will just deflate the period 0 quantity vector x^0 onto the utility possibility set indexed by $\bar{x}$, and Q_M is the ratio of these two deflation factors.

Note that the assumption of cost minimizing behavior is *not* required in order to define the Malmquist quantity index Q_M.

THEOREM 11. (Malmquist [1953; 231], Pollak [1971a; 62]): *If F satisfies conditions I, then (i) $\lambda > 0$, $x^0 \gg 0_N$, $\bar{x} \gg 0_N$ implies $Q_M(x^0, \lambda x^0, \bar{x}) = \lambda$ and (ii) $x^0 \gg 0_N$, $x^1 \gg 0_N$, $x^2 \gg 0_N$, $\bar{x} \gg 0_N$ implies $Q_M(x^0, x^1, \bar{x})Q_M(x^1, x^2, \bar{x}) = Q_M(x^0, x^2, \bar{x})$.*

Proof: (i) If F is merely continuous from above and increasing, then $D[F(\bar{x}), x]$ is well defined for all $\bar{x} \gg 0_N$ and $x \gg 0_N$. Moreover, D has the following homogeneity property (recall property (v) of conditions IV on D): for $\lambda > 0$, $D[F(\bar{x}), \lambda x] = \lambda D[F(\bar{x}), x]$. Thus $Q_M(x^0, \lambda x^0, \bar{x}) \equiv D[F(\bar{x}), \lambda x^0]/D[F(\bar{x}), x^0] = \lambda D[F(\bar{x}), x^0]/D[F(\bar{x}), x^0] = \lambda$. (ii) follows directly from definition (32).

Q.E.D.

Property (ii) in the above theorem is a desirable transitivity property of Q_M. $\widetilde{Q}_K$, Q_A, P_A and P_K all possess the analogous transitivity property (or circularity property as it is sometimes called in the index number literature).

THEOREM 12. *If F satisfies conditions I, $x^0 \gg 0_N$, $x^1 \gg 0_N$, $\bar{x} \gg 0_N$ and $F(\bar{x})$ is between $F(x^0)$ and $F(x^1)$, then the Malmquist quantity index $Q_M(x^0, x^1, \bar{x})$ will correctly indicate whether the aggregate has remained constant, increased or decreased from period 0 to period 1.*

Proof: (i) Suppose $F(x^0) = F(\bar{x}) = F(x^1)$. Then $Q_M(x^0, x^1, \bar{x}) = D[F(\bar{x}), x^1]/D[F(\bar{x}), x^0] = 1/1 = 1$. (ii) Suppose $F(x^0) \leq F(\bar{x}) \leq F(x^1)$ with $F(x^0) < F(x^1)$. Then $Q_M(x^0, x^1, \bar{x}) = k^1/k^0$ where $F(x^1/k^1) = F(\bar{x}) \leq F(x^1)$ which implies $k^1 \geq 1$ and $F(x^0/k^0) = F(\bar{x}) \geq F(x^0)$ which implies $0 < k^0 \leq 1$. Since at least one of the inequalities $F(\bar{x}) \leq F(x^1)$ and $F(\bar{x}) \leq F(x^0)$ is strict; at least one of the inequalities $k^1 \geq 1$ and $k^0 \leq 1$ must also be strict. Thus $Q_M(x^0, x^1, \bar{x}) = k^1/k^0 > 1$. The remaining case is similar. Q.E.D.

If F is nonhomothetic, then the restriction that the reference indifference surface indexed by $F(\bar{x})$ lie between the indifference surfaces indexed by $F(x^0)$ and $F(x^1)$ is necessary in order to prove Theorem 12; e.g. if $F(x^0) < F(x^1) < F(\bar{x})$, then it *need not* be the case that $Q_M(x^0, x^1, \bar{x}) > 1$.

The following result shows that the Malmquist quantity index satisfies the analogue to the Joseph–Pollak bounds for the Konüs price index.

THEOREM 13. *If F satisfies conditions I and $x^0 \gg 0_N$, $x^1 \gg 0_N$, $\bar{x} \gg 0_N$, then*

$$(33)\quad \min_i\{x_i^1/x_i^0 : i = 1,\ldots,N\} \le Q_M(x^0, x^1, \bar{x}) \le \max_i\{x_i^1/x_i^0 : i = 1,\ldots,N\}.$$

Proof: If F satisfies conditions I, then the deflation function D satisfies conditions IV. Thus $D(u,x)$ satisfies the same mathematical regularity properties with respect to x as $C(u,p)$ satisfies with respect to p. Since $C[F(\bar{x}), p^1]/C[F(\bar{x}), p^0] \equiv P_K(p^0, p^1, \bar{x})$ satisfies the inequalities in (9), $D[F(\bar{x}), x^1]/D(F(\bar{x}), x^0] \equiv Q_M(x^0, x^1, \bar{x})$ will satisfy the analogous inequalities (33).[16] Q.E.D.

In general, the Malmquist quantity index will depend on the reference indifference surface indexed by $\bar{x}$. As usual, two natural choices for $\bar{x}$ are x^0 or x^1, the observed quantity choices during period 0 or 1. Thus the *Laspeyres–Malmquist quantity index* is defined as

$$Q_M(x^0, x^1, x^0) \equiv D[F(x^0), x^1]/D[F(x^0), x^0] = D[F(x^0), x^1]$$

since $D[F(x^0), x^0] = 1$ if F is continuous from above and increasing, and the *Paasche–Malmquist quantity index* is defined as

$$Q_M(x^0, x^1, x^1) \equiv D[F(x^1), x^1]/D[F(x^1), x^0] = 1/D[F(x^1), x^0]$$

since $D[F(x^1), x^1] = 1$ if F is continuous from above and increasing.

THEOREM 14. (Malmquist [1953; 231]): *Suppose F satisfies conditions I and (14) holds. Then*

$$(34)\quad Q_M(x^0, x^1, x^0) \le p^{0T}x^1/p^{0T}x^0 \equiv Q_L \quad \text{and}$$

$$(35)\quad Q_M(x^0, x^1, x^1) \ge p^{1T}x^1/p^{1T}x^0 \equiv Q_P.$$

[16]More explicitly, $C[F(\bar{x}), p]$ is the support function for the set $L[F(\bar{x})] \equiv \{x : p^T x \ge C[F(\bar{x}), p]$ for every $p \gg 0_N\}$ and the sets $\{x : p^{0T}x \ge p^{0T}x^0,\ x \ge 0_N\}$ and $\{x : p^{1T}x \ge p^{1T}x^1, x \ge 0_N\}$ form outer approximations to this set where $x^0 \in \partial_p C[F(\bar{x}), p^0]$ and $x^1 \in \partial_p C[F(\bar{x}), p^1]$. $\partial_p C(u, p^0)$ denotes the set of supergradients to the concave function of p, $C(u,p)$, evaluated at the point p^0. Analogously, $D[F(\bar{x}), x]$ is the support function for the set $L^*[F(\bar{x})] \equiv \{p : p^T x \ge D[F(\bar{x}), x]$ for every $x \gg 0_N\}$ and the sets $\{p : p^T x^0 \ge p^{0T}x^0,\ p \ge 0_N\}$ and $\{p : p^T x^1 \ge p^{1T}x^1, p \ge 0_N\}$ form outer approximations to this set where $p^0 \in \partial_x D[F(\bar{x}), x^0]$ and $p^1 \in \partial_x D[F(\bar{x}), x^1]$.

Proof:

$$\begin{aligned} Q_M(x^0, x^1, x^0) &\equiv D[F(x^0), x^1] \\ &\equiv \max_k \{k : F(x^1/k) \geq F(x^0),\ k > 0\} \\ &= k^1 \quad \text{where } F(x^1/k^1) = F(x^0). \end{aligned}$$

Now

$$\begin{aligned} p^{0T} x^0 &= C[F(x^0), p^0] \\ &\equiv \min_x \{p^{0T} x : F(x) \geq F(x^0)\} \\ &\leq p^{0T} x^1 / k^1 \end{aligned}$$

since x^1/k^1 is feasible for the cost minimization problem. Thus

$$k^1 = Q_M(x^0, x^1, x^0) \leq p^{0T} x^1 / p^{0T} x^0 \equiv Q_L,$$

which proves (34). The proof of (35) is similar. Q.E.D.

THEOREM 15. *Suppose F satisfies conditions I and (14) holds. Then there exists a λ^* such that $0 \leq \lambda^* \leq 1$ and $Q_M(x^0, x^1, \lambda^* x^1 + (1 - \lambda^*)x^0)$ lies between Q_L and Q_P.*

Proof: Define $h(\lambda) \equiv Q_M[x^0, x^1, \lambda x^1 + (1 - \lambda)x^0] \equiv D\Big[F[\lambda x^1 + (1 - \lambda)x^0], x^1\Big] / D\Big[F[\lambda x^1 + (1 - \lambda)x^0], x^0\Big]$. Since $F[\lambda x^1 + (1 - \lambda)x^0]$ is continuous with respect to λ and $D(u, x)$ is continuous with respect to u (recall property (i) of conditions IV on D, $h(\lambda)$ is continuous for λ between 0 and 1. Moreover, $h(0) = Q_M(x^0, x^1, x^0)$ and $h(1) = Q_M(x^0, x^1, x^1)$. From Theorem 14, $h(0) \leq Q_L$ and $Q_P \leq h(1)$. Now repeat the proof of Theorem 10. Q.E.D.

It should be noted that λ^* can be chosen so that $0 \leq \lambda^* \leq 1$ and $Q_M[x^0, x^1, \lambda^* x^1 + (1 - \lambda^*)x^0]$ lies between Q_L and Q_P, and in addition, $F[\lambda^* x^1 + (1 - \lambda^*)x^0]$ lies between $F(x^0)$ and $F(x^1)$. Thus the Paasche and Laspeyres quantity indexes provide bounds for the Malmquist quantity index for some reference indifference surface which lies between the period 0 and period 1 indifference surfaces.

The following theorem relates the Paasche and Laspeyres Malmquist quantity indexes to the Paasche and Laspeyres implicit Konüs and Allen quantity indexes.

THEOREM 16. (Malmquist [1953; 233]): *Suppose F satisfies conditions I and (14) holds. Then*

$$(36)\qquad Q_M(x^0,x^1,x^0) \le \tilde{Q}_K(p^0,p^1,x^0,x^1,x^0) = Q_A(x^0,x^1,p^1) \quad \text{and}$$

$$(37)\qquad Q_M(x^0,x^1,x^1) \ge \tilde{Q}_K(p^0,p^1,x^0,x^1,x^1) = Q_A(x^0,x^1,p^0).$$

Proof:

$$\begin{aligned} Q_M(x^0,x^1,x^0) &= D[F(x^0),x^1] \\ &= k^1 \quad \text{say where } F(x^1/k^1) = F(x^0). \end{aligned}$$

Also

$$\begin{aligned} Q_A(x^0,x^1,p^1) &= p^{1T}x^1/C[F(x^0),p^1] \quad \text{using (26) and (14)} \\ &= \tilde{Q}_K(p^0,p^1,x^0,x^1,x^0) \quad \text{using (23)} \\ &= p^{1T}x^1/\min_x\{p^{1T}x : F(x) \ge F(x^0)\} \\ &\le p^{1T}x^1/p^{1T}(x^1/k^1) \quad \text{since } x^1/k^1 \text{ is} \\ &\qquad \text{feasible but not necessarily optimal} \\ &= k^1 \end{aligned}$$

which establishes (36). (37) follows in a similar manner. Q.E.D.

It is obvious that an *implicit Malmquist price index* $\tilde{P}_M$ can be defined as the expenditure ratio for the two periods deflated by Q_M: i.e. define

$$(38)\qquad \tilde{P}_M(p^0,p^1,x^0,x^1,\bar{x}) \equiv p^{1T}x^0/p^{0T}x^0 Q_M(x^0,x^1,\bar{x}).$$

However, the resulting price index does not have the desirable homogeneity property $\tilde{P}_M(p^0,\lambda p^0,x^0,x^1,\bar{x}) = \lambda$. Thus $\tilde{P}_M$ has properties analogous to the implicit Konüs quantity index $\tilde{Q}_K$, except that the role of prices and quantities is reversed.

Now that we have studied price and quantity indexes separately, it is time to observe that it is essential to study them together. For empirical work, it is highly desirable that the product of the price index P and the quantity index Q equal the actual expenditure ratio for the two periods under consideration, $p^{1T}x^1/p^{0T}x^0$. If P and Q satisfy this property, then we say that P and Q satisfy the *weak factor reversal test*[17] or the *product test.*[18] We have seen that the Konüs price index P_K is a desirable price index and that the Malmquist quantity index Q_M is a desirable quantity index since they each have a desirable homogeneity property. The following result shows that there exists at least one reference indifference surface such that P_K and Q_M satisfy the product test.

[17]The concept is associated with Irving Fisher [1922].

[18]This terminology is due to Frisch [1930].

THEOREM 17. (Malmquist [1953; 234]): *Suppose the aggregator function F satisfies conditions I and (14) holds. Then there exists a λ^* such that $0 \le \lambda^* \le 1$ and*

$$(39)\quad P_K[p^0, p^1, \lambda^* x^1 + (1-\lambda^*)x^0] Q_M[x^0, x^1, \lambda^* x^1 + (1-\lambda^*)x^0] = p^{1T}x^1/p^{0T}x^0.$$

Proof: For $0 \le \lambda \le 1$, define the continuous function

$$h(\lambda) \equiv P_K[p^0, p^1, \lambda x^1 + (1-\lambda)x^0] Q_M[x^0, x^1, \lambda x^1 + (1-\lambda)x^0].$$

Thus

$$\begin{aligned}
h(0) &\equiv P_K(p^0, p^1, x^0) Q_M(x^0, x^1, x^0) \\
&\equiv \Big[C[F(x^0), p^1]/C[F(x^0), p^0]\Big]\Big[D[F(x^0), x^1]/D[F(x^0), x^0]\Big] \\
&\quad \text{by (5) and (32)} \\
&\le \frac{C[F(x^0), p^1]}{C[F(x^0), p^0]}\frac{C[F(x^1), p^1]}{C[F(x^0), p^1]} \quad \text{using (36) and (26)} \\
&= p^{1T}x^1/p^{0T}x^0 \quad \text{using (14)} \\
&= \Big[C[F(x^1), p^1]/C[F(x^1), p^0]\Big]\Big[C[F(x^1), p^0]/C[F(x^0), p^0]\Big] \\
&\le \frac{C[F(x^1), p^1]}{C[F(x^1), p^0]}\frac{D[F(x^1), x^1]}{D[F(x^1), x^0]} \quad \text{using (37), (26) and (32)} \\
&= P_K(p^0, p^1, x^1) Q_M(x^0, x^1, x^1) \quad \text{using (5) and (32)} \\
&\equiv h(1).
\end{aligned}$$

Since $h(\lambda)$ is continuous over $[0,1]$ and since $h(0) \le p^{1T}x^1/p^{0T}x^0 \le h(1)$, there exists $0 \le \lambda^* \le 1$ such that $h(\lambda^*) = p^{1T}x^1/p^{0T}x^0$ and thus (39) is satisfied. Moreover, since $h(\lambda) \equiv \big(C\big[F[\lambda x^1 + (1-\lambda)x^0], p^1\big]/C\big[F[\lambda x^1 + (1-\lambda)x^0], p^0\big]\big)\big(D\big[F[\lambda x^1 + (1-\lambda)x^0], x^1\big]/D\big[F[\lambda x^1 + (1-\lambda)x^0], x^0\big]\big)$, we can choose λ^* so that $F[\lambda^* x^1 + (1-\lambda^*)x^0]$ lies between $F(x^0)$ and $F(x^1)$. Q.E.D.

Thus the reference indifference surface indexed by $\lambda^* x^1 + (1-\lambda^*)x^0$ which occurs in the above theorem lies between the surfaces indexed by x^0 and x^1, the quantity vectors observed during periods 0 and 1.

The final result in this section shows that all three quantity indexes that we have considered coincide (and are independent of reference price or quantity vectors) if the aggregator function is homothetic.

THEOREM 18. (Pollak [1971a; 65]): *If F is homothetic (so that there exists a continuous, monotonically increasing function of one variable such that $G[F(x)]$ is neoclassical) and (14) holds, then for every $x \gg 0_N$ and $p \gg 0_N$*

$$\begin{aligned} Q_M(x^0,x^1,x) &= \widetilde{Q}_K(p^0,p^1,x^0,x^1,x) = Q_A(x^0,x^1,p) \\ &= G[F(x^1)]/G[F(x^0)]. \end{aligned} \tag{40}$$

Proof:

$$\begin{aligned} Q_M(x^0,x^1,x) &\equiv D[F(x),x^1]/D[F(x),x^0] \\ &\equiv \max_{k>0}\{k : F(x^1/k) \geq F(x)\}/\max_{k>0}\{k : F(x^0/k) \geq F(x)\} \\ &= \frac{\max_k\{k : G[F(x^1/k)] \geq G[F(x)], k>0\}}{\max_k[k : G[F(x^0/k)] \geq G[F(x)], k>0\}} \\ &= k^1/k^0 \quad \text{say} \end{aligned}$$

where $G[F(x^1/k^1)] = G[F(x)]$ and $G[F(x^0/k^0)] = G[F(x)]$. Since $G[F(x)]$ is linearly homogeneous in x, the last two equations imply $k^1 = G[F(x^1)]/G[F(x)]$ and $k^0 = G[F(x^0)]/G[F(x)]$ which in turn implies $k^1/k^0 = Q_M(x^0,x^1,x) = G[F(x^1)]/G[F(x^0)]$. The other two equalities in (40) now follow from (29) and (30). Q.E.D.

COROLLARY 18.1.

$$Q_P \leq Q_M(x^0,x^1,x) = \widetilde{Q}_K(p^0,p^1,x^0,x^1,x) = Q_A(x^0,x^1,p) \leq Q_L.$$

Proof: Follows from (40) and (31). Q.E.D.

COROLLARY 18.2. *If $Q_M(x^0,x^1,x)$ is independent of $x \gg 0_N$ for all $x^0 \gg 0_N$ and $x^1 \gg 0_N$ and F satisfies conditions I, then F must be homothetic.*

Proof: If $Q_M(x^0,x^1,x)$ is independent of x, then $D[F(x),x^1]/D[F(x),x^0]$ is independent of x for all $x^0 \gg 0_N$ and $x^1 \gg 0_N$. Thus we must have $D[F(x),x^0] = f(x^0)/G[F(x)]$ for some functions f and G. Since F satisfies conditions I, D must satisfy conditions IV and it is evident that f can be taken to be neoclassical and G can be taken to be a monotonically increasing, continuous function of one variable with $G(u) > 0$ if $u > \bar{u} \equiv F(0_N)$. Since $D[F(x),x] = 1 = f(x)/G[F(x)]$ for every $x \gg 0_N$, we have $G[F(x)] = f(x)$, a positive, increasing, concave, linearly homogeneous and continuous function for $x \gg 0_N$. Thus F is homothetic. Q.E.D.

Finally, we note that if F is neoclassical and (14) holds, then: (i) all quantity indexes coincide and equal the value of the aggregator function evaluated at the period 1 quantities x^1 divided by the value of F evaluated at the period 0 quantities x^0; i.e., we have

$$(41) \quad Q_M(x^0, x^1, x) = \tilde{Q}_K(p^0, p^1, x^0, x^1, x) = Q_A(x^0, x^1, p) = F(x^1)/F(x^0)$$

for all $x \gg 0_N$ and $p \gg 0_N$; (ii) all price indexes coincide and equal the ratio of unit costs for the two periods; i.e., we have

$$(42) \qquad P_K(p^0, p^1, x) = \tilde{P}_M(p^0, p^1, x^0, x^1, x) = c(p^1)/c(p^0)$$

for all $x \gg 0_N$; and (iii) the expenditure ratio for the two periods is equal to the product of the price index times the quantity index:

$$(43) \qquad p^{1T}x^1/p^{0T}x^0 = [c(p^1)/c(p^0)][F(x^1)/F(x^0)].$$

4. Other Approaches to Index Number Theory

During the period 1875–1925, perhaps the main approach to index number theory was what Frisch [1936] called the 'atomistic' or 'statistical' approach. This approach assumed that all prices are affected proportionately (except for random errors) by the expansion of the money supply. Therefore, it does not matter which price index was used to measure the common factor of proportionality, as long as the index number contains a sufficient number of statistically independent price ratios. Proponents of this approach were Jevons and Edgeworth but the approach was rather successfully attacked by Bowley [1928] and Keynes. For references to this literature, see Frisch [1936; 2–5].

A 'neostatistical' approach has been initiated by Theil [1960]. For the case of two observations, *Theil's best linear price and quantity indexes* P_0, P_1, Q_0, Q_1 are the solution to the following constrained least squares problem:

$$\min_{P_0,P_1,Q_0,Q_1,e_1,e_2,e_3,e_4} \sum_{i=1}^{4} e_i^2 \quad \text{subject to}$$

$$(44) \quad \begin{array}{ll} \text{(i)} \; p^{0T}x^0 = P_0Q_0 + e_1, & \text{(ii)} \; p^{0T}x^1 = P_0Q_1 + e_2 \\ \text{(iii)} \; p^{1T}x^0 = P_1Q_0 + e_3, & \text{(iv)} \; p^{1T}x^1 = P_1Q_1 + e_4 \end{array}$$

and one other normalization such as $P_0 = 1$ is required. As usual, p^0 and p^1 are the price vectors for the two periods while x^0 and x^1 are the corresponding quantity vectors. P_0 and P_1 are scalars which are interpreted as the price level in periods 0 and 1 respectively while Q_0 and Q_1 are the quantity levels for the two periods. Finally, the e_i are regarded as errors. Kloek and de Wit [1961] suggested a number of modifications to Theil's approach; they suggested (44) for the case of two observations, but with the following three sets of additional normalizations: (1) $P_0 = 1$, $e_1 = 0$, (2) $P_0 = 1$, $e_1 + e_4 = 0$, and (3) $P_0 = 1$, $e_1 = 0$, $e_4 = 0$. Stuvel [1957] and Banerjee [1975] have suggested similar 'neostatistical' index number formulae: Stuvel's index numbers P_1/P_0 and Q_1/Q_0 can be generated by solving (44) subject to the additional normalizations $P_0 = 1$, $e_1 = 0$, $e_4 = 0$ and $e_2 = e_3$.

The other major approach to index number theory is the test or axiomatic approach, initiated by Irving Fisher [1911] [1922]. The test approach assumes that the price and quantity indexes are functions of the price and quantity vectors pertaining to two periods, say $P(p^0, p^1, x^0, x^1)$ and $Q(p^0, p^1, x^0, x^1)$. Tests are a prior 'reasonable' properties that the functions P and Q should possess. However, several researchers (e.g. Frisch [1930], Wald [1937], Samuelson [1974a], Eichhorn [1976] [1978a], Eichhorn and Voeller [1976]) have shown that not all a priori reasonable properties for P and Q can be consistent with each other; i.e. there are various impossibility theorems. Moreover, if one works with a restricted set of tests which are consistent, the resulting family of index number formulae is often not uniquely determined.

However, it turns out that the economic and test approaches to index number theory can be partially reconciled. In the following two sections, we shall assume explicit functional forms for the underlying aggregator function plus the assumption of cost minimizing behavior on the part of the consumer or producer. We shall show that certain functional forms for the aggregator function can be associated with certain functional forms for index number formulae. Many of the resulting index number formulae (e.g. Fisher's [1922] ideal formula) have been suggested as desirable in the literature on the test approach to index number theory.

5. Exact Index Number Formulae

Suppose we are given price and quantity data for two periods, p^0, p^1, x^0 and x^1. A *price index* P is defined to be a function of prices and quantities, $P(p^0, p^1, x^0, x^1)$, while a *quantity index* Q is defined to be another function of the observable prices and quantities for the two periods, $Q(p^0, p^1, x^0, x^1)$. Given either a price index or a quantity index, the other function can be defined implicitly by the following equation (Fisher's [1922] weak factor reversal test):

$$(45) \qquad P(p^0, p^1, x^0, x^1)Q(p^0, p^1, x^0, x^1) = p^{1T}x^1/p^{0T}x^0;$$

i.e., the product of the price index times the quantity index should equal the expenditure ratio between the two periods.

Assume that the producer or consumer is maximizing a neoclassical[19] aggregator function f subject to a budget constraint during the two periods. Under these conditions, it can be shown that the consumer (or producer) is also minimizing cost subject to a utility (or output) constraint and that the cost function C which corresponds to f can be written as

$$C[f(x), p] = f(x)c(p) \tag{46}$$

for $x \geq 0_N$ and $p \gg 0_N$ where $c(p) \equiv \min_x\{p^T x : f(x) \geq 1,\ x \geq 0_N\}$ is f's unit cost function.[20]

A quantity index $Q(p^0, p^1, x^0, x^1)$ is defined to be *exact* for a neoclassical aggregator function f if, for every $p^0 \gg 0_N$, $p^1 \gg 0_N$,[21] $x^r \gg 0_N$ a solution to the aggregator maximization problem $\max_x\{f(x) : p^{rT}x \leq p^{rT}x^r,\ x \geq 0_N\} = f(x^r) > 0$ for $r = 0, 1$, we have

$$Q(p^0, p^1, x^0, x^1) = f(x^1)/f(x^0). \tag{47}$$

Thus in (47), the price and quantity vectors (p^0, p^1, x^0, x^1) are *not* regarded as completely independent variables — on the contrary, we assume that (p^0, x^0) and (p^1, x^1) satisfy the following restrictions in order for the price and quantity vectors to be consistent with 'utility' maximizing behavior during the two periods:

$$p^r \gg 0_N,\ x^r \gg 0_N,\ f(x^r) = \max_x\{f(x) : p^{rT}x \leq p^{rT}x^r,\ x \geq 0_N\} > 0; \quad r = 0, 1. \tag{48}$$

If f is neoclassical, then, using (46), it can be verified that (48) implies (49) and vice versa:

$$p^r \gg 0_N,\ x^r \gg 0_N,\ p^{rT}x^r = f(x^r)c(p^r) = C(f(x^r), p^r) > 0; \quad r = 0, 1. \tag{49}$$

Now we are ready to define the notion of an exact price index.

A *price index* $P(p^0, p^1, x^0, x^1)$ is defined to be *exact* for a neoclassical aggregator function f which has the dual unit cost function c, if for every (p^0, x^0) and (p^1, x^1) which satisfies (48) or (49), we have

$$P(p^0, p^1, x^0, x^1) = c(p^1)/c(p^0). \tag{50}$$

---

[19] f is positive, linearly homogeneous and concave over the positive orthant and is extended to the nonnegative orthant Ω by continuity.

[20] Recall (6) with $G(u) \equiv u$. The function c is also neoclassical.

[21] Sometimes p^0 and p^1 are restricted to a subset of the positive orthant.

Note that if Q is exact for a neoclassical aggregator function f, then Q can be interpreted as a Malmquist, Allen or implicit Konüs quantity index (recall (41)), and the corresponding price index P defined implicitly by Q via (45) can be interpreted as a Konüs or implicit Malmquist price index (recall (42)).

Some examples of exact index number formulae are presented in the following theorems. Before proceeding with these theorems, it is convenient to develop some implications of (48) and (49). If f is neoclassical, (48) is satisfied, and f is differentiable at x^0 and x^1, then

$$(51) \qquad p^r/p^{rT}x^r = \nabla f(x^r)/x^{rT}\nabla f(x^r) = \nabla f(x^r)/f(x^r); \quad r = 0, 1.$$

The first equality in (51) follows from the Hotelling [1935; 71], Wold [1944; 69–71], [1953; 145] identity[22] while the second equality follows from Euler's Theorem on linearly homogeneous functions, $f(x^r) = x^{rT}\nabla f(x^r)$. Also if f is neoclassical, (49) holds and f's unit cost function c is differentiable at p^0 and p^1, then

$$(52) \qquad x^r/p^{rT}x^r = \nabla_p C[f(x^r), p^r]/C[f(x^r), p^r] = \nabla c(p^r)/c(p^r); \quad r = 0, 1.$$

The first equality in (52) follows from Shephard's [1953; 11] Lemma while the second equality follows from (49).

THEOREM 19. (Konüs and Byushgens [1926; 162], Pollak [1971a], Samuelson and Swamy [1974; 574]): *The Paasche and Laspeyres price indexes,* $P_P(p^0, p^1, x^0, x^1) \equiv p^{1T}x^1/p^{0T}x^1$ *and* $P_L(p^0, p^1, x^0, x^1) \equiv p^{1T}x^0/p^{0T}x^0$, *and the Paasche and Laspeyres quantity indexes,* $Q_P(p^0, p^1, x^0, x^1) \equiv p^{1T}x^1/p^{1T}x^0$ *and* $Q_L(p^0, p^1, x^0, x^1) \equiv p^{0T}x^1/p^{0T}x^0$, *are exact for a Leontief [1941] aggregator function,* $f(x) \equiv \min_i\{x_i/b_i : i = 1, \ldots, N\}$, *where* $x \equiv (x_1, \ldots, x_N)^T \geq 0_N$ *and* $b \equiv (b_1, \ldots, b_N)^T \gg 0_N$ *is a vector of positive constants.*

Proof: If f is the Leontief or fixed coefficients aggregator function defined above, then its unit cost function is $c(p) \equiv p^T b$ for $p \gg 0_N$. Now assume (49). Then

$$\begin{aligned} P_L &\equiv p^{1T}x^0/p^{0T}x^0 \\ &= p^{1T}[\nabla c(p^0)/c(p^0)] \text{ using (52)} \\ &= p^{1T}b/c(p^0) \text{ since } \nabla c(p^0) = b \\ &\equiv c(p^1)/c(p^0). \end{aligned}$$

[22]Alternatively, the first equality in (51) is implied by the Kuhn–Tucker conditions for the concave programming problem in (48) upon eliminating the Lagrange multiplier for the binding constraint $p^{rT}x \leq p^{rT}x^r$. The nonnegativity constraints $x \geq 0_N$ are not binding because we assume the solution $x^r \gg 0_N$.

Similarly,

$$\begin{aligned}P_P &\equiv p^{1T}x^1/p^{0T}x^1 = 1/(p^{0T}x^1/p^{1T}x^1)\\ &= 1/\left[p^{0T}[\nabla c(p^1)/c(p^1)]\right] \quad \text{using (52)}\\ &= c(p^1)/p^{0T}b \text{ since } \nabla c(p^1) = b\\ &\equiv c(p^1)/c(p^0).\end{aligned}$$

Thus P_L and P_P are exact price indexes for f, and thus the corresponding quantity indexes, Q_P and Q_L, defined implicitly by the weak factor reversal test (45), are exact quantity indexes for f. Q.E.D.

THEOREM 20. (Pollak [1971a] Samuelson and Swamy [1974; 574]): *The Paasche and Laspeyres price and quantity indexes are also exact for a linear aggregator function, $f(x) \equiv a^T x$ where $a^T \equiv (a_1, \ldots, a_N) \gg 0_N$ is a vector of fixed constants.*

Proof: Assume (48).[23] Then

$$\begin{aligned}Q_L &\equiv p^{0T}x^1/p^{0T}x^0\\ &= x^{1T}[\nabla f(x^0)/f(x^0)] \text{ using (51)}\\ &= x^{1T}a/f(x^0) \text{ since } \nabla f(x) = a\\ &\equiv f(x^1)/f(x^0).\end{aligned}$$

Similarly, $Q_P = f(x^1)/f(x^0)$ and so Q_L and Q_P are exact for the linear aggregator function f defined above. Thus the corresponding price indexes, P_P and P_L, defined implicitly by the weak factor reversal test (45) are exact price indexes for f and its corresponding unit cost function, $c(p) \equiv \min_x\{p^T x : a^T x \geq 1,\ x \geq 0_N\} = \min_i\{p_i/a_i : i = 1, \ldots, N\}$. Q.E.D.

The above theorems show that more than one index number formula can be exact for the same aggregator function, and one index number formula can be exact for quite different aggregator functions.

THEOREM 21. (Konüs and Byushgens [1926; 163–166], Afriat [1972b; 46], Pollak [1971a], Samuelson and Swamy [1974; 574]): *The family of geometric price indexes defined by $P_G(p^0, p^1, x^0, x^1) \equiv \prod_{i=1}^N (p_i^1/p_i^0)^{s_i}$ (where for $i = 1, 2, \ldots, N$, $s_i \equiv m_i(s_i^0, s_i^1)$, $s_i^0 \equiv p_i^0 x_i^0/p^{0T}x^0$, $s_i^1 \equiv p_i^1 x_i^1/p^{1T}x^1$ and*

[23]Note that the definition of exactness requires $x^r \gg 0_N$ and x^r is a solution to the appropriate aggregator maximization problem. Thus it can be seen that p^0 must be proportional to a.

m_i is any function which has the property $m_i(s,s) \equiv s$) is exact for a Cobb–Douglas [1928] aggregator function f defined by

$$(53)\quad f(x) \equiv \alpha_0 \prod_{i=1}^{N} x_i^{\alpha_i}, \ \text{where } \alpha_0 > 0,\ \alpha_1 > 0, \ldots, \alpha_N > 0,\ \sum_{i=1}^{N} \alpha_i = 1.$$

The family of geometric quantity indexes,

$$Q_G(p^0,p^1,x^0,x^1) \equiv \prod_{i=1}^{N} (x_i^1/x_i^0)^{s_i}, \qquad s_i \equiv m_i(s_i^0, s_i^1)$$

is also exact for the aggregator function defined by (53).

Proof: If f is Cobb–Douglas and (48) holds, then for $r = 0,1$, differentiating (53) yields

$$x_i^r \frac{\partial f(x^r)}{\partial x_i} / f(x^r) = \alpha_i = x_i^r p_i^r / p^{rT} x^r \quad \text{using (51)}$$
$$\equiv s_i^r.$$

Thus $s_i^0 = s_i^1 = \alpha_i = s_i \equiv m_i(s_i^0, s_i^1)$ and

$$P_G(p^0,p^1,x^0,x^1) \equiv \prod_{i=1}^{N} (p_i^1/p_i^0)^{s_i} = \prod_{i=1}^{N} (p_i^1/p_i^0)^{\alpha_i}$$
$$= k\prod_{i=1}^{N} (p_i^0)^{\alpha_i} / k\prod_{i=1}^{N} (p_i^0)^{\alpha_i} = c(p^1)/c(p^0)$$

since it can be verified by Lagrangian techniques that the Cobb–Douglas function defined by (53) has the unit cost function

$$c(p) \equiv k \prod_{i=1}^{N} p_i^{\alpha_i} \ \text{where } k \equiv 1/\alpha_0 \prod_{i=1}^{N} \alpha_i^{\alpha_i}.$$

Thus P_G is exact for f. Similarly

$$Q_G(p^0,p^1,x^0,x^1) \equiv \prod_{i=1}^{N} (x_i^1/x_i^0)^{s_i} = \prod_{i=1}^{N} (x_i^1/x_i^0)^{\alpha_i}$$
$$= \alpha_0 \prod_{i=1}^{N} (x_i^1)^{\alpha_i} / \alpha_0 \prod_{i=1}^{N} (x_i^0)^{\alpha_i} = f(x^1)/f(x^0)$$

and so Q_G is also exact for f defined by (53). Q.E.D.

THEOREM 22. (Byushgens [1925], Konüs and Byushgens [1926; 1971], Frisch [1936; 30], Wald [1939; 331], Afriat [1972b; 45] [1977], Pollak [1971a] and Diewert [1976a; 132]):[24] *Irving Fisher's [1922] ideal quantity index*

$$Q_F(p^0, p^1, x^0, x^1) \equiv (p^{1T}x^1/p^{1T}x^0)^{1/2}(p^{0T}x^1/p^{0T}x^0)^{1/2} = (Q_P Q_L)^{1/2}$$

and the corresponding price index

$$P_F(p^0, p^1, x^0, x^1) \equiv (p^{1T}x^1/p^{0T}x^1)^{1/2}(p^{1T}x^0/p^{0T}x^0)^{1/2}$$
$$= (P_P P_L)^{1/2} = p^{1T}x^1/p^{0T}x^0 Q_F(p^0, p^1, x^0, x^1)$$

are exact for the homogeneous quadratic function f defined by

$$f(x) \equiv (x^T A x)^{1/2}, \; x \in S \tag{54}$$

where A is a symmetric $N \times N$ matrix of constants and S is any open, convex subset of the nonnegative orthant Ω such that f is positive, linearly homogeneous and concave over this subset.[25]

Proof: We suppose that the following modified version of (48) holds:[26]

$$p^r \gg 0_N, \; x^r \gg 0_N, \; f(x^r) = \max_x\{f(x) : p^{rT}x \leq p^{rT}x^r, \; x \in S\}; \quad r = 0, 1. \tag{55}$$

Since only the budget constraints $p^{rT}x \leq p^{rT}x^r$ will be binding in the concave programming problems defined in (55), the Hotelling–Wold relations (51) will also hold, since the f defined by (54) is differentiable. Thus

$$p^r/p^{rT}x^r = \nabla f(x^r)/f(x^r) \text{ for } r = 0, 1 \text{ by (51)}$$
$$= \frac{1}{2}(x^{rT}Ax^r)^{-1/2}2Ax^r/(x^{rT}Ax^r)^{1/2} \text{ differentiating (54)}$$
$$= Ax^r/x^{rT}Ax^r, \tag{56}$$

[24]Samuelson [1947; 155] states that S. Alexander also derived this result in an unpublished Harvard paper.

[25]f can be extended to the nonnegative orthant as follows. Because $(x^TAx)^{1/2}$ is linearly homogeneous, S can be taken to be a convex cone. Extend f to $\overline{S}$, the closure of S, by continuity. Now define the free disposal level sets of f by $L(u) \equiv \{x : x \geq x', f(x') \geq u, x' \in \overline{S}\}$ for $u \geq 0$. The extended f is defined as $f(x) \equiv \max_u\{u : x \in L(u), u \geq 0\}$ for $x \geq 0_N$.

[26]The nonnegativity constraints $x \geq 0_N$ have been replaced by $x \in S$. Because we assume that S is an open set and we assume that $x^r \in S$, the constraints $x \in S$ are not binding in (55).

and

$$
\begin{aligned}
Q_F(p^0,p^1,x^0,x^1) &\equiv [x^{1T}(p^0/p^{0T}x^0)/x^{0T}(p^1/p^{1T}x^1)]^{1/2} \\
&= [x^{1T}(Ax^0/x^{0T}Ax^0)/x^{0T}(Ax^1/x^{1T}Ax^1)]^{1/2} \text{ using (56)} \\
&= (x^{1T}Ax^1)^{1/2}/(x^{0T}Ax^0)^{1/2} \text{ since } x^{1T}Ax^0 = x^{0T}Ax^1 \\
&\equiv f(x^1)/f(x^0) \text{ using (54).}
\end{aligned}
$$

Thus Q_F and the corresponding implicit price index

$$
\begin{aligned}
P_F(p^0,p^1,x^0,x^1) &= p^{1T}x^1/p^{0T}x^0 Q_F(p^0,p^1,x^0,x^1) \\
&= f(x^1)c(p^1)/f(x^0)c(p^0)[f(x^1)/f(x^0)] \text{ using (49)} \\
&= c(p^1)/c(p^0)
\end{aligned}
$$

are exact for the aggregator function f defined by (54) where c is the unit cost function which is dual to f. Q.E.D.

The set S which occurs in (54) will be nonempty if we take A to be a symmetric matrix with one positive eigenvalue (and the corresponding eigenvector is positive) while the other eigenvalues of A are zero or negative. For example, take $A = aa^T$ where $a \gg 0_N$ is a vector of positive constants. In this case, S can be taken to be the positive orthant and $f(x) \equiv (x^T aa^T x)^{1/2} = a^T x$, a linear aggregator function. Thus the Fisher price and quantity indexes are also exact for a linear aggregator function.

The above example shows that the matrix A in (54) does not have to be invertible. However if A^{-1} does exist, then, using Lagrangian techniques, it can be shown[27] that $c(p) \equiv (p^T A^{-1} p)^{1/2}$ for $p \in S^*$ where S^* is the set of positive prices where $c(p)$ is positive, linearly homogeneous and concave.

6. Superlative Index Number Formulae

The last example of an exact index number formula is very important for the following reason: unlike the linear aggregator function $a^T x$ or the geometric aggregator function defined by (53), the homogeneous quadratic aggregator function $f(x) \equiv (x^T Ax)^{1/2}$ can provide a second order differential approximation to an arbitrary, linearly homogeneous, twice continuously differentiable

[27]See Pollak [1971a] and Afriat [1972b; 45].

aggregator function, i.e. $(x^T Ax)^{1/2}$ is a *flexible functional form.*[28] Thus if the true aggregator function can be approximated closely by a homogeneous quadratic, and the producer or consumer is engaging in competitive maximizing behavior during the two periods, then the Fisher price and quantity indexes will closely approximate the true ratios of unit and output (or utility). Note that it is not necessary to econometrically estimate the (generally unknown) coefficients which occur in the A matrix, *only the observable price and quantity vectors are required.*

Diewert [1976a; 117] defined a quantity index Q to be *superlative*[29] if it is exact for an aggregator function f which is capable of providing a second order differential approximation to an arbitrary twice continuously differentiable linearly homogeneous aggregator function. Thus Theorem 22 implies that Fisher's ideal index number formula Q_F is superlative.

THEOREM 23. (Konüs and Byushgens [1926; 167–172], Pollak [1971a], Diewert [1976a; 133–134]): *Irving Fisher's ideal price and quantity indexes, P_F and Q_F, are exact for the aggregator function which is dual to the unit cost function c defined by*

$$c(p) \equiv (p^T Bp)^{1/2} \tag{57}$$

where B is a symmetric matrix of constants and S^* is any convex subset of Ω such that c is positive, linearly homogeneous and concave over S^*.[30]

Proof: Assume that (49) is satisfied where $p^0, p^1 \in S^*$, c is defined by (57) and f is the aggregator function dual to this c. Then, since c is differentiable,

---

[28] f is a flexible functional form if it can provide a second order (differential) approximation to an arbitrary twice continuously differentiable function f^* at a point x^*. f differentially approximates f^* at x^* iff (i) $f(x^*) = f^*(x^*)$, (ii) $\nabla f(x^*) = \nabla f^*(x^*)$ and (iii) $\nabla^2 f(x^*) = \nabla^2 f^*(x^*)$, where both f and f^* are assumed to be twice continuously differentiable at x^* (and thus the two Hessian matrices in (iii) will be symmetric). Thus a general flexible functional form f must have at least $1 + N + N(N+1)/2$ free parameters. If f and f^* are both linearly homogeneous, then $f^*(x^*) = x^{*T}\nabla f^*(x^*)$ and $\nabla^2 f^*(x^*)x^* = 0_N$, and thus a flexible linearly homogeneous functional form f need have only $N + N(N-1)/2 = N(N+1)/2$ free parameters. The term 'differential approximation' is in Lau [1974; 184]. Diewert [1974b; 125] or [1976a; 130] shows that $(x^T Ax)^{1/2}$ is a flexible linearly homogeneous functional form.

[29] The term is due to Fisher [1922; 247] who defined a quantity index Q to be superlative if it was numerically close to his ideal index, Q_F.

[30] The aggregator function f which is dual to c defined by (57) can be constructed using the local duality techniques explained in Blackorby and Diewert [1979].

(52) also holds. Thus we have

$$\begin{aligned}
P_F(p^0,p^1,x^0,x^1) &\equiv (p^{1T}x^1/p^{0T}x^1)^{1/2}(p^{1T}x^0/p^{0T}x^0)^{1/2} \\
&= [p^{0T}\nabla c(p^1)/c(p^1)]^{-1/2}[p^{1T}\nabla c(p^0)/c(p^0)]^{1/2} \\
&\quad \text{using (52)} \\
&= (p^{0T}Bp^1/p^{1T}Bp^1)^{-1/2}(p^{1T}Bp^0/p^{0T}Bp^0)^{1/2} \\
&\quad \text{differentiating (57)} \\
&= (p^{1T}Bp^1)^{1/2}/(p^{0T}Bp^0)^{1/2} \text{ since } p^{0T}Bp^1 = p^{1T}Bp^0 \\
&\equiv c(p^1)/c(p^0) \text{ using (57).}
\end{aligned}$$

Thus P_F and the corresponding implicit quantity index

$$\begin{aligned}
Q_F(p^0,p^1,x^0,x^1) &= p^{1T}x^1/p^{0T}x^0 P_F(p^0,p^1,x^0,x^1) \\
&= f(x^1)c(p^1)/f(x^0)c(p^0)[c(p^1)/c(p^0)] \\
&\quad \text{using (49)} \\
&= f(x^1)/f(x^0)
\end{aligned}$$

are exact for the unit cost function defined by (57). Q.E.D.

The set S^* which occurs in (57) will be nonempty if we take B to be a symmetric matrix with one positive eigenvalue (and the corresponding eigenvector is a vector with positive components) while the other eigenvalues of B are zero or negative. For example, take $B \equiv bb^T$ where $b \gg 0_N$ is a vector of positive constants. In this case, S^* can be taken to be the positive orthant and $c(p) = (p^Tbb^Tp)^{1/2} = p^Tb$, a Leontief unit cost function. Thus the Fisher price and quantity indexes are also exact for a Leontief aggregator function.[31] This example shows that the f and c defined by Theorem 23 do not have to coincide with the f and c defined in Theorem 22. However, Q_F and P_F are exact for both classes of functions. Of course, if B^{-1} or A^{-1} exist, then the f and c defined in Theorem 22 coincide with the f and c defined in Theorem 23 (for a subset of prices and quantities at least).

A price index P is defined to be *superlative* if it is exact for a unit cost function c which can provide a second order differential approximation to an arbitrary twice continuously differentiable unit cost function. Since the c defined by (57) can provide such an approximation, Theorem 23 implies that P_F is a superlative price index.

[31]This fact was first noted by Pollak [1971a].

If P is a superlative price index and $\widetilde{Q}$ is the corresponding quantity index defined implicitly by the weak factor reversal test (45), then we define the pair of index number formulae $(P, \widetilde{Q})$ to be *superlative*. Similarly, if Q is a superlative quantity index and $\widetilde{P}$ is the corresponding implicit price index defined by (45), then the pair of index number formulae $(\widetilde{P}, Q)$ is also defined to be *superlative*.

Before defining some additional pairs of superlative indexes, it is necessary to note the following result. If

$$f^*(z_1, \ldots, z_N) \equiv \alpha_0 + \sum_{i=1}^{N} \alpha_i z_i + \frac{1}{2}\sum_{i=1}^{N}\sum_{j=1}^{N} \alpha_{ij} z_i z_j$$

is a quadratic function defined over an open convex set S, then for every $z^0, z^1 \in S$, the following identity is true:

$$(58) \qquad f^*(z^1) - f^*(z^0) = \frac{1}{2}[\nabla f^*(z^1) + \nabla f^*(z^0)]^T (z^1 - z^0)$$

where $\nabla f^*(z^r)$ is the gradient vector of f^* evaluated at z^r, $r = 0, 1$. The above identity follows simply by differentiating f^* and substituting the partial derivatives into (58).[32]

Now define the Törnqvist [1936] price and quantity indexes, P_0 and Q_0:

$$(59) \qquad P_0(p^0, p^1, x^0, x^1) \equiv \prod_{i=1}^{N} (p_i^1/p_i^0)^{(s_i^0+s_i^1)/2}$$

$$(60) \qquad Q_0(p^0, p^1, x^0, x^1) \equiv \prod_{i=1}^{N} (x_i^1/x_i^0)^{(s_i^0+s_i^1)/2}$$

where $p^0 \gg 0_N$, $p^1 \gg 0_N$, $x^0 \gg 0_N$, $x^1 \gg 0_N$, $s_i^0 \equiv p_i^0 x_i^0 / p^{0T} x^0$ and $s_i^1 \equiv p_1^1 x_1^1 / p^{1T} x^1$ for $i = 1, 2, \ldots, N$.

THEOREM 24. (Diewert [1976a; 119]): *Q_0 is exact for the homogeneous translog aggregator function f defined as*[33]

$$(61) \qquad \ln f(x) \equiv \alpha_0 + \sum_{i=1}^{N} \alpha_i \ln x_i + \frac{1}{2}\sum_{i=1}^{N}\sum_{j=1}^{N} \alpha_{ij} \ln x_i \ln x_j, \quad x \in S$$

where $\sum_{i=1}^{N} \alpha_i = 1$, $\alpha_{ij} = \alpha_{ji}$ for all i, j, $\sum_{j=1}^{N} \alpha_{ij} = 0$ for $i = 1, \ldots, N$ and S is an open convex subset of Ω such that f is positive and concave over S (the above restrictions on the α's ensure that f is linearly homogeneous).

Proof: Assume that the producer or consumer is engaging in maximizing behavior during periods 0 and 1 so that (55) holds. Now define $z_i \equiv \ln x_i^r$

[32]On the other hand if f^* satisfies (58) for all $z^0, z^1 \in S$, then Diewert [1976a; 138] (assuming that f^* is thrice differentiable) and Lau [1979] (assuming that f^* is once differentiable) show that f^* must be a quadratic function.

[33]This functional form is due to Christensen, Jorgenson and Lau [1971] and Sargan [1971].

for $r = 0, 1$ and $i = 1, 2, \ldots, N$. If we define $f^*(z) \equiv \alpha_0 + \sum_{i=1}^N \alpha_i z_i + (1/2)\sum_{i=1}^N \sum_{j=1}^N \alpha_{ij} z_i z_j$ where the α's are as defined in (61), then, since f^* is quadratic in z, we can apply the identity (58). Since

$$\partial f^*(z^r)/\partial z_j \equiv \partial \ln f(x^r)/\partial \ln x_j = [x_j^r/f(x^r)][\partial f(x^r)/\partial x_j]$$

for $r = 0, 1$ and $j = 1, \ldots, N$, (58) translates into the following identity involving the partial derivatives of the f defined by (61):

$$\begin{aligned}\ln f(x^1) - \ln f(x^0) &= \frac{1}{2}\sum_{i=1}^N \left[\frac{x_i^1}{f(x^1)}\frac{\partial f(x^1)}{\partial x_i} + \frac{x_i^0}{f(x^0)}\frac{\partial f(x^0)}{\partial x_i}\right](\ln x_i^1 - \ln x_i^0) \\ &= \frac{1}{2}[\nabla_{\ln x} \ln f(x^1) + \nabla_{\ln x} \ln f(x^0)](\ln x^1 - \ln x^0)\end{aligned}$$

or

$$\ln f(x^1)/f(x^0) = \frac{1}{2}\sum_{i=1}^N \left[\frac{x_i^1 p_i^1}{p^{1T}x^1} + \frac{x_i^0 p_i^0}{p^{0T}x^0}\right] \ln(x_i^1/x_i^0) \text{ using (51).}$$

Therefore

$$f(x^1)/f(x^0) = \prod_{i=1}^N (x_i^1/x_i^0)^{(s_i^1+s_i^0)/2} \equiv Q_0(p^0, p^1, x^0, x^1).$$

Q.E.D.

Define the implicit Törnqvist price index, $\widetilde{P}_0(p^0, p^1, x^0, x^1) \equiv p^{1T}x^1/[p^{0T}x^0 \times Q_0(p^0, p^1, x^0, x^1)]$. Since Q_0 is exact for the homogeneous translog f defined by (61), and since the homogeneous translog f is a flexible functional form (it can provide a second order differential approximation to an arbitrary twice continuously differentiable linearly homogeneous aggregator function), $(\widetilde{P}_0, Q_0)$ is a superlative pair of index number formulae.

THEOREM 25. (Diewert [1976a; 121]):[34] *P_0 defined by (59) is exact for the translog unit cost function c defined as*

$$\text{(62)} \quad \ln c(p) \equiv \alpha_0^* + \sum_{i=1}^N \alpha_i^* \ln p_i + \frac{1}{2}\sum_{i=1}^N \sum_{j=1}^N \alpha_{ij}^* \ln p_i \ln p_j, \quad p \in S^*$$

[34]Theil [1965; 71–72] virtually proved this theorem; however, he did not impose linear homogeneity on $c(p)$ defined by (62), which is required in order for (52) to be valid.

where $\sum_{i=1}^{N}\alpha_i^* = 1$, $\alpha_{ij}^* = \alpha_{ji}^*$ *for all* i, j, $\sum_{j=1}^{N}\alpha_{ij}^* = 0$ *for* $i = 1,\ldots,N$ *and* S^* *is an open, convex subset of* Ω *such that* c *is positive and concave over* S^*.

Proof: Assume that the producer or consumer is engaging in cost minimizing behavior during periods 0 and 1 and thus we assume that (49) and its consequence (52) hold, with $p^0, p^1 \in S^*$. Since $\ln c(p)$ is quadratic in the variables $z_i \equiv \ln p_i$, we can again apply the identity (58) which translates into the following identity involving the partial derivatives of the c defined by (62):

$$\ln c(p^1) - \ln c(p^0) = \frac{1}{2}\sum_{i=1}^{N}\left[\frac{p_i^1}{c(p^1)}\frac{\partial c(p^1)}{\partial p_i} + \frac{p_i^0}{c(p^0)}\frac{\partial c(p^0)}{\partial p_i}\right](\ln p_i^1 - \ln p_i^0)$$

or

$$\ln c(p^1)/c(p^0) = \frac{1}{2}\sum_{i=1}^{N}\left[\frac{p_i^1 x_i^1}{p^{1T}x^1} + \frac{p_i^0 x_i^0}{p^{0T}x^0}\right]\ln(p_i^1/p_i^0) \text{ using (52).}$$

Therefore

$$c(p^1)/c(p^0) = P_0(p^0,p^1,x^0,x^1) \quad \text{using definition (59).}$$

Q.E.D.

Now define the implicit Törnqvist quantity index, $\widetilde{Q}_0(p^0,p^1,x^0,x^1) \equiv p^{1T}x^1/p^{0T}x^0 P_0(p^0,p^1,x^0,x^1)$. Since P_0 is exact for the flexible functional form defined by (62), $(P_0,\widetilde{Q}_0)$ is also a superlative pair of index number formulae. It should be noted that the translog unit cost function is in general *not* dual to the homogeneous translog aggregator function defined by (61) (except when all $\alpha_{ij} = 0 = \alpha_{ij}^*$ and $\alpha_i = \alpha_i^*$, in which case (61) and (62) reduce to the Cobb–Douglas functional form).

Thus far, we have found three pairs of superlative index number formulae: (P_F,Q_F), $(P_0,\widetilde{Q}_0)$ and $(\widetilde{P}_0,Q_0)$. In turns out that there are many more such formulae. For $r \neq 0$, define the *quadratic mean of order r aggregator function*[35] f_r as

$$(63) \qquad f_r(x) \equiv \left(\sum_{i=1}^{N}\sum_{j=1}^{N} a_{ij}x_i^{r/2}x_j^{r/2}\right)^{1/r}, \qquad x \in S$$

[35]An ordinary mean of order r (see Hardy, Littlewood and Polya [1934]) is defined as $F_r(x) \equiv (\sum_{i=1}^{N} a_i x_i^r)^{1/r}$ for $x \gg 0_N$ where $a_i \geq 0$ and $\sum_{i=1}^{N} a_i = 1$. Note that $kF_r(x)$ where $k > 0$ is the constant elasticity of substitution functional form (see Arrow, Chenery, Minhas and Solow [1961]) so that f_r defined by (63) contains this functional form as a special case.

where S is an open subset of Ω where f_r is neoclassical, and define the *quadratic mean order r unit cost function*[36] c_r as

$$c_r(p) \equiv \left(\sum_{i=1}^{N}\sum_{j=1}^{N} b_{ij}p_i^{r/2}p_j^{r/2}\right)^{1/r}, \qquad p \in S^* \tag{64}$$

where S^* is an open subset of Ω where c_r is neoclassical. For $r \neq 0$, define the following price and quantity indexes:

(65)

$$P_r(p^0,p^1,x^0,x^1) \equiv \left[\sum_{i=1}^{N} s_i^0(p_i^1/p_i^0)^{r/2}\right]^{1/r}\left[\sum_{j=1}^{N} s_j^1(p_j^1/p_j^0)^{-r/2}\right]^{-1/r}$$

$$Q_r(p^0,p^1,x^0,x^1) \equiv \left[\sum_{i=1}^{N} s_i^0(x_i^1/x_i^0)^{r/2}\right]^{1/r}\left[\sum_{j=1}^{N} s_j^1(x_j^1/x_j^0)^{-r/2}\right]^{-1/r}$$

where p^0, p^1, x^0, $x^1 \gg 0_N$, $s_i^0 \equiv p_i^0x_i^0/p^{0T}x^0$ and $s_i^1 \equiv p_i^1x_i^1/p^{1T}x^1$ for $i = 1,2,\ldots,N$.

It can be shown[37] (in a manner analogous to the proof of Theorem 22), that for each $r \neq 0$, Q_r defined by (65) is exact for f_r defined by (63). Similarly, it can be shown[38] (in a manner analogous to the proof of Theorem 23), that P_r defined by (65) is exact for c_r defined by (64). Since it is easy to show (cf. Diewert [1976a; 130] that f_r and c_r are flexible functional forms for each $r \neq 0$, it can be shown that $(P_r, \widetilde{Q}_r)$ and $(\widetilde{P}_r, Q_r)$ are pairs of superlative index number formulae for each $r \neq 0$, where $\widetilde{Q}_r \equiv p^{1T}x^1/p^{0T}x^{0T}P_r$ and $\widetilde{P}_r \equiv p^{1T}x^1/p^{0T}x^0Q_r$. Note that $P_2 = P_F$ (Fisher's ideal price index) and $Q_2 = Q_F$ (Fisher's ideal quantity index) so that $(P_2, \widetilde{Q}_2) = (\widetilde{P}_2, Q_2) = (P_F, Q_F)$. Moreover, it can be shown that the homogeneous translog aggregator function defined by (61) is a limiting case of f_r defined by (63) as r tends to zero (similarly, the translog unit cost function defined by (62) is a limiting case of c_r as r tends to zero)[39] and that Q_0 defined by (60) is a limiting case of Q_r as r tends to 0 while P_0 defined by (59) is a limiting case of P_r as r tends to 0.[40]

Given such a multiplicity of superlative indexes, the question arises: which index number formula should be used in empirical applications? The answer appears to be that it doesn't matter, provided that the variation in prices and quantities is not too great going from period 0 to period 1. This is

[36]See Denny [1974] who introduced c_r to the economics literature.
[37]See Diewert [1976a; 132].
[38]See Diewert [1976a; 133–134].
[39]See Diewert [1980; 451].
[40]See Khaled [1978; 95–96].

because it has been shown[41] that the functions P_r and P_s differentially approximate each other to the second order for all r and s, provided that the derivatives are evaluated at any point where $p^0 = p^1$ and $x^0 = x^1$: i.e. we have $P_r(p^0, p^1, x^0, x^1) = \widetilde{P}_s(p^0, p^1, x^0, x^1)$, $\nabla P_r(p^0, p^1, x^0, x^1) = \nabla \widetilde{P}_s(p^0, p^1, x^0, x^1)$ and $\nabla^2 P_r(p^0, p^1, x^0, x^1) = \nabla^2 \widetilde{P}_s(p^0, p^1, x^0, x^1)$ for all r and s, provided that $p^0 = p^1 \gg 0_N$ and $x^0 = x^1 \gg 0_N$. ∇P_r stands for the $4N$ dimensional vector of first order partials of P_r, $\nabla^2 P_r$ stands for the $4N$ matrix of second order partials of P_r, etc. The quantity indexes Q_r and $\widetilde{Q}_s$ similarly differentially approximate each other to the second order for all r and s, provided that prices and quantities are the same for the two periods. These results are established by straightforward but tedious calculations — moreover, the assumption of optimizing behavior on the part of the consumer or producer is not required in order to derive these results.

Diewert [1978b] also shows that the Paasche and Laspeyres price indexes, P_P and P_L, differentially approximate each other and the superlative indexes, P_r and $\widetilde{P}_s$, to the *first* order for all r and s, provided that prices and quantities are the same for the two periods. Thus if the variation in prices and quantities is relatively small between the two periods, the indexes P_L, P_P, P_r and $\widetilde{P}_s$ will all yield approximately the same answer.

Diewert [1978b] argues that the above results provide a reasonably strong justification for using the *chain principle* when calculating official indexes such as the consumer price index or the GNP deflator, rather than using a fixed base, since in using the chain principle the base is changed every year, and thus the changes between p^0 and p^1 and x^0 and x^1 will be minimized, leading to smaller discrepancies between P_L and P_P, and even smaller discrepancies between the superlative indexes P_r and $\widetilde{P}_s$.[42]

However, in some situations (e.g. in cross country comparisons or when decennial census data are being used) there can be considerable variation in the price and quantity data going from period (or observation) 0 to period (or observation) 1, in which case the indexes P_r and $\widetilde{P}_s$ can differ considerably. In this situation, it is sometimes useful to compare the variation in the N quantity ratios (x_i^1/x_i^0) to the variation in the N price ratios (p_i^1/p_i^0). If there is less variation in the quantity ratios than in the price ratios, then the quantity indexes Q_r defined by (66) are share weighted averages of the quantity ratios and will tend to be more stable than the implicit indexes $\widetilde{Q}_r$. On the other hand, if there is less variation in the price ratios than in the quantity ratios (the more typical case), then the price indexes P_r defined by (65) are share weighted averages of the price ratios (p_i^1/p_i^0) and will tend to be in closer

[41]See Diewert [1978b] who utilizes the work of Vartia [1976a] [1976b]. Vartia [1978] provides an alternative proof.

[42]The chain principle can also be justified from the viewpoint of Divisia indexes; see Wold [1953; 134–139] and Jorgenson and Griliches [1967].

agreement with each other than the implicit price indexes $\widetilde{P}_r$. Thus, in the first situation, we would recommend the use of $(\widetilde{P}_r, Q_r)$ for some r,[43] while in the second situation we would recommend the use of $(P_r, \widetilde{Q}_r)$ for some r.[44] Notice that the Fisher index, $(P_F, Q_F) = (P_2, \widetilde{Q}_2) = (\widetilde{P}_2, Q_2)$ can be used in either situation. A further advantage for the Fisher formulae (P_F, Q_F) is that Q_F is consistent with revealed preference theory: i.e., even if the true aggregator function f is nonhomothetic, under the assumption of maximizing behavior, Q_F will correctly indicate the direction of change in the aggregate when revealed preference theory tells us that the aggregate is decreasing, increasing or remaining constant (cf. Diewert [1976a; 137]). Recall also that Q_F is consistent both with a linear aggregator function (perfect substitutability) and a Leontief aggregator function (no substitutability). No other superlative index number formula Q_r or $\widetilde{Q}_r$, $r \neq 2$, has the above rather nice properties.

We conclude this section by showing that some of the above superlative index number formulae are also exact for nonhomothetic aggregator functions.

THEOREM 26. (Diewert [1976a; 122]): *Let the functional form for the cost function $C(u,p)$ be a general translog defined by*

$$(66)\quad \ln C(u,p) \equiv \alpha_0 + \sum_{i=1}^N \alpha_i \ln p_i + \frac{1}{2}\sum_{i=1}^N \sum_{j=1}^N \gamma_{ij} \ln p_i \ln p_j + \delta_0 \ln u + \sum_{i=1}^N \delta_i \ln p_i \ln u + \frac{1}{2}\varepsilon_0 (\ln u)^2$$

where the parameters satisfy the following restrictions:

$$(67)\quad \begin{aligned} &\sum_{i=1}^N \alpha_i = 1; \ \gamma_{ij} = \gamma_{ji} \text{ for all } i,j; \\ &\sum_{j=1}^N \gamma_{ij} = 0 \qquad \text{for } i = 1,2,\ldots,N, \quad \text{and} \sum_{i=1}^N \delta_i = 0. \end{aligned}$$

Let (u^0, p^0) and (u^1, p^1) belong to a (u,p) region where $C(u,p)$ satisfies conditions II where $u^0 > 0$, $u^1 > 0$, $p^0 \gg 0_N$, $p^1 \gg 0_N$ and the corresponding quantity vectors are $x^0 \equiv \nabla_p C(u^0, p^0) > 0_N$ and $x^1 \equiv \nabla_p C(u^1, p^1) > 0_N$ respectively. Then

$$(68)\qquad P_0(p^0, p^1, x^0, x^1) = C(u^*, p^1)/C(u^*, p^0)$$

---

[43]If $(x_i^1/x_i^0) = k > 0$ for all i, then $(\widetilde{P}_r, Q_r) = (p^{1T}x^1/p^{0T}x^0 k, k)$ for all r, and the use of $(\widetilde{P}_r, Q_r)$ can be theoretically justified using Leontief's [1936; 54–57] Aggregation Theorem.

[44]If $(p_i^1/p_i^0) = k > 0$ for all i, then $(P_r, \widetilde{Q}_r) = (k, p^{1T}x^1/p^{0T}x^0 k)$ for all r, and the use of $(P_r, \widetilde{Q}_r)$ can be theoretically justified using Hicks' [1946; 312–313] Composite Commodity Theorem. See also Wold [1953; 102–110], Gorman [1953; 76–77] and Diewert [1978a; 23].

where P_0 is the Törnqvist price index defined by (59) and the reference utility level $u^* \equiv (u^0u^1)^{1/2}$.

Proof: For a fixed u^*, $\ln C(u^*, p)$ is quadratic in the variables $z_i \equiv \ln p_i$ and thus we may apply the identity (53) to obtain

$$\ln C(u^*,p^1) - \ln C(u^*,p^0)$$

$$= \frac{1}{2}\sum_{i=1}^{N}\Big[[p_i^1 \ln C(u^*,p^1)/\partial p_i] + [p_i^0 \ln C(u^*,p^0)/\partial p_i]\Big](\ln p_i^1 - \ln p_i^0)$$

$$= \frac{1}{2}\sum_{i=1}^{N}\Big[[p_i^1 \partial \ln C(u^1,p^1)/\partial p_i] + [p_i^0 \partial \ln C(u^0,p^0)/\partial p_i]\Big](\ln p_i^1 - \ln p_i^0)$$

where the equality follows upon evaluating the derivatives of C and noting that $2\ln u^* = \ln u^1 + \ln u^0$,

$$= \ln P_0(p^0,p^1,x^0,x^1)$$

using the definitions of x^0, x^1 and P_0 and equations (52). Q.E.D.

Note that the right hand side of (68) is the true Konüs price index which corresponds to the general translog cost function defined by (66), evaluated at the reference utility level of u^*, the square root of the product of the period 0 and 1 utility levels, u^0 and u^1. We note that the translog cost function can provide a second order differential approximation to an arbitrary twice continuously differentiable cost function.

THEOREM 27. (Diewert [1976a; 123–124]): *Let the aggregator function F be such that F's distance function D is the translog distance function defined by $\ln D(u,x) \equiv \ln C(u,x)$ where C is defined by (66) and (67). Let (u^0,x^0) and (u^1,x^1) belong to a (u,x) region where $D(u,x)$ satisfies conditions IV where $u^0 > 0$, $u^1 > 0$, $x^0 \gg 0_N$, $x^1 \gg 0_N$, $D(u^0,x^0) = 1$, $D(u^1,x^1) = 1$ and the corresponding vectors of normalized prices are $p^0/p^{0T}x^0 \equiv \nabla_x D(u^0,x^0) > 0_N$ and $p^1/p^{1T}x^1 \equiv \nabla_x D(u^1,x^1) > 0_N$ respectively.*[45] *Then*

$$Q_0(p^0,p^1,x^0,x^1) = D(u^*,x^1)/D(u^*,x^0) \tag{69}$$

where Q_0 is the Törnqvist quantity index defined by (60) and the reference utility level $u^* \equiv (u^0u^1)^{1/2}$.

Proof: For a fixed u^*, $\ln D(u^*,x)$ is quadratic in the variables $z_i \equiv \ln x_i$ and thus we may apply the identity (58) to obtain

$$\ln D(u^*,x^1) - \ln D(u^*,x^0)$$

$$= \frac{1}{2}\sum_{i=1}^{N}\Big[[x_i^1 \partial \ln D(u^*,x^1)/\partial x_i] + [x_i^0 \partial \ln D(u^*,x^0)/\partial x_i]\Big](\ln x_i^1 - \ln x_i^0)$$

[45]These assumptions imply that x^r is a solution to the aggregator maximization problem $\max_x\{F(x) : p^{rT}x = p^{rT}x^r, x \geq 0_N\} = F(x^r) \equiv u^r$ for $r = 0,1$ where F is locally dual (cf. Blackorby and Diewert [1979]) to the translog distance function D defined above.

$$= \frac{1}{2}\sum_{i=1}^{N} [[x_i^1 \ln D(u^1, x^1)/\partial x_i] + [x_i^0 \partial \ln D(u^0, x^0)/\partial x_i]] \ln(x_i^1/x_i^0)$$

where the equality follows upon evaluating the derivatives of D and noting that $2 \ln u^* = \ln u^1 + \ln u^0$

$$= \frac{1}{2}\sum_{i=1}^{N} [[x_i^1 p_i^1/p^{1T} x^1 D(u^1, x^1)] + [x_i^0 p_i^0/p^{0T} x^0 D(u^0, x^0)]] \ln(x_i^1/x_i^0)$$

using $p^r/p^{rT} x^r = \nabla_x D(u^r, x^r), \quad r = 0, 1,$[46]

$$= \ln Q_0(p^0, p^1, x^0, x^1)$$

using $D(u^1, x^1) = 1$, $D(u^0, x^0) = 1$ and the definition of Q_0. Q.E.D.

Note that the right hand side of (69) is the Malmquist quantity index which corresponds to the translog distance function, evaluated at the reference utility level $u^* = (u^0 u^1)^{1/2}$. Theorem 27 provides a fairly strong justification for the use of Q_0 in empirical applications, since the translog distance function can differentially approximate an arbitrary twice continuously differentiable distance function to the second order.[47] However, the Fisher ideal index Q_2 can be given a similar strong justification in the context of nonhomothetic aggregator functions.[48]

7. Historical notes and additional related topics

Our survey of the economic theory of index numbers is based on the work of Konüs [1924], Frisch [1936], Allen [1949], Malmquist [1953], Pollak [1971a], Afriat [1972a] [1972b] [1977] and Samuelson and Swamy [1974]. The results noted in Sections 2 and 3 are either taken directly from or are straightforward modifications of results obtained by the above authors, except that in many cases we have weakened the original author's regularity conditions.[49]

[46]This identity is due to Shephard [1953; 10–13] and Hanoch [1978a; 116].

[47]Let D be a distance function which satisfies certain local regularity properties and let F be the corresponding local aggregator function, and C be the corresponding local cost function. Blackorby and Diewert [1979] show that if D differentially approximates D^* to the second order, then F differentially approximates F^*, and C differentially approximates C^* to the second order where F^* and C^* are dual to D^*.

[48]See Diewert [1976b; 149].

[49]Our regularity conditions can be further weakened: for all of the results in Sections 2 and 3 which do not involve the Malmquist quantity index, we need only assume that F be continuous and be subject to local nonsatiation (it turns out that the corresponding C will still satisfy conditions II). Also Theorems 11, 12, 14 and 16 can be proven provided that F be only continuous from above and increasing.

The reader will have noted that many of the proofs in Sections 2 and 3 use arguments that are used in revealed preference theory. For further material on the interconnections between revealed preference theory and index number theory, see Leontief [1936], Samuelson [1947; 146–163], Allen [1949], Diewert [1976b], Vartia [1976b; 144] and Afriat [1977].

There is extensive literature on the measurement of real output or real value added that is analogous to our discussion on the measurement of utility or real input: see Samuelson [1950a], Bergson [1961], Moorsteen [1961], Fisher and Shell [1972b; 49–113] (the last three references make use of a quantity index analogous to the Malmquist index), Samuelson and Swamy [1974; 588–592], Sato [1976b], Archibald [1977] and Diewert [1980].

Background material on the duality between cost, production or utility, and distance or deflation functions can be found in Shephard [1953] [1970], McFadden [1978a], Hanoch [1978a], Blackorby, Primont and Russell [1978], Diewert [1974a] [1978c], Deaton [1979] and Weymark [1980].

Turning now to Sections 5 and 6, for theorems which prove converses to Theorems 19 to 25 under various regularity conditions, see Byushgens [1925], Konüs and Byushgens [1926], Pollak [1971a], Diewert [1976a] and Lau [1979].

Sato [1976a] shows that a certain index number formula (which was defined independently by Vartia [1974]) is exact for the CES aggregator function defined by (63) with $a_{ij} \equiv 0$ for $i \neq j$ for all r, while Lau [1979] develops a partial converse theorem.

In Theorem 22, preferences were assumed to be represented by the transformed quadratic function, $(x^T Ax)^{1/2}$. The assumption that preferences can be represented, at least locally, by a general quadratic function of the form $a_0 + a^T x + 1/2 x^T Ax$ has a long history in economics, perhaps starting with Bennet [1920]. Other authors who have approximated preferences quadratically, in addition to those mentioned in Theorem 22, include Bowley [1928], Hotelling [1938], Hicks [1946; 331–333], Kloek [1967], Theil [1967; 200–212] [1968], and Harberger [1971].

Kloek and Theil utilize quadratic approximations in the logarithms of prices and quantities and they obtain results which are related to Theorems 25 and 26 above. Kloek [1967] shows that the Törnqvist price index $P_0(p^0, p^1, x^0, x^1)$ approximates the true Konüs price index $P_K(p^0, p^1, u^m)$ to the second order where u^m, an intermediate utility level, is defined implicitly by the equation $C(u^m, p^0)/C(u^0, p^0) = C(u^1, p^1)/C(u^m, p^1)$ and C is the true cost function. On the quantity side, Kloek [1967] shows that the implicit Törnqvist quantity index $\widetilde{Q}_0(p^0, p^1, x^0, x^1)$ approximates the true Allen quantity index $Q_A(x^0, x^1, p^m) \equiv C[F(x^1), p^m]/C[F(x^0), p^m]$ to the second order where $p^m \equiv (p_1^m, p_2^m, \ldots, p_N^m)^T$, an intermediate price vector, is defined by $p_i^m \equiv (p_i^0 p_i^1)^{1/2}$, $i = 1, \ldots, N$ and F is the aggregator function dual to the true cost function C. On the other hand, Theil [1968] shows that $P_0(p^0, p^1, x^0, x^1)$ approximates the true Konüs price index $P_K(p^0, p^1, \bar{u})$ to the second order where $\bar{u}$, an intermediate utility level, is defined as $\bar{u} \equiv G(p^m/y^m)$ where G is the indirect utility function dual to the true cost function C,[50] p^m is Kloek's intermediate price vector defined above and $y^m \equiv (p^{0T}x^0 p^{1T}x^1)^{1/2}$ is an intermediate expenditure. Finally, on the quantity side, Theil [1967] [1968] proves Kloek's result (i.e. that $\widetilde{Q}_0(p^0,p^1,x^0,x^1)$ approximates $Q_A(x^0,x^1,p^m)$ to the second order) and in addition, shows that the direct Törnqvist quantity index $Q_0(p^0,p^1,x^0,x^1)$ also approximates $Q_A(x^0,x^1,p^m)$ to the second order.

It should be noted that index number theory and consumer surplus analysis are closely related. Thus the Paasche–Allen quantity index $Q_A(x^0,x^1,p^1) \equiv C[F(x^1),p^1]/C[F(x^0),p^1]$, is closely related to Hicks' [1941–42; 128] [1946; 40–41] *compensating variation in income*,[51] $C[F(x^1),p^1] - C[F(x^0),p^1]$, and the Laspeyres–Allen quantity index, $Q_A(x^0,x^1,p^0) \equiv C[F(x^1),p^0]/C[F(x^0),p^0]$, is closely related to Hicks' [1941–42; 128] [1946; 331] *equivalent variation in income*, $C[F(x^1),p^0] - C[F(x^0),p^0]$. Thus the various bounds we developed for index numbers in the previous section have counterparts in consumer surplus analysis. Hicks [1941–42] and Samuelson [1947; 189–202] emphasized the interconnection between index number theory and consumer surplus measures. For additional results and references to the literature on consumer surplus, see Hotelling [1938], Samuelson [1942], Harberger [1971], Silberberg [1972], Hause [1975], Chipman and Moore [1976] and Diewert [1976b]. The attractiveness of the Malmquist quantity index $Q_M(x^0,x^1,x)$ does not seem to have been noted in the applied welfare economics literature, although the closely related concept inherent in Debreu's [1951] coefficient of resource utilization has been recognized. Perhaps in the future there will be more applications of the Kloek–Theil approximation results, or of Theorem 27 above which shows that the Törnqvist quantity index Q_0 is numerically equal to a certain Malmquist index.

Another type of price and quantity index which we must mention is the Divisia [1925] [1926; 40] index (which is perhaps due to Bennet [1920; 461]). The Bennet–Divisia justification for these indexes proceeds as follows. Regard $(x_1,\ldots,x_N)^T \equiv x$ and $(p_1,\ldots,p_N)^T \equiv p$ as functions of time, $x(t)$ and $p(t)$ for $i = 1,\ldots,N$. Now differentiate expenditure with respect to time and we obtain.[52]

$$(70) \quad \partial\Big[\sum\nolimits_{i=1}^{N} p_i(t)x_i(t)\Big]\Big/\partial t = \sum\nolimits_{i=1}^{N} p_i(t)\partial x_i(t)/\partial t + \sum\nolimits_{i=1}^{N} x_i(t)\partial p_i(t)/\partial t.$$

[50] $G(p^m/y^m) \equiv \max_u\{u : C(u,p^m/y^m) \le 1\} \equiv \max_x\{F(x) : (p^m/y^m)^T x \le 1, x \ge 0_N\}$ where C is the cost function and F is the aggregator function.

[51] Hicks' verbal definition of the compensating variation can be interpreted to mean $C[F(x^0),p^1] - C[F(x^0),p^0]$, and this interpretation is related to the Laspeyres–Konüs cost of living index.

[52] "The fundamental idea is that over a short period the rate of increase of expenditure of a family can be divided into two parts x and I, where x measures the increase due to change of prices and I measures the increase due to increase of consumption; x is the total of the various quantities consumed, each multiplied by the appropriate rate of increase of price, and I is the total of the

Now divide both sides of the above equation through by $\sum_{i=1}^N p_i(t)x_i(t) \equiv p(t)^T x(t)$ and we obtain the identity:

$$(71)\quad \partial \ln[p(t)^T x(t)]/\partial t = \sum_{i=1}^N s_i(t)\partial \ln x_i(t)/\partial t + \sum_{i=1}^N s_i(t)\partial \ln p_i(t)/\partial t$$

where $s_i(t) \equiv p_i(t)x_i(t)/p(t)^T x(t)$ for $i = 1,2,\ldots,N$. The term on the left hand side of (70) is the rate of change of expenditures, which is decomposed into a share weighted rate of change of quantities plus a share weighted rate of change of prices. Denote $\dot{x}_i(t) \equiv \partial x_i(t)/\partial t$ and $\dot{p}_i(t) \equiv \partial p_i(t)/\partial t$ and integrate both sides of (70) to obtain

$$(72)\quad \ln p(1)^T x(1)/p(0)^T x(0) = \int_0^1 \left[\sum_{i=1}^N s_i(t)\dot{x}_i(t)/x_i(t)\right] dt + \int_0^1 \left[\sum_{i=1}^N s_i(t)\dot{p}_i(t)/p_i(t)\right] dt.$$

The first term on the right hand side of the above equation is defined to be the natural logarithm of the ***Divisia quantity index***, $\ln[X(1)/X(0)]$, while the second term is the logarithm of the ***Divisia price index***, $\ln[P(1)/P(0)]$.

The above derivation of the Divisia indexes, $X(1)/X(0)$ and $P(1)/P(0)$, is devoid of any economic interpretation. However, Ville [1951–52], Malmquist [1953; 227], Wold [1953; 134–147], Solow [1957], Gorman [1959; 479] [1970], Jorgenson and Griliches [1967; 253] and Hulten [1973] show that if the consumer or producer is continuously maximizing a well behaved linearly homogeneous aggregator function subject to a budget constraint between $t = 0$ and $t = 1$, then $P(1)/P(0) = P_K(p(0), p(1), \bar{x})$ (i.e. the Divisia price index equals the true Konüs price index for any reference quantity vector $\bar{x} \gg 0_N$) and we can deduce that $X(1)/X(0) = Q_M(x(0), x(1), \bar{x}) = Q_A(x(0), x(1), \bar{p}) = \widetilde{Q}_K(p(0), p(1), x(0), x(1), \bar{x})$ (i.e. the Divisia quantity index equals the Malmquist, Allen, and implicit Konüs quantity indexes for all reference vectors $\bar{x} \gg 0_N$ and $\bar{p} \gg 0_N$). On the other hand, Ville [1951–52; 127], Malmquist [1953; 226–227], Gorman [1970; 7], Silberberg [1972; 944] and Hulten [1973; 1021–1022] show that if the aggregator function is not homothetic, then the line integrals defined on the right hand side of (72) are not independent of the path of integration and thus the Divisia indexes are also path dependent.

We have not stressed the Divisia approach to index numbers in this survey since economic data typically are not collected on a continuous time basis. Since there are many ways of approximating the line integrals in (72) using discrete

---

prices of commodities, each multiplied by the rate of increase in its consumption' (Bennet [1920; 455]). I is the first term on the right hand side of (70) while x is the second term.

data points, the Divisia approach to index number theory does not significantly narrow down the range of discrete type index number formulae, $P(p^0, p^1, x^0, x^1)$ and $Q(p^0, p^1, x^0, x^1)$, that are consistent with the Divisia approach.

The line integral approach also occurs in consumer surplus analysis; see Samuelson [1942] [1947; 189–202], Silberberg [1972], Rader [1976] and Chipman and Moore [1976].

Divisia indexes and exact index number formulae also play a key role in another area of economics which has a vast literature, namely the *measurement of total factor productivity.* A few references to this literature are Solow [1957], Domar [1961], Richter [1966], Jorgenson and Griliches [1967] [1972], Gorman [1970], Ohta [1974], Star [1974], Usher [1974], Christensen, Cummings and Jorgenson [1980], Diewert [1976a; 124–129] [1980; 487–498] and Allen [1981]. To see the relationship of this literature to superlative index number formulae, consider the following example: Let $u^r \equiv f(x^r) > 0$, $r = 0, 1$ be 'intermediate' output produced by a competitive (in input markets) cost minimizing firm where $x^r \gg 0_N$ is a vector of inputs utilized during period r, and f is the homogeneous translog production function defined by (61). Letting $w^0 \gg 0_N$ and $w^1 \gg 0_N$ be the vectors of input prices the producer faces during periods 0 and 1, Theorem 24 tells us that

$$(73) \qquad f(x^1)/f(x^0) = Q_0(w^0, w^1, x^0, x^1)$$

where Q_0 is the Törnqvist quantity index defined by (60). Using (49), we also have

$$(74) \qquad c(w^r)f(x^r) = w^{rT}x^r, \qquad r = 0, 1$$

where $c(w)$ is the unit cost function which is dual to $f(x)$. Suppose now that 'final' output is $y^r \equiv a^r f(x^r)$, $r = 0, 1$ where $a^r > 0$ is defined to be a technology index for period r. The ratio a^1/a^0 can be defined to be a measure of Hicks neutral technical progress.[53] Using (73),

$$(75) \qquad a^1/a^0 \equiv (y^1/y^0)/[f(x^1)/f(x^0)] = y^1/y^0 Q_0(w^0, w^1, x^0, x^1).$$

Thus a^1/a^0 can be calculated using observable data.[54] The unit cost function for y in period r is $c(w)/a^r$. Now suppose the producer behaves monopolistically on his output market and sells his period r output y^r at a price p^r equal to unit cost times a markup factor $m^r > 0$, i.e.

$$(76) \qquad p^r \equiv m^r c(w^r)/a^r, \qquad r = 0, 1.$$

[53]See Blackorby, Lovell and Thursby [1976] for a discussion of the various types of neutral technological change.

[54]This part of the analysis is due to Diewert [1976a; 124–129].

Using (76),

$$(77)\quad m^1/m^0 = (p^1/p^0)(a^1/a^0)/[c(w^1)/c(w^0)] = (p^1y^1/p^0y^0)/(w^{1T}x^1/w^{0T}x^0)$$

using (74) and (75). Thus the rate of markup change m^1/m^0 can be calculated by (77), the value of output ratio deflated by the value of inputs ratio, using observable data.[55] However, if pure profits are zero in each period, then $p^r y^r = w^{rT}x^r = [m^r c(w^r)/a^r][a^r f(x^r)]$ (using (76)) $= m^r w^{rT}x^r$ (using (74)) so that $m^r = 1$ for $r = 0, 1$.

Another area of research which somewhat surprisingly is closely related to index number theory is the measurement of inequality; see Blackorby and Donaldson [1978] [1980] [1981].

Typically, a price or quantity index is not constructed in a single step. For example, in constructing a cost of living index, first food, clothing, transportation and other subindexes are constructed and then they are combined to form a single cost of living index. Vartia [1974; 39–42] [1976a; 124] [1976b; 84–89] defines an index number formula $P(p^0, p^1, x^0, x^1)$ to be ***consistent in aggregation*** if the numerical value of the index constructed in two (or more) stages necessarily coincides with the value of the index calculated in a single stage. Vartia [1976b; 90] stresses the importance of the consistency in aggregation property for national income accounting and notes that the Paasche and Laspeyres indexes have this property (as do the geometric indexes P_G and Q_G defined in Theorem 21 above). Vartia [1976b; 121–140] exhibits many other index number formulae that are consistent in aggregation. Unfortunately, the two families of superlative indexes, $(P_r, \tilde{Q}_r)$ and $(\tilde{P}_s, Q_s)$, are ***not*** consistent in aggregation for any r or s. However, Diewert [1978b] using some of Vartia's results shows that the superlative indexes are ***approximately*** consistent in aggregation (to the second order in a certain sense). Additional results are contained in Blackorby and Primont [1980]. Related to the consistency in aggregation property for an index number formula are the following issues which have been considered by Pollak [1975], Primont [1977], Blackorby and Russell [1978] and Blackorby, Primont and Russell [1978; Chapter 9]: (i) under what conditions do well defined Konüs cost of living subindexes exist for a subset of the commodity space and (ii) under what conditions can the subindexes be combined into the true overall Konüs cost of living index P_K? Finally, a related result is due to Gorman [1970; 3] who shows that the line integral Divisia indexes defined above 'aggregate conformably' or are consistent in aggregation, to use Vartia's term.

If we are given more than two price and quantity observations, then some ideas due to Afriat [1967] can be utilized in order to construct ***nonparametric***

---

[55]This argument is essentially due to Allen [1981]. Allen also generalized his results to many outputs and to nonneutral measures of technical change.

index numbers. Let there be I given price-quantity vectors (p^i, x^i) where $p^i \gg 0_N$, $x^i > 0_N$, $i = 1, 2, \ldots, I$. Use the given data in order to define Afriat's ijth cross coefficient, $D_{ij} \equiv (p^{iT}x^j/p^{iT}x^i) - 1$ for $1 \leq i, j \leq I$. Now consider the following linear programming problem in the $2I + 2I^2$ variables λ_i, ϕ_i, s_{ij}^+, s_{ij}^-, $i, j = 1, \ldots, I$:

$$\text{(78)} \qquad \text{minimize} \sum_{i=1}^{I} \sum_{j=1}^{I} s_{ij}^- \quad \text{subject to}$$

$$\text{(i)} \qquad \lambda_i D_{ij} = \phi_j - \phi_i + s_{ij}^+ - s_{ij}^-; \quad i, j = 1, 2, \ldots, I,$$

$$\text{(ii)} \qquad \lambda_i \geq 1; \quad i = 1, 2, \ldots, I, \quad \text{and}$$

$$\text{(iii)} \qquad \phi_i \geq 0, \ s_{ij}^+ \geq 0, \ s_{ij}^- \geq 0; \quad i, j = 1, 2, \ldots, I.$$

Diewert [1973b][56] shows that if x^i is a solution to

$$\text{(79)} \qquad \max_x \{F(x) : p^{iT}x \leq p^{iT}x^i, \quad x \geq 0_N\}$$

for $i = 1, 2, \ldots, I$ where F is a continuous from above aggregator function which is subject to local nonsatiation (so that the budget constraint $p^{iT}x \leq p^{iT}x^i$ will always hold as an equality for an x which maximizes $F(x)$ subject to the budget constraint), then the objective function in the programming problem (78) will attain its lower bound of zero. On the other hand, Afriat [1967] shows that if the objective function in (78) attains its lower bound of 0 so that $\lambda_i^* D_{ij} \geq \phi_j^* - \phi_i^*$ for all i and j where λ_i^*, ϕ_i^* denote solution variables to (78), then the given quantity vector x^i is a solution to the utility maximization problem (79) for $i = 1, 2, \ldots, I$. Moreover Afriat [1967; 73–74] shows that a utility function F^* which is consistent with the given data in the sense that $F^*(x^i) = \max_x\{F^*(x) : p^{iT}x \leq p^{iT}x^i;\ x \geq 0_N\}$ for $i = 1, 2, \ldots, I$ can be defined as $F^*(x) \equiv \min_i\{F_i^*(x) : i = 1, \ldots, I\}$ where

$$\text{(80)} \qquad F_i^*(x) \equiv \phi_i^* + \lambda_i^*[(p^{iT}x/p^{iT}x^i) - 1], \quad i = 1, 2, \ldots, I,$$

and where the number ϕ_i^* and λ_i^* are taken from the solution to (78). Afriat notes that this F^* is continuous, increasing and concave over the nonnegative orthant and that $F^*(x^i) = \phi_i^*$ for $i = 1, \ldots, I$. Thus if the observed data are consistent with a decision maker maximizing a continuous from above, locally

---

[56] Afriat [1967] has essentially this result. However, there is a slight error in his proof and he does not phrase the problem as a linear programming problem. (78) corrects some severe typographical errors in Diewert's [1973b; 421] equation (3.2).

nonsatiated aggregator function $F(x)$ subject to I budget constraints, then the solution to the linear programming problem (78) can be used in order to construct an approximation F^* to the true F, and this F^* will satisfy much stronger regularity conditions. Diewert [1973b; 424] notes that we can test whether the given data are consistent with the additional hypothesis that the true aggregator function is homothetic or linearly homogeneous by adding the following restrictions to (78): (iv) $\lambda_i = \phi_i$, $1 = 1, \ldots, I$. Geometrically, these additional restrictions force all of the hyperplanes defined by (61) through the origin; i.e. $F_i^*(0_N) = 0$ for all i. Once the linear program (78) is solved, either with or without the additional normalizations (iv), we can calculate $F^*(x^i) = \phi_i^*$ for all i and thus the quantity indexes $F^*(x^{i+1})/F^*(x^i)$ can readily be calculated. Diewert and Parkan [1978] calculated these nonparametric quantity indexes using some Canadian time series data[57] and compared them with the superlative indexes Q_2, Q_0 and $\widetilde{Q}_0$. The differences among all of these indexes turned out to be small.[58] The above method for constructing nonparametric indexes is of course closely related to revealed preference theory.

Finally, we mention that there is an analogous 'revealed production theory' which allows one to construct nonparametric index numbers and nonparametric approximations to production functions and production possibility sets by solving various linear programming problems:[59] see Farrell [1957], Afriat [1972a], Hanoch and Rothschild [1972] and Diewert and Parkan [1983].

[57]However, slightly different but equivalent normalizations were used. In particular, when the general nonhomothetic problem (78 (i), (ii) and (iii)) was solved, ((78) (iii)) was replaced by $\lambda_i \geq 0$ for $i = 1, \ldots, I$, $\phi_1 \equiv 1$ and $\phi_I \equiv Q_2(p^1, p^I, x^1, x^I)$ in order to make the nonhomothetic nonparametric quantity indexes, ϕ_{i+1}^*/ϕ_i^*, comparable to $Q_2(p^i, p^{i+1}, x^i, x^{i+1})$ for $i = 1, 2, \ldots, I-1$.
[58]Diewert and Parkan [1978] also investigated empirically the consistency in aggregation issue. Price indexes were constructed residually using (45).
[59]In the context of production theory, the (output) aggregate $F(x)$ is observable, in contrast to the utility theory context where $F(x)$ is unobservable.

Essays in Index Number Theory, Volume I
W.E. Diewert and A.O. Nakamura (Editors)

Chapter 8
EXACT AND SUPERLATIVE INDEX NUMBERS*

W.E. Diewert

1. Introduction

One of the most troublesome problems facing national income accountants and econometricians who are forced to construct some data series, is the question of which functional form for an index number should be used. In the present paper, we consider this question and relate functional forms for the underlying production or utility function (or aggregator function, to use a neutral terminology) to functional forms for index numbers.

First, define a *quantity index* between periods 0 and 1, $Q(p^0, p^1, x^0, x^1)$, as a function of the prices in periods 0 and 1, $p^0 > 0_N$ and $p^1 > 0_N$ (where 0_N is an N dimensional vector of zeros), and the corresponding quantity vectors, $x^0 > 0_N$ and $x^1 > 0_N$. A *price index* between periods 0 and 1, $P(p^0, p^1, x^0, x^1)$, is a function of the same price and quantity vectors. Given either a price index or a quantity index, the other function can be defined implicitly by the following equation (Fisher's [1922] weak factor reversal test):

$$P(p^0, p^1, x^0, x^1)Q(p^0, p^1, x^0, x^1) = p^1 \cdot x^1/p^0 \cdot x^0; \tag{1.1}$$

i.e., the product of the price index times the quantity index should yield the expenditure ratio between the two periods. (We indicate the inner product of two vectors as $p \cdot x$ or $p^T x$.)

Examples of price indexes are

$$P_{La}(p^0, p^1, x^0, x^1) \equiv p^1 \cdot x^0/p^0 \cdot x^0 \qquad \text{(Laspeyres price index)},$$
$$P_{Pa}(p^0, p^1, x^0, x^1) \equiv p^1 \cdot x^1/p^0 \cdot x^1 \qquad \text{(Paasche price index)}.$$

---

*This article was first published in the *Journal of Econometrics* 4(2), 1976, pp. 115–145. A preliminary version was presented at Stanford in August 1973. The author is indebted to L.J. Lau, D. Aigner, K.J. Arrow, E.R. Berndt, C. Blackorby, L.R. Christensen and K. Lovell for helpful comments. This research was partially supported by National Science Foundation Grant GS-3269-A2 at the Institute for Mathematical Studies in the Social Sciences at Stanford University, and by the Canada Council.

The geometric mean of the Paasche and Laspeyres indexes has been suggested as a price index by Bowley [1928] and Pigou [1912], but it is Irving Fisher [1922] who termed the resulting index *ideal*:

$$(1.2) \qquad P_{Id}(p^0,p^1,x^0,x^1) \equiv (p^1 \cdot x^0 p^1 \cdot x^1 / p^0 \cdot x^0 p^0 \cdot x^1)^{1/2}.$$

The Laspeyres, Paasche and ideal quantity indexes are defined in a similar manner — quantities and prices are interchanged in the above formulae. In particular, the ideal quantity index is defined as

$$(1.3) \qquad Q_{Id}(p^0,p^1,x^0,x^1) \equiv (p^1 \cdot x^1 p^0 \cdot x^1 / p^1 \cdot x^0 p^0 \cdot x^0)^{1/2}.$$

Notice that $P_{Id}Q_{Id} = p^1 \cdot x^1/p^0 \cdot x^0$; i.e., the ideal price and quantity indexes satisfy the 'adding up' property (1.1). The following theorem shows that the ideal quantity index may be used to compute the quantity aggregates $f(x^r)$ provided that the aggregator function f has a certain functional form.

THEOREM 1.4. (Byushgens [1925], Konüs and Byushgens [1926], Frisch [1936; 30], Wald [1939; 331], Afriat [1972b; 45] and Pollak [1971a]): *Let* $p^r \gg 0_N$ *for periods* $r = 0,1,2,\ldots,R$, *and suppose that* $x^r > 0_N$ *is a solution to* $\max_x\{f(x) : p^r \cdot x \leq p^r \cdot x^r,\ x \geq 0_N\}$, *where* $f(x) \equiv (x^T A x)^{1/2} \equiv (\sum_{j=1}^N \sum_{k=1}^N x_j a_{jk} x_k)^{1/2}$, $a_{jk} = a_{kj}$, *and the maximization takes place over a region where* $f(x)$ *is concave and positive (which means* A *must have* $N-1$ *zero or negative eigenvalues and one positive eigenvalue). Then*

$$(1.5) \qquad f(x^r)/f(x^0) = Q_{Id}(p^0,p^r,x^0,x^r), \qquad r = 1,2,\ldots,R.$$

Thus given the *base period normalization* $f(x^0) = 1$, the ideal quantity index may be used to calculate the aggregate $f(x^r) = (x^{rT} A x^r)^{1/2}$ for $r = 1,2,\ldots,R$, and *we do not have to estimate the unknown coefficients in the A matrix.* This is the major advantage of this method for determining the aggregates $f(x^r)$ (as opposed to the econometric methods suggested by Arrow [1974]), and it is particularly important when N (the number of goods to be aggregated) is large compared to R (the number of observations in addition to the base period observation p^0, x^0).

If a quantity index $Q(p^0,p^r,x^0,x^r)$ and a functional form for the aggregator function f satisfy (1.5) then we say that Q is *exact* for f. Konüs and Byushgens [1926] show that the geometric quantity index $\prod_{i=1}^N (x_i^1/x_i^0)^{s_i}$ (where $s_i \equiv p_i^0 \cdot x_i^0/p^0 \cdot x^0$) is exact for a Cobb–Douglas aggregator function, while Afriat [1972b], Pollak [1971a] and Samuelson–Swamy [1974] present other examples of exact index numbers. However, it appears that out of all the exact index numbers thus far exhibited, *only* the ideal index corresponds to a functional form for f which is capable of providing a second order approximation to an arbitrary twice differentiable linearly homogeneous function. For a proof that the functional form $(x^T Ax)^{1/2}$ can provide such a second order approximation, see Diewert [1974b].

Let us call a quantity index Q *'superlative'* (see Fisher [1922; 247] for an undefined notion of a superlative index number) if it is *exact* for an f which can provide a second order approximation to a linearly homogeneous function.

In the following section, we show that the Törnqvist [1936], Theil [1965] [1967] and Kloek [1966] [1967] quantity index (which has been used by Christensen and Jorgenson [1970], Star [1974], Jorgenson and Griliches [1972; 83], Star and Hall [1976] as a discrete approximation to the Divisia [1926] index) is also a superlative index number. In Section 3, we use the results of Section 2 to provide a rigorous interpretation of the Jorgenson–Griliches method of measuring technical progress for discrete data.

In Section 4, we introduce an entire family of superlative index numbers. Section 5 presents some conclusions which tend to support the use of Fisher's ideal quantity index in empirical applications and the final section is an appendix which sketches the proofs of various theorems developed in the following sections.

2. The Törnqvist–Theil 'Divisia' Index and the Translog Function

Before stating our main results, it will be necessary to state a preliminary result which is extremely useful in its own right. Let z be an N dimensional vector and define the *quadratic function* $f(z)$ as

$$\begin{aligned} f(z) &\equiv a_0 + a^T z + \frac{1}{2} z^T A z \\ &= a_0 + \sum_{j=1}^{N} a_i z_i + \frac{1}{2} \sum_{i=1}^{N} \sum_{j=1}^{N} a_{ij} z_i z_j, \end{aligned} \tag{2.1}$$

where the a_i, a_{ij} are constants and $a_{ij} = a_{ji}$ for all i, j.

The following lemma is a global version of the Theil [1967; 222–223] and Kloek [1966] local result.

LEMMA 2.2. Quadratic Approximation Lemma. *If and only if the quadratic function f is defined by (2.1), then*

$$f(z^1) - f(z^0) = \frac{1}{2}[\nabla f(z^1) + \nabla f(z^0)]^T (z^1 - z^0), \tag{2.3}$$

where $\nabla f(z^r)$ is the gradient vector of f evaluated at z^r.

The above result should be contrasted with the usual Taylor series expansion for a quadratic function,

$$f(z^1) - f(z^0) = [\nabla f(z^0)]^T (z^1 - z^0) + \frac{1}{2}(z^1 - z^0)^T \nabla^2 f(z^0)(z^1 - z^0),$$

where $\nabla^2 f(z^0)$ is the matrix of second order partial derivatives of f evaluated at an initial point z^0. In the expansion (2.3), a knowledge of $\nabla^2 f(z^0)$ is not required, but a knowledge of $\nabla f(z^1)$ is required. It must be emphasized that (2.3) holds as an equality for all z^1, z^0 belonging to an open set if and only if f is a quadratic function.

Actually, the Quadratic Approximation Lemma (2.2) is closely related to the following result which we will prove as a corollary to (2.2):

LEMMA 2.4. (Bowley [1928; 224–225]): *If a consumer's preferences can be represented by a quadratic function f defined by (2.1), if $x^1 \gg 0_N$ is a solution to the utility maximization problem*

$$\max_z\{f(z) : p^1 \cdot z = Y^1,\ z \geq 0_N\}, \tag{2.5}$$

where $p^1 \gg 0_N$ and $Y^1 \equiv p^1 \cdot x^1$, and if $x^0 \gg 0_N$ (i.e., each component of x^0 is positive) is a solution to the utility maximization problem

$$\max_z\{f(z) : p^0 \cdot z = Y,\ z \geq 0_N\}, \tag{2.6}$$

where $p^0 \gg 0_N$ and $Y^0 \equiv p^0 \cdot x^0$, then the change in utility between periods 0 and 1 is

$$f(x^1) - f(x^0) = \frac{1}{2}(\lambda_1^* p^1 + \lambda_0^* p^0) \cdot (x^1 - x^0), \tag{2.7}$$

where λ_i^* is the marginal utility of income in period i for $i = 0, 1$. That is, λ_1^* is the optimal value of the Lagrange multiplier for the maximization problem (2.5), and λ_0^* is the Lagrange multiplier for (2.6).

Bowley's Lemma is frequently used in applied welfare economics and cost-benefit analysis, while the Quadratic Approximation Lemma is frequently used in index number theory, which indicates the close connection between the two fields.

Suppose that we are given a homogeneous translog aggregator function (Christensen, Jorgenson and Lau [1971]) defined by

$$\ln f(x) \equiv \alpha_0 + \sum_{n=1}^N \alpha_n \ln x_n + \frac{1}{2}\sum_{j=1}^N \sum_{k=1}^N \gamma_{jk} \ln x_j \ln x_k,$$

where $\sum_{n=1}^N \alpha_n = 1$, $\gamma_{jk} = \gamma_{kj}$ and $\sum_{k=1}^N \gamma_{jk} = 0$ for $j = 1, 2, \ldots, N$.

Jorgenson and Lau have shown that the homogeneous translog function can provide a second order approximation to an arbitrary twice continuously differentiable linearly homogeneous function. Let us use the parameters which occur in the translog functional form in order to define the following function, f^*:

$$(2.8) \qquad f^*(z) \equiv \alpha_0 + \sum_{j=1}^{N} \alpha_i z_i + \frac{1}{2}\sum_{i=1}^{N}\sum_{j=1}^{N} \gamma_{ij} z_i z_j.$$

Since the function f^* is quadratic, we can apply the Quadratic Approximation Lemma (2.2), and we obtain

$$(2.9) \qquad f^*(z^1) - f^*(z^0) = \frac{1}{2}[\nabla f^*(z^1) + \nabla f^*(z^0)] \cdot (z^1 - z^0).$$

Now we relate f^* to the translog function f. We have

$$(2.10) \qquad \begin{aligned} \partial f^*(z^r)/\partial z_j &= \partial \ln f(x^r)/\partial \ln x_j \\ &= [\partial f(x^r)/\partial x_j][x_j^r/f(x^r)], \\ f^*(z^r) &= \ln f(x^r), \\ z_j^r &= \ln x_j^r, \quad \text{for } r = 0,1 \text{ and } j = 1,2,\ldots,N. \end{aligned}$$

If we substitute relation (2.10) into (2.9), we obtain

$$(2.11) \qquad \ln f(x^1) - \ln f(x^0) = \frac{1}{2}\left[\hat{x}^1 \frac{\nabla f(x^1)}{f(x^1)} + \hat{x}^0 \frac{\nabla f(x^0)}{f(x^0)}\right] \cdot (\ln x^1 - \ln x^0),$$

where $\ln x^1 \equiv (\ln x_1^1, \ln x_2^1, \ldots, \ln x_N^1)$, $\ln x^0 \equiv (\ln x_1^0, \ln x_2^0, \ldots, \ln x_N^0)$, $\hat{x}^1 \equiv$ the vector x^1 diagonalized into a matrix, and $\hat{x}^0 \equiv$ the vector x^0 diagonalized into a matrix.

Assume that $x^r \gg 0_N$ is a solution to the aggregator maximization problem $\max_x\{f(x) : p^r \cdot x = p^r \cdot x^r, x \geq 0_N\}$, where $p^r \gg 0_N$ for $r = 0, 1$, and f is the homogeneous translog function. The first order conditions for the two maximization problems, after elimination of the Lagrange multipliers (Konüs and Byushgens [1926; 155], Hotelling [1935; 71–74], Wold [1944; 69–71] and Pearce [1964; 59]), yield the relations $p^r/p^r \cdot x^r = \nabla f(x^r)/x^r \cdot \nabla f(x^r)$ for $r = 0, 1$. Since f is linearly homogeneous, $x^r \cdot \nabla f(x^r)$ may be replaced by $f(x^r)$ in the above, and substitution of these last two relations into (2.11) yields

$$\begin{aligned} \ln[f(x^1)/f(x^0)] &= \frac{1}{2}\left[\frac{\hat{x}^1 p^1}{p^{1T} x^1} + \frac{\hat{x}^0 p^0}{p^{0T} x^0}\right] \cdot (\ln x^1 - \ln x^0) \\ &= \sum_{n=1}^{N} \frac{1}{2}(s_n^1 + s_n^0) \ln(x_n^1/x_n^0), \end{aligned}$$

or

(2.12)

$$f(x^1)/f(x^0) = \prod_{n=1}^{N} (x_n^1/x_n^0)^{(s_n^1+s_n^0)/2} \equiv Q_0(p^0, p^1, x^0, x^1),$$

where $s_n^r \equiv p_n^r x_n^r / p^r \cdot x^r$, the nth share of cost in period r.

The right hand side of (2.12) is the quantity index which corresponds to Irving Fisher's [1922] price index number 124, using (1.1). It has also been advocated as a quantity index by Törnqvist [1936] and Theil [1965] [1967] [1968]. It has been utilized empirically by Christensen and Jorgenson [1969] [1970] as a discrete approximation to the Divisia [1926] index and by Star [1974] and Star and Hall [1976] in the context of productivity measurement. The above argument shows that this quantity index is *exact* for a homogeneous translog aggregator function, and in view of the second order approximation property of the homogeneous translog function, we see that the right hand side of (2.12) is a *superlative* quantity index.

It can also be seen (using the if and only if nature of the Quadratic Approximation Lemma (2.2)) that the homogeneous translog function is the only differentiable linear homogeneous function which is exact for the Törnqvist–Theil quantity index.

The above argument can be repeated (with some changes in notation) if the unit cost function for the aggregator function is the translog unit cost function defined by

$$\ln c(p) \equiv \alpha_0^* + \sum_{j=1}^{N} \alpha_n^* \ln p_n + \frac{1}{2} \sum_{j=1}^{N} \sum_{k=1}^{N} \gamma_{jk}^* \ln p_j \ln p_k,$$

where $\sum_{n=1}^{N} \alpha_n^* = 1$, $\gamma_{jk}^* = \gamma_{kj}^*$ and $\sum_{k=1}^{N} \gamma_{jk}^* = 0$ for $j = 1, 2, \ldots, N$. We also need the following results.

LEMMA 2.13. (Shephard [1953; 11], Samuelson [1947]): *If f is positive, linearly homogeneous and concave; if*

$$p^r \cdot x^r = \min_x \{p^r \cdot x : f(x) \geq f(x^r)\} \equiv c(p^r) f(x^r) \quad \text{for } r = 0, 1,$$

and if the unit cost function c is differentiable at p^r, then

$$x^r = \nabla c(p^r) f(x^r) \qquad \text{for } r = 0, 1.$$

COROLLARY[1] 2.14.

$$x^r / p^r \cdot x^r = \nabla c(p^r) / c(p^r) \qquad \text{for } r = 0, 1.$$

Now under the assumption of cost minimizing behavior in periods 0 and 1 (which implies (2.14)), we have upon applying the Quadratic Approximation Lemma (2.2) to the translog unit cost function,

$$\ln c(p^1) - \ln c(p^0) = \frac{1}{2}\left[\hat{p}^1 \frac{\nabla c(p^1)}{c(p^1)} + \hat{p}^0 \frac{\nabla c(p^0)}{c(p^0)}\right] \cdot (\ln p^1 - \ln p^0)$$

$$= \frac{1}{2}\left[\hat{p}^1 \frac{x^1}{p^1 \cdot x^1} + \hat{p}^0 \frac{x^0}{p^0 \cdot x^0}\right] \cdot (\ln p^1 - \ln p^0) \quad \text{(using (2.14)}$$

$$= \sum\nolimits_{n=1}^{N} (s_n^1 + s_n^0) \ln(p_n^1/p_n^0),$$

or

$$c(p^1)/c(p^0) = \prod\nolimits_{n=1}^{N} (p_n^1/p_n^0)^{(s_n^1+s_n^0)/2}, \tag{2.15}$$

where $s_n^r = p_n^r x_n^r / p^r \cdot x^r$ (the nth share of cost in period r), $p^r \gg 0_N$ (period r prices, $r = 0, 1$), $x^r \geq 0_N$ (period i quantities, $i = 0, 1$), and $c(p)$ = the translog unit cost function.[2]

The right hand side of (2.15) corresponds to Irving Fisher's [1922] price index 123. The above argument shows that this price index is *exact* for a translog unit cost function, and that this is the only differentiable unit cost function which is exact for this price index.

Let us denote the right hand side of (2.15) as the price index function $P_0(p^0, p^1, x^0, x^1)$, and denote the right hand side of (2.12) as the quantity index $Q_0(p^0, p^1, x^0, x^1)$. It can be verified that $P_0(p^0, p^1, x^0, x^1)Q_0(p^0, p^1, x^0, x^1) \neq p^1 \cdot x^1/p^0 \cdot x^0$ in general; i.e., the price index P_0 and the quantity index Q_0 do not satisfy the weak factor reversal test (1.1). This is perfectly reasonable, since the quantity index Q_0 is consistent with a homogeneous translog (direct) aggregator function, while the price index P_0 is consistent with an aggregator function which is dual to the translog unit cost function, and *the two aggregator functions do not in general coincide*; i.e., they correspond to different (aggregation) technologies. Thus, given Q_0, the corresponding price index, which satisfies (1.1), is defined by $\widetilde{P}_0(p^0, p^1, x^0, x^1) \equiv p^1 \cdot x^1/[p^0 \cdot x^0 Q^0(p^0, p^1, x^0, x^1)]$. The quantity index Q_0 and the corresponding (implicit) price index P_0 were used by Christensen and Jorgenson [1969] [1970] in order to measure U.S.

---

[1] *Proof:* divide the equation in Lemma 2.13 by $p^r \cdot x^r = c(p^r)f(x^r)$.

[2] Note that the validity of (2.15) depends crucially on the validity of (2.14), which will be valid if p^0 and p^1 belong to an open convex set of prices P, such that the translog $c(p)$ satisfies the regularity conditions of positivity, linear homogeneity and concavity over P.

real input and output. On the other hand, given P_0, the corresponding (implicit) quantity index, which satisfies (1.1), is defined by $\tilde{Q}_0(p^0,p^1,x^0,x^1) \equiv p^1 \cdot x^1/[p^0 \cdot x^0 P_0(p^0,p^1,x^0,x^1)]$. The price-quantity index pair $(P_0,\tilde{Q}_0)$ was advocated by Kloek [1967; 2] over the pair $(\tilde{P}_0,Q_0)$ on the following grounds: as we disaggregate more and more, we can expect the individual consumer or producer to utilize positive amounts of fewer and fewer goods (i.e., as N grows, components of the vectors x^0 and x^1 will tend to become zero), but the prices which the producer or consumer faces are generally positive irrespective of the degree of disaggregation. Since the logarithm of zero is not finite, Q_0 will tend to be indeterminate as the degree of disaggregation increases, but P_0 will still be well defined (provided that all prices are positive).

Theil [1968] and Kloek [1967] provided a somewhat different interpretation of the indexes P_0 and Q_0, an interpretation which does not require the aggregator function to be linear homogeneous. Let the aggregate u be defined by $u = f(x)$, where f is a not necessarily homogeneous aggregator function which satisfies for example the Shephard [1970] or Diewert [1971a] regularity conditions for a production function. For $p \gg 0_N$, $Y > 0$, define the total cost function by $C(u,p) \equiv \min_x\{p \cdot x : f(x) \geq u; x \geq 0_N\}$ and the indirect utility function by $g(p/Y) \equiv \max_x\{f(x) : p \cdot x \leq Y, x \leq 0_N\}$. The true cost of living price index evaluated at 'utility' level u is defined as $P(p^0,p^1;u) \equiv C(u,p^1)/C(u,p^0)$ and the Theil index of quantity (or 'real income') evaluated at prices p is defined as $Q_T(p;u^0,u^1) \equiv C(u^1,p)/C(u^0,p)$. The Theil-Kloek results are that: (i) $P_0(p^0,p^1,x^0,x^1)$ is a second order approximation to $P(p^0,p^1;g(v^*))$, where the nth component of v^* is $v_n^* \equiv (p_n^0 p_n^1/p^0 \cdot x^0 p^1 \cdot x^1)^{\frac{1}{2}}$, for $n = 1,2,\ldots,N$, and (ii) $Q_0(p^0,p^1,x^0,x^1)$ is a second order approximation to $Q_T[p^*, g(p^0/p^0 \cdot x^0), g(p^1/p^1 \cdot x^1)]$, where the nth component of p^* is $p_n^* \equiv (p_n^0 p_n^1)^{1/2}$.

In view of the Theil–Kloek approximation results, we might be led to ask whether the index number P_0 is exact for any general (nonhomothetic) functional forms for the cost function $C(u,p)$. The following theorem answers this question in the affirmative:

THEOREM 2.16. *Let the functional form for the cost function be a general translog of the form*

$$\ln C(u,p) \equiv \alpha_0^* + \sum_{i=1}^N \alpha_i^* \ln p_i + \frac{1}{2}\sum_{j=1}^N \sum_{k=1}^N \gamma_{jk}^* \ln p_j \ln p_k$$

$$+ \beta^* \ln u + \frac{1}{2}\delta^*(\ln u)^2 + \sum_{k=1}^N \varepsilon_i^* \ln u \ln p_i,$$

where $\sum_{i=1}^N \alpha_i^* = 1$, $\gamma_{jk}^* = \gamma_{kj}^*$, $\sum_{k=1}^N \gamma_{jk}^* = 0$, *for* $j = 1,2,\ldots,N$, *and* $\sum_{i=1}^N \varepsilon_j^* = 0$.[3] *Let* (u^0,p^0) *and* (u^1,p^1) *belong to a* (u,p) *region where* $C(u,p)$

[3]These restrictions ensure the linear homogeneity of $C(u,p)$ in p.

satisfies the appropriate regularity conditions for a cost function (e.g., see Shephard [1970], Hanoch [1970] or Diewert [1971a]) and define the quantity vectors $x^0 \equiv \nabla_p C(u^0, p^0)$ *and* $x^1 \equiv \nabla_p C(u^1, p^1)$. *Then*

$$P_0(p^0, p^1, x^0, x^1) = C(u^*, p^1)/C(u^*, p^0),$$

where $u^* \equiv (u^0 u^1)^{1/2}$ *and* P_0 *is defined by the right hand side of (2.15).*

In contrast to the case of a linearly homogeneous aggregator function where the cost function takes the simple form $C(u,p) = c(p)u$, Theorem (2.16) is not an if and only if result; that is, the index number $P_0(p^0, p^1; x^0, x^1)$ is exact for functional forms for $C(u,p)$ other than the translog. In fact, Theorem (2.16) remains true if: (i) we define C as $\ln C(u,p) \equiv \alpha_0(u) + \sum_{i=1}^N [\alpha_i + \varepsilon_i h(u)] \ln p_i + \frac{1}{2} \sum_{j=1}^N \sum_{k=1}^N \gamma_{jk} \ln p_j \ln p_k$, where $\sum_{i=1}^N \alpha_i = 1$, $\sum_{i=1}^N \varepsilon_i = 0$, $\gamma_{jk} = \gamma_{kj}$, $\sum_{k=1}^N \gamma_{jk} = 0$, for $j = 1, 2, \ldots, N$, and h is a monotonically increasing function of one variable, and (ii) define the reference utility level u^* as the solution to the equation $2h(u^*) = h(u^1) + h(u^0)$. (In the translog case, $h(u) \equiv \ln u$.)

Thus the same price index P_0 is exact for more than one functional form (and reference utility level) for the true cost of living.

We can also provide a justification for the quantity index Q_0 in the context of an aggregator function f which is not necessarily linearly homogeneous. In order to provide this justification, it is necessary to define the quantity index which has been proposed by Malmquist [1953] and Pollak [1971a] in the context of consumer theory, and by Bergson [1961] and Moorsteen [1961] in the context of producer theory.

Given an aggregator function f and an aggregate $u \equiv f(x)$, define f's *distance function* as $D(u,x) \equiv \max_k\{k : f(x/k) \geq u\}$. To use the language of utility theory, the distance function tells us by what proportion one has to deflate (or inflate) the given consumption vector x in order to obtain a point on the utility surface indexed by u. It can be shown that if f satisfies certain regularity conditions, then f is completely characterized by D (see Shephard [1970], Hanoch [1970] and McFadden [1970]). In particular, $D(u,x)$ is linearly homogeneous, nondecreasing and concave in the vector of variables x and nonincreasing in u in Hanoch's formulation.

Now define the Malmquist quantity index as $Q_M(x^0, x^1, u) \equiv D(u, x^1)/D(u, x^0)$. Note that the index depends on x^0 (the base period quantities), x^1 (the current period quantities) and on the base indifference surface (which is indexed by u) onto which the points x^0 and x^1 are deflated. The following theorem relates the translog functional form to the Malmquist quantity index.

THEOREM 2.17. *Let an aggregator function f satisfying the Hanoch [1970] and Diewert [1971a] regularity conditions be given such that f's distance function D is a general translog of the form*

$$\ln D(u,x) = \alpha_0 + \sum_{i=1}^{N} \alpha_i \ln x_i + \frac{1}{2}\sum_{j=1}^{N}\sum_{k=1}^{N} \gamma_{jk} \ln x_j + \beta \ln u + \frac{1}{2}\delta(\ln u)^2 + \sum_{i=1}^{N} \varepsilon_i \ln u \ln x_i,$$

where $\sum_{i=1}^{N}\alpha_i = 1$, $\gamma_{jk} = \gamma_{kj}$, $\sum_{k=1}^{N}\gamma_{jk} = 0$, for $j = 1,2,\ldots,N$, and $\sum_{i=1}^{N}\varepsilon_i = 0$. Suppose that the quantity vector x^0 is a solution to the aggregator maximization problem $\max_x\{f(x) : p^0 \cdot x = p^0 \cdot x^0\}$, while x^1 is a solution to $\max_x\{f(x) : p^1 \cdot x = p^1 \cdot x^1\}$ and $u^0 \equiv f(x^0)$, $u^1 \equiv f(x^1)$. Then

$$Q_0(p^0,p^1,x^0,x^1) = D(u^*,x^1)/D(u^*,x^0) \equiv Q_M(x^0,x^1,u^*),$$

where $u^* \equiv (u^0u^1)^{1/2}$ and Q_0 is defined in (2.12).

As was the case with the price index P_0, the quantity index Q_0 is equal to Malmquist quantity indexes which are defined by nontranslog distance functions; i.e., Theorem (2.17) is not an if and only if result.

However, Theorems (2.16) and (2.17) do provide a rather strong justification for the use of P_0 or Q_0 since the translog functional form provides a second order approximation to a general cost or distance function (which in turn are dual to a general nonhomothetic aggregator function).

Finally, note that Theorems (2.16) and (2.17) have a 'global' character to them in contrast to the Theil–Kloek 'local' results.

3. Productivity Measurement and 'Divisia' Indexes

Jorgenson and Griliches [1972; 83–84] have advocated the use of the indexes $\widetilde{P}_0$, Q_0 in the context of *productivity measurement.* It is perhaps appropriate to review their procedure in the light of the results of the previous section.

First, we note (by a straightforward computation) that it is not in general true that 'a discrete Divisia index of discrete Divisia indexes is a discrete Divisia index of the components' (Jorgenson and Griliches [1972; 83]), where the 'Divisia' quantity index is defined to be Q_0. In view of the one-to-one nature of the index number Q_0 with the translog functional form for the aggregator function f in the linearly homogeneous case, it can be seen that the Jorgenson–Griliches assertion will be true if the producer or consumer is maximizing an aggregator function f subject to an expenditure constraint, where f is *both* a homogeneous translog function *and* a translog of micro translog aggregator functions. The set of such translog functions is not empty since it contains the set of Cobb–Douglas functions. Thus if cost shares are approximately constant (which corresponds to the Cobb–Douglas case), then the Jorgenson–Griliches assertion will be approximately true.

It can be similarly shown that, in general, it is not true that a discrete 'Divisia' price index of discrete 'Divisia' indexes is a discrete 'Divisia' price index of the components, where the 'Divisia' price index is defined to be P_0: the first method of constructing a price index is justified if the aggregator function has a unit cost function dual of the form $\hat{c}[c^1(p^1), c^2(p^2), \ldots, c^J(p^J)]$, where $(p^1, p^2, \ldots, p^J)$ represents a partition of the price vector p and the functions $\hat{c}$, $c^1, c^2, \ldots, c^J$ are all translog unit cost functions, while the second method of constructing a price index is justified if the aggregator function has a unit cost function dual, $c(p)$, which is translog.

Jorgenson and Griliches [1972] use the index number formula $Q_0(p^0, p^1, x^0, x^1)$ defined by the right hand side of (2.12) not only to form an index of real input, but also to form an index of real output. Just as the aggregation of inputs into a composite input rests on the duality between unit cost and homogeneous production functions, the aggregation of outputs into a composite output can be based on the duality between unit revenue and homogeneous factor requirements functions.[4] We briefly outline this latter duality.

Suppose that a producer is producing M outputs, $(y_1, y_2, \ldots, y_M) \equiv y$, and the technology of the producer can be described by a *factor requirements function*, g, where $g(y)$ = the minimum amount of aggregate input required to produce the vector of outputs y.[5] The producer's *unit* (aggregate input) *revenue function*[6] is defined for each price vector $w \geq 0_M$ by

$$(3.1) \qquad r(w) \equiv \max_y \{w \cdot y : g(y) \leq 1,\ y \geq 0_M\}.$$

Thus given a factor requirements function g, (3.1) may be used to define a unit revenue function. On the other hand, given a unit revenue function $r(w)$ which is a positive, linearly homogeneous, convex function for $w \gg 0_M$, a

[4]See Diewert [1969] [1974a], Fisher and Shell [1972a] and Samuelson and Swamy [1974] on this topic.

[5]Assume g is defined for $y \geq 0_M$, and has the following properties: (i) $g(y) > 0$ for $y \gg 0_M$ (*positivity*), (ii) $g(\lambda y) = \lambda g(y)$ for $\lambda \geq 0$, $y \geq 0$ (*linear homogeneity*), and (iii) $g[\lambda y^1 + (1-\lambda)y^2] \leq \lambda g(y^1) + (1-\lambda)g(y^2)$ for $0 \leq \lambda \leq 1$, $y^1 \geq 0_M$, $y^2 \geq 0_M$ (*convexity*).

[6]If g satisfies the three properties listed in footnote 5, then r also has those three properties.

factor requirements functions g^* consistent with r may be defined for $y \gg 0_M$ by[7]

(3.2)

$$\begin{aligned} g^*(y) &\equiv \min_\lambda\{\lambda : w\cdot y \le r(w)\lambda \quad \text{for every} \quad w \ge 0_M\} \\ &= \min_\lambda\{\lambda : 1 \le r(w)\lambda \quad \text{for every} \quad w \ge 0_M \text{ such that } w\cdot y = 1\} \\ &= 1/\min_w\{r(w) : w\cdot y = 1,\ w \ge 0_M\}. \end{aligned}$$

The translog functional form may be used to provide a second order approximation to an arbitrary twice differentiable factor requirements function. Thus assume that g is defined (at least over the relevant range of y's) by

(3.3)

$$\ln g(y^r) \equiv a_0 + \sum_{m=1}^{M} a_m \ln y_m^r + \frac{1}{2}\sum_{j=1}^{M}\sum_{k=1}^{M} c_{jk} \ln y_j^r \ln y_k^r, \quad \text{for } r = 0, 1,$$

where

$$\sum_{m=1}^{M} a_m = 1, \quad c_{jk} = c_{kj}, \quad \sum_{k=1}^{M} c_{jk} = 0, \text{ for } j = 1, 2, \ldots, M.$$

Now assume that $y^r \gg 0_M$ is a solution to the aggregate input minimization problem $\min_y\{g(y) : w^r\cdot y = w^r\cdot y^r,\ y \ge 0_M\}$, where $w^r \gg 0_M$ for $r = 0, 1$, and g is the translog function defined by (3.3). Then the first order necessary conditions for the minimization problems along with the linear homogeneity of g yield the relations $w^r/w^r\cdot y^r = \nabla g(y^r)/g(y^r)$, for $r = 0, 1$, and using these two relations in Lemma (2.2) applied to (3.3), we deduce that

$$\text{(3.4)} \qquad \begin{aligned} g(y^1)/g(y^0) &= \prod_{m=1}^{M} (y_m^1/y_m^0)^{[(w_m^1 y_m^1/w^1\cdot y^1)+(w_m^0 y_m^0/w^0\cdot y^0)]/2} \\ &\equiv Q_0^*(w^0, w^1; y^0, y^1). \end{aligned}$$

Thus again the Törnqvist formula can be used to aggregate quantities consistently, provided that the underlying aggregator function is a homogeneous translog.

Similarly if the revenue function $r(w)$ is translog over the relevant range of data and if the producer is in fact maximizing revenue, then we can show that $r(w^1)/r(w^0) = P^*(w^0, w^1, y^0, y^1) \equiv Q_0^*(y^0, y^1, w^0, w^1)$, the Törnqvist price index. (Note that in $Q_0^*(y^0, y^1, w^0, w^1)$, prices and quantities are interchanged compared to $Q_0^*(w^0, w^1, y^0, y^1)$ which appeared in (3.4) above.)

[7]The proof is analogous to the proof of the Samuelson–Shephard duality theorem presented in Diewert [1974a]; alternatively, see Samuelson and Swamy [1974].

Using the above material, we may now justify the Jorgenson–Griliches [1972] method of measuring technical progress. Assume that the production possibilities efficient set can be represented as a set of outputs y and inputs x such that

$$g(y) = f(x), \tag{3.5}$$

where g is the homogeneous translog factor requirements function defined by (3.3), and f is the homogeneous translog production function defined in Section 2. Let $w^r \gg 0_M$, $p^r \gg 0_N$, $r = 0, 1$ be vectors of output and input prices during periods 0 and 1, and assume that $y^0 \gg 0_M$ and $x^0 \gg 0_N$ is a solution to the period 0 profit maximization problem,

$$\max_{y,x}\{w^0 \cdot y - p^0 \cdot x : g(y) = f(x)\}. \tag{3.6}$$

Suppose that 'technical progress' occurs between periods 0 and 1 which we assume to be a parallel outward shift of the 'isoquants' of the aggregator function f; i.e., we assume that the equation which defines the efficient set of outputs and inputs in period 1 is $g(y) = (1+\tau)f(x)$ where τ represents the amount of 'technical progress' if $\tau > 0$ or 'technical regress' if $\tau < 0$. Finally, assume that $y^1 \gg 0_M$ and $x^1 \gg 0_N$ is a solution to the period 1 profit maximization problem,

$$\max_{y,x}\{w^1 \cdot y - p^1 \cdot x : g(y) = (1+\tau)f(x)\}. \tag{3.7}$$

Thus we have $g(y^0) = f(x^0)$ and $g(y^1) = (1+\tau)f(x^1)$. It is easy to see that $y^r \gg 0_M$ is a solution to the aggregate input minimization problem $\min_y\{g(y) : w^r \cdot y = w^r \cdot y^r,\ y \geq 0_M\}$, for $r = 0, 1$, and thus (3.4) holds. Similarly, $x^r \gg 0_N$ is a solution to the aggregator maximization problem $\max_x\{f(x) : p^r \cdot x = p^r \cdot x^r,\ x \geq 0_N\}$, for $r = 0, 1$, and thus (2.12) holds. Substitution of (2.12) and (3.4) into the identity $g(y^1)/g(y^0) = (1+\tau)f(x^1)/f(x^0)$ yields the following expression for $(1+\tau)$ in terms of observable prices and quantities:

$$(1+\tau) = \prod_{m=1}^{M} \left(y_m^1/y_m^0\right)^{[(w_m^1 y_m^1/w^1 \cdot y^1)+(w_m^0 y_m^0/w^0 \cdot y^0)]/2} \Big/ \prod_{n=1}^{n} \left(x_n^1/x_n^0\right)^{[(p_n^1 x_n^1/p^1 \cdot x^1)+(p_n^0 x_n^0/p^0 \cdot x^0)]/2}. \tag{3.8}$$

Thus the Jorgenson–Griliches method of measuring technical progress can be justified if: (i) the economy's production possibilities set can be represented by a *separable* transformation surface defined by $g(y) = f(x)$, where the input aggregator function f and the output aggregator function g are both homogeneous translog functions, (ii) producers are maximizing profits, and (iii) technical progress takes place in the *'neutral'* manner postulated above.

Since the separability assumption $g(y) = f(x)$ is somewhat restrictive from an a priori theoretical point of view, it would be of some interest to devise a measure of technical progress which did not depend on this separability assumption. This can be done, but only at a cost as we shall see below.

Suppose that technology can be represented by a transformation function,[8] where $y_1 = t(y_2, y_3, \ldots, y_M; x_1, x_2, \ldots, x_N) \equiv t(\tilde{y}; x) \equiv t(z)$ is the maximum amount of output one that can be produced given that the vector of other outputs $\tilde{y} = (y_2, y_3, \ldots, y_M)$ is to be produced by the vector of inputs $x = (x_1, x_2, \ldots, x_N)$. Assume that t is a positive, linearly homogeneous, concave function over a convex set of the nonnegative orthant S in $K \equiv M - 1 + N$ dimensional space. Assume also that $t(\tilde{y}; x)$ is nonincreasing in the components of the other outputs vector $\tilde{y}$ and nondecreasing in the components of the input vector x. Suppose that the transformation function t is defined for z belonging to S by

$$(3.9) \qquad \ln t(z) \equiv \alpha_0 + \sum_{k=1}^{K} \ln z_k + \frac{1}{2} \sum_{j=1}^{K} \gamma_{jk} \ln z_j \ln z_k,$$

where $\sum_{k=1}^{K} \alpha_k = 1$, $\gamma_{jk} = \gamma_{kj}$ and $\sum_{k=1}^{K} \gamma_{jk} = 0$, for $j = 1, 2, \ldots, K$; i.e., t is a translog transformation function over the set S.

Suppose that $y^r \equiv (y_1^r, y_2^r, \ldots, y_M^r) \gg 0_M$ (output vectors), $x^r \equiv (x_1^r, x_2^r, \ldots, x_N^r) \gg 0_N$ (input vectors), $w^r \gg 0_M$ (output price vectors), $p^r \gg 0_N$ (input price vectors), and $w^r \cdot y^r = p^r \cdot x^r$ (value of outputs equals value of inputs) for periods $r = 0, 1$. Assume that $z^0 \equiv (y_2^0, \ldots, y_m^0, x_1^0, \ldots, x_N^0) \equiv (\tilde{y}^0, x^0)$ is a solution to the following output maximization subject to an expenditure constraint problem in period 0:

$$(3.10) \qquad \max_z \{t(z) : q^0 \cdot z = q^0 \cdot z^0, \quad z \text{ belongs to } S\}$$

where t is the translog transformation function defined by (3.9), $q^0 \equiv (-w_2^0, -w_3^0, \ldots, -w_M^0; p_1^0, p_2^0, \ldots, p_N^0) \equiv (-\tilde{w}^0; p^0)$, and

$$(3.11) \qquad y_1^0 = t(z^0).^9$$

The first order conditions for the maximization problem (3.10) imply that (Konüs–Byushgens [1926] Lemma)

$$(3.12) \qquad q^0 / q^0 \cdot z^0 = \nabla t(z^0)/t(z^0).$$

[8]For a more detailed discussion of transformation functions and their properties, see Diewert [1973a].
[9]Note that $q^0 \cdot z^0 = {w_1}^0 \cdot {y_1}^0 > 0$, since $w^0 \cdot y^0 = p^0 \cdot x^0$.

As before, we assume that 'neutral' input augmenting technical progress takes place between periods 0 and 1; i.e., if $(y;x)$ was an efficient vector of outputs and inputs in period 0, then $[y;(1+\tau)^{-1}x]$ is on the efficiency surface in period 1. Thus the efficiency surface in period 1 can be defined as the set of $(y_1, y_2, \ldots, y_M; x_1, x_2, \ldots, x_N)$ which satisfy the following equation:

$$(3.13) \qquad y_1 = t[y_2, y_3, \ldots, y_M; (1+\tau)x_1, (1+\tau)x_2, \ldots, (1+\tau)x_N].$$

Assume that $(y_2^1, y_3^1, \ldots, y_M^1; x_1^1, x_2^1, \ldots, x_N^1) \equiv (\tilde{y}^1; x^1)$ is a solution to the period 1 output maximization subject to an expenditure constraint problem $\max_{y,x}\{t[\tilde{y};(1+\tau)x] : -\widetilde{w}^1 \cdot y + p^1 \cdot x = -\widetilde{w}^1 \cdot y^1 + p^1 \cdot x^1; [\tilde{y} : (1+\tau)x]$ belongs to $S\}$. Then $z^1 \equiv [\tilde{y}^1;(1+\tau)x^1]$ is a solution to the following output maximization problem:

$$(3.14) \qquad \max_z\{t(z) : q^1 \cdot z = q^1 \cdot z^1, \quad z \text{ belongs to } S\},$$

where t is the translog function defined by (3.9), $q^1 \equiv (-\widetilde{w}^1; p^1)$, and

$$(3.15) \qquad y_1^1 = t(z^1) = t[\tilde{y}^1;(1+\tau)x^1].$$

Again, the Konüs-Byushgens-Hotelling Lemma applied to the maximization problem (3.14), using the linear homogeneity of t, implies that[10]

$$(3.16) \qquad q^1/q^1 \cdot z^1 = \nabla t(z^1)/t(z^1).$$

Now substitute (3.12) and (3.16) into the identity (2.11), except that t replaces f and z replaces x, and we obtain

$$(3.17) \qquad \begin{aligned} t(z^1)/t(z^0) &= \sum\nolimits_{k=1}^{K} (z_k^1/z_k^0)^{[(q_k^1 z_k^1/q^1\cdot z^1)+(q_k^0 z_k^0/q^0\cdot z^0)]/2} \\ &= t[\tilde{y}^1;(1+\tau)x^1]/t(y^0;x^0). \end{aligned}$$

Combining (3.13), (3.15) and (3.17), we obtain the following equation in τ:

$$(3.18) \qquad \begin{aligned} y_1^1/y_1^0 = &\prod\nolimits_{n=1}^{N} [(1+\tau)(x_n^1/x_n^0)]^{[p_n^1(1+\tau)x_n^1/V^1(\tau)+p_n^0 x_n^0/V^0]/2} \\ &\Big/ \prod\nolimits_{m=2}^{M} (y_m^1/y_m^0)^{[w_m^1 y_m^1/V^1(\tau)+w_m^0 y_m^0/V^0]/2}, \end{aligned}$$

where $V^0 \equiv -\sum_{m=2}^{M} w_m^0 y_m^0 + \sum_{n=1}^{N} p_n^0 x_n^0$ = net cost of producing output y_1 in period 0, and $V^1(\tau) \equiv -\sum_{m=2}^{M} w_m^1 y_m^1 + \sum_{n=1}^{N} p_n^1 (1+\tau) x_n^1$.

[10]We assume that τ is small so that $q^1 \cdot z^1 \equiv -\widetilde{w}^1 \cdot \tilde{y}^1 + p^1 \cdot (1+\tau)x^1 > 0$.

Given data on outputs, inputs and prices, equation (3.18) can be solved for the unknown rate of technical progress τ. Note that equation (3.18) is quite different from the Jorgenson–Griliches equation for τ defined by (3.8) (except that the two equations are equivalent when $M = 1$; i.e., when there is only one output).

However, it should be pointed out that our more general measure of technical progress, which is obtained by solving (3.18) for τ, suffers from some disadvantages: (i) our procedure is computationally more difficult,[11] and (ii) our procedure is not symmetric in the outputs (that is, the first output y_1 is asymmetrically singled out in (3.18)). Thus different orderings of the outputs could give rise to different measures of technical progress. This is because each ordering of the outputs corresponds to a *different* translog assumption about the underlying technology and thus different measures of τ can be obtained. However, all of these measures should be close in empirical applications since the different translog functions are all approximating the same technology to the second order.

4. Quadratic Means of Order r and Exact Index Numbers

For $r \neq 0$, the (homogeneous) *quadratic mean of order r aggregator function* is defined by

$$(4.1) \qquad f_r(x) \equiv \left[\sum_{i=1}^{N} \sum_{j=1}^{N} a_{ij} x_i^{r/2} x_j^{r/2}\right]^{1/r},$$

where $a_{ij} = a_{ji}$, $1 \leq i, j \leq N$, are parameters, and the domain of definition of f_r, is restricted to $x \equiv (x_1, x_2, \ldots, x_N) \gg 0_N$ such that $\sum\sum a_{ij} x_i^{r/2} x_j^{r/2} > 0$, and f_r is concave. The above functional form is due to McCarthy [1967], Kadiyala [1971–72], Denny [1972] [1974] and Hasenkamp [1973]. Denny also defined the *quadratic mean of order r unit cost function*,

$$(4.2) \qquad c_r(p) \equiv \left[\sum_{i=1}^{N} \sum_{j=1}^{N} b_{ij} p_i^{r/2} p_j^{r/2}\right]^{1/r}, \qquad b_{ij} = b_{ji}, \quad r \neq 0.$$

Denny noted that if $r = 1$, then (4.1) reduces to the *generalized linear* functional form (Diewert [1969] [1971a]), (4.2) reduces to the *generalized Leontief* functional form (Diewert [1969] [1971a]), and if all $a_{ij} = 0$ for $i \neq j$, then

[11]Furthermore, we cannot a priori rule out the possibility that equation (3.18) will have either multiple solutions for τ or no solutions at all. The Fisher measure of technical progress, to be introduced in Section 5, overcomes these difficulties.

(4.1) reduces to the CES functional form (Arrow, Chenery, Minhas and Solow [1961]), while if all $b_{ij} = 0$ for $i \neq j$, then (4.2) reduces to the CES unit cost function.

We may also note that when $r = 2$, (4.1) reduces to the Konüs–Byushgens [1926] homogeneous quadratic production or utility function, while (4.2) reduces to the Konüs–Byushgens unit cost function. This functional form has also been considered by Afriat [1972b; 72] and Pollak [1971a] in the context of utility functions and by Diewert [1969] [1974b] in the context of revenue and factor requirements functions.

Lau [1973] has shown that the limit as r tends to zero of the quadratic mean of order r aggregator function (4.1) is the homogeneous translog aggregator function and similarly that the limit as r tends to zero of (4.2) is the translog unit cost function.

This completes our discussion of special cases of the above family of functional forms. The following theorem shows that the functional form is 'flexible'.

THEOREM 4.3. *Let f be any linearly homogeneous, twice continuously differentiable, positive function defined over an open subset of the positive orthant in N dimensional space. Then for any $r \neq 0$, f_r defined by (4.1) can provide a second order differential approximation to f.*

By a second order differential approximation to f at a point $x^* \gg 0_N$,[12] we mean that there exists a set of a_{ij} parameters for f_r defined by (4.1), such that $f_r(x^*) = f(x^*)$, $\nabla f_r(x^*) = \nabla f(x^*)$, and $\nabla^2 f_r(x^*) = \nabla^2 f_r(x^*)$; i.e., the values of f_r and f and their first and second order partial derivatives at x^* all coincide.

Define the *quadratic mean of order r quantity index* Q_r for $x^0 \gg 0_N$, $x^1 \gg 0_N$, $p^0 > 0_N$, $p^1 > 0_N$, for $r \geq 0$, as

(4.4)
$$Q_r(p^0, p^1; x^0, x^1) \equiv \left[\frac{\sum_{i=1}^N (x_i^1/x_i^0)^{r/2}(p_i^0 x_i^0/p^0 \cdot x^0)}{\sum_{k=1}^N (x_k^0/x_k^1)^{r/2}(p_k^1 x_k^1/p^1 \cdot x^1)}\right]^{1/r}$$
$$= \left[\sum_{i=1}^N (x_i^1/x_i^0)^{r/2} s_i^0\right]^{1/r} \left[\sum_{k=1}^N (x_k^1/x_k^0)^{-r/2} s_k^1\right]^{-1/r}.$$

Thus for any $r \neq 0$, Q_r may be calculated as a function of observable prices and quantities in two periods. Note that Q_r can be expressed as the product of a mean of order r[13] in the square roots of the quantity relatives

[12]This terminology follows Lau [1974].

[13]Ordinary, as opposed to quadratic, means of order r were defined by Hardy, Littlewood and Polya [1934].

$(x_i^1/x_i^0)^{1/2}$ (using base period cost shares as weights) times a mean of order $-r$ in the square roots of the quantity relatives $(x_i^1/x_i^0)^{1/2}$ (using period one cost shares as weights).

It is perhaps of some interest to note which of Irving Fisher's [1911] [1922] *tests* are satisfied by the quantity index Q_r. It can be verified that Q_r satisfies: (i) the *commodity reversal test* (i.e., the value of the index number does not change if the ordering of the commodities is changed); (ii) the *identity test* (i.e., $Q_r(p^0,p^0;x^0,x^0) \equiv 1$); (in fact $Q_r(p^0,p^1;x^0,x^0) \equiv 1$ with the quantity index equal to one so long as all quantities remain unchanged); (iii) the *commensurability test* (i.e., $Q_r(D^{-1}p^0,D^{-1}p^1,Dx^0,Dx^1) = Q_r(p^0,p^1,x^0,x^1)$ where D is a diagonal matrix with positive elements down the main diagonal so that the quantity index remains invariant to changes in units of measurement); (iv) the *determinateness test* (i.e., $Q_r(p^0,p^1,x^0,x^1)$ does not become zero, infinite or indeterminate if an individual price becomes zero for any $r \neq 0$ and $Q_r(p^0,p^1;x^0,x^1)$ does not become zero, infinite or indeterminate if an individual *quantity* becomes zero if $0 < r \leq 2$);[14] (v) the *proportionality test* (i.e., $Q_r(p^0,p^1;x^0,\lambda x^0) = \lambda$ for every $\lambda > 0$); and (vi) the *time or point reversal test* (i.e., $Q_r(p^0,p^1;x^0,x^1)Q_r(p^1,p^0;x^1,x^0) \equiv 1$).

Define the *quadratic mean of order r price index* P_r for $p^0 \gg 0_N$, $p^1 \gg 0_N$, $x^0 > 0_N$, $x^1 > 0_N$, for $r \neq 0$, as

$$(4.5) \qquad P_r(p^0,p^1,x^0,x^1) \equiv \left[\frac{\sum_{i=1}^N (p_i^1/p_i^0)^{r/2}(p_i^0x_i^0/p^0\cdot x^0)}{\sum_{k=1}^N (p_k^0/p_k^1)^{r/2}(p_k^1x_k^1/p^1\cdot x^1)}\right]^{1/r}$$
$$= Q_r(x^0,x^1;p^0,p^1).$$

It is easy to see that P_r will also satisfy Fisher's tests (i) to (vi). The only Fisher tests *not* satisfied by the indexes P_r and Q_r are: (vii) the *circularity test* (i.e., $P_r(p^0,p^1;x^0,x^1)P_r(p^1,p^2;x^1,x^2) \not\equiv P(p^0,p^2;x^0,x^2)$); and (viii) the *factor reversal test* (i.e., $P_r(p^0,p^1;x^0,x^1)Q_r(p^0,p^1;x^0,x^1) \not\equiv p^1\cdot x^1/p^0\cdot x^0$ except that P_2 and Q_2, the 'ideal' price and quantity indexes, satisfy the factor reversal test).

For $r \neq 0$ define the *implicit quadratic mean of order r price index* $\widetilde{P}_r$ as

$$(4.6) \qquad \widetilde{P}_r(p^0,p^1,x^0,x^1) \equiv p^1\cdot x^1/[p^0\cdot x^0 Q_r(p^0,p^1,x^0,x^1)],$$

and define the *implicit quadratic mean of order r quantity index* $\widetilde{Q}_r$ as

$$(4.7) \qquad \widetilde{Q}_r(p^0,p^1,x^0,x^1) \equiv p^1\cdot x^1/[p^0\cdot x^0 P_r(p^0,p^1,x^0,x^1)].$$

Thus the two pairs of indexes $(Q_r,\widetilde{P}_r)$ and $(\widetilde{Q}_r,P_r)$ will satisfy the weak factor reversal test (1.1).

The following theorem relates the aggregator function f_r to the quantity index Q_r:

[14]Thus the quantity indexes Q_r, for $0 < r \leq 2$, are somewhat more satisfactory than the Törnqvist–Theil index Q_0 defined by (2.12).

THEOREM 4.8. *Suppose that (i) $f_r(x)$ is defined by (4.1), where $r \neq 0$; (ii) $x^0 \gg 0_N$ is a solution to the maximization problem* $\max_x\{f_r(x) : p^0 \cdot x \leq p^0 \cdot x^0$, *$x$ belongs to $S\}$, where S is a convex subset of the nonnegative orthant in $\mathbf{R}^N$, $f_r(x^0) > 0$ and the price vector p^0 is such that $p^0 \cdot x^0 > 0$; and (iii) $x^1 \gg 0_N$ is a solution to the maximization problem* $\max_x\{f_r(x) : p^1 \cdot x \leq p^1 \cdot x^1$, *$x$ belongs to $S\}$, $f_r(x^1) > 0$ and the price vector p^1 is such that $p^1 \cdot x^1 > 0$; then*

$$f_r(x^1)/f_r(x^0) = Q_r(p^0, p^1, x^0, x^1). \tag{4.9}$$

Thus the quadratic mean of order r quantity index Q_r is exact for the quadratic mean of order r aggregator function, which in view of Theorem (4.3) implies that Q_r is a *superlative* index number.

Suppose that $x^s \gg 0_N$ is a solution to $\max_x\{f_r(x) : p^s \cdot x \leq p^s \cdot x^s$, x belongs to $S\}$, where $f_r(x) > 0$, $p^s \cdot x^s > 0$ for $s = 0, 1, 2$. Then using (4.9) three times, we find that

$$\begin{aligned} Q_r(p^0, p^1, x^0, x^1)Q_r(p^1, p^2, x^1, x^2) &= f_r(x^1)[f_r(x^0)]^{-1}f_r(x^2)[f_r(x^1)]^{-1} \\ &= f_r(x^2)/f_r(x^0) \\ &= Q_r(p^0, p^2, x^0, x^2). \end{aligned}$$

Thus under the assumption that the producer or consumer is maximizing $f_r(x)$ subject to an expenditure constraint each period, we find that Q_r will satisfy the *circularity test* in addition to the other Fisher tests which it satisfies. A similar proposition is true for any exact index number, a fact which was first noted by Samuelson and Swamy [1974]. Since the circularity test is capable of empirical refutation, we see that we can empirically refute the hypothesis that an economic agent is maximizing $f_r(x)$ subject to an expenditure constraint. Thus violations of the circularity test could mean either that the economic agent was not engaging in maximizing behavior or that his aggregator function was not $f_r(x)$.

We note that Theorem (4.8) did not require that all prices be nonnegative; only that quantities be positive. f_r can also be a *transformation* function (recall Section 2) which is nondecreasing in inputs and nonincreasing in outputs. Theorem (4.8) will still hold except that prices of other outputs must be indexed negatively while prices of inputs are taken to be positive. The quantity index Q_r may be used in the context of productivity measurement just as we used the index Q_0 in Section 3. We will return to this topic in Section 5.

Theorem (4.8) tells us that f_r defined by (4.1) is exact for Q_r defined by (4.4). However, could there exist a linearly homogeneous functional form f different from f_r which is also exact for Q_r? The answer is no, as the following theorem shows:

THEOREM 4.10. (Generalization of Byushgens [1925], Konüs and Byushgens [1926]): *Let S be an open subset of the positive orthant in $\mathbf{R}^N$ which is also a convex cone. Suppose f is defined over S and is (i) positive, (ii) once-differentiable, (iii) linearly homogeneous, and (iv) concave. Suppose that f is exact for the quantity index Q_r defined by (4.4) for $r \neq 0$ (i.e., if x^s is a solution to $\max_x\{f(x) : p^s \cdot x \leq p^s \cdot x^s$, x belongs to $S\}$ for $s = 0, 1$, then $Q_r(p^0, p^1; x^0, x^1) = f(x^1)/f(x^0)$). Then f is a quadratic mean of order r defined by (4.1) for some a_{ij}, $1 \leq i \leq j \leq N$.*

We note that the functional form f_r defined by (4.1) may also be used as a factor requirements function, and that the quantity index Q_r defined by (4.4) will still be exact for f_r; i.e., Theorems (4.8) and (4.10) will still hold except that the maximization problems $\max_x\{f_r(x) : p^s \cdot x \leq p^s \cdot x^s$, x belongs to $S\}$ are replaced by the minimization problems $\min_x\{f_r(x) : p^s \cdot x \geq p^s \cdot x^s$, x belongs to $S\}$ for $s = 0, 1$, and condition (iv) is changed from concavity to convexity. Thus the quadratic mean of order r quantity indexes Q_r can be used to aggregate either inputs or outputs provided that the functional form for the aggregator function is a quadratic mean of order r.

The above theorems have their counterparts in the dual space.

THEOREM 4.11. *Suppose that (i) $c_r(p) \equiv \left(\sum_i \sum_j b_{ij} p_i^{r/2} p_j^{r/2}\right)^{1/r}$, where $b_{ij} = b_{ji}$ for all i, j, $r \neq 0$ and $(p_1, p_2, \ldots, p_N) \equiv p$ belongs to S where S is an open, convex cone which is a subset of the positive orthant in $\mathbf{R}^N$; (ii) $c_r(p)$ is positive, linearly homogeneous and concave over S; (iii) $x^0/p^0 \cdot x^0 = \nabla c_r(p^0)/c_r(p)$, where $p^0 \gg 0_N$ so that (using the corollary (2.14) to Shephard's lemma) x^0 is a solution to the aggregator maximization problem $\max_x\{\tilde{f}_r(x) : p^0 \cdot x \leq p^0 \cdot x^0, x \geq 0_N\}$, where $\tilde{f}_r$[15] is the direct aggregator function which is dual to $c_r(p)$; and (iv) $x^1/p^1 \cdot x^1 = \nabla c_r(p^1)/c_r(p^1)$, $p^1 \gg 0_N$ so that x^1 is a solution to the aggregator maximization problem $\max_x\{\tilde{f}_r(x) : p^1 \cdot x \leq p^1 \cdot x^1, x \geq 0_N\}$. Then*

$$c_r(p^1)/c_r(p^0) = P_r(p^0, p^1, x^0, x^1), \tag{4.12}$$

where P_r is the quadratic mean of order r price index defined by (4.5).

The proof of Theorem (4.11) is analogous to the proof of Theorem (4.8), except that p replaces x, c_r replaces f_r, and Corollary (2.14) is used instead of the Konüs–Byushgens–Hotelling Lemma.

Thus the quadratic mean of order r unit cost function c_r is *exact* for the price index P_r. Since, by Theorem (4.3), c_r can provide a second order approximation to an arbitrary twice differentiable unit cost function, we see

---

[15]Define $\tilde{f}_r(x) \equiv 1/\max_p\{c_r(p) : p \cdot x = 1$, p belongs to $\tilde{S}\}$, where $\tilde{S}$ is the closure of S.

that P_r is a *superlative* price index for each $r \neq 0$. We note also that there is an analogue to Theorem (4.10) for P_r; i.e., c_r is essentially the only functional form which is exact for the price index function P_r.

However, if we relax the assumption that the underlying aggregator function be linearly homogeneous, then the index numbers P_r and Q_r can be exact for a number of true cost of living price indexes and Malmquist quantity indexes, respectively; i.e., analogues to Theorems (2.16) and (2.17) hold.

We have obtained two families of price and quantity indexes: P_r, $\widetilde{Q}_r$ and $\widetilde{P}_r$, Q_r defined by (4.6) and (4.4) for any $r \neq 0$. The first price-quantity family corresponds to an aggregator function $\tilde{f}_r$ which has the unit cost function c_r defined by (4.2) as its dual, and the second price-quantity family corresponds to an aggregator function f_r defined by (4.1). Recall also that the price-quantity indexes P_0, $\widetilde{Q}_0$ correspond to a translog unit cost function, while $\widetilde{P}_0$, Q_0 correspond to a homogeneous translog aggregator function.

For various values of r, some of the indexes P_r or $\widetilde{P}_r$ have been considered in the literature. For $r = 2$, $P_2 \equiv \widetilde{P}_2$ becomes the Pigou [1912] and Fisher [1922] *ideal* price index which corresponds to the Konüs–Byushgens [1926] homogeneous quadratic aggregator function $f_2(x) \equiv (x^T Ax)^{1/2}$, where $A = A^T$ is a symmetric matrix of coefficients and it also corresponds to the unit cost function $c_2(p) \equiv (p^T Bp)^{1/2}$, where $B = B^T$ is a symmetric matrix of coefficients. If A^{-1} exists, then it is easy to show that the unit cost function which is dual to f_2 is $\tilde{c}_2(p) = (p^T A^{-1} p)^{1/2}$ (at least for a range of prices). However, if $f_2(x) \equiv (x^T aa^T x)^{1/2} = a^T x$, where $a \gg 0_N$ is a vector of coefficients (*linear aggregator function*), then $\tilde{c}_2(p) = \tilde{c}_2(p_1, p_2, \ldots, p_N) \equiv \min_i\{p_i/a_i : i = 1, 2, \ldots, N\}$ which is *not* a member of the family of unit cost functions defined by $c_2(p) \equiv (p^T Bp)^{1/2}$. On the other hand, if $c_2(p) = (p^T bb^T p)^{1/2} = b^T p$, where $b \gg 0_N$ is a vector of coefficients (*Leontief unit cost function*), then the dual aggregator function is $\tilde{f}_2(x_1, x_2, \ldots, x_N) = \min_i\{x_i/b_i : i = 1, 2, \ldots, N\}$, which is a *Leontief aggregator function*. Thus P_2 is exact for a Leontief aggregator function (a fact which was noted by Pollak [1971]) and, since $P_2 \equiv \widetilde{P}_2$, it is also exact for a linear aggregator function. This is an extremely useful property for an index number formula, since the two types of aggregator functions correspond to *zero* substitutability between the commodities to be aggregated and *infinite* substitutability, respectively.

For $r = 1$, the price index P_1 has been recommended by Walsh [1901; 105]. P_1 is exact for the unit cost function c_1, whose dual $\tilde{f}_1$ is the *generalized Leontief aggregator function* (see Diewert [1971a] for the definition of this function) which has the Leontief aggregator function as a special case. Walsh also recommended the price index $\widetilde{P}_1$, which is exact for the *generalized linear aggregator function* f_1 (see Diewert [1971a] for the definition of this function) which of course has the linear aggregator function as a special case.

If, in fact, a producer or consumer were maximizing a linear homogeneous function subject to an expenditure constraint for a number of time periods, we would expect (in view of the Approximation Theorem (4.3)) that the price indexes P_r and $\widetilde{P}_r$ should more or less coincide, particularly if the variation in relative prices were small. However, since real world data are not necessarily consistent with this maximization hypothesis, let us consider some empirical evidence on this point.

Table 1. Comparison of Some Index Numbers Tabulated by Fisher

Price index	Fisher number	1913	1914	1915	1916	1917	1918
P_{Pa}	54	100	100.3	100.1	114.4	161.1	177.4
P_{La}	53	100	99.9	99.7	114.1	162.1	177.9
P_0	123	100	100.1	99.9	113.8	162.1	177.8
$\widetilde{P}_0$	124	100	100.16	99.85	114.25	161.74	178.16
$\widetilde{P}_1$	1153	100	100.13	99.89	114.20	161.70	177.83
P_1	1154	100	100.12	99.90	114.24	161.73	177.76
P_2	353	100	100.12	99.89	114.21	161.56	177.65

Irving Fisher [1922; 489] tabled the wholesale prices and the quantities marketed for 36 primary commodities in the U.S. during the war years (1913–1918), a time of very rapid price and quantity changes. Fisher calculated and compared 134 different price indexes using this data. Table 1 reproduces Fisher's [1922; 244–247] computations for the Paasche and Laspeyres price indexes, P_{Pa} and P_{La}, as well as for P_0, $\widetilde{P}_0$, P_1, $\widetilde{P}_1$ and $P_2 \equiv \widetilde{P}_2 \equiv P_{Id}$. Fisher's identification number is given in column 2 of the table; e.g. P_2 or the 'ideal' price index was identified as number 353 by Fisher. All index numbers were calculated using 1913 as a base.

Note that the Paasche and Laspeyres indexes coincide to about two significant figures, while the last four indexes mostly lie between the Paasche and Laspeyres indexes and coincide to three significant figures. Fisher [1922; 278] also calculated P_2 (and some of the other 'very good' index numbers) using different years as the base year and then he compared how the various series differed; that is, he tested for 'circularity'. Fisher found that the average discrepancy was only about 1/3 percent between any two bases. Thus as far as Fisher's *time series* data are concerned, it appears that any one of the price indexes, P_r or $\widetilde{P}_r$, gives the same answer to three significant figures, and that violations of circularity are only about 1/3 percent so that the choice of base year is not too important.[16]

To determine how the price indexes P_r compare for different r's in the context of *cross section* data, one may look at Ruggles' [1967; 189–190] paper which compares the consumer price indexes P_{Pa}, P_{La}, P_0 and P_2 for 19 Latin American countries for the year 1961. The indexes P_0 and P_2, using Argentina as a base, differed by about one percent per observation, while P_0 and P_2, using Venezuela as a base (the relative prices in the two countries differed markedly), differed by about 1.5 percent per observation. P_2 failed the circularity test (comparing values with Venezuela and Argentina as the base country) by an average of about two percent per observation, while P_0 failed the circularity test by about three percent per observation.[17] Thus it appears that the indexes P_r differ more and violate circularity more in the context of cross section analysis than in time series analysis. However, the agreement between P_0 and P_2 in the cross section context is still remarkable since the Paasche and Laspeyres indexes differed by about 50 percent per observation.

5. Concluding Remarks

We have obtained two families of *superlative* price and quantity indexes, $(P_r, \tilde{Q}_r)$ and $(\tilde{P}_r, Q_r)$; that is, each of these index numbers is exact for a homogeneous aggregator function which is capable of providing a second order approximation to an arbitrary twice continuously differentiable aggregator function (or its dual unit cost function). Moreover, $(P_r, \tilde{Q}_r)$ and $(\tilde{P}_r, Q_r)$ satisfy many of the Irving Fisher tests for index numbers in addition to their being consistent with a homogeneous aggregator function. Note also that if prices are varying proportionately, then the aggregates $\tilde{Q}_r$ and Q_r are consistent with Hicks' [1946] aggregation theorem.

---

[16]However, as a matter of general principle, it would seem that the chain method of calculating index numbers would be preferable, since over longer periods of time, the underlying functional form for the aggregator function may gradually change, so that, for example, (1.5) will only be approximately satisfied, with the degree of approximation becoming better as r approaches 0.

[17]This failure of the circularity test should not be too surprising from the viewpoint of economic theory since we do not expect the aggregator function for the 270 consumer goods and services to be representable as a linearly homogeneous function; that is, we do not expect all 'income' or expenditure elasticities to be unitary.

Although any one of the index number pairs, $(P_r, \widetilde{Q}_r)$ or $(\widetilde{P}_r, Q_r)$ could be used in empirical applications, we would recommend the use of

$$(P_2, \widetilde{Q}_2) \equiv (\widetilde{P}_2, Q_2) \equiv [(p^1 \cdot x^1 p^1 \cdot x^0 / p^0 \cdot x^1 p^0 \cdot x^0)^{1/2}, \\ (x^1 \cdot p^1 x^1 \cdot p^0 / x^0 \cdot p^1 x^0 \cdot p^0)^{1/2}],$$

Irving Fisher's [1922] ideal index numbers, as the preferred pair of index numbers. There are at least three reasons for this selection.

(i) The functional form for the Fisher-Konüs-Byushgens ideal index number is *particularly simple* and this leads to certain simplifications in applications. For example, recall equation (3.18) which we used in order to measure technical progress, $(1+\tau)$, in an economy whose transformation function could be represented by a (nonseparable) homogeneous translog transformation function t. If we assume $t(z) = f_2(z)$, where f_2 is defined by (4.1) for $r = 2$, then the analogue to (3.18) is

$$(5.1) \qquad \frac{y_1^1}{y_1^0} = \frac{[-\widetilde{w}^1 \cdot \widetilde{y}^1 + p^1 \cdot (1+\tau)x^1][-\widetilde{w}^0 \cdot \widetilde{y}^1 + p^0 \cdot (1+\tau)x^1]^{1/2}}{(-\widetilde{w}^1 \cdot \widetilde{y}^0 + p^1 \cdot x^0)(-\widetilde{w}^0 \cdot y^0 + p^0 \cdot x^0)}.$$

If we square both sides of (5.1), the resulting quadratic equation in $(1+\tau)$ can easily be solved, given market data.

(ii) The indexes $P_2(p^0, p^1; x^0, x^1)$ and $\widetilde{Q}_2(p^0, p^1; x^0, x^1)$ are functions of $p^0 \cdot x^1 / p^0 \cdot x^0$ and $p^1 \cdot x^0 / p^1 \cdot x^1$, which are 'sufficient statistics' for revealed preference theory,[18] and moreover $\widetilde{Q}_2$ is *consistent with revealed preference theory* in the following sense: (a) if $p^0 \cdot x^1 < p^0 \cdot x^0$ and $p^1 \cdot x^0 \geq p^1 \cdot x^1$ (i.e., x^0 revealed preferred to x^1), then $\widetilde{Q}_2(p^0, p^1; x^0, x^1) < 1$; (b) if $p^0 \cdot x^1 \geq p^0 \cdot x^0$ and $p^1 \cdot x^0 < p^1 \cdot x^1$ (i.e., x^1 revealed preferred to x^0), then $\widetilde{Q}_2(p^0, p^1; x^0, x^1) > 1$ (i.e., the quantity index indicates an increase in the aggregate); and (c) if $p^0 \cdot x^1 = p^0 \cdot x^0$ and $p^1 \cdot x^0 = p^1 \cdot x^1$ (i.e., x^0 and x^1 revealed to be equivalent or indifferent), then $\widetilde{Q}_2(p^0, p^1; x^0, x^1) = 1$ (i.e., the quantity index remains unchanged). Thus even if the true aggregator function f is nonhomothetic, the quantity index $\widetilde{Q}_2$ will correctly indicate the direction of change in the aggregate when revealed preference theory tells us that the aggregate is decreasing, increasing or remaining constant.

(iii) The index number pair (P_2, Q_2) is consistent with both a *linear aggregator function* (infinite substitutability between the goods to be aggregated) and a *Leontief aggregator function* (zero substitutability between the commodities to be aggregated). No other $(P_r, \widetilde{Q}_r)$ or $(\widetilde{P}_r, Q_r)$ has this very useful property.

---

[18] See Samuelson [1947], Houthakker [1950] and Afriat [1967]. We should also mention the *nonparametric method* of price and quantity index number determination pioneered by Afriat [1967] which depends only on the R^2 inner products of the rth price vector p^r and the sth quantity vector x^s, $p^r \cdot x^s$, if there are R observations. See Diewert [1973b; 424, footnote 2] for an algorithm which would enable one to calculate a polyhedral approximation $\phi(x)$ to the 'true' linearly homogeneous aggregator function $f(x)$.

6. Proofs of Theorems

Proof of (2.2).

$$\begin{aligned} f(z^1) - f(z^0) &= a^T z^1 + \frac{1}{2} z^{1T} A z^1 - a^T z^0 - \frac{1}{2} z^{0T} A z^0 \\ &= a^T (z^1 - z^0) + \frac{1}{2} z^{1T} A (z^1 - z^0) + \frac{1}{2} z^{0T} A (z^1 - z^0) \\ &= \frac{1}{2} (a + Az^1 + a + Az^0)^T (z^1 - z^0), \quad \text{since } A = A^T \\ &= \frac{1}{2} [\nabla f(z^1) + \nabla f(z^0)]^T (z^1 - z^0). \end{aligned}$$

Assume f is thrice differentiable and satisfies the functional equation $f(x) - f(y) = \frac{1}{2}[\nabla f(x) + \nabla f(y)]^T (x - y)$, for all x and y, in an open neighborhood. We wish to find the function that is characterized by the fact that its average slope between any two points equals the average of the endpoint slopes in the direction defined by the difference between the two points. If f is a function of one variable, the functional equation becomes $f(x) - f(y) = \frac{1}{2}[f'(x) + f'(y)](x - y)$. If we differentiate this last equation twice with respect to x, we obtain the differential equation $\frac{1}{2} f'''(x)(x - y) = 0$, which implies that $f(x)$ is a polynomial of degree two. The general case follows in an analogous manner using the directional derivative concept.

Proof of (2.4). λ_1^* and x^1 will satisfy the first order necessary conditions for an interior maximum for the maximization problem (2.5),

$$\nabla f(x^1) = \lambda_1^* p^1; \qquad p_1 \cdot x_1 = Y^1. \tag{6.1}$$

Similarly, λ_0^* and x^0 will satisfy the first order conditions for the constrained maximization problem (2.6),

$$\nabla f(x^0) = \lambda_0^* p^0; \qquad p_0 \cdot x_0 = Y^0.$$

Now substitute the first part of (6.1) into the right hand side of the identity (2.3), and obtain (2.7).

Proof of (2.16). For a fixed u^*, $\ln C(u^*; p)$ is quadratic in the vector of variables $\ln p$ and we may apply the quadratic approximation Lemma (2.2) to obtain

$$\ln C(u^*,p^1) - \ln C(u^*,p^0) = \frac{1}{2}\left[\hat{p}^1\frac{\nabla_p C(u^*,p^1)}{C(u^*,p^1)} + \hat{p}^0\frac{\nabla_p C(u^*,p^0)}{C(u^*,p^0)}\right] \cdot (\ln p^1 - \ln p^0)$$

$$= \frac{1}{2}\left[\hat{p}^1\frac{\nabla_p C(u^1,p^1)}{C(u^1,p^1)} + \hat{p}^0\frac{\nabla_p C(u^0,p^0)}{C(u^0,p^0)}\right] \cdot (\ln p^1 - \ln p^1)$$

where the equality follows upon evaluating the derivatives of C and noting that $2\ln u^* = \ln u^1 + \ln u^0$,

$$= \ln P_0(p^0,p^1,x^0,x^1)$$

using the definitions of $x^0 \equiv \nabla_p C(u^0,p^0)$, $x^1 \equiv \nabla_p C(u^1,p^1)$ and P_0.

Proof of (2.17). It is first necessary to express the partial derivatives of D with respect to the components of x, $\nabla_x D(u^r,x^r)$, $r = 0,1$, in terms of the partial derivatives of f. We have $D(u^r,x^r) \equiv \max_k\{k : f(x^r/k) \geq u^r\} = 1$, for $r = 0,1$, since each x^r is on the u^r 'utility' surface. To find out how the distance $D(u^0,x^0)$ changes as the components of x^0 change, apply the implicit function theorem to the equation $f(x^0/k) = u^0$ (where $k = 1$ initially). We find that

$$\partial k/\partial x_j \equiv \partial D(u^0,x^0)/\partial x_j = f_j(x^0) \Big/ \sum_{k=1}^N x_k^0 f_k(x^0), \quad j = 1,2,\ldots,N.$$

Similarly

$$\partial D(u^1,x^1)/\partial x_j = f_j(x^1) \Big/ \sum_{k=1}^N x_k^1 f_k(x^1), \quad j = 1,2,\ldots,N.$$

Furthermore, the first order conditions for the two aggregator maximization problems after elimination of the Lagrange multipliers yield the relations

$$p_j^0/p^0 \cdot x^0 = f_j(x^0) \Big/ \sum_{k=1}^N x_k^0 f_k(x^0), \qquad j = 1,2,\ldots,N,$$

and

$$p_j^1/p^1 \cdot x^1 = f_j(x^1) \Big/ \sum_{k=1}^N x_k^1 f_k(x^1), \qquad j = 1,2,\ldots,N.$$

Upon noting that the right hand sides of the last set of relations are identical to the right hand sides of the earlier relations, we obtain

(6.2) $$\nabla_x D(u^0,x^0) = p^0/p^0 \cdot x^0 \quad \text{and} \quad \nabla_x D(u^1,x^1) = p^1/p^1 \cdot x^1.$$

Now for a fixed u^*, $\ln D(u^*, x)$ is quadratic in the vector of variables $\ln x$ and we may again apply the Quadratic Approximation Lemma (2.2) to obtain the following equality:

$$\ln D(u^*, x^1) - \ln D(u^*, x^0) = \frac{1}{2}\left[\hat{x}^1 \frac{\nabla_x D(u^*, x^1)}{D(u^*, x^1)} + \hat{x}^0 \frac{\nabla_x D(u^*, x^0)}{D(u^*, x^0)}\right]$$
$$\cdot (\ln x^1 - \ln x^0)$$
$$= \ln Q_0(p^0, p^1, x^0, x^1),$$

where the equality follows upon evaluating the derivatives of D, noting that $2\ln u^* = \ln u^1 + \ln u^0$, using (6.2), the equalities $D(u^1, x^1) = 1$, $D(u^0, x^0) = 1$ and the definition of Q_0.

Proof of (4.3). Since both f and f_r are twice continuously differentiable, their Hessian matrices evaluated at x^*, $\nabla^2 f(x^*)$ and $\nabla^2 f_r(x^*)$, are both symmetric. Thus we need only show that $\partial^2 f(x^*)/\partial x_i \partial x_j = \partial^2 f_r(x^*)/\partial x_i \partial x_j$, for $1 \le i \le j \le N$. Furthermore, by Euler's theorem on linearly homogeneous functions, $f(x^*) = x^{*T}\nabla f(x^*)$ and $f_r(x^*) = x^{*T}\nabla f_r(x^*)$. Since the partial derivative functions $\partial f(x)/\partial x_i$ are homogeneous of degree zero, application of Euler's theorem on homogeneous functions yields, for $i = 1, 2, \ldots, N$,

$$\sum_{j=1}^{N} x_j^* \partial^2 f(x^*)/\partial x_i \partial x_j = 0 = \sum_{j=1}^{N} x_j^* \partial^2 f_r(x^*)/\partial x_i \partial x_j. \tag{6.3}$$

Thus the above material implies that $f_r(x^*) = f(x^*)$, $\nabla f_r(x^*) = \nabla f(x^*)$ and $\nabla^2 f_r(x^*) = \nabla^2 f(x^*)$ will be satisfied under our present hypotheses if and only if

$$\partial f_r(x^*)/\partial x_i = f_i^* \equiv \partial f(x^*)/\partial x_i, \quad \text{for } i = 1, 2, \ldots, N, \tag{6.4}$$

$$\partial^2 f_r(x^*)/\partial x_i \partial x_j = f_{ij}^* \equiv \partial^2 f(x^*)/\partial x_i \partial x_j, \quad \text{for } 1 \le i < j \le N. \tag{6.5}$$

Thus we need to choose the $N(N+1)/2$ independent parameters $a_{ij} (1 \le i \le j \le N)$, so that the $N + N(N-1)/2 = N(N+1)2$ equations (6.4) and (6.5) are satisfied. Recall that $x^* \equiv (x_1^*, x_2^*, \ldots, x_N^*) \gg 0_N$ and that $y^* \equiv x^{*T}\nabla f(x^*) = f(x^*) > 0$, since f is assumed to be positive over its domain of definition. Thus since $y^* > 0$, $x_i^* > 0$ and $r \neq 0$, the numbers a_{ij}^*, for $1 \le i < j \le N$, can be defined by solving the following equations for a_{ij}^*:

$$f_{ij}^* = \frac{1-r}{y^*} f_i^* f_j^* + \frac{r}{2} y^{*(1-r)} a_{ij}^* x_i^{*r/2-1} x_j^{*r/2-1}, \quad 1 \le i < j \le N. \tag{6.6}$$

The system of equation (6.6) is equivalent to (6.5) if we also make use of (6.4). Now define $a^*_{ji} = a^*_{ij}$, for $i \neq j$, and then a^*_{ii} is defined as the solution to the following equation:

$$(6.7) \qquad \sum_{j=1}^{N} a^*_{ij} x_i^{*r/2-1} x_j^{*r/2} y^{*(1-r)} = f_i^*, \qquad i = 1, 2, \ldots, N.$$

Now define $f_r(x) \equiv \left(\sum_{i=1}^{N} \sum_{j=1}^{N} a^*_{ij} x_i^{r/2} x_j^{r/2}\right)^{1/r}$, and it can be verified readily that equations (6.4) and (6.5) are satisfied by f_r as defined.

Proof of (4.8). Using assumptions (ii) and (iii) of (4.8) yields

$$(6.8) \qquad v^0 \equiv p^0/p^0 \cdot x^0 = \nabla f_r(x^0)/f_r(x^0),$$

$$(6.9) \qquad v^1 \equiv p^1/p^1 \cdot x^1 = \nabla f_r(x^1)/f_r(x^1).$$

Upon differentiating $f_r(x^0)$, the ith equation in (6.8) becomes

$$v_i^0 \equiv p_i^0/p^0 \cdot x^0 = (x_i^0)^{(r/2)-1} \sum_{j=1}^{N} a_{ij} x_j^{0r/2} \Big/ \sum_{k=1}^{N} \sum_{m=1}^{N} a_{km} x_k^{0r/2} x_m^{0\ r/2},$$

and therefore

$$(6.10) \qquad \sum_{i=1}^{N} x_i^{1r/2} v_i^0 x_i^{01-(r/2)} = \sum_i \sum_j x_i^{1r/2} a_{ij} x_j^{0r/2} \Big/ \sum_k \sum_m a_{km} x_k^{0r/2} x_m^{0\ r/2}.$$

Similarly, using equation (6.9), we obtain

$$(6.11) \qquad \sum_{i=1}^{N} x_i^{0r/2} v_i^{1-(r/2)} = \sum_i \sum_j x_i^{0r/2} a_{ij} x_j^{1r/2} \Big/ \sum_k \sum_m a_{km} x_k^{1r/2} x_m^{1\ r/2}.$$

Upon noting that $a_{ij} = a_{ji}$, take the ratio of (6.10) to (6.11),

$$(6.12) \qquad \frac{\sum_i (x_i^1/x_i^0)^{r/2} v_i^0 x_i^0}{\sum_j (x_j^0/x_j^1)^{r/2} v_j^1 x_j^1} = \frac{\sum_k \sum_m a_{km} x_k^{1r/2} x_m^{1\ r/2}}{\sum_k \sum_m a_{km} x_k^{0r/2} x_m^{0\ r/2}} = \left[\frac{f_r(x^1)}{f_r(x^0)}\right]^r.$$

Take the rth root of both sides of (6.12) and obtain (4.9).

Proof of (4.10). Let x, y be any two points belonging to S such that

$$(6.13) \qquad 1 = f(x) = f(y) = x \cdot \nabla f(x) = y \cdot \nabla f(y),$$

where the last two equalities follow from the linear homogeneity of f. Since f is a concave function over S, for every z belonging to S, $f(z) \leq f(x) + \nabla f(x) \cdot (z - x) = f(x) + \nabla f(x) \cdot z - f(x) = \nabla f(x) \cdot z$, and similarly $f(z) \leq \nabla f(y) \cdot z$. Thus x is a solution to $\max_z \{f(z) : \nabla f(x) \cdot z \leq \nabla f(x) \cdot x,\ z$ belongs to $S\}$, and

y is a solution to $\max_z\{f(z) : \nabla f(y) \cdot z \leq \nabla f(y) \cdot y,\ z$ belongs to $S\}$. Since f is exact for Q_r for some $r \neq 0$ by assumption, we must have, using (6.13),

$$Q_r(\nabla f(x), \nabla f(y); x, y) = f(y)/f(x) = 1,$$

or

$$\sum_{i=1}^{N}(x_i/y_i)^{r/2} f_i(y) y_i / \nabla f(y) \cdot y = \sum_{n=1}^{N}(y_n/x_n)^{r/2} f_n(x) x_n / \nabla f(x) \cdot x,$$

or

$$\sum_{n=1}^{N} y_n^{r/2}[f_n(x) x_n^{1-(r/2)}] = \sum_{n=1}^{N} y_n^{1-(r/2)} f_n(y) x_n^{r/2}, \tag{6.14}$$

where $f_n(y) \equiv \partial f(y)/\partial y_n$, $f_n(x) \equiv \partial f(x)/\partial x_n$, and $x \cdot \nabla f(x) = 1 = y \cdot \nabla f(y)$. Replace the vector $y \equiv (y_1, y_2, \ldots, y_N)$ which occurs in (6.14) with the vector y^n belonging to S, where $f(y^n) = 1$, for $n = 1, 2, \ldots, N$. Regard the resulting system of N equations as N linear equations in the N unknowns,

$$f_1(x) x_1^{1-(r/2)},\ f_2(x) x_2^{1-(r/2)}, \ldots, f_N(x) x_N^{1-(r/2)},$$

and since we can choose the vectors $y^1, y^2, \ldots, y^N$ to be such that the coefficient matrix on the left hand side of the system of N equations is nonsingular, we may invert the coefficient matrix and obtain the solution

$$f_n(x) x_n^{1-(r/2)} = \sum_{j=1}^{N} A_{nj} x_j^{r/2}, \quad n = 1, 2, \ldots, N, \tag{6.15}$$

for some constants, A_{ij}, $1 \leq i, j \leq N$. Equation (6.15) is valid for any x belonging to S, such that $f(x) = 1$; in particular, (6.15) is true for $x = y$,

$$f_n(y) y_n^{1-(r/2)} = \sum_{j=1}^{N} A_{nj} y_j^{r/2}, \quad n = 1, 2, \ldots, N. \tag{6.16}$$

Now substituting (6.15) into the left hand side of (6.14) and (6.16) into the right hand side of (6.14), we obtain

$$\sum_{n=1}^{N} y_n^{r/2} \sum_{j=1}^{N} A_{nj} x_j^{r/2} = \sum_{n=1}^{N} x_n^{r/2} \sum_{j=1}^{N} A_{nj} y_n^{r/2},$$

or

$$\sum_n \sum_j y_n^{r/2} A_{nj} x_j^{r/2} = \sum_n \sum_j x_n^{r/2} A_{nj} y_j^{r/2}. \tag{6.17}$$

Since (6.17) is true for every x, y, such that $f(x) = 1 = f(y)$, we must have

$$A_{nj} = A_{jn}, \quad \text{for} \quad 1 \leq n, j \leq N. \tag{6.18}$$

Now take $x_n^{r/2}$ times (6.15) and sum over n,

$$\sum_{n=1}^{N} x_n^{r/2} f_n(x) x_n^{1-(r/2)} = \sum_{n=1}^{N} \sum_{j=1}^{N} A_{nj} x_n^{r/2} x_j^{r/2} = 1,$$

since $x \cdot \nabla f(x) = f(x) = 1$.

Thus if $f(x) = 1$, then x satisfies the equation $\sum_n \sum_j A_{nj} x_n^{r/2} x_j^{r/2} = 1$, where $A_{nj} = A_{jn}$. Since f is linearly homogeneous by assumption, we must have for x belonging to S,

$$f(x) = \left[\sum_{n=1}^{N} \sum_{j=1}^{N} A_{nj} x_n^{r/2} x_j^{r/2}\right]^{1/r}. \tag{6.19}$$

Essays in Index Number Theory, Volume I
W.E. Diewert and A.O. Nakamura (Editors)

Chapter 9
SUPERLATIVE INDEX NUMBERS AND CONSISTENCY IN AGGREGATION*

W.E. Diewert

1. Introduction

Recently, Vartia [1974] [1976a] proposed a discrete approximation to the continuous time Divisia[1] price or quantity index which has the following two remarkable properties: (i) the price index and the corresponding quantity index (which is defined by the same formula except that prices and quantities are interchanged) satisfy Fisher's [1922] factor reversal test (i.e., the product of the price and quantity indexes equals the expenditure ratio for the two periods under consideration) and (ii) the Vartia I price or quantity index has the property of *consistency in aggregation*.

Vartia defines an index number formula to be consistent in aggregation if the value of the index calculated in two stages necessarily coincides with the value of the index as calculated in an ordinary way; i.e., in a single stage.

The economic theory of index numbers (see Afriat [1972b], Pollak [1971a], Samuelson and Swamy [1974]) is concerned with rationalizing functional forms for index numbers with functional forms for the underlying aggregator function. In Section 2, we show that the Vartia I price and quantity indexes are consistent only with a Cobb–Douglas aggregator function. This is perhaps not a surprising result, since thus far, the only way in which the two stage method of calculating index numbers has been justified from the viewpoint of the economic theory of index numbers, is to assume that the underlying aggregator function is weakly separable in the same partition that corresponds to the two

---

*This article was first published in 1978 in *Econometrica* 46(4), pp. 883–900. The author wishes to thank C. Blackorby, Z. Griliches, C. Sims, and a referee for helpful comments and Keith Wales for programming help. Correspondence with P.A. Samuelson has also been very helpful. This research was supported by a Canada Council grant.

[1]See Divisia [1926], Wold [1953; Ch. 8], Jorgenson and Griliches [1967], and Hulten [1973].

stages.[2] Thus to justify the two stage method of constructing index numbers for any partition of variables, one thus far has had to assume that the aggregator function is weakly separable in any partition of its variables, but then the results of Leontief [1947] and Gorman [1968a] imply that the aggregator function is strongly separable in the coordinate-wise partition of its variables. If we also assume that the aggregator function is linearly homogeneous, then using Bergson's [1936] results, it can be seen that the aggregator function must be a mean of order r (Hardy, Littlewood and Polya [1934]); i.e., a CES function. The mean of order r functions that are consistent in aggregation include the Cobb–Douglas aggregator function (i.e., $r = 0$) and the linear aggregator function (i.e., $r = 1$). For a more comprehensive discussion of the difficulties involved in combining subindexes to form a complete index, see Pollak [1975] and Blackorby, Primont, and Russell [1978].

In spite of the rather negative result that the Vartia I price and quantity indexes are exact only for a Cobb–Douglas aggregator function, we show in Section 3 that the Vartia index approximates to the second order any *superlative* index, and that all superlative indexes closely approximate each other. Moreover, using the consistency in aggregation property of the Vartia index, we can show that all superlative indexes are *approximately* consistent in aggregation. A superlative quantity index is an index number formula which is consistent with a consumer or producer maximizing a "flexible" aggregator function subject to a budget constraint. A "flexible" functional form is one which can provide a second order approximation to an arbitrary function (c.f. Diewert [1976a]). Thus a practical objection to the use of superlative index number formulae (i.e., they are not consistent in aggregation) loses its force, since they will be approximately consistent in aggregation. Moreover, the degree of approximation will become closer if the *chain principle* for constructing indexes is used rather than the fixed base method (if we are dealing with time series data where changes in prices and quantities between successive periods are generally smaller than changes relative to a fixed base).

In Section 4, we apply the results of Section 3 to some issues in the theory of aggregation, to the measurement of productivity change, and to applied welfare economics. Proofs of our theorems are relegated to an appendix.

A second appendix numerically compares various index number formulae, constructed in one and two stages and also constructed using the chain principle versus a fixed base. This may help to illustrate more concretely some of the rather abstract approximation theorems presented in the paper.

An index number formula like Vartia's could be termed *pseudo superlative* since it approximates a superlative index number formulae to the second order. A final appendix proves some theorems about pseudo superlative index numbers and relates these results to some results due to Samuelson and Swamy [1974].

[2]See Shephard [1953; 61–71], Solow [1955–56], Gorman [1959], Blackorby, Primont, and Russell [1978], and Geary and Morishima [1973].

2. Exact Index Numbers and the Vartia Indexes

Define an *aggregator function* $f(x)$ as a function of N nonnegative variables $x \geq 0_N$ which has the following three properties: (i) f is positive for positive arguments (i.e., $f(x) > 0$ for $x \gg 0_N$); (ii) f is linearly homogeneous (i.e., $f(\lambda x) = \lambda f(x)$ for $\lambda \geq 0$, $x \geq 0_N$); and (iii) f is concave (i.e., $f[\lambda x^1 + (1-\lambda)x^2] \geq \lambda f(x^1) + (1-\lambda)f(x^2)$ for $0 \leq \lambda \leq 1$, $x^1 \geq 0_N$, $x^2 \geq 0_N$).

Generally, aggregator functions are taken to be either utility functions in the consumer context or production functions in the producer context.

An aggregator function f satisfying the above conditions has a *total cost function* defined by $C(u,p) \equiv \min\{p \cdot x : f(x) \geq u\} = u \min_{x/u}\{p \cdot x/u : f(x/u) \geq 1\}$ (using the linear homogeneity of f) $= u \min_z\{p \cdot z : f(z) \geq 1\} \equiv uc(p)$ where $p \cdot x \equiv \sum_{i=1}^N p_i x_i$ and $c(p)$ denotes the *unit cost function* which corresponds to f.

It turns out that the unit cost function $c(p)$ satisfies the same regularity conditions as f; i.e., $c(p)$ is positively linearly homogeneous and concave for $p \gg 0_N$. Moreover, given a unit cost function satisfying these regularity conditions, its aggregator function dual can be defined by $f(x) \equiv 1/[\max_p\{c(p) : p \cdot x = 1, \; p \geq 0_N\}]$.[3]

Thus if the functional form for the aggregator function is known (or the functional form for its unit cost function dual $c(p)$ is known) and the economic agent is engaging in cost minimizing behavior, then the quantity aggregate can be defined as $u \equiv f(x)$ (or $u \equiv p \cdot x/c(p)$) and the price of the aggregate may be defined as $p_0 \equiv p \cdot x/f(x)$ (or $p_0 \equiv c(p)$). Since the functional form for f (or c) is *not* generally known, the above results are not particularly useful in empirical applications. However, as Afriat [1972b], Pollak [1971a], Samuelson and Swamy [1974], and Diewert [1976a] have shown, it is sometimes possible to relate known functional forms for index numbers to functional forms for aggregator functions. We now outline this theory of exact index numbers.

First define a *quantity index* between periods 0 and 1 as a function, $Q(p^0, p^1; x^0, x^1)$, of the price vectors in periods 0 and 1, p^0 and p^1 respectively, and the corresponding quantity vectors x^0 and x^1 while a *price index* between periods 0 and 1 is a function, $P(p^0, p^1; x^0, x^1)$, of the same price and quantity vectors. Given either a price index or a quantity index, the other function can be defined implicitly by the following equation (Fisher's [1922] weak factor reversal test):

$$(1) \qquad P(p^0, p^1, x^0, x^1)Q(p^0, p^1, x^0, x^1) = p^1 \cdot x^1/p^0 \cdot x^0;$$

i.e., the product of the price index times the quantity index should yield the expenditure ratio between the two periods.

[3]See Shephard [1953], Samuelson [1953–54] [1972], Chipman [1970], and Diewert [1974a].

The positive linearly homogeneous, concave aggregator function f is defined to be *exact* for the quantity index $Q(p^0, p^1, x^0, x^1)$ if for every $p^0 \gg 0_N$, $p^1 \gg 0_N$, x^0 a solution to the aggregator maximization problem $\max_x\{f(x) : p^0 \cdot x \leq p^0 \cdot x \leq p^0 \cdot x^0, x \geq 0_N\}$, and x^1 a solution to $\max_x\{f(x) : p^1 \cdot x \leq p^1 \cdot x^1, x \geq 0_N\}$, we have

$$(2) \qquad Q(p^0, p^1, x^0, x^1) = f(x^1)/f(x^0).$$

Similarly, the positive, linearly homogeneous, concave aggregator function f with unit cost function dual c is defined to be *exact* for the price index $P(p^0, p^1, x^0, x^1)$ if for every $p^0 \gg 0_N$, $p^1 \gg 0_N$, x^0 a solution to $\max_x\{f(x) : p^0 \cdot x \leq p^0 \cdot x^0, x \geq 0_N\}$, and x^1 a solution to $\max_x\{f(x) : p^1 \cdot x \leq p^1 \cdot x^1, x \geq 0_N\}$, we have

$$(3) \qquad P(p^0, p^1; x^0, x^1) = c(p^1)/c(p^0).$$

With the above preliminaries disposed of, define the Vartia[4] [1974] [1976a] price index $P_V(p^0, p^1; x^0, x^1)$ as

$$(4) \quad \ln P_V(p^0, p^1; x^0, x^1) \equiv \sum_{i=1}^{N} [L(p_i^1 x_i^1, p_i^0 x_i^0)/L(p^1 \cdot x^1, p^0 \cdot x^0)] \ln(p_i^1/p_i^0)$$

where the logarithmic mean function L, introduced into the economics literature by Vartia [1974] and Sato [1976a], is defined by $L(a,b) \equiv (a-b)/(\ln a - \ln b)$ for $a \neq b$ and $L(a,a) \equiv a$.

The Vartia quantity index $Q_V(p^0, p^1, x^0, x^1)$ is defined by

$$(5) \quad \begin{aligned} \ln Q_V(p^0, p^1, x^0, x^1) &\equiv \sum_{i=1}^{N} [L(p_i^1 x_i^1, p_i^0 x_i^0)/L(p^1 \cdot x^1, p^0 \cdot x^0)] \ln(x_i^1/x_i^0) \\ &= \ln P_V(x^0, x^1, p^0, p^1); \end{aligned}$$

i.e., the price and quantity indexes have the same functional form except that the role of prices and quantities are interchanged. Vartia shows that P_V and Q_V satisfy the factor reversal test (1) and have the property of consistency in aggregation.

THEOREM 1. *The only once differentiable, positive, linearly homogeneous and concave unit cost function which is exact for the Vartia price index defined by (4) is the Cobb–Douglas unit cost function.*

THEOREM 2. *The only once differentiable, positively linearly homogeneous and concave aggregator function which is exact for the Vartia quantity index defined by (5) is the Cobb–Douglas aggregator function.*

Since the Cobb–Douglas aggregator function is extremely restrictive, one might suppose that the Vartia indexes would not be very useful. However, the results of the next section indicate that the Vartia indexes are useful in establishing certain general theorems about superlative indexes.

[4]We consider only the Vartia I indexes since they are the indexes which have the consistency in aggregation property. Sato [1976a] showed that the Vartia II indexes were exact for a CES aggregator function.

3. Approximation Properties of the Vartia Indexes

A frequently[5] used discrete approximation to the Divisia price index is defined by

$$(6)\quad \ln P_0(p^0,p^1,x^0,x^1) \equiv \frac{1}{2}\sum_{i=1}^{N}[(p_i^1x_i^1/p^1\cdot x^1)+(p_i^0x_i^0/p^0\cdot x^0)]\ln(p_i^1/p_i^0)$$

while a quantity index Q_0 is defined by

(7)

$$\ln Q_0(p^0,p^1,x^0,x^1) \equiv \frac{1}{2}\sum_{i=1}^{N}[(p_i^1x_i^1/p^1\cdot x^1)+(p_i^0x_i^0/p^0\cdot x^0)]\ln(x_i^1/x_i^0)$$

$$= \ln P_0(x^0,x^1,p^0,p^1).$$

Törnqvist [1936] [1937] and [1971; 47] urged the use of the above indexes, while Kloek [1966] [1967] and Theil [1967] [1968] showed that the indexes had some good local approximation properties. In addition, Diewert [1976a] showed that the translog[6] unit cost function is exact for P_0 defined by (6) and that a linearly homogeneous translog aggregator function is exact for Q_0 defined by (7).

Diewert [1976a] defined a price index (quantity index) to be *superlative* if a unit cost function c (aggregator function f), capable of providing a second order differential approximation to an arbitrary twice differentiable linearly homogeneous function, is exact for it. Since a linearly homogeneous translog function can provide a second order approximation to an arbitrary twice differentiable linearly homogeneous function (see Lau [1974]), it can be seen that P_0 defined by (6) is a superlative price index and Q_0 defined by (7) is a superlative quantity index. In general, "superlative" indexes are consistent with "flexible" functional forms for the underlying aggregator function.

Since the price index P_0 defined by (6) resembles somewhat the Vartia price index P_V defined by (4), the following result may not be too surprising.

THEOREM 3. *The Vartia price index differentially approximates*[7] *the superlative price index P_0 to the second order at any point where the prices and quantities for the two periods are equal; i.e., $P_V(p^0,p^1,x^0,x^1) = P_0(p^0,p^1,x^0,x^1)$ and the first and second order partial derivatives of the two functions coincide provided that $p^0 = p^1 \gg 0_N$ and $x^0 = x^1 \gg 0_N$.*

The proof of the above theorem makes use of the following two lemmas which are also proved in an appendix.

---

[5]See, for example, the empirical work by Christensen and Jorgenson [1969] [1970].

[6]See Christensen, Jorgenson and Lau [1971].

[7]The term is due to Lau [1974]: it means that the approximating function has the same level and first and second order partial derivatives at a point as the function that it is approximating: i.e., $c(p^*) = c^*(p^*)$, $\nabla_p c(p^*) = \nabla_p c^*(p^*)$, and $\nabla_{pp}^2 c(p^*) = \nabla_{pp}^2 c^*(p^*)$ for some p^*.

LEMMA 1. *Define the functions f and g of the two variables λ and γ by $f(\lambda,\gamma) \equiv \frac{1}{2}[(\lambda/\gamma)+1]$ and $g(\lambda,\gamma) \equiv L(\lambda,1)/L(\gamma,1)$ where L is the Vartia mean function. Then f differentially approximates g to the first order at the point $\lambda = 1$, $\gamma = 1$; i.e., $f(1,1) = g(1,1)$, $f_1(1,1) = g_1(1,1)$, and $f_2(1,1) = g_2(1,1)$.*

LEMMA 2. *Suppose that: (i) the function of n variables g differentially approximates the function g^* to the second order at the point y^*; i.e., $g(y^*) = g^*(y^*)$, $\nabla_y g(y^*) = \nabla_y g^*(y^*)$, and $\nabla^2_{yy} g(y^*) = \nabla^2_{yy} g^*(y^*)$; (ii) the function of n variables f differentially approximates the function f^* to the first order at y^*; i.e., $f(y^*) = f^*(y^*)$ and $\nabla_y f(y^*) = \nabla_y f^*(y^*)$ and (iii) $g(y^*) = g^*(y^*) = 0$. Then $h(y) \equiv f(y)g(y)$ differentially approximates $h^*(y) \equiv f^*(y)g^*(y)$ to the second order at the point y^*.*

The following theorem may be proved in a manner analogous to the proof of Theorem 3.

THEOREM 4. *The Vartia quantity index differentially approximates the superlative quantity index Q_0 to the second order at any point where the prices and quantities for the two periods are equal.*

Thus $P_V(p^0,p^1,x^0,x^1)$ will be close to $P_0(p^0,p^1,x^0,x^1)$ provided that p^0 is close to p^1 and x^0 is close to x^1. If we call an index which can approximate a superlative index differentially to the second order at any point where $p^0 = p^1$ and $x^0 = x^1$ a *pseudo superlative* index, it can be seen that the Vartia price and quantity indexes are pseudo superlative.

It turns out that the Vartia price and quantity indexes closely approximate other superlative indexes and that all of the superlative indexes that are known thus far approximate each other to the second order for small changes in prices and quantities.

For $r \neq 0$, define the quadratic mean of order r price index P_r as

$$(8) \qquad P_r(p^0,p^1,x^0,x^1) \equiv \left[\frac{\sum_{i=1}^N (p_i^0 x_i^0/p^0\cdot x^0)(p_i^1/p_i^0)^{r/2}}{\sum_{k=1}^N (p_k^1 x_k^1/p^1\cdot x^1)(p_k^0/p_k^1)^{r/2}}\right]^{1/r}.$$

It can be shown (Diewert [1976a]) that P_r is exact for the quadratic mean of order r unit cost function, $c_r(p) \equiv \left(\sum_{i=1}^N \sum_{j=1}^N b_{ij} p_i^{r/2} p_j^{r/2}\right)^{1/r}$. Since c_r can approximate an arbitrary unit cost function to the second order, P_r is a superlative price index.

For $r \neq 0$, define the quadratic mean of order r quantity index Q_r as

$$(9) \qquad Q_r(p^0,p^1,x^0,x^1) \equiv \left[\frac{\sum_{i=1}^N (p_i^0 x_i^0/p^0\cdot x^0)(x_i^1/x_i^0)^{r/2}}{\sum_{j=1}^N (p_j^1 x_j^1/p^1\cdot x^1)(x_j^0/x_j^1)^{r/2}}\right]^{1/r}.$$

It can similarly be shown that Q_r is exact for the quadratic mean of order r aggregator function, $f_r(x) \equiv \left(\sum_{i=1}^N \sum_{j=1}^N a_{ij} x_i^{r/2} x_j^{r/2}\right)^{1/r}$, and that Q_r is a superlative quantity index.

THEOREM 5. *For any $r \neq 0$, $P_r(p^0, p^1, x^0, x^1) = P_0(p^0, p^1, x^0, x^1)$ and the first and second order partial derivatives of the two functions coincide, provided that $p^0 = p^1 \gg 0_N$ (all price components are positive) and $x^0 = x^1 > 0_N$ (at least one quantity component is positive).*

THEOREM 6. *For any $r \neq 0$, the quantity index Q_r differentially approximates Q_0 to the second order at any point where the prices and quantities for the two periods are equal; i.e., $Q_r(p^0, p^1, x^0, x^1) = Q_0(p^0, p^1, x^0, x^1)$ and the first and second order partial derivatives of the two functions coincide, provided that $p^0 = p^1 > 0_N$ and $x^0 = x^1 \gg 0_N$.*

Theorems 3 and 5 imply that the Vartia price index P_V approximates all of the superlative indexes P_0 and P_r while Theorems 4 and 6 imply that the Vartia quantity index Q_V differentially approximates all of the superlative indexes Q_0 and Q_r, provided that price and quantity changes are small between the two periods.[8] It is worth emphasizing that these theorems hold *without* the assumption of optimizing behavior on the part of economic agents; i.e., they are theorems in numerical analysis rather than economics.

4. Applications and Conclusions

For many years, it was thought that the indexes P_0 and Q_0 had the property of consistency in aggregation;[9] i.e., it was thought that a discrete "Divisia" index of discrete "Divisia" indexes was the discrete "Divisia" index of the components. Although P_0 and Q_0 are not consistent in aggregation, the results of the previous section show why they are approximately consistent in aggregation: each P_0 subindex is approximated to the second order by a Vartia index of the same size, while the "macro" P_0 index is approximated to the second order by a "macro" Vartia index. Thus the macro index of the subindexes is approximated to the second order by a Vartia macro index of Vartia subindexes which is identically equal to a Vartia index of the original micro components, which in turn approximates to the second order a P_0 index in the micro components. Therefore, given time series data where indexes are constructed by chaining observations in successive periods, we would expect P_0 and Q_0 to be approximately consistent in aggregation.

The same conclusion holds for the quadratic mean of order r price indexes P_r and quantity indexes Q_r: they will be approximately consistent in aggregation since each P_r approximates P_V and each Q_r approximates Q_V.

---

[8]Unfortunately, we cannot specify exactly how small is "small" without performing extensive computations involving the third order partial derivatives of the index number formulae; cf. the discussion in Lau [1974a; 183].

[9]See, for example, Jorgenson and Griliches [1972; 83] and Theil [1973; 498].

Some empirical evidence which tends to support the theoretical results above is available. Parkan [1975a] compared the price indexes P_0, P_2 (Fisher's [1922] ideal price index) and $\widetilde{P}_0$ (defined implicitly by (1), the weak factor reversal test, using Q_0 as the quantity index) and the quantity indexes Q_0, Q_2 (Fisher's ideal quantity index) and $\widetilde{Q}_0$ (defined implicitly by (1) using P_0 as the price index) using some Canadian post war consumption data on 13 goods constructed by Gussman [1972] and Cummings and Meduna [1973]. He also calculated the nonparametric price and quantity indexes defined by Diewert [1973b; 424]. Parkan [1975a] then computed all four of the price indexes and all four of the quantity indexes in two stages, calculating subaggregates in each case and then aggregating these subaggregates using the same index number formula. It was found that the resulting total of eight price indexes generally coincided to three significant figures, and the eight quantity indexes similarly closely approximated each other. The theoretical results above provide an explanation for this rather puzzling empirical phenomenon. In Appendix 2, we replicate portions of Parkan's computations.

The above results also have an application to the measurement of total factor productivity. Let u be the quantity of some output, let x be a vector of other outputs and inputs (all indexed positively), let p be the corresponding vector of output and input prices (where price components which correspond to outputs are indexed negatively while components which correspond to inputs are indexed with a plus sign), and suppose that the technology of the economy can be represented by a homogeneous translog transformation function t, where $u = t(x)$. Then if producers are maximizing $t(x)$ subject to an expenditure constraint for periods 0 and 1, it can be shown that $u^1/u^0 = Q_0(p^0, p^1, x^0, x^1)$ provided that no technical change has taken place between the two periods. If technical change takes place in a factor augmenting fashion, then the equation $u^1/u^0 = Q_0(p^0, p^1, x^0, x^1)$ (which depends only on observable prices and quantities) may be modified in a manner which will enable one to compute the amount of technical progress (cf. Diewert [1976a]). It turns out that Theorem 4 above can be extended to cover the case where a subset of the prices are negative, so that the Vartia quantity index will still approximate Q_0. Thus we may use the same argument as before to show that the "Divisia" index $Q_0(p^0, p^1, x^0, x^1)$ may be approximated to the second order by taking a "Divisia" index of "Divisia" subindexes or in fact taking any superlative quantity index of superlative subindexes.[10]

[10]This argument still does not quite provide a rigorous justification for the Jorgenson–Griliches [1967] [1972] method of measuring technical progress in the case of discrete data. Their method can be justified if we assume that the economy's transformation surface can be represented by an equation like $g(u, x_1, \ldots, x_M) = f(x_{M+1}, x_{M+2}, \ldots, x_N)$ where $u, x_1, \ldots, x_M$ are outputs and $x_{M+1}, \ldots, x_N$ are inputs so that outputs are separable from inputs which

The results of the previous section may also be applied to the problem of measuring changes in the welfare of a consumer. Let u be utility, x a consumption vector, and p the corresponding vector of rental prices. If the consumer is maximizing a homogeneous translog utility function subject to an expenditure constraint for periods 0 and 1, then since the homogeneous translog is exact for Q_0,

$$u^1/u^0 = Q_0(p^0, p^1, x^0, x^1); \tag{10}$$

i.e., the relative change in utility is equal to the value of the "Divisia" quantity index Q_0. If changes in prices and quantities are small between the two periods, then the theorems in the previous section show that the change in welfare can be approximated to the second order by evaluating the "Divisia" index Q_0, or by the Vartia index Q_V, or by taking a "Divisia" index of "Divisia" subindexes, or by taking any known superlative index of superlative indexes. This same string of equivalences can be applied in a more general situation, since it is possible to justify the quantity index Q_0 in the context of an aggregator function f which is not necessarily linearly homogeneous. Such justifications for Q_0 have been provided by Kloek [1967] and Theil [1968].

In order to provide a somewhat different justification for Q_0, it is necessary to define the Malmquist [1953] quantity index. Given an aggregator function f and an aggregate $u \equiv f(x)$, define f's *distance function* as $D(u, x) \equiv \max_k\{k : f(x/k) \geq u\}$. The distance function tells us by what proportion one has to deflate the given consumption vector x in order to obtain a point on the indifference surface indexed by u. Now define the Malmquist quantity index as $Q_M(x^0, x^1, u) \equiv D(u, x^1)/D(u, x^0)$ and note that it depends on the two quantity vectors x^0 and x^1, and the base indifference surface (indexed by u) onto which the points x^0 and x^1 are deflated. Then it can be shown (Diewert [1976a]) that if the consumer's preferences can be represented by a translog distance function (which can approximate general nonhomothetic preferences to the second order) and the consumer is maximizing utility subject to a budget constraint for the two periods, then

$$Q_M(x^0, x^1, u^*) = Q_0(p^0, p^1, x^0, x^1), \tag{11}$$

where the reference utility level $u^* \equiv (u^0 u^1)^{1/2}$ is the square root of the product of the base and current period utility levels. As before, the right hand side of (11) can be approximated for small changes in prices and quantities by any of the indexes Q_r or by the Vartia index Q_V.

---

is more restrictive than assuming $u = t(x_1, \ldots, x_N)$. However, it turns out that the Jorgenson–Griliches method of measuring total factor productivity *can* be rigorously justified, at least approximately, even in the general (nonseparable) case; cf. Diewert [1980].

To summarize, the above arguments show that constructing aggregate price and quantity indexes by aggregating two (or more) stages will give approximately the same answer that a one stage index would, provided that either a superlative index or the Vartia index is used.[11]

The present author has argued elsewhere (Diewert [1976a]) that the Fisher price and quantity indexes, P_2 and Q_2, are probably the best of the superlative indexes to use in empirical applications. But what about using the pseudo superlative Vartia indexes, which have the very attractive property of consistency in aggregation (whereas the superlative indexes are only approximately consistent in aggregation)? Unfortunately, the Vartia quantity index has the property that rescaling the prices in either period will generally change the index (i.e., in general $Q_V(\lambda p^0, p^1, x^0, x^1) \neq Q_V(p^0, p^1, x^0, x^1)$ for $\lambda \neq 1$) while the Vartia price index has the property that rescaling the period 1 prices does not in general change the value of the price index by the same scale factor (i.e., in general $P_V(p^0, \lambda p^1, x^0, x^1) \neq \lambda P_V(p^0, p^1, x^0, x^1)$ for $\lambda \neq 1$). These defects of the Vartia indexes will probably preclude their use in empirical situations, but as we have seen, the Vartia indexes have proven to be very useful from a theoretical point of view.

Appendix 1: Proofs of Theorems

Proof of Theorem 1. If x^s is a solution to $\max_x\{f(x) : p^s \cdot x \leq p^s \cdot x^s,\ x \geq 0_N\}$ for $s = 0, 1$ where f is positively linearly homogeneous and concave over the positive orthant, then it is easy to see that x^s is also a solution to the expenditure minimization problem $\min_x\{p^s \cdot x : f(x) \geq f(x^s)\} = c(p^s)u^s$ for $s = 0, 1$ where c is the unit cost function which corresponds to f and $u^s \equiv f(x^s)$ for $s = 0, 1$. Moreover, if $c(p)$ is once differentiable, then by Shephard's Lemma [1953; 11], $x^s = \nabla_p c(p^s)u^s$ for $s = 0, 1$ where $\nabla_p c(p^s)$ is the gradient vector of c evaluated at the price vector p^s.

Thus using Shephard's Lemma to eliminate quantities in the index number formula, if c is once differentiable, positively linearly homogeneous, concave and exact for the Vartia price index defined by (4), we must have for every $p^0 \gg 0_N$, $p^1 \gg 0_N$, x^s a solution to $\max_x\{f(x) : p^s \cdot x \leq p^s \cdot x^s,\ x \geq 0_N\}$ for $s = 0, 1$ where f is dual to c,

$$(12) \qquad \ln\left[\frac{c(p^1)}{c(p^0)}\right] \equiv \sum_{i=1}^{N} \frac{L[p_i^1 c_i(p^1)u^1, p_i^0 c_i(p^0)u^0]}{L[p^1 \cdot \nabla_p c(p^1)u^1, p^0 \cdot \nabla_p c(p^0)u^0]} \ln\left[\frac{p_i^1}{p_i^0}\right],$$

[11]However, note that our arguments do not justify the existence of well behaved *subaggregates* which satisfy the usual regularity conditions of microeconomic theory.

where $u^0 \equiv f(x^0)$, $u^1 \equiv f(x^1)$, and $c_i(p^s) \equiv \partial c(p^s)/\partial p_i$ for $i = 1, 2, \ldots, N$. Thus (12) is to hold for some functional form for c with the appropriate regularity properties for all $p^s \gg 0_N$ and scalars $u^s > 0$ for $s = 0, 1$.

Let the last $N-1$ components of the vectors p^0 and p^1 be identical; then (12) becomes

$$(13) \qquad \ln\left[\frac{c(p^1)}{c(p^0)}\right] = \frac{L[p_1^1 c_1(p^1)u^1, p_1^0 c_1(p^0)u^0]}{L[p^1 \cdot \nabla_p^1 c(p^1)u^1, p^0 \cdot \nabla_p c(p^0)u^0]} \ln\left[\frac{p_1^1}{p_1^0}\right].$$

Since (13) must hold for all positive u^1 and u^0, and the left hand side is independent of u^0 and u^1, the right hand side must also be independent of u^0 and u^1. This will be the case only if $p_1^1 c_1(p^1) = \alpha_1 p^1 \cdot \nabla_p c(p^1)$ and $p_0^1 c_1(p^0) = \alpha_1 p^0 \cdot \nabla_p c(p^0)$ where α_1 is a constant; i.e., the expenditure share on the first good must be constant. We can similarly show that expenditure shares on all goods must be constant; i.e., we must have for $j = 1, 2, \ldots, N$ and all $p^0 \gg 0_N$,

$$(14) \qquad \alpha_j = p_j^0 c_j(p^0)/p^0 \cdot \nabla_p c(p^0) = p_j^0 c_j(p^0)/c(p^0),$$

where the second equality follows from the assumption that c is linearly homogeneous. From (14) and the linear homogeneity of $c(p)$, it follows that $c(p)$ must be proportional to $\prod_{j=1}^N p_j^{\alpha_j}$ where $\sum_{j=1}^N \alpha_j = 1$. Since we require $c(p)$ to be concave, we must also have $\alpha_j \geq 0$. Thus in order to be exact for the Vartia price index, $c(p)$ must be of the Cobb–Douglas functional form.

Proof of Theorem 2. If x^s is a solution to $\max_x\{f(x) : p^s \cdot x \leq p^s \cdot x^s, x \geq 0_N\}$ for $s = 1, 0$ where f is positively linearly homogeneous and concave over the positive orthant, then it is well known that if f is once differentiable, elimination of the Lagrange multipliers for the maximization problems yields the identities $p^s/p^s \cdot x^s = \nabla_x f(x^s)/x^s \cdot \nabla_x f(x^s) = \nabla_x f(x^s)/f(x^s)$ for $s = 0, 1$ where the second equality follows from the linear homogeneity of f. Thus

$$(15) \qquad p^s = p^s \cdot x^s \nabla_x f(x^s)/f(x^s), \qquad s = 0, 1.$$

Thus using (15) to eliminate prices from the index number formula, if f is once differentiable, positively linearly homogeneous, concave, and exact for the Vartia quantity index Q_V defined by (5), we must have for every $x^0 \gg 0_N$, $x^1 \gg 0_N$, scalars $e^0 > 0$, $e^1 > 0$ and price vectors $p^s \equiv e^s \nabla_x f(x^s)/f(x^s)$ for $s = 0, 1$,

$$(16) \quad \ln\left[\frac{f(x^1)}{f(x^0)}\right] = \sum_{i=1}^N \frac{L[e^1 f_i(x^1)x_i^1/f(x^1), e^0 f_i(x^0)x_i^0/f(x^0)]}{L(e^1, e^0)} \ln\left[\frac{x_i^1}{x_i^0}\right].$$

Since the left hand side of (16) is independent of the scalars e^0 and e^1, the right hand side must also be, and this will only be the case if for all $x^1 \gg 0_N$ and $x^0 \gg 0_N$, there exist constants α_j such that

$$(17) \qquad \alpha_j = f_j(x^1)x_j^1/f(x^1) = f_j(x^0)x_j^0/f(x^0) \qquad (j = 1, 2, \ldots, N).$$

Upon integrating the partial differential equations (17), we find that $f(x)$ must be proportional to $\prod_{j=1}^N x_j^{\alpha_j}$ and in order for f to be linearly homogeneous and concave, we further require that $\alpha_j \geq 0$ and $\sum_{j=1}^N \alpha_j = 1$. Thus in order to be exact for the Vartia quantity index, f must be of Cobb–Douglas functional form.

Proof of Lemma 1. The proof is a straightforward computation if one makes repeated use of l'Hospital's rule (see Rudin [1953; 82–83]) and the definition of the Vartia mean function L.

Proof of Lemma 2. Obviously $h(y^*) = h^*(y^*)$. We have

$$\begin{aligned} \nabla_y h(y^*) &= f(y^*)\nabla_y g(y^*) + \nabla_y f(y^*)g(y^*) \\ &= f^*(y^*)\nabla_y g^*(y^*) + \nabla_y f^*(y^*)g^*(y^*) \\ &= \nabla_y h^*(y^*) \end{aligned}$$

and

$$\begin{aligned} \nabla_{yy}^2 h(y^*) = f(y^*)\nabla_{yy}^2 g(y^*) &+ \nabla_y f(y^*)\nabla_y^T g(y^*) \\ &+ \nabla_y g(y^*)\nabla_y^T f(y^*) + \nabla_{yy}^2 f(y^*)g(y^*) \\ = \nabla_{yy}^2 h^*(y^*), \end{aligned}$$

since $g(y^*) = g^*(y^*) = 0$ where $\nabla_y^T f(y^*)$ is the transpose of the column vector $\nabla_y f(y^*)$, etc.

Proof of Theorem 3. We show that $\ln P_V$ differentially approximates $\ln P_0$. Define the values $v_j^s = p_i^s x_j^s$ for $s = 0, 1$ and $j = 1, 2, \ldots, N$. Then it is easy to see that our result will be true if the first and second order partial derivatives of the log of the Vartia price index regarded as a function of p^0, p^1, v^0, and v^1,

$$\sum_{i=1}^N \left[\frac{v_i^1 - v_i^0}{\ln v_i^1 - \ln v_i^0}\right] \left[\frac{\ln(\sum_{j=1}^N v_j^1) - \ln(\sum_{j=1}^N v_j^0)}{\sum_{j=1}^N v_j^1 - \sum_{j=1}^N v_j^0}\right] \ln\left[\frac{p_i^1}{p_i^0}\right],$$

are equal to the first and second order partial derivatives of the log of P_0 regarded as a function of p^0, p^1, v^0, and v^1,

$$\sum_{i=1}^N \frac{1}{2}\left[\frac{v_i^1}{\sum_{j=1}^N v_j^1} + \frac{v_i^0}{\sum_{j=1}^N v_j^0}\right] \ln\left[\frac{p_i^1}{p_i^0}\right],$$

evaluated at any point such that $p^1 = p^0 \gg 0_N$, $v^1 = v^0 \gg 0_N$. It turns out that matters are simplified if we introduce the additional variables $V^1 \equiv \sum_{j=1}^N v_j^1$ and $V^0 \equiv \sum_{j=1}^N v_j^0$ and regard the index numbers as functions of p^1, p^0, v^1, v^0, V^1, and V^0. Let us also define the jth component λ_j of the vector of variables λ by $v_j^1 = \lambda_j v_j^0$, and the jth component δ_j of the vector of variables δ by $p_j^1 = \delta_j p_j^0$, $j = 1, 2, \ldots, N$. Define the scalar γ by $V^1 = \gamma V^0$. It can be seen that our theorem will be true if the first and second order partial derivatives of the log of the Vartia price index regarded as a function of p^0, δ, v^0, λ, V^0, and γ,

$$\sum_{i=1}^N \frac{v_i^0}{V^0}\left[\frac{\lambda_i - 1}{\ln \lambda_i}\right]\left[\frac{\ln \gamma}{\gamma - 1}\right] \ln \delta_i,$$

are equal to the first and second order partial derivatives of the log of P_0 regarded as a function of p^0, δ, v^0, λ, V^0, and γ,

$$\sum_{i=1}^N \frac{1}{2}\left[\frac{\lambda_i}{\gamma} + 1\right]\left[\frac{v_i^0}{V^0}\right] \ln \delta_i,$$

evaluated at any point such that $p^0 \gg 0_N$, $v^0 \gg 0_N$, $V^0 > 0$, and where $\lambda = 1_N$ (a vector of ones), $\delta = 1_N$, and $\gamma = 1$. Thus we need only show that for each i, the level, the first order partial derivatives and the second order partial derivatives of the function $h^i(\lambda_i, \gamma, \delta_i) \equiv f^i(\lambda_i, \gamma, \delta_i) g^i(\lambda_i, \gamma, \delta_i)$ where $f^i(\lambda, \gamma, \delta_i) \equiv (\lambda_i - 1)(\ln \lambda_i)^{-1}(\ln \gamma)(\gamma - 1)^{-1}$ and $g^i(\lambda_i, \gamma, \delta_i) \equiv \ln \delta_i$ are equal to the corresponding level, first and second order partial derivatives of the function $h^{i^*}(\lambda_i, \gamma, \delta_i) \equiv f^{i^*}(\lambda_i, \gamma, \delta_i) g^{i^*}(\lambda_i, \gamma, \delta_i)$ where $f^{i^*}(\lambda_i, \gamma, \delta_i) \equiv \frac{1}{2}[(\lambda_i/\gamma) + 1]$ and $g^{i^*}(\lambda_i, \gamma, \delta_i) \equiv \ln \delta_i$, evaluated at $\lambda_i = 1$, $\gamma = 1$, and $\delta_i = 1$.

By Lemma 1, the level and the first order partial derivatives of f^i and f^{i^*} coincide at $\lambda_i = 1$, $\gamma = 1$, and $\delta_i = 1$. Since $g^i \equiv g^{i^*}$, the levels and all partial derivatives of g^i and g^{i^*} coincide. Moreover, $g^i(1,1,1) = g^{i^*}(1,1,1) = 0$ and thus our result follows using Lemma 2.

Proof of Theorem 5. Straightforward but tedious calculations show that if $p^0 = p^1 \equiv p \gg 0_N$, $x^0 = x^1 \equiv x > 0_N$, then:

(i) $$\ln P_0(p^0, p^1, x^0, x^1) = \ln P_r(p^0, p^1, x^0, x^1) = 0;$$

(ii) $$\partial \ln P_0(p^0, p^1, x^0, x^1)/\partial x_i^1 = \partial \ln P_r(p^0, p^1, x^0, x^1)/\partial x_i^1 = 0,$$

$$\partial \ln P_0/\partial x_i^0 = \partial \ln P_r/\partial x_i^0 = 0$$

dropping the arguments p^0, p^1, x^0, x^1 for brevity,

$$\partial \ln P_0/\partial p_i^1 = \partial \ln P_r/\partial p_i^1 = x_i/p \cdot x,$$

$$\partial \ln P_0/\partial p_i^0 = \partial \ln P_r/\partial p_i^0 = -x_i/p \cdot x$$

for $i = 1, 2, \ldots, N$;

(iii)
$$\begin{aligned}
\partial^2 \ln P_0/\partial x_i^1 \partial x_j^1 &= \partial^2 \ln P_r/\partial x_i^1 \partial x_j^1 = 0,\\
\partial^2 \ln P_0/\partial x_i^1 \partial x_j^0 &= \partial^2 \ln P_r/\partial x_i^1 \partial x_j^0 = 0,\\
\partial^2 \ln P_0/\partial x_i^1 \partial p_j^1 &= \partial^2 \ln P_r/\partial x_i^1 \partial p_j^1\\
&= \left(\frac{1}{2}\delta_{ij}/p\cdot x\right) - \left[\frac{1}{2}p_i x_i/(p\cdot x)^2\right]
\end{aligned}$$

where δ_{ij} equals 1 if $i = j$ and is zero otherwise,

$$\begin{aligned}
\partial^2 \ln P_0/\partial x_i^1 \partial p_j^0 &= \partial^2 \ln P_r/\partial x_i^1 \partial p_i^0\\
&= -\left(\frac{1}{2}\delta_{ij}/p\cdot x\right) + \left[\frac{1}{2}p_i x_j(p\cdot x)^2\right],\\
\partial^2 \ln P_0/\partial x_i^0 \partial x_j^0 &= \partial^2 \ln P_r/\partial x_i^0 \partial x_j^0 = 0,\\
\partial^2 \ln P_0/\partial x_i^0 \partial p_j^1 &= \partial^2 \ln P_r/\partial x_i^0 \partial p_j^0\\
&= \left(\frac{1}{2}\delta_{ij}/p\cdot x\right) - \left[\frac{1}{2}p_i x_j/(p\cdot x)^2\right],\\
\partial \ln P_0/\partial x_i^0 \partial p_j^0 &= \partial^2 \ln P_r/\partial x_i^0 \partial p_j^0\\
&= -\left(\frac{1}{2}\delta_{ij}/p\cdot x\right) + \left[\frac{1}{2}p_i x_j/(p\cdot x)^2\right],\\
\partial^2 \ln P_0/\partial p_i^1 \partial p_j^1 &= \partial^2 \ln P_r/\partial p_i^1 \partial p_j^1\\
&= -x_i x_j/(p\cdot x)^2,\\
\partial^2 \ln P_0/\partial p_i^1 \partial p_j^0 &= \partial^2 \ln P_r/\partial p_i^1 \partial p_j^0 = 0,
\end{aligned}$$

and

$$\begin{aligned}
\partial^2 \ln P_0/\partial p_i^0 \partial p_j^0 &= \partial^2 \ln P_r/\partial p_i^0 \partial p_j^0\\
&= x_i x_j/(p\cdot x)^2
\end{aligned}$$

for $1 \leq i, j \leq N$.

Proof of Theorem 6. Proof is the same as Theorem 5 except that prices and quantities are interchanged.

Appendix 2: Empirical Comparison of Index Numbers

In this appendix, we use the Canadian consumer data constructed by Cummings and Meduna [1973] in order to compare empirically various index number formulae.

Table 1. Comparison of Single State Chained Index Numbers

Year	Vartia P_V	Törnqvist P_0	Implicit Törnqvist $\tilde{P}_0$	Fisher Ideal P_2	Laspeyres P_L	Paasche P_P
1947	1.0000	1.0000	1.0000	1.0000	1.0000	1.0000
1951	1.3851	1.3851	1.3850	1.3851	1.3857	1.3845
1956	1.4874	1.4875	1.4874	1.4875	1.4888	1.4861
1961	1.6556	1.6558	1.6556	1.6557	1.6578	1.6536
1966	1.8795	1.8797	1.8795	1.8796	1.8831	1.8761
1971	2.3228	2.3230	2.3227	2.3228	2.3285	2.3172

The primary data are price indexes for 13 components of Canadian consumer expenditures for the years 1947–1971 and the corresponding per capita quantity series. These data were constructed by Cummings and Meduna by aggregating (using Törnqvist price indexes) Canadian national accounts data pertaining to expenditure on over 40 consumer goods categories plus some series constructed by Gussman [1972]. Rental prices for each category of consumer durables were constructed. These data are available on request.

Table 1 lists Vartia (P_V), Törnqvist (P^0), implicit Törnqvist ($\tilde{P}_0$), Fisher (P_2), Laspeyres ($P_L(p^0,p^1,x^0,x^1) \equiv (x^0 \cdot p^1/x^0 \cdot p^0)$, and Paasche ($P_P(p^0,p^1, x^0,x^1) \equiv x^1 \cdot p^1/x^1 \cdot p^0$) price indexes for the 13 goods using the chain principle. Only selected years are reported in order to conserve space. Note that all of the indexes coincide to three significant figures and that P_V, P_0, $\tilde{P}_0$, P_2, and P_L coincide to four significant figures (after rounding off) over the entire period. The close correspondence of P_V, P_0, $\tilde{P}_0$, and P_2 is not unexpected since all of these indexes differentially approximate each other to the *second* order at a point where prices and quantities are equal and, of course, using the chain principle, prices and quantities are approximately equal going from one year to the next. What is somewhat unexpected is the close correspondence of P_L and P_P to the other indexes. However, in the following appendix we show that P_L and P_P differentially approximate the other four indexes to the *first* order at any point where prices and quantities are equal. Thus, for our data, it appears that annual changes in prices and quantities are small enough so that chained Paasche and Laspeyres price indexes approximate reasonably closely any chained superlative price index.

The index numbers in Table 1 were constructed using a double precision Fortran program. As a check on our computations, we calculated a chained Vartia quantity index, used the weak factor reversal test (1) to construct an

Table 2. Comparison of Single Stage Fixed Base Index Number

Year	P_V	P_0	$\tilde{P}_0$	P_2	P_L	P_P
1947	1.0000	1.0000	1.0000	1.0000	1.0000	1.0000
1951	1.3825	1.3833	1.3822	1.3834	1.3833	1.3834
1956	1.4814	1.4834	1.4804	1.4831	1.4847	1.4815
1961	1.6487	1.6532	1.6483	1.6518	1.6564	1.6472
1966	1.8772	1.8843	1.8782	1.8818	1.9070	1.8568
1971	2.3121	2.3224	2.3158	2.3188	2.3621	2.2763

implicit price index $\tilde{P}_V$, and we found $\tilde{P}_V$ coincided with P_V to five decimal places.

In order to ascertain the effect of chaining, see Table 2 which compares the six index number formulae using a fixed base (1947=1) throughout. The fixed base Vartia, Törnqvist, and Fisher indexes turn out to be very close to each other as well as to their chained counterparts (all coincide to within one half percent). However, the general effect of using a fixed base is to increase the differences between the various index number formulae: in the fixed base case, P_L differs from P_V by about two percent in 1971, while P_L differs from P_P by four percent in 1971.

The relatively large divergence between the Paasche and Laspeyres indexes when a fixed base year is used is a source for some concern. In many national accounting systems (such as Canada's), the consumer price index is constructed using a fixed base Laspeyres index, while the implicit deflator for the consumption expenditures component of GNP is a fixed base Paasche index. Given the importance of consumer price indexes in indexing wages and cost of living supplements, it is important that official price indexes be consistent with each other.

The above considerations suggest that government agencies should use chained rather than fixed base indexes in order to deflate expenditures into constant dollar quantities. More formally, there are at least three reasons why chained rather than fixed base indexes should be used in the context of time series data where period to period changes in prices and quantities are small:

(i) All superlative, pseudo superlative, Paasche, and Laspeyres index numbers should coincide quite closely if they are constructed using the chain principle. (The degree of coincidence should be somewhat greater for the superlative and pseudo superlative indexes, since their index number formula derivatives coincide to the second order, but only to the first order with the Paasche and Laspeyres formulae.)

(ii) The Paasche, Laspeyres, or any superlative index number can be regarded as discrete approximations to the continuous line integral Divisia index, which has some useful optimality properties from the viewpoint of economic theory.[12] These discrete approximations will be closer to the Divisia index if the chain principle is used.

(iii) The use of chained indexes avoids the vexing problems which arise when the base year in the fixed base indexes is changed. For example, the base year in the Canadian system of national accounts has just been changed from 1961 to 1971, and new Paasche price indexes for the various components of GNP using 1971 as the base year have been computed for the years 1971–1974. Unfortunately the new price and quantity indexes are not proportional to the old indexes which used 1961 as the base year. Thus most econometric models using national accounts data will have to be reestimated using the new indexes. The use of chained indexes would avoid the discontinuities introduced by these periodic changes in the base year.

We now turn to a comparison of index numbers constructed in one stage versus two stages. We formed the subaggregate "nondurables" by aggregating the following five goods: food, alcohol, tobacco, energy, and other nondurables. The subaggregate "services" consisted of two goods: medical services, and other services; the subaggregate "semidurables" consisted of two goods: clothing, and other semidurables; and finally the subaggregate "durables" consisted of the following four goods: motor vehicles, housing, land, and other durables. Two stage price indexes were constructed by using a price index number formula (along with its corresponding quantity index defined by the weak factor reversal test) in order to construct price and quantity indexes for the four subaggregates listed above. The same index number formula was then used in order to construct an aggregate price index. Of course, for the Vartia, Laspeyres, and Paasche indexes, the two stage procedure gave rise to precisely the same aggregate indexes as the usual one stage procedure; i.e., P_V, P_L, and P_P are consistent in aggregation.

Table 3. Comparison of Two Stage Chained Index Numbers

Year	P_0	$\tilde{P}_0$	P_2
1947	1.0000	1.0000	1.0000
1951	1.3851	1.3851	1.3851
1956	1.4875	1.4874	1.4875
1961	1.6558	1.6556	1.6557
1966	1.8797	1.8796	1.8796
1971	2.3230	2.3228	2.3228

[12]See Malmquist [1953], Wold [1953], and Hulten [1973].

Table 3 lists the aggregate price indexes which resulted when the two stage procedure was applied using the chain principle with the index number formulae P_0, $\widetilde{P}_0$, and P_2. Comparison of Tables 1 and 3 shows that the Törnqvist indexes, P_0 and $\widetilde{P}_0$, and the Fisher index, P_2, are approximately consistent in aggregation to a very high degree of approximation indeed.

Recall that in equation (4) of the text, the logarithm of the ith price relative, $\ln(p_i^1/p_i^0)$, is weighted by $w_i(p^0,p^1,x^0,x^1) \equiv L(p_i^1x_i^1, p_i^1x_i^0)/L(p^1 \cdot x^1, p^0 \cdot x^0)$ where L is the logarithmic mean function. Vartia [1974; 91–93] shows that $\sum_{i=1}^N w_i(p^0,p^1,x^0,x^1) < 1$ (unless $p_i^1x_i^1/p_i^0x_i^0$ = constant for $i = 1,2,\ldots,N$ in which case the weights w_i sum to precisely 1). The sum of the weights for selected years is tabled in Table 4; the first column of Table 4 is the sum of the chained weights, $\sum_{i=1}^{13} w_i(p^{t-1},p^t,x^{t-1},x^t)$, while the second column is the sum of the fixed base weights, $\sum_{i=1}^{13} w_i(p^{1947},p^t,x^{1947},x^t)$.

Table 4. Sum of Vartia Weights

Year	Chained Sum	Fixed Base Sum
1948	.99979	.99979
1951	.99977	.99746
1956	.99995	.99419
1961	.99994	.99112
1966	.99997	.98929
1971	.99998	.98631

Note that it is not correct to say that the Vartia price and quantity indexes are biased downwards due to the fact that the weights $w_i(p^0,p^1,x^0,x^1)$ generally sum to a number less than one. This would be the case if the weights were constant, but they are not. Moreover, the Vartia quantity index has the same weights as the Vartia price index, and since $P_V(p^0,p^1,x^0,x^1)$ $Q_V(p^0,p^1,x^0,x^1) = p^1 \cdot x^1/p^0 \cdot x^0$, it cannot be the case that both the Vartia price and quantity indexes are biased downwards.

Appendix 3: More on Pseudo Superlative Index Numbers

In this section, we prove some additional theorems about pseudo superlative index numbers.

THEOREM 7. *If $P(p^0, p^1, x^0, x^1)$ is a pseudo superlative price index, then the corresponding quantity index defined by using the weak factor reversal test, $Q(p^0, p^1, x^0, x^1) \equiv p^1 \cdot x^1/[p^0 \cdot x^0 P(p^0, p^1, x^0, x^1)]$ is a pseudo superlative quantity index.*

Proof. The Vartia quantity index can be defined as

$$Q_V(p^0, p^1, x^0, x^1) \equiv p^1 \cdot x^1/[p^0 \cdot x^0 P_V(p^0, p^1, x^0, x^1)] \tag{18}$$

where P_V is the Vartia price index. Since P is pseudo superlative, P and P_V have the same first and second order partial derivatives when evaluated at an equal price and equal quantity point, and thus comparison of the definition of Q with the definition of Q_V given by (18) shows that Q and Q_V will have the same first and second order partial derivatives. Q.E.D.

Of course, a similar theorem holds if the roles of P and Q are interchanged in Theorem 7.

The following theorem shows that the Paasche and Laspeyres price indexes are *not* pseudo superlative.

THEOREM 8. *Let $P(p^0, p^1, x^0, x^1)$ be any pseudo superlative price index (i.e., P has the first and second order partial derivatives defined in Theorem 5). Then $P(p^0, p^1, x^0, x^1) = P_L(p^0, p^1, x^0, x^1) \equiv p^1 \cdot x^0/p^0 \cdot x^0 = P_P(p^0, p^1, x^0, x^1) \equiv p^1 \cdot x^1/p^0 \cdot x^1$ and the first order partial derivatives of the pseudo superlative index P, the Laspeyres index P_L, and the Paasche index P_P coincide, provided that $p^0 = p^1 \equiv p \gg 0_N$ and $x^0 = x^1 \equiv x > 0_N$. However, the second order partial derivatives of P, P_L, and P_P evaluated under the same conditions do not coincide.*

Proof. Again, straightforward but tedious calculations show that if $p^0 = p^1 \equiv p \gg 0_N$, $x^0 = x^1 \equiv x > 0_N$, then $\ln P_L = \ln P_P = 0 = \ln P$, and the first order partial derivatives of $\ln P_L(p^0, p^1, x^0, x^1)$, $\ln P_P(p^0, p^1, x^0, x^1)$ and $\ln P(p^0, p^1, x^0, x^1)$ all coincide. The second order partial derivatives of $\ln P_L(p^0, p^1, x^0, x^1)$ are (dropping the arguments p^0, p^1, x^0, x^1 for brevity):

$$\partial^2 \ln P_L/\partial x_i^1 \partial x_j^1 = 0,$$

$$\partial^2 \ln P_L/\partial x_i^1 \partial x_j^0 = 0,$$

$$\partial^2 \ln P_L/\partial x_i^1 \partial p_j^1 = 0, \quad \neq \partial^2 \ln P/\partial x_i^1 \partial p_j^1,$$

$$\partial^2 \ln P_L/\partial x_i^1 \partial p_j^0 = 0, \quad \neq \partial^2 \ln P/\partial x_i^1 \partial p_j^0,$$

$$\partial^2 \ln P_L/\partial x_i^0 \partial x_j^0 = 0,$$

$$\partial^2 \ln P_L/\partial x_i^0 \partial p_j^1 = (\delta_{ij}/p \cdot x) - [p_i x_j/(p \cdot x)^2] \neq \partial^2 \ln P/\partial x_i^0 \partial p_j^1,$$

$$\partial^2 \ln P_L/\partial x_i^0 \partial p_j^0 = -(\delta_{ij}/p \cdot x) + [p_i x_j/(p \cdot x)^2] \neq \partial^2 \ln P/\partial x_i^0 \partial p_j^0,$$

$$\partial^2 \ln P_L/\partial p_i^1 \partial p_j^1 = -x_i x_j/(p \cdot x)^2,$$

$$\partial^2 \ln P_L/\partial p_i^1 \partial p_j^0 = 0,$$

$$\partial^2 \ln P_L/\partial p_i^0 \partial p_j^0 = x_i x_j/(p \cdot x)^2,$$

where $\delta_{ij} = 1$ if $i = j$ and is 0 otherwise.

The second order partial derivatives of $\ln P_P(p^0, p^1, x^0, x^1)$ can be obtained from the second order partial derivatives of $\ln P_L(p^0, p^1, x^0, x^1)$ listed above if x^0 and x^1 are interchanged. Q.E.D.

A *symmetric mean* $m(x, y)$ of two nonnegative numbers x and y can be defined as any function which satisfies the following three conditions:

(19-i) $$m(x, y) = m(y, x),$$

(19-ii) $$m(x, x) = x,$$

(19-iii) $$\min(x, y) \leq m(x, y) \leq \max(x, y).$$

Samuelson and Swamy [1974; 582] assert a theorem which states that in the case of a homothetic aggregator function, any symmetric mean of the Laspeyres and Paasche index numbers will approximate the true index number up to the third order in accuracy. Although the following theorem does not make the assumptions about optimizing behavior on the part of economic agents that Samuelson and Swamy make in their theorem, the following theorem in numerical analysis does provide a counterpart to the Samuelson–Swamy result.

THEOREM 9. *Any twice continuously differentiable symmetric mean of the Laspeyres and Paasche price indexes, $m[P_L(p^0, p^1, x^0, x^1), P_P(p^0, p^1, x^0, x^1)]$, is a pseudo superlative price index.*

Proof. If $m(P_L, P_P) = P_L^{\frac{1}{2}} P_P^{\frac{1}{2}}$, then the resulting index is P_2, Fisher's ideal index, which is pseudo superlative by Theorem 5. Note that when $p^0 = p^1$ and $x^0 = x^1$, $m(P_L, P_P) = m(1, 1) = 1 = P_L^{\frac{1}{2}} P_P^{\frac{1}{2}}$ where we have used property (19-ii) of m. Thus we need only show that if $p^0 = p^1 \equiv p \gg 0_N$, $x^0 = x^1 \equiv x > 0_N$, then the first and second order partials of $m[P_L(p^0, p^1, x^0, x^1), P_P(p^0, p^1, x^0, x^1)]$ equal the corresponding partial derivatives of $P_2(p^0, p^1, x^0, x^1)$ for any twice differentiable symmetric mean function m. Thus we need to know the first and second order partials of $m(x, y)$ evaluated at $(x, y) = (1, 1)$.

By partially differentiating (19-i) with respect to x, we obtain

(20) $$m_1(x, y) = m_2(y, x)$$

where m_i denotes partial differentiation with respect to the ith argument of m, $i = 1, 2$. Partial differentiation of (19-ii) with respect to x yields the following identity:

(21) $$m_1(x,x) + m_2(x,x) = 1.$$

When $x = y$, (20) and (21) imply the following relations:

(22) $$m_1(x,x) = m_2(x,x) = \frac{1}{2}.$$

By partially differentiating (21) with respect to x, we obtain the following identity (using also $m_{12} = m_{21}$ which follows from the assumption that m is twice continuously differentiable):

(23) $$m_{11}(x,x) + 2m_{12}(x,x) + m_{22}(x,x) = 0.$$

Now partially differentiate (20) with respect to x, set $x = y$, and obtain the following identity:

(24) $$m_{11}(x,x) = m_{22}(x,x).$$

Relations (23), (24), and $m_{12} = m_{21}$ imply that:

(25) $$m_{11}(x,x) + m_{12}(x,x) = 0 = m_{21}(x,x) + m_{22}(x,x).$$

The magnitude of $m_{11}(x,x) = m_{22}(x,x) = -m_{12}(x,x)$ cannot be determined in general, but it turns out that we only need equations (22) and (25) in order to evaluate the first and second order partial derivatives of $m[P_L(p^0,p^1,x^0,x^1), P_P(p^0,p^1,x^0,x^1)]$ and to show that they are equal to the corresponding partial derivatives of $P_2(p^0,p^1,x^0,x^1)$ when $p^0 = p^1$ and $x^0 = x^1$. (Recall that the first order partials of P_L and P_P are equal using Theorem 8.)

Q.E.D.

Note that the proof of the above theorem did not require property (19-iii) on the symmetric mean m. The proof of the following theorem is analogous to the proof of Theorem 9.

THEOREM 10. *Any twice continuously differentiable symmetric mean of two pseudo superlative indexes is also a pseudo superlative index.*

Vartia [1974] [1976a] presents a geometric proof that the logarithmic mean $L(x,y) \equiv (x-y)(\ln x - \ln y)$ lies between the geometric mean and the arithmetic mean of the nonnegative numbers x and y; i.e., he shows that $M_0(x,y) \equiv x^{\frac{1}{2}}y^{\frac{1}{2}} \le L(x,y) \le M_1(x,y) \equiv \frac{1}{2}x + \frac{1}{2}y$.

We conclude this appendix by stating that

(26) $$M_0(x,y) \le L(x,y) \le M_{r^*}(x,y), \qquad x > 0, \quad y > 0,$$

where $r^* \equiv 1/3$ and the mean of order r for $r \neq 0$ is defined by $M_r(x,y) \equiv (\frac{1}{2}x^r + \frac{1}{2}y^r)^{1/r}$. Since

$$\min(x,y) \le M_0(x,y) \le M_{1/3}(x,y) \le M_1(x,y) \le \max(x,y)$$

(see Hardy, Littlewood and Polya [1934; Theorems 5 and 16]), it can be seen that $L(x,y)$ is a symmetric mean (recall definition (19)) and that the bounds (26) on $L(x,y)$ are tighter than Vartia's bounds.

Essays in Index Number Theory, Volume I
W.E. Diewert and A.O. Nakamura (Editors)

Chapter 10
DIRECT VERSUS IMPLICIT SUPERLATIVE INDEX NUMBER FORMULAE*

Robert C. Allen and W. Erwin Diewert

1. Introduction

Economists and statisticians who construct estimates of total factor productivity or who estimate production functions or systems of consumer demand functions are often forced to aggregate subsets of their data. In order to perform this aggregation, an index number formula is generally used. A price index $P(p^0, p^1, x^0, x^1)$ is defined to be a function P of the prices of the N commodities to be aggregated in periods 0 and 1, $p^0 \equiv (p_1^0, \ldots, p_N^0)$ and $p^1 \equiv (p_1^1, \ldots, p_N^1)$, respectively, and of the corresponding quantities utilized during periods 0 and 1, $x^0 \equiv (x_1^0, \ldots, x_N^0)$ and $x^1 \equiv (x_1^1, \ldots, x_N^1)$, respectively. A quantity index $Q(p^0, p^1, x^0, x^1)$ is defined to be another function Q of the price and quantity vectors for the two periods. Generally, we assume that P and Q satisfy Fisher's [1922] weak factor reversal test:

(1) $$P(p^0, p^1, x^0, x^1)Q(p^0, p^1, x^0, x^1) \equiv p^1 \cdot x^1 / p^0 \cdot x^0$$

where

$$p^t \cdot x^t \equiv \sum_{i=1}^{N} p_i^t x_i^t \quad \text{for } t = 0, 1.$$

Thus P is to be interpreted as the ratio of the price level in period 1 to the price level in period 0 while Q is the ratio of the quantity levels.

At this stage, a problem arises: Which functional form for P and Q should be chosen in order to aggregate the data?

---

*First published in *The Review of Economics and Statistics*, Vol. 63, No. 3, 1981, pp. 430–435. Robert C. Allen is also a professor in the Department of Economics at the University of British Columbia. This research was supported by a Canada Council grant. The authors are indebted to Don Andrews, Robert Feenstra and Russell Uhler for research assistance and to an unknown editor, C. Blackorby, D. Donaldson and the members of the UBC Theory Workshop for helpful comments.

Diewert [1976a] has suggested the use of a *superlative* index number formula. A quantity index Q is defined to be superlative if it is *exact* for a neoclassical aggregator function[1] f which is capable of providing a second order differential approximation to a twice continuously differentiable linearly homogeneous aggregator function. A price index P is defined to be *superlative* if it is exact for a unit cost function c which can provide a second order differential approximation to an arbitrary twice continuously differentiable unit cost function. For formal definitions and examples of exact index number formulae, see Konüs and Byushgens [1926], Pollak [1971a], Afriat [1972b], Samuelson and Swamy [1974], Sato [1976a], Diewert [1976a] [1978b] and Lau [1979].

If either a price or quantity index is defined, then a corresponding quantity or price index can be defined implicitly by using the weak factor reversal test (1). Moreover, if P is a superlative price index and $\tilde{Q}$ is the corresponding quantity index defined implicitly by (1), then as Diewert [1978b] has shown, $\tilde{Q}$ is also superlative. Similarly, if Q is a superlative quantity index and $\tilde{P}$ is the corresponding implicit price index defined by (1), then $\tilde{P}$ is also superlative.

Unfortunately, there are many superlative index number formulae so the practical problem arises: which index number formulae should be used? If the prices and quantities do not vary very much between the two periods in question, then Diewert [1978b] and Vartia [1978] have shown that all the superlative index number formulae approximate each other rather closely, so that the choice of index number formula is relatively unimportant. However, in many applications involving the use of cross section data or decennial census data, there can be a tremendous amount of variation in prices or in quantities between the two periods so that the superlative index number formulae can generate quite different aggregates.

In Section 2 we note some easily checked theoretical and numerical bounds for the commonly used superlative index number formulae. If these bounds are narrow, then the choice of index number formula will not matter much.

In Section 3 we suggest a criterion for choosing between directly defined and implicitly defined superlative index number formulae based on Leontief's [1936; 54–57] and Hicks' [1946; 312–313] Aggregation Theorems.

In Section 4 we present a numerical example based on the 1889 and 1909 censuses of the U.S. steel industry which illustrates the bounds developed in Section 2 and the criterion suggested in Section 3.

---

[1]A neoclassical aggregator function is a function which is (i) positive, (ii) linearly homogeneous and (iii) concave over the positive orthant $\{x : x \gg 0_N\}$. f will also be continuous and nondecreasing over the positive orthant and can be extended to the nonnegative orthant by continuity (see Diewert [1974a]). An aggregator function could be either a utility or production function.

2. Numerical and Theoretical Bounds for the Indexes P_r and $\widetilde{P}_s$

Define the period t cost shares as $s_i^t \equiv p_i^t x_i^t / p^t \cdot x^t = 0, 1$; $i = 1, \ldots, N$. Define the *mean of order r price index* using period t shares as

$$(2) \qquad P_{r,t}(p^0, p^1, x^0, x^1) \equiv \begin{cases} \left[\sum_{i=1}^N s_i^t (p_i^1/p_i^0)^r\right]^{1/r} & \text{for } r \neq 0 \\ \prod_{i=1}^N (p_i^1/p_i^0)^{s_i^t} & \text{for } r = 0. \end{cases}$$

Some well known price indexes are special cases of $P_{r,t}$:

$$P_{1,0} \equiv \Sigma_i s_i^0 (p_i^1/p_i^0) = p^1 \cdot x^0 / p^0 \cdot x^0$$

is the *Laspeyres* price index,

$$P_{-1,1} \equiv [\Sigma_i s_i^1 (p_i^1/p_i^0)^{-1}]^{-1} = p^1 \cdot x^1 / p^0 \cdot x^1$$

is the *Paasche* price index,

$$P_{1,1} \equiv \Sigma_i s_i^1 (p_i^1/p_i^0)$$

is the *Palgrave* index, and

$$P_{-1,0} \equiv [\Sigma_i s_i^0 (p_i^1/p_i^0)^{-1}]^{-1}$$

is what Vartia [1978] calls the *Harmonic Laspeyres* price index.

The mean of order r, t price indexes $P_{r,t}$ can also be used to define the *quadratic mean of order r price indexes* P_r:

$$(3) \qquad \begin{aligned} P_r(p^0, p^1, x^0, x^1) \equiv {} & [P_{r/2,0}(p^0, p^1, x^0, x^1) \\ & \times P_{-r/2,1}(p^0, p^1, x^0, x^1)]^{1/2}. \end{aligned}$$

Note that $P_2 \equiv (P_{1,0})^{1/2}(P_{-1,1})^{1/2}$ is Irving Fisher's [1922] *ideal price* index, and

$$P_0(p^0, p^1, x^0, x^1) \equiv \prod_{i=1}^N (p_i^1/p_i^0)^{(s_i^0 + s_i^1)/2}$$

is the *Törnqvist* price index.

It can be shown (Diewert [1976a]) that P_r for $r \neq 0$ is exact for the quadratic mean of order r cost function,

$$c_r(p) \equiv (\Sigma_i \Sigma_j b_{ij} p_i^{r/2} p_j^{r/2})^{1/r}$$

and P_0 is exact for the translog[2] unit cost function,

$$\ln c_0(p) \equiv \alpha_0 + \Sigma_i \alpha_i \ln p_i + \frac{1}{2}\Sigma_i \Sigma_j \alpha_{ij} \ln p_i \ln p_j,$$

where

$$\alpha_{ij} = \alpha_{ji}, \quad \Sigma_i \alpha_i = 1$$

and $\Sigma_i \alpha_{ij} = 0$ for $j = 1, \ldots, N$.[3] Since c_r and c_0 can approximate an arbitrary unit cost function to the second order, the price indexes P_r are superlative for every r.

If prices and quantities are positive for the two periods under consideration (an assumption we make for the remainder of this paper), then using some theorems in Hardy, Littlewood and Polya [1934; 14–15], it can be shown that the superlative indexes P_r satisfy the following numerical bounds:

$$\min_i \{p_i^1/p_i^0\} \leq P_r(p^0, p^1, x^0, x^1)$$

$$\leq \max_i \{p_i^1/p_i^0\}. \tag{4}$$

Mean of order r, t and *quadratic mean of order r quantity indexes*, $Q_{r,t}$ and Q_r, respectively, can be defined in a manner analogous to the definitions of $P_{r,t}$ and P_r: simply interchange the role of prices and quantities in the definitions; e.g., $Q_r(p^0, p^1, x^0, x^1) \equiv P_r(x^0, x^1, p^0, p^1)$.

It can also be shown that Q_r for $r \neq 0$ is exact for the quadratic mean of order r aggregator function $f_r(x)$ (defined analogously to c_r) and Q_0 is exact for the homogeneous translog aggregator function f_0 (defined analogously to c_0). Since f_r and f_0 can approximate an arbitrary linearly homogeneous function to the second order, the quantity indexes Q_r are superlative for every r. Moreover, the following numerical bounds, which are analogous to (4), are valid:

$$\min_i \{x_i^1/x_i^0\} \leq Q_r(p^0, p^1, x^0, x^1)$$

$$\leq \max_i \{x_i^1/x_i^0\}. \tag{5}$$

Define the *implicit mean of order r, t* and *quadratic mean of order r price indexes* as

$$\tilde{P}_{r,t}(p^0, p^1, x^0, x^1) \equiv \frac{p^1 \cdot x^1}{p^0 \cdot x^0 Q_{r,t}(p^0, p^1, x^0, x^1)} \tag{6}$$

---

[2]The translog functional form was independently introduced by Christensen, Jorgenson and Lau [1971] and Sargan [1971].

[3]This last result is almost in Theil [1965; 71–72].

and

$$\widetilde{P}_r(p^0,p^1,x^0,x^1) \equiv \frac{p^1 \cdot x^0}{p^1 \cdot x^0 Q_r(p^0,p^1,x^0,x^1)}, \tag{7}$$

respectively. From (5) and (7), it can be seen that the superlative price indexes $\widetilde{P}_r$ satisfy the following numerical bounds for every r:

$$\begin{aligned} p^1 \cdot x^1/p^0 \cdot x^0 \max_i\{x_i^1/x_i^0\} &\le \widetilde{P}_r(p^0,p^1,x^0,x^1) \\ &\le p^1 \cdot x^1/p^0 \cdot x^0 \min_i\{x_i^1/x_i^0\}. \end{aligned} \tag{8}$$

For arbitrary positive price and quantity vectors, it need not be the case that the P_r satisfy the bounds in (8) or that the $\widetilde{P}_r$ satisfy the bounds in (4). However, given that the producer or consumer is engaging in cost minimizing behavior for the two periods under consideration and that the true functional form for his or her unit cost function is $c_r(p)$ for some r, then $P_r(p^0,p^1,x^0,x^1)$ will satisfy not only the bounds in (4), but also the bounds in (8).[4] Similarly, if the consumer or producer has the utility or production function $f_r(x)$ for some r, then $\widetilde{P}_r(p^0,p^1,x^0,x^1)$ will satisfy not only the bounds in (8), but also the bounds in (4).[5]

In fact if $c_r(p)$ is the true unit cost function and the economic agent is engaging in minimizing behavior, P_r will satisfy the following bounds:

$$\begin{aligned} P_{-1,1} \equiv p^1 \cdot x^1/p^0 \cdot x^1 &\le P_r(p^0,p^1,x^0,x^1) \\ &\le p^1 \cdot x^0/p^0 \cdot x^0 \equiv P_{1,0}. \end{aligned} \tag{9}$$

Similarly, if $f_r(x)$ is the true aggregator function and the producer or consumer minimizes costs in the two periods, $\widetilde{P}_r$ will also satisfy the bounds in (9).[6]

Thus numerically, P_r will satisfy the bounds in (4) while $\widetilde{P}_r$ will satisfy the bounds in (8). Furthermore, under the usual assumptions when P_r is exact for c_r and $\widetilde{P}_r$ is exact for f_r, P_r should also satisfy (8), $\widetilde{P}_r$ should satisfy (4), and both P_r and $\widetilde{P}_r$ should satisfy the bounds in (9).

---

[4]This follows from the fact that P_r is exact for c_r (see Diewert [1976a; 133–34]) and from Theorem 13 in Diewert [1981a].

[5]This follows from the fact that Q_r is exact for f_r (see Diewert [1976a; 132]) and from a theorem due to Samuelson [1947; 159] and Pollak [1971; 12].

[6]These last two bounds follow from the fact that the Konüs [1924] cost of living index is bounded from below by the Paasche index and from above by the Laspeyres index when the aggregator function is linearly homogeneous. This result is probably due to Frisch [1936; 25].

3. Hicks' Composite Commodity Theorem Versus Leontief's Aggregation Theorem

Unfortunately, the following example shows that the bounds in (4) or (8) can be very wide indeed. Suppose that there are only two commodities and $p^0 \equiv (1,1)$, $p^1 \equiv (k,k)$ where $k > 0$, $x^0 \equiv (1,0)$ and $x^1 \equiv (0,1)$. Then it can be verified that the upper and lower bounds in (4) both equal k and thus $P_r = k$ for $-2 \le r \le 2$. However, the upper and lower bounds for the implicit indexes $\widetilde{P}_r$ in (8) are $+\infty$ and 0, respectively.[7]

The reason why the direct price indexes $P_{r,t}$ and P_r all equal k in our example is that the prices are *proportional* in the two periods; i.e., $p^1 = kp^0$. On the other hand, if the quantity vectors x^0 and x^1 were proportional, then the direct quantity indexes $Q_{r,t}$ and Q_r would all be equal and thus the corresponding implicit price indexes $\widetilde{P}_{r,t}$ and $\widetilde{P}_r$ would also all be equal. In our example, the two quantity vectors are highly nonproportional (in fact, they are orthogonal), and thus the indexes $Q_{r,t}$ exhibit great variability.

Thus it is useful to compare the variation in the N quantity ratios (x_i^1/x_i^0) to the variation in the N price ratios (p_i^1/p_i^0). If there is less variation in the price ratios than in the quantity ratios (the typical case), then the direct price indexes P_r are essentially share weighted averages of the price ratios (p_i^1/p_i^0) and will tend to be in closer agreement with each other than the implicit price indexes $\widetilde{P}_r$. Thus in this situation, we would recommend the use of a superlative direct price index and the corresponding implicit quantity index, $(P_r, \widetilde{Q}_r)$, for some r. In fact, if $(p_i^1/p_i^0) = k > 0$ for all i, then $(P_r, \widetilde{Q}_r) = (k, p^1 \cdot x^1/p^0 \cdot x^0 k)$ for all r, and the use of $(P_r, \widetilde{Q}_r)$ can be theoretically justified using Hicks' [1946; 312–313] Composite Commodity Theorem.

On the other hand, if there is less variation in the quantity ratios than in the price ratios, then the quantity indexes Q_r are essentially share weighted averages of the quantity ratios and will tend to be more stable than the implicit quantity indexes $\widetilde{Q}_r$. In fact, if $(x_i^1/x_i^0) = k$ for all i, then $(\widetilde{P}_r, Q_r) = (p^1 \cdot x^1/p^0 \cdot x^0 k, k)$ for all r, and the use of $(\widetilde{P}_r, Q_r)$ can be theoretically justified using Leontief's [1936; 54–57] Aggregation Theorem.

A practical problem now arises: How can we decide whether prices are more highly proportional than quantities? A simple procedure is the following: regress $\log(p_i^1/p_i^0)$ (or $\log(x_i^1/x_i^0)$) on a constant and let $D(p^0, p^1)$ (or $D(x^0, x^1)$) be the sum of squared residuals of the regression; this will be our measure of nonproportionality of the vectors p^0 and p^1 (x^0 and x^1). Assuming

[7] Actually, in this case the indexes $\widetilde{P}_r \equiv k/(0 \times \infty)^{1/2}$ are undefined. However, if we changed the quantity vectors to $x^0 \equiv (1, \varepsilon)$ and $x^1 \equiv (\varepsilon, 1)$ for small $\varepsilon > 0$, then as ε approached zero, the upper and lower bounds would approach $+\infty$ and 0, respectively, and the indexes $\widetilde{P}_r$ would vary considerably as r changed.

that the vectors p^0 and p^1 are strictly positive, it can be verified that D has the following desirable properties: (i) $D(p^0, kp^0) = 0$ for every scalar $k > 0$ so that if prices are proportional during the two periods, the distance or deviation from proportionality is zero, (ii) $D(p^0, p^1) > 0$ if $p^0 \neq kp^1$ for any $k > 0$, (iii) $D(p^0, p^1) = D(p^1, p^0)$; i.e., the deviation from proportionality of prices p^0 from p^1 is the same as the deviation of prices p^1 from p^0 (symmetry with respect to time), (iv) $D(p^0, kp^1) = D(p^0, p^1)$ for all scalars $k > 0$, (v) $D(Ap^0, Ap^1) = D(p^0, p^1)$ where A is a diagonal matrix with positive diagonal elements; i.e., D is invariant to scale changes in the units of measurement, and (vi) $D(Bp^0, Bp^1) = D(p^0, p^1)$ where B is a permutation matrix (symmetry with respect to goods).[8] Moreover, it can be shown that D increases as the Euclidean distance between the logarithmic deviations of prices from their means in each period, p^{0^*} and p^{1^*}, increases where the ith component of p^{t^*} is defined as $p_i^{t^*} \equiv \ln p_i^t - (\Sigma_j \ln p_j^t)/N$ for $t = 0, 1$. We will say that prices are less proportional than quantities if $D(p^0, p^1) > D(x^0, x^1)$ and we would recommend the use of the superlative index number pair $(\tilde{P}_r, Q_r)$ for some r in order to aggregate the data. On the other hand, if $D(p^0, p^1) < D(x^0, x^1)$, then we would recommend the use of the superlative index number pair $(P_r, \tilde{Q}_r)$ for some r.

4. An Empirical Example

This investigation grew out of an attempt to measure productivity growth in U.S. steelworks and rolling mills between the 1889 and 1909 census years.[9] Productivity growth may be measured as an index of output divided by an index of inputs. The choice of a formula to index inputs had no practical consequence since all the usual formulae gave very similar results — the direct and implicit Törnqvist input indexes, for instance, were within 0.025% of each other.[10] This result is hardly surprising in the light of Section 3 of this paper. The same inputs — pig iron, ferromanganese, scrap, iron ore, fuel, capital, and labor — were used in both years. While there was some variation in the factor mix, it was not extreme. Likewise, the prices of the inputs tended to move in harmony. With the conditions for both Hicks and Leontief aggregation approximately satisfied, all of the usual price and quantity indexes move in unison. Inputs grew by a factor of 2.77 between 1889 and 1909.

[8] These properties are not sufficient to determine the functional form of D.
[9] For a discussion of the implications of this question, see Allen [1979].
[10] The data are tabled and discussed in Allen [1979].

No such happy congruence of results characterized the indexes of output.[11] In this case, even common superlative indexes of quantity differ appreciably — the direct Törnqvist index, for instance, is five percent greater than the implicit Törnqvist index. The need to discriminate among these quantity indexes prompted the research embodied in this paper.

Inspection of the data suggests that prices were much more proportional than quantities. Almost all prices were lower in 1909 than in 1889, but the output of some products had increased enormously while the production of others had declined. The highest and lowest price ratios were 1.0437 and 0.4465 while the highest and lowest quantity ratios were 16.0963 and 0.2454, respectively. Our formal criterion developed in the previous section also indicates that prices were more highly proportional than quantities:[12] regressing $\ln p_i^1 / \ln p_i^0$ and $\ln x_i^1 / \ln x_i^0$ on a constant produced the following sums of squared residuals:

$$D(p^0, p^1) = .601$$

and

$$D(x^0, x^1) = 18.386.$$

With prices more proportional than quantities, one would expect less variation among the direct price indexes and among the implicit quantity indexes than among the implicit price indexes and the direct quantity indexes. This expectation is borne out by an examination of Table 1. Note that the common superlative direct price indexes P_r are all closely clustered around the value 0.77 while there is more variation in the common implicit superlative price indexes $\widetilde{P}_r$. Their divergent behavior becomes more pronounced as r approaches

[11]There are 10 outputs: rails, bars, structural steel, skelp, plates and sheets, rerolled rails, cut nails, wire, other rolled, and miscellaneous. The price vectors for 1889 and 1909 were p^0: 32.606, 43.608, 55.953, 42.415, 60.303, 56.497, 47.169, 61.958, 58.593, 45.333; p^1: 28.372, 31.893, 30.876, 31.467, 38.567, 25.228, 49.229, 49.068, 54.891, 40.277. The corresponding quantity vectors were x^0: 1,867,600, 1,572,400, 276,360, 428,100, 652,690, 349,460, 261,470, 129,170, 640,420, 137,810; x^1: 2,844,100, 3,707,800, 2,096,800, 403,970, 2,833,000, 106,350, 45,059, 1,459,700, 5,572,200, 644,090. Quantities were measured in long tons, prices were measured in dollars per long ton.

[12]However, it is interesting to note that prices are not more highly *correlated* than quantities: the partial correlation coefficient between the two price vectors is only 0.4241 compared to 0.5293 for the quantities.

$\pm\infty$.[13] P_r is confined within the narrow limits $0.6642 \leq P_r \leq 0.7743$ while $\widetilde{P}_r$ ranges over the interval $0.7731 \leq \widetilde{P}_r \leq 1.9875$. P_r lies between the Paasche and Laspeyres indexes for $-4 < r < 13$ while $\widetilde{P}_r$ satisfies this property only in the range $0 < r < 3$ (recall (9)).[14]

5. Conclusions

Table 1 lists the estimates of the growth in total factor productivity implied by each of the indexes of output. The range is broad. Diewert [1976a] has argued that superlative indexes should be preferred to other indexes. The application of that criterion eliminates the most extreme values, but the range of the remaining estimates of productivity growth is still large. In this paper we have argued that in situations like the present one — namely, where prices are more highly proportional than quantities — implicit superlative quantity indexes are to be preferred to direct superlative quantity indexes. This preference is due to the property that the price indexes P_r are essentially based on share weighted averages of relative prices and thus the aggregates generated by these indexes should all be numerically close and, moreover, they can be (approximately)

---

[13]It can be shown that

$$\lim_{r\to\infty}(P_{r/2,t})^{1/2} = \max_i (p_i^1/p_i^0)^{1/2},$$

$$\lim_{r\to\infty}(P_{-r/2,t})^{1/2} = \min_i (p_i^1/p_i^0)^{1/2},$$

and that

$$\lim_{r\to\infty} P_r = [\max_i (p_i^1/p_i^0)^{1/2}][\min_j (p_j^1/p_j^0)^{1/2}] = \lim_{r\to-\infty} P_r .$$

See Hardy, Littlewood and Polya [1934; 15].

[14]When aggregating outputs, we assume that the producer maximizes revenue subject to a factor requirements constraint. The fact that a revenue maximization problem has replaced a cost minimization problem leads to a reversal of the inequalities in (9), but not in (4) or (8). For our data, the Laspeyres output price is less than the Paasche index, which is theoretically correct. For additional material on the output aggregation problem, see Samuelson and Swamy [1974] or Diewert [1976a; 125–26].

Table 1. Price, Quantity, and Efficiency Indexes

Price Indexes			Quantity Indexes		
Symbol	Name	Value	Symbol	Value	Efficiency Index
Direct Price Index, Implicit Quantity Index					
	$\min_i\{p_i^1/p_i^0\}$	0.4465	—	6.2114	2.24
P_{-91}	Lowest P_r	0.6642[a]	$\widetilde{Q}_{-91}$	4.1756[a]	1.57
$P_{-1,0}$	Harmonic Laspeyres	0.7322	$\widetilde{Q}_{-1,0}$	3.7877	1.37
$P_{1,0}$	Laspeyres	0.7651	$Q_{-1,1}$	3.6248	1.31
P_{-2}	$(P_{-1,0})^{1/2}(P_{1,1})^{1/2}$	0.7692[a]	$\widetilde{Q}_{-2}$	3.6055[a]	1.30
P_0	Törnqvist	0.7724[a]	$\widetilde{Q}_0$	3.5906[a]	1.30
P_1	Generalized Leontief[b]	0.7735[a]	$\widetilde{Q}_1$	3.5854[a]	1.29
P_2	Fisher Ideal	0.7740[a]	$\widetilde{Q}_2$	3.5831[a]	1.29
P_3	Highest P_r	0.7743[a]	$\widetilde{Q}_3$	3.5818[a]	1.29
$P_{-1,1}$	Paasche	0.7830	$Q_{1,0}$	3.5420	1.28
$P_{1,1}$	Palgrave	0.8080	$\widetilde{Q}_{1,1}$	3.4324[a]	1.24
	$\max_i\{p_i^1/p_i^0\}$	1.0437	—	2.6573	0.96
Implicit Price Index, Direct Quantity Index					
	$p^1 \cdot x^1/p^0 \cdot x^0 \min_i\{x_i^1/x_i^0\}$	16.0963	$\min_i\{x_i^1/x_i^0\}$	0.1723	0.06
$\widetilde{P}_{-1,0}$	Implicit Harmonic Laspeyres	2.3301	$Q_{-1,0}$	1.1902	0.43
$\widetilde{P}_{\pm\infty}$	Highest $\widetilde{P}_r$	1.9875	$Q_{\pm\infty}$	1.3954[a]	0.50
$\widetilde{P}_{-2}$	$(\widetilde{P}_{-1,0})^{1/2}(\widetilde{P}_{1,1})^{1/2}$	1.0159[a]	Q_{-2}	2.7299[a]	0.99
$\widetilde{P}_0$	Implicit Törnqvist	0.8105[a]	Q_0	3.4218[a]	1.24
$\widetilde{P}_2$	$(P_{-1,1})^{1/2}(\widetilde{P}_{1,0})^{1/2} = P_2$	0.7740[a]	Q_2	3.5831[a]	1.29
$\widetilde{P}_1$	Generalized Linear[b] (lowest $\widetilde{P}_r$)	0.7731[a]	Q_1	3.5873[a]	1.30
$\widetilde{P}_{-1,1}$	Implicit Palgrave	0.4428	$Q_{1,1}$	6.2632	2.26
	$p^1 \cdot x^1/p^0 \cdot x^0 \max_i\{x_i^1/x_i^0\}$	0.2454	$\max_i\{x_i^1/x_i^0\}$	11.3006	4.08

Note: The efficiency index equals the corresponding quantity index divided by 2.77, the relative level of inputs. The quantity index equals the expenditure ratio, 2.7734, divided by the price index on the same line.

[a] Denotes superlative indexes.

[b] Diewert [1976a; 135] shows that P_1 is exact for the Generalized Leontief and $\widetilde{P}_1$ is exact for the Generalized Linear functional forms defined in Diewert [1971a].

justified using Hicks' Aggregation Theorem. Furthermore, under these conditions, it is likely that the chosen P_r will satisfy the theoretically valid bounds (8) and (9) in addition to (4).

On the other hand, if quantities are more proportional than prices, then the use of one of the direct superlative quantity indexes Q_r and corresponding implicit price indexes $\tilde{P}_r$ is recommended, since these indexes are based on share weighted averages of relative quantities and thus the aggregates generated by these indexes should all be numerically close and can be justified using Leontief's Aggregation Theorem. Furthermore, under these conditions, it is likely that the chosen $\tilde{P}_r$ will satisfy the theoretically valid bounds (4) and (9) in addition to (8).

Finally, suppose that our proportionality criterion cannot distinguish whether prices are more proportional than quantities. Under these circumstances, which index number formula should be used? A reasonable suggestion under these circumstances is the following: use Fisher's ideal formula since $P_2 = \tilde{P}_2$ and thus the formula is (approximately) consistent with both Hicks' and Leontief's Aggregation Theorems (since P_2 satisfies the bounds in (4) and $\tilde{P}_2$ satisfies the bounds in (8)). Moreover, P_2 will also lie between the Paasche and Laspeyres indexes, since it is the geometric mean of the two indexes (so that P_2 will also satisfy the bounds in (9)). P_2 is the only superlative index number which has this desirable property.

Essays in Index Number Theory, Volume I
W.E. Diewert and A.O. Nakamura (Editors)

Chapter 11
GROUP COST OF LIVING INDEXES: APPROXIMATIONS AND AXIOMATICS*

W.E. Diewert

1. Introduction

Following Frisch [1936], at least three approaches to constructing a cost of living or consumer price index can be distinguished: (i) the *"statistical"* approach, (ii) the *test* or *axiomatic* approach and (iii) the *preference field* (cf. Wold [1953; 135]) or *economic* (cf. Samuelson and Swamy [1974]) approach.

In the statistical approach, price ratios of the form p_i^{t+1}/p_i^t (where p_i^t is the price of commodity i in period t, $i = 1, \ldots, N$) are assumed to be stochastically distributed around a common mean. In this approach, the index number problem reduces to finding an appropriate statistical estimator for the common mean. See Frisch [1936] for a review of this approach and Diewert [1981a; 179–180] for a discussion of related "neostatistical" approaches.

In the test or axiomatic approach, the price index is regarded as a function of $4N$ variables, $P(p^0, p^1, x^0, x^1)$, which satisfies certain properties or axioms or tests, where $p^t \equiv (p_1^t, p_2^t, \ldots, p_N^t)$ is a vector of period t prices and $x^t \equiv (x_1^t, x_2^t, \ldots, x_N^t)$ is a vector of period t commodity consumptions for $t = 0$ and 1. If $N = 1$ so that there is only one commodity, then the price index function is usually defined to be $P(p_1^0, p_1^1, x_1^0, x_1^1) \equiv p_1^1/p_1^0$. The test approach to index number theory was popularized by Irving Fisher [1922] and perfected by Eichhorn [1976] [1978b] and Eichhorn and Voeller [1976] [1983].

The economic approach to the construction of a consumer price index is based on the assumption of utility maximizing behavior on the part of the individual or household during the two periods under consideration. The resulting theoretical price index depends on the (unknown) household preferences. The concept is due to Konüs [1924] and it is known as the Konüs cost of living index. In order to obtain an operational approximation to the theoretical cost of living index, it is usually necessary to make additional assumptions. Thus

---

*First published in *Methods of Operations Research*, Vol. 48, 1984, pp. 23–45. The author thanks the Social Sciences and Humanities Research Council of Canada for financial support.

the economic approach to the construction of a consumer price index may be broken down into a number of sub approaches: (1) the *bounds* or *nonparametric* approach, (2) the *exact* approach, and (3) the *econometric* approach.

In the bounds approach, we attempt to find the tightest possible bounds for the theoretical index that can be computed using observable price and quantity data. Of course, we also use the assumption of utility maximizing behavior in all of the economic approaches. Contributors to this approach include Konüs [1924], Samuelson [1947; 159], Malmquist [1953], Pollak [1971a], and Afriat [1972b], [1973d], [1977].

In the econometric approach, we assume the household's utility function is a specific functional form. Then the observed price and quantity data are used in order to estimate econometrically the parameters of the system of consumer demand equations. Once these parameters have been estimated, the Konüs cost of living index may be calculated. Contributors to this approach include Goldberger and Gamaletsos [1970], Christensen and Manser [1975], Braithwait [1980] and Jorgenson and Slesnick [1983].

In the exact approach, it is again necessary to assume that consumer preferences can be adequately approximated by a certain class of preferences. However, no econometric estimation is necessary in this approach. Theorem 2 below illustrates this approach (while Theorem 1 illustrates the bounds approach). Contributors to the exact approach include Konüs and Byushgens [1926], Afriat [1972b], Samuelson and Swamy [1974] and Diewert [1976a].

This completes our brief review of the alternative approaches to the construction of a consumer price index.

It should be noted that with the exceptions of Pollak [1981], Jorgenson and Slesnick [1983] and Diewert [1983a], the economic approach has centered around the single household case. Thus one of the main purposes of this paper will be to extend the theory to the many households case.

The other main purpose of this paper will be to provide an axiomatic approach to the various concepts of group cost of living indexes. Several years ago, Professor Eichhorn suggested that this axiomatization must be possible. Thus it seems appropriate to dedicate this paper to him.

In Section 2, we define the Konüs cost of living index for a single household. We also list some theorems that indicate how approximations to the theoretical index may be computed; i.e., we illustrate the bounds and the exact approaches to the economic theory of index numbers.

In Sections 3 to 5, we define several different theoretical group cost of living indexes and we develop some theorems which indicate how they may be numerically approximated by observable data. We also characterize each theoretical index axiomatically.

Finally, in Section 6 we define a group welfare index, characterize it axiomatically and develop numerical approximations.

2. The Single Household Konüs Cost of Living Index

We assume that the household has preferences over combinations of N goods that may be represented by a utility function F where $u = F(x)$ is the utility level or standard of living that can be attained if the individual consumes the consumption vector $x \equiv (x_1, x_2, \ldots, x_N)^T \geq 0_N$.

We assume that the utility function F defined over $\{x : x \geq 0_N\}$ satisfies the following conditions I:

(i) continuity,

(ii) increasingness; i.e., if $x'' \gg x' \geq 0_N$, then $F(x'') > F(x')$,

(iii) quasiconcavity; i.e., for each utility level u, the upper level set $L(u) \equiv \{x : F(x) \geq u\}$ is convex,

(iv) $F(0_N) = 0$, and

(v) $F(x)$ tends to $+\infty$ as the components of x all tend to $+\infty$.

We shall assume that the household maximizes the utility function $F(x)$ subject to a budget constraint of the form $p \cdot x = \sum_{n=1}^{N} p_n x_n \leq y$ where $p \gg 0_N$ is a positive vector of commodity (rental) prices and $y > 0$ is expenditure on the N commodities.[1]

The household's utility maximization problem can be decomposed into two stages. In the first stage, the household attempts to minimize the cost of achieving a given utility level, and in the second stage, the maximal utility level that is just consistent with the budget constraint is chosen.

The solution to the first stage problem defines the household's *cost function* C: for $u \geq 0$, $p \gg 0_N$,

$$C(u, p) \equiv \min_x \{p \cdot x : F(x) \geq u, \; x \geq 0_N\}. \tag{1}$$

Given that F satisfies conditions I, C defined for $u \geq 0$, $p \gg 0_N$ will satisfy the following conditions II:

(i) C is continuous,

(ii) $C(0, p) = 0$ for every $p \gg 0_N$,

(iii) for every $p \gg 0_N$, $C(u, p)$ is increasing in u and $C(u, p)$ tends to $+\infty$ as u tends to $+\infty$,

(iv) $C(u, p)$ is positively linearly homogeneous in p for fixed u; i.e., for $u \geq 0$, $p \gg 0_N$ and $\lambda > 0$, $C(u, \lambda p) = \lambda C(u, p)$,

(v) $C(u, p)$ is concave in p for fixed u,

(vi) $C(u, p)$ is increasing in p for fixed $u > 0$; i.e., if $p'' \gg p' \gg 0_N$, $u > 0$, then $C(u, p'') > C(u, p')$, and

[1]Notation: x^T denotes the transpose of the column vector x, $p^T x = p \cdot x \equiv \sum_{n=1}^{N} p_n x_n$ denotes the inner product of the vectors p and x, $x \geq 0_N$ means each component of the vector x is nonnegative, $x \gg 0_N$ means each component is positive, and $x > 0_N$ means $x \geq 0$ but $x \neq 0_N$.

(vii) C is such that the function $F^*(x) \equiv \max_u\{u : p \cdot x \geq C(u,p)$ for every $p \gg 0_N$, $u \geq 0\}$ is continuous for $x \geq 0_N$.

Moreover, if we are given a cost function C satisfying conditions II, C may be used in order to construct the underlying preference function F which will satisfy conditions I.[2]

Our interest in C stems from the fact that it may be used to define the Konüs [1924] *cost of living index* P_K: for $p^0 \gg 0_N$, $p^1 \gg 0_N$ and $u > 0$ define

$$(2) \qquad P_K(p^0, p^1, u) \equiv C(u, p^1)/C(u, p^0).$$

Thus P_K depends on three variables: (i) p^0, a vector of period 0 or base period prices, (ii) p^1, a vector of period 1 or current period prices, and (iii) u, a number that indexes the reference indifference surface. Thus $P_K(p^0, p^1, u)$ is the minimum cost of achieving the standard of living indexed by u when the household faces period 1 prices p^1 *relative* to the minimum cost of achieving the same standard of living when the household faces period 0 prices p^0. If there is only one good, then it can be seen that $P_K(p_1^0, p_1^1, u) = p_1^1/p_1^0$ for all $u > 0$. In this case, there is obviously no index number problem.

In the general case when there is more than one good, the functional form for the cost of living index P_K obviously depends on the functional form for the household's cost function C, which in turn is determined by the form of the household's preference function F. Our fundamental problem is that we do not know what the functional forms for F or C, and hence P_K, are. Theorems 1 and 2 below indicate how empirically implementable approximations to P_K may be computed. However, first it is useful to introduce the concept of a *mechanistic price index formula*. This is simply a function P of the observable price and quantity vectors for time periods 0 and 1, (p^0, p^1, x^0, x^1), where P has a known functional form. Two examples of such formulae are the *Laspeyres price index* P_L defined by

$$(3) \qquad P_L(p^0, p^1, x^0, x^1) \equiv p^1 \cdot x^0 / p^0 \cdot x^0$$

and the *Paasche price index* P_P defined by

$$(4) \qquad P_P(p^0, p^1, x^0, x^1) \equiv p^1 \cdot x^1 / p^0 \cdot x^1.$$

Irving Fisher [1922] gives hundreds of examples of mechanistic price index formulae. The axiomatic characterization of these indexes may be found in Eichhorn [1976] [1978b] and Eichhorn and Voeller [1976] [1983]. Note that these indexes are functions of $4N$ arguments whereas the price index that appears in the following Theorem 1 has only $2N + 1$ arguments.

---

[2]This is a version of the Shephard [1953] Duality Theorem; see Diewert [1982].

Since the Paasche and Laspeyres indexes are often rather close to each other numerically, Theorem 1 below is a very useful result. However, in order to prove the theorem, we require cost minimizing behavior on the part of the consumer during periods 0 and 1. We shall also assume that we can observe the consumer's quantity choices $x^0 > 0_N$ and $x^1 > 0_N$ made during periods 0 and 1 in addition to the corresponding price vectors $p^0 \gg 0_N$ and $p^1 \gg 0_N$. Thus we assume:

$$(5) \qquad p^0 \cdot x^0 = C[F(x^0), p^0]; \; p^1 \cdot x^1 = C[F(x^1), p^1].$$

THEOREM 1. (Konüs [1924; 20–21]: *Let the consumer's utility functions F satisfy conditions I and suppose that the observed data for periods 0 and 1, (p^0, x^0) and (p^1, x^1) respectively, satisfy the cost minimization assumptions (5). Then there exists a reference utility level u^* that lies between the base utility level $u^0 \equiv F(x^0)$ and the period 1 utility level $u^1 \equiv F(x^1)$ such that the consumer's true cost of living index for this reference utility level, $P_K(p^0, p^1, u^*)$, lies between $P_L \equiv p^1 \cdot x^0/p^0 \cdot x^0$ and $P_P \equiv p^1 \cdot x^1/p^0 \cdot x^1$.*

Thus if P_L and P_P are close to each other numerically, an average of P_L and P_P, such as Irving Fisher's [1922] ideal index number P_F, will closely approximate the Konüs cost of living index $P_K(p^0, p^1, u^*)$ defined in Theorem 1. P_F is defined as the geometric average of P_L and P_P; i.e.,

$$(6) \qquad P_F(p^0, p^1, x^0, x^1) \equiv (p^1 \cdot x^0/p^0 \cdot x^0)^{1/2}(p^1 \cdot x^1/p^0 \cdot x^1)^{1/2}.$$

Theorem 1 above illustrates the bounds approach in the economic theory of index numbers. Theorem 2 below illustrates the exact approach. Before stating Theorem 2, it is first necessary to define the translog cost function C_T by:

$$(7) \qquad \begin{aligned} \ln C_T(u,p) \equiv a_0 + \sum_{i=1}^N a_i \ln p_i + (1/2)\sum_{i=1}^N\sum_{j=1}^N a_{ij} \ln p_i \ln p_j + a_{00} \ln u \\ + \sum_{i=1}^N a_{0i} \ln p_i \ln u + (1/2) a_{00} (\ln u)^2, \end{aligned}$$

where the parameters a_{ij} satisfy the following restrictions:

$$(8) \qquad \sum_{i=1}^N a_i = 1; \; a_{ij} = a_{ji}; \quad \sum_{j=1}^N a_{ij} = 0 \text{ for } i = 1, \ldots, N; \quad \sum_{i=1}^N a_{0i} = 0.$$

It is also necessary to define the base period expenditure shares $s_i^0 \equiv p_i^0 x_i^0/p^0 \cdot x^0$ and the period 1 expenditure shares $s_i^1 \equiv p_i^1 x_i^1/p^1 \cdot x^1$ for $i = 1, \ldots, N$.

Define the *translog*[3] *price index* P_T by

$$(9) \qquad P_T(p^0, p^1, x^0, x^1) \equiv \prod_{i=1}^N (p_i^1/p_i^0)^{(s_i^0+s_i^1)/2}.$$

[3]This corresponds to the terminology used in Christensen, Cummings and Jorgenson [1980]. Diewert [1981a; 187] called P_T the Törnqvist price index, but the term translog price index seems to be more descriptive.

THEOREM 2. (Diewert [1976a; 122]): *Suppose the household's cost function C equals the translog cost function C_T defined by (7) and suppose the observed data (p^0, x^0) and (p^1, x^1) satisfy the cost minimization assumptions (5) where F is the utility function dual to C. Define $u^0 \equiv F(x^0)$, $u^1 \equiv F(x^1)$ and $u^* \equiv (u^0u^1)^{1/2}$. The true cost of living index $P_K(p^0, p^1, u^*)$ evaluated at the intermediate utility level u^* may be calculated by evaluating the translog price index P_T; i.e., we have*

$$P_K(p^0, p^1, u^*) \equiv C_T(u^*, p^1)/C_T(u^*, p^0) = P_T(p^0, p^1, x^0, x^1). \tag{10}$$

We note that the translog function C_T defined by (7) can provide a second order approximation to an arbitrary twice continuously differentiable cost function, so that Theorem 2 is not restricted to the homothetic case. Thus Theorem 2 provides a very strong economic justification for the use of the translog price index P_T to approximate the true cost of living $P_K(p^0, p^1, u^*)$.

Theorem 1 provided a strong justification for the use of the Fisher index P_F while Theorem 2 justified the use of the translog index P_T. The following theorem tells us that it does not matter very much whether we use P_F or P_T in empirical applications: the two index number functions approximate each other to the second order around any equal price and quantity point.

THEOREM 3. (Diewert [1978b]): *The functions $P_F(p^0, p^1, x^0, x^1)$ and $P_T(p^0, p^1, x^0, x^1)$ differentially approximate each other to the second order around any point where $p^0 = p^1 \gg 0_N$ and $x^0 = x^1 \gg 0_N$; i.e., we have:*

$$P_F(p^0, p^0, x^0, x^0) = P_T(p^0, p^0, x^0, x^0) \tag{11}$$

$$\nabla P_F(p^0, p^0, x^0, x^0) = \nabla P_T(p^0, p^0, x^0, x^0) \text{ and} \tag{12}$$

$$\nabla^2 P_F(p^0, p^0, x^0, x^0) = \nabla^2 P_T(p^0, p^0, x^0, x^0) \tag{13}$$

where ∇P stands for the $4N$ dimensional vector of first order partial derivatives of the function P and $\nabla^2 P$ is the $4N \times 4N$ matrix of second order partial derivatives of P.

This completes our review of the single household theory of the cost of living index. We now turn our attention to the many household case.

3. The Plutocratic Cost of Living Index

Assume that there are H distinct households, where household h has utility function $f^h(x)$ satisfying conditions I noted in Section 2 with a dual cost function $C^h(u_h, p)$.

Let $p^0 \gg 0_N$ and $p^1 \gg 0_N$ be the period 0 and 1 price vectors as usual, let $x_h^t \equiv (x_{1h}^t, x_{2h}^t, \ldots, x_{Nh}^t)^T$ be household h's observed quantity vector for period t, $t = 0, 1$ and for $h = 1, \ldots, H$, let $u_h^0 \equiv F^h(x_h^0)$ and $u_h^1 \equiv F^h(x_h^1)$ be the period 0 and 1 utility levels attained by household h, and define the utility vectors $u^1 \equiv (u_1^1, \ldots, u_H^1)^T$ and $u^0 \equiv (u_1^0, \ldots, u_H^0)^T$. Assume cost minimizing behavior for each household during both periods; i.e.,

$$p^0 \cdot x_h^0 = C^h(u_h^0, p^0) = C^h[F^h(x_h^0), p^0] \quad \text{for } h = 1, \ldots, H$$

(14) and

$$p^1 \cdot x_h^1 = C^h(u_h^1, p^1) = C^h[F^h(x_h^1), p^1] \quad \text{for } h = 1, \ldots, H.$$

Pollak [1981; 328] defines what he calls a Scitovsky–Laspeyres cost of living index for a group of households as the ratio of the total expenditure required to enable each household to attain its reference indifference curve at period 1 prices to that required at period 0 prices. This same concept of a group cost of living index was suggested by Prais [1959] in less precise language. Prais referred to his concept as a *plutocratic price index*. Formally, this Prais–Pollak *plutocratic cost of living index* may be defined as:

$$P_{PP}(p^0, p^1, u) \equiv \sum_{h=1}^{H} C^h(u_h, p^1) / \sum_{h=1}^{H} C^h(u_h, p^0) \tag{15}$$

$$= \sum_{h=1}^{H} s^h(u, p^0) C^h(u_h, p^1) / C^h(u_h, p^0) \tag{16}$$

where the household h share of total expenditure at the vector of reference utility levels $u \equiv (u_1, \ldots, u_H)$ and prices $p \equiv (p_1, \ldots, p_N)$ is defined as

$$s^h(u, p) \equiv C^h(u_h, p) / \sum_{j=1}^{H} C^j(u_j, p) \text{ for } h = 1, \ldots, H. \tag{17}$$

The *Laspeyres–Pollak cost of living index* is defined as $P_{PP}(p^0, p^1, u^0)$ while the *Paasche–Pollak cost of living index* is defined as $P_{PP}(p^0, p^1, u^1)$.

It is convenient to define aggregate consumption vectors $\bar{x}^t$ as follows:

$$\bar{x}^0 \equiv \sum_{h=1}^{H} x_h^0; \qquad \bar{x}^1 \equiv \sum_{h=1}^{H} x_h^1. \tag{18}$$

Armed with the above definitions, the bounds approach to approximating the plutocratic cost of living index may be illustrated by stating the following two theorems.

THEOREM 4. (Diewert [1983a]): *Suppose each consumer's utility function F^h satisfies conditions I and suppose the cost minimization assumption (14) holds. Then*

$$\min_i\{p_i^1/p_i^0\} \le P_{PP}(p^0,p^1,u^0) \le p^1 \cdot \bar{x}^0/p^0 \cdot \bar{x}^0 \equiv P_L \tag{19}$$

$$P_P \equiv p^1 \cdot \bar{x}^1/p^0 \cdot \bar{x}^1 \le P_{PP}(p^0,p^1,u^1) \le \max_i\{p_i^1/p_i^0\}. \tag{20}$$

Note that P_L in (19) is the usual Laspeyres price index involving the *aggregate* base period consumption vector $\bar{x}^0$, while P_P in (20) is the usual Paasche price index involving the *aggregate* period 1 consumption vector $\bar{x}^1$. The consumer price index constructed by statistical agencies is usually an approximation to P_L. All that is required to construct P_L is: the current vector of prices p^1, the base price vector p^0, and the base period aggregate consumption vector $\bar{x}^0 \equiv \Sigma_h x_h^0$.

THEOREM 5. (Diewert [1983a]): *Under the conditions of Theorem 4, there exists a reference utility vector $u^* \equiv (u_1^*, \ldots, u_H^*)$ such that each component u_h^* lies between u_h^0 and u_h^1 and the Prais–Pollak group cost of living index evaluated at this reference utility vector, $P_{PP}(p^0,p^1,u^*)$, lies between $P_L \equiv p^1 \cdot \bar{x}^0/p^0 \cdot \bar{x}^0$ and $P_P \equiv p^1 \cdot \bar{x}^1/p^0 \cdot \bar{x}^1$.*

Note that the bounds in Theorems 4 and 5 depend only on aggregate price and quantity data. Furthermore, we would expect P_L and P_P to be very close to each other provided that p^0 and p^1 are not "too" different, and hence $P_{PP}(p^0,p^1,u^*)$ may be closely approximated by either P_L or P_P (or, say, by the Fisher ideal index $P_F \equiv (P_L P_P)^{1/2}$) under these circumstances.

We now turn to an axiomatic characterization of the plutocratic cost of living index.

THEOREM 6. *Let $P(p^0,p^1,u)$ be a function of $2N + H$ variables defined for $p^0 \gg 0_N$, $p^1 \gg 0_N$ and $u > 0_H$ satisfying the following properties:*

(i) $P(p^0,p^1,u) > 0$ for $p^0 \gg 0_N$, $p^1 \gg 0_N$, $u > 0_H$ (positivity);

(ii) $P(p^0,p^1,u) = 1/P(p^1,p^0,u)$ for $p^0 \gg 0_N$, $p^1 \gg 0_N$, $u > 0_H$ (time reversal);

(iii) $P(p^0,p^2,u) = P(p^0,p^1,u)P(p^1,p^2,u)$ for $p^0 \gg 0_N$, $p^1 \gg 0_N$, $p^2 \gg 0_N$, $u > 0_H$ (circularity);

(iv) $P(p^*,p,u_1,u_2,\ldots,u_H) = \sum_{h=1}^H u_h P(p^*,p,u_h e_h)$ for some $p^* \gg 0_N$ and for all $p \gg 0_N$, $u = (u_1,\ldots,u_H) > 0_H$ where e_h is a unit vector with a one in component h and $u_h P(p^*,p,u_h e_h) \equiv 0$ if $u_h = 0$ (an additivity across consumers property); and

(v) for the same vector of prices $p^* \gg 0_N$ that occurred in (iv) above, the functions $u_h P(p^*,p,u_h e_h) \equiv C^h(u_h,p)$ regarded as functions of $u_h \ge 0$ and $p \gg 0_N$ satisfy conditions II for cost functions for $k = 1,2,\ldots,H$.

Then P is a plutocratic cost of living index of the form (15) that corresponds to the household preferences that are dual to the functions C^h defined in (v) for $h = 1, \ldots, H$; i.e., $P \equiv P_{PP}$ satisfies (15). Moreover, each C^h satisfies the following money metric[4] *scaling of utility property for the reference prices p^*:*

$$(21) \qquad C^h(u_h, p^*) = u_h \text{ for } u_h > 0, \quad h = 1, \ldots, H.$$

Conversely, given cost functions C^h satisfying conditions II and the money metric property (21), then $P \equiv P_{PP}$ defined by (15) satisfies properties (i) to (v) listed above in the theorem.

Proof: Let $p^0 \gg 0_N$, $p^1 \gg 0_N$, $u \equiv (u_1, u_2, \ldots, u_H) > 0_H$ and let p^* be the vector of prices that occurs in (iv) and (v). Then

$$\begin{aligned} P(p^0, p^1, u) &= P(p^0, p^*, u)P(p^*, p^1, u) && \text{by (iii)} \\ &= P(p^*, p^1, u)/P(p^*, p^0, u) && \text{by (ii)} \\ &= \Sigma_h u_h P(p^*, p^1, u_h e_h)/\Sigma_h u_h P(p^*, p^0, u_h e_h) && \text{by (iv)} \\ &= \Sigma_h C^h(u_h, p^1)/\Sigma_h C^h(u_h, p^0) && \text{by (v).} \end{aligned}$$

We need to show that the functions C^h defined in property (v) satisfy the money metric property (21). Thus for $h = 1, \ldots, H$ and $u_h > 0$,

$$\begin{aligned} C^h(u_h, p^*) &\equiv u_h P(p^*, p^*, u_h e_h) \\ &= u_h P(p^*, p^*, u_h e_h)P(p^*, p^*, u_h e_h) && \text{by (iii)} \\ &= u_h P(p^*, p^*, u_h e_h)/P(p^*, p^*, u_h e_h) && \text{by (ii)} \\ &= u_h. && \text{by (i).} \end{aligned}$$

The converse part of the theorem is straightforward. Q.E.D.

Prais [1959] and Muellbauer [1974] criticized the plutocratic cost of living index because it implicitly gives too much weight to high expenditure households (recall equation (16) above). Thus Prais and Muellbauer suggested that we take a simple average of the individual household Konüs cost of living indexes, $C^h(u_h, p^1)/C^h(u_h, p^0)$, and this is what they called a *democratic price index*. We shall find it convenient to consider slightly more general averages, since we may want to weight different households differently, depending on their demographic characteristics (family size, age, etc.). We also find it convenient to consider both arithmetic type averages and geometric type averages. Thus we consider additive type averages in the following section and geometric type averages in the subsequent section.

---

[4]The term is due to Samuelson [1974b].

4. Additive Democratic Cost of Living Indexes

Let $u \equiv (u_i, \ldots, u_H)$ denote a vector of reference utility levels, let $p^0 \gg 0_N$ denote a vector of positive base period prices, let $p^1 \gg 0_N$ denote a vector of current period prices, and let $\alpha > 0_H$ denote a vector of demographic weights such that $\sum_{h=1}^{H} \alpha_h = 1_H \cdot \alpha = 1$ where 1_H is a vector of ones of dimension H. Then we define the *additive democratic cost of living* P_D as

$$(22) \qquad P_D(p^0, p^1, u, \alpha) \equiv \sum\nolimits_{h=1}^{H} \alpha_h \frac{C^h(u_h, p^1)}{C^h(u_h, p^0)}.$$

Assume cost minimizing behavior for all households during both periods; i.e., assume (14) (and recall the notation immediately above (14)). Recall also Theorem 4 and note that it is valid if $H = 1$. Making repeated use of Theorem 4 applied to individual households, we can obtain the following bounds on the Laspeyres–Democratic index $P_D(p^0, p^1, u^0)$ and on the Paasche–Democratic index $P_D(p^0, p^1, u^1, \alpha)$:

$$(23) \qquad \min_i\{p_i^1/p_i^0\} \le P_D(p^0, p^1, u^0, \alpha) \le \sum\nolimits_{h=1}^{H} \alpha_h \frac{p^1 \cdot x_h^0}{p^0 \cdot x_h^0},$$

$$(24) \qquad \sum\nolimits_{h=1}^{H} \alpha_h \frac{p^1 \cdot x_h^1}{p^0 \cdot x_h^1} \le P_D(p^0, p^1, u^1, \alpha) \le \max_i\{p_i^1/p_i^0\}.$$

Note that the right hand side of (23) is an arithmetic α_h weighted average of the individual household Laspeyres price indexes, $\Sigma_h \alpha_h P_L(p^0, p^1, x_h^0, x_h^1)$, which we denote by P_L^α. The left hand side of (24) is an arithmetic average of the individual Paasche indexes, $\Sigma_h \alpha_h P_P(p^0, p^1, x_h^0, x_h^1)$, which we denote by P_P^α.

The following theorem provides a group counterpart to Theorem 1.

THEOREM 7. *Let each consumer's utility function F^h satisfy conditions I. Suppose that the observed data for periods 0 and 1, (p^0, x_h^0) and (p^1, x_h^1) for $h = 1, \ldots, H$, satisfy the cost minimization assumption (14). Then there exists a reference utility vector $u^* \equiv (u_1^*, \ldots, u_H^*)$ such that each component u_h^* lies between u_h^0 and u_h^1 and the additive democratic group cost of living index evaluated at this reference utility vector $P_D(p^0, p^1, u^*, \alpha)$ lies between P_L^α and P_P^α.*

Proof: Define $h(\lambda) \equiv P_D[p^0, p^1, (1-\lambda)u^0 + \lambda u^1]$ for $0 \le \lambda \le 1$. Note that $h(0) = P_D(p^0, p^1, u^0, \alpha)$ and $h(1) = P_D(p^0, p^1, u^1, \alpha)$. There are 24 a priori inequality relations that are possible between the four numbers $h(0)$, $h(1)$, P_L^α and P_P^α. However, (23) and (24) imply that $h(0) \le P_L^\alpha$ and $P_P^\alpha \le h(1)$. This means that there are really only six possible inequalities between the four numbers: (1) $h(0) \le P_L^\alpha \le P_P^\alpha \le h(1)$, (2) $h(0) \le P_P^\alpha \le P_L^\alpha \le h(1)$, (3) $h(0) \le$

$P_P^\alpha \leq h(1) \leq P_L^\alpha$, (4) $P_P^\alpha \leq h(0) \leq P_L^\alpha \leq h(1)$, (5) $P_P^\alpha \leq h(1) \leq h(0) \leq P_L^\alpha$ and (6) $P_P^\alpha \leq h(0) \leq h(1) \leq P_L^\alpha$. Since the individual cost functions $C^h(u_h, p^t)$ are continuous in u_h, it can be seen that $P_D(p^0, p^1, u, \alpha)$ defined by (22) is continuous in the vector of utility variables u. Hence $h(\lambda)$ is a continuous function for $0 \leq \lambda \leq 1$ and assumes all intermediate values between $h(0)$ and $h(1)$. By inspecting cases (1) to (6) above, it can be seen that we can choose λ between 0 and 1 (call this number λ^*) so that $P_L^\alpha \leq h(\lambda^*) \leq P_P^\alpha$ for case (1) or so that $P_P^\alpha \leq h(\lambda^*) \leq P_L^\alpha$ for cases (2) to (6). Now define $u^* \equiv (1-\lambda^*)u^0 + \lambda^* u^1$ and the proof is complete. Q.E.D.

The method of proof used in Theorem 7 (and in Theorems 1 and 4 as well) is due to Konüs [1924].

For typical data, we would expect P_L^α and P_P^α to be very close to each other (closer than the individual indexes P_L^h and P_P^h since P_L^α and P_P^α are averages of the individual P_L^h and P_P^h), so $P_D(p^0, p^1, u^*, \alpha)$ may be very closely approximated by either P_L^α or P_P^α. The practical difficulty with using the democratic price index $P_D(p^0, p^1, u^*, \alpha)$ as a general measure of inflation between periods 0 and 1 is that P_L^α and P_P^α can only be constructed if we have individual household data. The group index defined in the previous section has bounds that could be constructed from aggregate data.

The following theorem provides an axiomatic characterization of the additive democratic price index that is analogous to the axiomatic characterization of the plutocratic index that was provided by Theorem 6.

THEOREM 8. *Let $p^0 \gg 0_N$, $p^1 \gg 0_N$, $p^2 \gg 0_N$, $u \gg 0_H$ and $\alpha > 0_H$ with $1_H \cdot \alpha = 1$. Let $P(p^0, p^1, u, \alpha)$ be a function of $2N + 2H$ variables that satisfies the following properties:*

(i) $P(p^0, p^1, u, \alpha) > 0$ (positivity);

(ii) for $h = 1, 2, \ldots, H$, $P(p^0, p^1, u, e_h) = 1/P(p^1, p^0, u, e_h)$ where e_h is a unit vector with a one in component h (time reversal);

(iii) $P(p^0, p^2, u, e_h) = P(p^0, p^1, u, e_h)P(p^1, p^2, u, e_h)$ for $h = 1, 2, \ldots, H$ (circularity);

(iv) $P(p^0, p^1, u, \alpha) = \sum_{h=1}^H \alpha_h P(p^0, p^1, u, e_h)$ (additivity); and

(v) for some positive vector of prices $p^* \gg 0_N$, the function $u_h P(p^*, p, u, e_h) \equiv C^h(u_h, p)$ depends only on u_h and $p \gg 0_N$ and C^h satisfies conditions II for cost functions for $h = 1, 2, \ldots, H$.

Then P is an additive democratic cost of living index of the form (22) that corresponds to the household preferences that are dual to the functions C^h defined in (v) for $h = 1, 2, \ldots, H$. Moreover, each C^h satisfies the money metric scaling of utility property (21) for the reference prices p^*. Conversely, given cost functions C^h satisfying conditions II and the money metric property (21) *for $h = 1, \dots, H$, then $P \equiv P_D$ where P_D defined by (22) satisfies properties (i) to (v) listed above.*

Proof: Let $p^0 \gg 0_N$, $p^1 \gg 0_N$, $u \equiv (u_1, u_2, \dots, u_H) \gg 0_H$, $\alpha > 0_H$ with $1_H \cdot \alpha = 1$, and let p^* be the vector of prices that occurs in (v). Then

$$\begin{aligned} P(p^0, p^1, u, \alpha) &= \sum_{h=1}^{H} \alpha_h P(p^0, p^1, u, e_h) && \text{by (iv)} \\ &= \Sigma_h \alpha_h P(p^0, p^*, u, e_h) P(p^*, p^1, u, e_h) && \text{by (iii)} \\ &= \Sigma_h \alpha_h u_h P(p^*, p^1, u, e_h) / u_h P(p^*, p^0, u, e_h) && \text{by (ii)} \\ &= \Sigma_h \alpha_h C^h(u_h, p^1) / C^h(u_h, p^0) && \text{by (v).} \end{aligned}$$

To show that C^h satisfies the money metric property (21), let $u_h > 0$. Then

$$\begin{aligned} C^h(u_h, p^*) &\equiv u_h P(p^*, p^*, u, e_h) \\ &= u_h P(p^*, p^*, u, e_h) P(p^*, p^*, u, e_h) && \text{by (iii)} \\ &= u_h P(p^*, p^*, u, e_h) / P(p^*, p^*, u, e_h) && \text{by (ii)} \\ &= u_h && \text{by (i).} \end{aligned}$$

The converse part of the theorem is straightforward. Q.E.D.

5. Multiplicative Democratic Cost of Living Indexes

Making the same notational assumptions as at the beginning of Section 4, we may define the ***multiplicative democratic cost of living index*** as

$$(25) \qquad P_M(p^0, p^1, u, \alpha) \equiv \prod_{h=1}^{H} [C^h(u_h, p^1)/C^h(u_h, p^0)]^{\alpha_h}.$$

Thus P_M is a generalized geometric mean of the individual household Konüs cost of living indexes where household h receives the (demographic) weight α_h. The following theorem provides a group counterpart to Theorem 2.

THEOREM 9. *Suppose household h has preferences F^h which are dual to a translog cost function C^h of the type defined by (7) for $h = 1, 2, \dots, H$. Suppose the observed price-quantity data $(p^t, x_1^t, \dots, x_H^t)$ for period $t = 0, 1$ satisfy the cost minimization assumptions (14). Define $u_h^t \equiv F^h(x_h^t)$ for $t = 0, 1$ and $h =$*

$1, 2, \ldots, H$ and $u_h^* \equiv (u_h^0 u_h^1)^{1/2}$ and $u^* \equiv (u_1^*, u_2^*, \ldots, u_H^*)$. *Let the demographic weight vector* α *satisfy* $\alpha > 0_H$ *and* $1_H \cdot \alpha = 1$. *Then the multiplicative democratic cost of living index* $P_M(p^0, p^1, u^*, \alpha)$ *evaluated at the intermediate utility vector* u^* *defined above may be calculated by evaluating the right hand side of (26) below:*

$$(26) \qquad P_M(p^0, p^1, u^*, \alpha) = \sum_{i=1}^{N} (p_i^1/p_i^0)^{\sum_{h=1}^{H} \alpha_h (s_i^{0h} + s_i^{1h})/2}$$

where the period t *expenditure share of commodity* i *for consumer* h *is defined by*

$$s_i^{th} \equiv p_i^t x_{ih}^t / p^t \cdot x^t \text{ for } t = 0, 1;\ h = 1, \ldots, H;\ i = 1, \ldots, N.$$

Proof: Follows directly from Theorem 2 and definition (25). Q.E.D.

COROLLARY. *If* $\alpha_h \equiv 1/H$ *for* $h = 1, \ldots, H$, *then (26) may be rewritten as*

$$(27) \qquad P_M(p^0, p^1, u^*, 1/H) = \prod_{i=1}^{N} (p_i^1/p_i^0)^{(S_i^0 + S_i^1)/2}$$

where the average expenditure share for commodity i *during period* t *is defined as*

$$(28) \qquad S_i^t \equiv \sum_{h=1}^{H} s_i^{th}/H; \quad t = 0, 1; \quad i = 1, \ldots, N.$$

It is interesting to contrast Theorem 5 with the Corollary to Theorem 9. If we know total (or average) commodity purchases by households of each commodity, then we may form an accurate approximation to the plutocratic index which appears in the statement of Theorem 5. On the other hand, if we know the average household expenditure shares defined by (28), then we may form an accurate approximation to the (multiplicative) democratic price index $P_M(p^0, p^1, u^*, 1/H)$.

We conclude this section with an axiomatic characterization of the multiplicative democratic index.

THEOREM 10. *Let* $p^0 \gg 0_N$, $p^1 \gg 0_N$, $p^2 \gg 0_N$, $u \gg 0_H$ *and* $\alpha > 0_H$ *with* $\alpha \cdot 1_H = 1$. *Let* $P(p^0, p^1, u, \alpha)$ *be a function of* $2N + 2H$ *variables that satisfies the following properties:*

(i) $P(p^0, p^1, u, \alpha) > 0$ *(positivity);*

(ii) $P(p^0, p^1, u, \alpha) = 1/P(p^1, p^0, u, \alpha)$ *(time reversal);*

(iii) $P(p^0, p^2, u, \alpha) = P(p^0, p^1, u, \alpha) P(p^1, p^2, u, \alpha)$ *(circularity);*

(iv) $P(p^0, p^1, u, \alpha) = \prod_{h=1}^{H} [P(p^0, p^1, u, e_h)]^{\alpha_h}$ *(multiplicative property); and*

(v) for some positive vector of prices $p^* \gg 0_N$, the function $u_h P(p^*, p, u, e_h) \equiv C^h(u_h, p)$ depends only on u_h and $p \gg 0_N$ and C^h satisfies conditions II for cost functions for $h = 1, 2, \ldots, H$.

Then P is a multiplicative democratic cost of living index of the form (25) that corresponds to the household preferences that are dual to the functions C^h defined in (v) for $h = 1, \ldots, H$. Moreover, each C^h satisfies the money metric property (21) for $h = 1, \ldots, H$ for the reference prices p^*. Conversely, given cost functions C^h satisfying conditions II and the money metric property (21), then $P \equiv P_M$ where P_M is defined by (25) satisfies properties (i) to (v) listed above.

Proof:

$$
\begin{aligned}
P(p^0, p^1, u, \alpha) &= \prod_{h=1}^{H} [P(p^0, p^1, u, e_h)]^{\alpha_h} && \text{by (iv)} \\
&= \prod_{h=1}^{H} [P(p^0, p^*, u, e_h) P(p^*, p^1, u, e_h)]^{\alpha_h} && \text{by (iii)} \\
&= \prod_{h=1}^{H} [P(p^*, p^1, u, e_h) / P(p^*, p^0, u, e_h)]^{\alpha_h} && \text{by (ii)} \\
&= \prod_{h=1}^{H} [u_h P(p^*, p^1, u, e_h) / u_h P(p^*, p^0, u, e_h)]^{\alpha_h} \\
&= \prod_{h=1}^{H} [C^h(u_h, p^1) / C^h(u_h, p^0)]^{\alpha_h} && \text{by (v).}
\end{aligned}
$$

The remainder of the proof is straightforward. Q.E.D.

6. Social Welfare Indexes

For a single household, the *Malmquist* [1953; 232] *quantity index* Q_M seems to be an appropriate concept for a utility index. In order to define Q_M, we must first define the *deflation* or *distance* function $D(u, x)$ that corresponds to the consumer's utility function F. For $u > 0$ and $x > 0_N$, define

$$
D(u, x) \equiv \max_k \{k : F(x/k) \geq u,\ k > 0\}. \tag{29}
$$

Thus $D(u, x^1)$ is the deflation factor k_1, say, that will just reduce the vector x^1 proportionately so that $F(x^1/k_1) = u$. If F satisfies conditions I, then D

will satisfy certain regularity conditions (conditions III, say) and a D satisfying these conditions will uniquely characterize F.[5]

For $u > 0$, $x^0 \gg 0_N$, $x^1 \gg 0_N$, define the *Malmquist Quantity Index* Q_M as

$$(30) \qquad Q_M(x^0, x^1, u) \equiv D(u, x^1)/D(u, x^0).$$

In general, the Malmquist quantity index $Q_M(x^0, x^1, u)$ will depend on the reference indifference surface indexed by u.

Turning now to the many consumer case, suppose that household h's preferences may be represented by the deflation function $D^h(u_h, x_h)$ that is dual to the utility function $F^h(x_h)$, where F^h satisfies conditions I as usual.

Define the $N \times H$ matrix of period 0 (1) consumer choices by $X^0(X^1)$; i.e.,

$$X^0 \equiv (x_1^0, x_2^0, \ldots, x_H^0), \qquad X^1 \equiv (x_1^1, x_2^1, \ldots, x_H^1)$$

where $x_h^t \equiv (x_{1h}^t, x_{2h}^t, \ldots, x_{Nh}^t)^T$ is consumer h's observed consumption vector during period t for $t = 0, 1$ and $h = 1, 2, \ldots, H$.

It is natural to try and aggregate individual welfare changes into a single scalar measure of overall welfare change. A simple *index of social welfare change* that respects individual preferences is

$$(31) \qquad W(X^0, X^1, u, \beta) \equiv \sum_{h=1}^{H} \beta_h \; D^h(u_h, x_h^1)/D^h(u_h, x_h^0)$$

where $u \equiv (u_1, u_2, \ldots, u_H)$ is a vector of individual reference utility levels and the weights β_h satisfy the following restriction:

$$(32) \qquad \sum_{h=1}^{H} \beta_h = 1.$$

The reader will note that our indicator of social welfare change defined by (31) is a quantity counterpart to the additive democratic cost of living index defined by (22) (which was an indicator of price change).

Our reason for imposing the restriction (32) is the following: if $x_h^1 = x_h^0$ for $h = 1, \ldots, H$ so that there is no change in consumption between periods 0 and 1, then we want our indicator of social welfare change (or our aggregate quantity index) to indicate that there has been no change; i.e., we want $W(X^0, X^0, u, \beta) = 1$. Hence we must have (32).

[5]The appropriate regularity conditions are listed in Diewert [1982; 560], and references to the literature on duality theorems between F and D may be found there also. Essentially, $D(u, x)$ has the same regularity properties as the cost function $C(u, p)$ where x replaces p, except that $D(u, x)$ decreases in u while $C(u, p)$ increases in u.

In general, the weights β_h do not have to be positive or even nonnegative. However, if the weights are nonnegative, then we obtain the following bounds for W, which are analogous to the bounds in (23) and (24) except that the role of prices and quantities is interchanged:[6]

$$(33)\quad \min_{i,h}\{x_{ih}^1/x_{ih}^0\} \leq W(X^0,X^1,u^0,\beta) \leq \sum_{h=1}^{H} \beta_h p^0\cdot x_h^1/p^0\cdot x_h^0 \equiv \sum_{h=1}^{H} \beta_h Q_L^h$$

and

$$(34)\quad \sum_{h=1}^{H} \beta_h Q_P^h \equiv \sum_{h=1}^{H} \beta_h p^1\cdot x_h^1/p^1\cdot x_h^0 \leq W(X^0,X^1,u^1,\beta) \leq \max_{i,h}\{x_{ih}^1/x_{ih}^0\}$$

where Q_L^h and Q_P^h are the Laspeyres and Paasche quantity indexes for consumer h, $u^0 \equiv (u_1^0,\ldots,u_H^0)$, $u^1 \equiv (u_1^1,\ldots,u_H^1)$ and $u_h^t \equiv F(x_h^t)$.

Thus the Laspeyres social welfare indicator $W(X^0,X^1,u^0,\beta)$ is bounded from above by a weighted average of the individual household Laspeyres quantity indexes Q_L^h, and the Paasche social welfare indicator $W(X^0,X^1,u^1,\beta)$ is bounded from below by a weighted average of the individual household Paasche quantity indexes Q_P^h.

A quantity counterpart to Theorem 7 also holds: if the weights β_b are nonnegative, then there exists a reference utility vector $u^* = (1-\lambda^*)u^0 + \lambda^* u^1$ for some λ^* between 0 and 1 such that the social welfare change index $W(X^0,X^1,u^*,\beta)$ lies between $\sum_{h=1}^{H}\beta_h Q_L^h$ (the upper bound in (33)) and $\sum_{h=1}^{H}\beta_h Q_P^h$ (the lower bound in (34).[7]

Finally, we have the following theorem which provides an axiomatic characterization for welfare indexes of the form (31).

THEOREM 11. *Let* $X^0 \equiv (x_1^0,\ldots,x_H^0)$, $X^1 \equiv (x_1^1,\ldots,x_H^1)$ *and* $X^2 \equiv (x_1^2,\ldots,x_H^2)$ *be* $N\times H$ *matrices of positive elements,* $u \gg 0_H$ *and* $\beta > 0_H$ *with* $1_H\cdot\beta = 1$. *Let* $W(X^0,X^1,u,\beta)$ *be a function of* $2NH+2H$ *variables that satisfies the following properties:*

(i) $W(X^0,X^1,u,\beta) > 0$ *(positivity);*

(ii) for $h = 1,\ldots,H$, $W(X^0,X^1,u,e_h) = 1/W(X^0,X^1,u,e_h)$ *(time reversal);*

(iii) for $h = 1,\ldots,H$, $W(X^0,X^2,u,e_h) = W(X^0,X^1,u,e_h)W(X^1,X^2,u,e_h)$ *(circularity);*

(iv) $W(X^0,X^1,u,\beta) = \sum_{h=1}^{H}\beta_h W(X^0,X^1,u,e_h)$ *(additivity); and*

[6]For a proof of these inequalities, see Diewert [1983a].

[7]See Diewert [1983a] for a proof.

(v) for some positive matrix of reference quantities $X^* \equiv (x_1^*, \ldots, x_H^*)$, *the function* $W(X^*, X^1, u, e_h)/u_h \equiv D^h(u_h, x_h)$ *depends only on* u_h *and* x_h^1 *(column* h *in* X^1*) and* D^h *satisfies conditions III for deflation functions for* $h = 1, 2, \ldots, H$.

Then W *is a social welfare index of the form (31) that corresponds to the household preferences that are dual to the deflation functions* D^h *defined in (v) for* $h = 1, \ldots, H$. *Moreover, each* D^h *satisfies the following scaling property for the reference quantities* x_h^* *that appear in (v):*

$$D^h(u_h, x_h^*) = 1/u_h \text{ for all } u_h > 0. \tag{35}$$

Conversely, given deflation functions D^h *satisfying conditions III and the scaling property (35) for* $h = 1, \ldots, H$, *then if we define* W *by (31),* W *satisfies properties (i) to (v) listed above.*

The proof is a straightforward modification of the proof of Theorem 8.

Essays in Index Number Theory, Volume I
W.E. Diewert and A.O. Nakamura (Editors)

Chapter 12
TEST APPROACHES TO INTERNATIONAL COMPARISONS*

W.E. Diewert

1. Introduction

For many purposes, it is useful to compare the real outputs or real incomes of a number of countries or regions at a single point in time. This information may be required to allocate international aid or interregional transfer payments between member countries or regions. In this paper, we consider test approaches to the problem of providing international or interregional comparisons. For a survey of the literature on international comparisons, see Kravis [1984]. For presentations of the axiomatic or test approach for making comparisons between two countries, see Voeller [1981] and Eichhorn and Voeller [1983].

In Section 2, we discuss how best to use bilateral real output indexes in order to make multilateral real output comparisons in a symmetric manner. We find that there are at least six plausible methods for combining the bilateral indexes into consistent multilateral indexes. In order to distinguish between these possibilities, we consider an axiomatic or test approach to multilateral quantity indexes in Sections 3 and 4. Using this test approach, we are able to eliminate some of our six methods.

2. Multilateral Comparisons of Real Output: Preliminary Approaches

Suppose that we have collected price and (net) output data on N commodities for I regions or countries, labor inputs and other primary resource commodities. Let the positive price vector for country i be $p^i \equiv (p_1^i, \ldots, p_N^i) \gg 0_N$, and let

*This paper was first published in *Measurement in Economics*, W. Eichhorn (ed.), Heidelberg: Physica-Verlag, 1988, pp. 67–86. Research support from the National Science Foundation, Grant #SES-8420937 is gratefully acknowledged. The author is indebted to W. Eichhorn, J. van Yzeren, Y. Vartia and a referee for helpful comments. None of the above individuals or institutions is responsible for any errors or opinions expressed in this report.

the corresponding quantity vector be $y^i \equiv (y_1^i, \ldots, y_N^i)$ for $i = 1, \ldots, I$. If commodity n is a net input into the production sector of country i (e.g., an imported good), then $y_n^i < 0$. However, we assume that the net value of outputs is positive for each country:

$$(1) \qquad p^i \cdot y^i \equiv \sum_{n=1}^{N} p_n^i y_n^i > 0 \quad \text{for} \quad i = 1, \ldots, I.$$

We assume that bilateral comparisons of real output can be made using an index number formula, $Q(p^j, p^i, y^j, y^i)$, which is supposed to represent the real output of country i relative to j. Microeconomic considerations[1] suggest that two good functional forms for Q are the Fisher [1922] ideal quantity index Q_F and the implicit translog quantity index $\tilde{Q}_T$ defined below by (2) and (3) respectively:

$$(2) \qquad Q_F(p^j, p^i, y^j, y^i) \equiv (p^i \cdot y^i p^j \cdot y^i / p^i \cdot y^j p^j \cdot y^j)^{1/2};$$

$$(3) \qquad \tilde{Q}_T(p^j, p^i, y^j, y^i) \equiv p^i \cdot y^i / p^j \cdot y^j \prod_{n=1}^{N} (p_n^i / p_n^j)^{(s_n^i + s_n^j)/2}$$

where $s_n^i \equiv p_n^i y_n^i / p^i \cdot y^i$ for $i = 1, \ldots, I$ and $n = 1, \ldots, N$.

The simplest way to make consistent multilateral comparisons of real output using a bilateral quantity index Q as a building block is to use what Kravis [1984; 10] calls the star system. More explicitly, the star system works as follows. First, pick an appropriate bilateral quantity index number formula, which gives the real output of country i relative to country j, $Q(p^j, p^i, y^j, y^i)$ say. Secondly, pick a numeraire country, say $j = 1$. Then country i's share of real world private product is defined by

$$(4) \qquad \sigma_i^j = Q(p^j, p^i, y^j, y^i) / \sum_{k=1}^{I} Q(p^j, p^k, y^j, y^k), \qquad i = 1, 2, \ldots, I,$$

where $j = 1$. Note that $\sum_{i=1}^{I} \sigma_i^j = 1$.

Of course, the problem with the star system is that the results depend on the choice of the numeraire country j; i.e., all countries are not treated in a symmetric manner in the star system. Thus we need to consider alternative methods for making multilateral comparisons.

Our first alternative method starts off with the following definition:

$$(5) \qquad \alpha_i \equiv \left[\sum_{j=1}^{I} [Q(p^j, p^i, y^j, y^i)]^{-1} \right]^{-1}, \quad i = 1, 2, \ldots, I.$$

Country i's share of real world private product is now defined as:

$$(6) \qquad s_i \equiv \alpha_i / \sum_{k=1}^{I} \alpha_k.$$

[1]See Diewert [1976a], [1981a] and Caves, Christensen and Diewert [1982b].

The intuitive interpretation of the above system for achieving multilateral comparisons of real output in a transitive and symmetric way can be grasped if we consider the case of one output good; i.e., $N = 1$. In this case, if Q is Q_F or $\tilde{Q}_T$ or any reasonable quantity index, (5) becomes

$$\alpha_i = y_1^i / \sum_{j=1}^{I} y_1^j, \quad i = 1, 2, \ldots, I; \tag{7}$$

i.e., α_i becomes country i's share of total world output. In this $N = 1$ case, we will have $\sum_{k=1}^{I} \alpha_k = 1$, so that $s_i = \alpha_i =$ country i's share of world private product.

In the general $N > 1$ case, α_i defined by (5) is country i's share of world product in the "metric" of country i. Since these metrics are not quite compatible in general, we make further adjustments via definition (6) to ensure that the shares sum to one. Let us call the resulting method for making multilateral comparisons of real output the *own share system.*

It turns out that the own share system, using the Fisher formula, is related to the Eltetö and Köves [1964] and Szulc [1964] (EKS) method for making multilateral comparisons. This can be seen as follows: using the Fisher formula, we have

$$\begin{aligned} \alpha_i &= I^{-1}\left[\sum_{j=1}^{I} I^{-1}[Q_F(p^j, p^i, y^j, y^i)]^{-1}\right]^{-1} \\ &\cong I^{-1}\left[\prod_{j=1}^{I} Q_F(p^j, p^i, y^j, y^i)\right]^{1/I} \end{aligned}$$

where we have approximated the harmonic mean by a geometric mean.

Thus

$$\begin{aligned} \alpha_i/\alpha_k &= s_i/s_k \\ &\cong \prod_{j=1}^{I} Q_F(p^j, p^i, y^j, y^i)^{1/I} \Big/ \prod_{j=1}^{I} Q_F(p^j, p^k, y^j, y^k)^{1/I} \qquad (8) \\ &= \text{the EKS output of country } i \text{ relative to } k. \end{aligned}$$

Thus we have provided an alternative justification for the use of the EKS method for making multilateral comparisons of real output.

Instead of using the Fisher formula Q_F in (8), Caves, Christensen and Diewert [1982a] advocated the use of the direct translog formula Q_T in place of Q_F. In the present context where some outputs can be negative, we would recommend the use of the implicit translog quantity index $\tilde{Q}_T$ in place of Q_T.

For any choice of a binary quantity index number formula Q, let us define the *generalized EKSCCD* method for making multilateral quantity comparisons as follows. Define

$$\beta_i \equiv \prod_{j=1}^{I} Q(p^j, p^i, y^j, y^i)^{1/I}, \quad i = 1, \ldots, I. \tag{9}$$

Country i's EKSCCD share of world private product is now defined as:

$$(10) \qquad S_i \equiv \beta_i / \sum_{k=1}^{I} \beta_k, \quad i = 1, \ldots, I.$$

We note that the own share method (6) satisfies the following *consistency in aggregation* property whereas the EKSCCD method does not: suppose any one country is split into two countries, each of which has the same price vector and one half of the original quantity vector. Then the relative outputs of the countries which are not split up should remain constant.[2]

We now consider another class of methods for making multilateral comparisons of real output.

Recall that the shares σ_i^j, $i = 1, \ldots, I$, defined by (4) allowed us to partition up world product using the bilateral metric of country j. What we now propose to do is to average over these country j metrics:

$$(11) \qquad \sigma_i(s_1, \ldots, s_I) \equiv \sum_{j=1}^{I} s_j \sigma_i^j, \quad i = 1, \ldots, I.$$

We still must choose the country weights s_j which occur in (11). A natural first choice would be to give each country's metric the same weight and choose *democratic weights* where $s_j \equiv 1/I$ for $j = 1, \ldots, I$. Fisher [1922; 305] recommended this method as being the very best (if it were practicable) and called it the *blend method.* However, this method seems to give tiny countries a comparatively large influence in the formation of the averages defined by (11). Moreover, the resulting method does not satisfy the consistency in aggregation (or invariance under country partitioning) property mentioned above. Thus it may be preferable to use *plutocratic weights,*[3] where $s_j \equiv p^j \cdot y^j / \sum_{k=1}^{I} p^k \cdot y^k$ is country j's share of world private product. This plutocratic averaging method satisfies the consistency in aggregation property but it has a drawback as well: the resulting shares defined by (11) are not invariant to scale changes in the prices of any one country. This is a very serious drawback, since we would not want country i's share of world real output to depend on the inflation rate in country j. A final variant for (11) would be to use *quantity weights,* where the s_j which appear in (11) are defined by (6), the own shares. The resulting method satisfies the consistency in aggregation property and is also invariant to scale changes in country prices.

In this section, we have considered six different methods for making multilateral comparisons of real output, using bilateral index number formulae based

[2]van Yzeren [1983] and Hill [1984; 131] informally discuss this consistency in aggregation property when they ask whether averages of bilateral indexes should be weighted according to the "importance" of the country.

[3]The terms democratic and plutocratic are due to Prais [1959].

on producer theory as building blocks. The six methods were: (i) the star system, (ii) the own share system, (iii) the EKSCCD method, (iv) the democratic weights method, (v) the plutocratic weights method, and (vi) the (own share) quantity weights method. However, we have not been able to recommend any of these methods as being clearly the best. Thus in the next two sections we shall turn to the test or axiomatic approach to index number theory and see whether this approach leads to any clear cut recommendations. In the following section, we momentarily digress and consider the axiomatic properties of bilateral index number formulae. Then in Section 4, we shall look for multilateral counterparts to the bilateral properties studied in Section 3.

3. Bilateral Comparisons of Real Output: The Test Approach

The test or axiomatic approach to bilateral index number theory has its origins in the work of Walsh [1901] and Fisher [1911][1922]. More recently, the test approach has been more definitively studied by Eichhorn and Voeller [1976] [1983] and Vartia [1985]. The reason why we are devoting a section of the present paper to this well studied topic is that the existing literature does not deal with the problems that arise from the fact that, in our framework, some quantities can be negative. The existing literature on the test approach assumes that all quantities are at least nonnegative. Thus it is necessary to review existing index number tests (or axioms or properties) and see if they are sensible when some quantities can be negative.

We regard a bilateral index number formula as a function, $Q(p^1, p^2, y^1, y^2)$, of $4N$ variables. The N dimensional vectors p^1 and p^2 represent prices in countries 1 and 2 and the vectors y^1 and y^2 represent the net output vectors in countries 1 and 2 (if $y_n^i < 0$, then the nth good is being utilized as an input into the private production sector of country i). We shall denote the tests or properties that we would like our bilateral quantity index Q to satisfy by BT1, BT2, etc. Our first test is:

BT1: *Positivity*: $Q(p^1, p^2, y^1, y^2) > 0$ for $p^i \gg 0_N$, y^i such that $p^i \cdot y^i > 0$, $i = 1, 2$. Moreover, Q is a continuous function.

Although the positivity test seems very reasonable, it can be verified that the Fisher ideal index Q_F does *not* satisfy it. (The implicit translog quantity index $\widetilde{Q}_T$ does satisfy it however.) In order to have the Fisher formula satisfy a positivity test, we must strengthen BT1 to the following property:

BT1′: *Strong Positivity*: Q is continuous and $Q(p^1, p^2, y^1, y^2) > 0$ for $p^1 \gg 0_N$, $p^2 \gg 0_N$ and y^i such that $p^j \cdot y^i > 0$ for $i, j = 1, 2$.

In words, we restrict the domain of definition of Q to strictly positive price vectors and quantity vectors which have a positive inner product with both price vectors.

The following test is weaker than the usual identity test (e.g., see Eichhorn and Voeller [1976] [1983]). It is a slight modification of a test due to Vartia [1985]:

BT2: *Identity*: $Q(\alpha p, \beta p, y, y) = 1$ for $\alpha > 0$, $\beta > 0$.

The above test says that if the price vectors in the two countries are proportional and the quantity vectors are identical, then the output of country 2 relative to 1 should equal unity. A stronger identity test arises if we drop the hypothesis of price proportionality; i.e., $Q(p^1, p^2, y, y) = 1$ for $p^i \gg 0_N$, $p^i \cdot y > 0$ for $i = 1, 2$. Q_F satisfies this stronger identity property but $\widetilde{Q}_T$ does not. However both $\widetilde{Q}_T$ and Q_F satisfy the weaker property BT2.

The following property (which may be found in Walsh [1901]) is a fundamental one for binary index number formulae.

BT3: *Proportionality*: $Q(p^1, p^2, y^1, \lambda y^2) = \lambda Q(p^1, p^2, y^1, y^2)$ for $\lambda > 0$. Thus if we double all output in country 2, the quantity index doubles.

If there is only one good, then it is easy to show that tests BT2 and BT3 determine the functional form for Q; i.e., if $N = 1$, then $Q(p^1, p^2, y^1, y^2) = y^2/y^1$ which is the output of country 2 divided by the output of country 1.

The next test is also a variant of a test due to Vartia [1985].

BT4: *Strong Monetary Unit Test*: $Q(\alpha p^1, \beta p^2, \gamma y^1, \gamma y^2) = Q(p^1, p^2, y^1, y^2)$ for $\alpha > 0$, $\beta > 0$ and $\gamma > 0$.

The above test says that if we simultaneously change the level of prices in each country by a (possibly different) proportional factor and we change all quantities in both countries by the same proportional factor γ, then the quantity index remains unchanged. This property implies that the quantity index does not depend on inflation rates in the two countries. Furthermore, if both countries experience output growth at the same rate, then the quantity index remains unchanged since it is supposed to represent the output of country 2 relative to that of country 1.

The following three tests are invariance or symmetry tests.

BT5: *Commensurability* (or *Invariance to Scale Changes in Units*): Let D be a diagonal $N \times N$ matrix with positive elements on its main diagonal. Then $Q(Dp^1, Dp^2, D^{-1}y^1, D^{-1}y^2) = Q(p^1, p^2, y^1, y^2)$ where D^{-1} denotes the inverse matrix of D.

BT6: *Country Reversal* (or *Symmetric Treatment of Countries*): $Q(p^2, p^1, y^2, y^1) = 1/Q(p^1, p^2, y^1, y^2)$.

The above test says that if we interchange the role of the countries in the index number formula Q, then the resulting value of the index equals the reciprocal of the original value of the index. It is the country counterpart to Irving Fisher's [1922] time reversal test. Test BT5 is in Fisher [1911]; however, Pierson [1896] informally suggested tests BT5 and BT6.

BT7: *Commodity Reversal Test* (or *Symmetric Treatment of Commodities*): Let P denote an $N \times N$ permutation matrix; i.e., each row and column of P contains one unit element and the remaining components are zeros. Then $Q(Pp^1, Pp^2, Py^1, Py^2) = Q(p^1, p^2, y^1, y^2)$.

This test was considered by Irving Fisher [1922; 81] and more recently by Vartia [1985].

It is straightforward to show that our two desirable bilateral index number formulae, Q_F and $\tilde{Q}_T$, that arose in our producer theory approaches to making bilateral output comparisons satisfy all of the above tests, except that Q_F does not satisfy BT1.

Some additional tests that have been considered in the bilateral context (see Eichhorn and Voeller [1976]) are:

BT8: *Strong Proportionality*: For $\lambda > 0$, $Q(p^1, p^2, y, \lambda y) = \lambda$;

BT9: *Monotonicity*: For $y^2 \geq y^3$, $Q(p^1, p^2, y^1, y^2) \geq Q(p^1, p^2, y^1, y^3)$;

BT10: *Mean Value Test*: $\min_n\{y_n^2/y_n^1\} \leq Q(p^1, p^2, y^1, y^2) \leq \max_n\{y_n^2/y_n^1\}$ where $p^i \gg 0_N$, and $y^i \equiv (y_1^i, y_2^i, \ldots, y_N^i)$, $i = 1, 2$.

The Fisher quantity index Q_F satisfies BT8 and BT9 but the implicit translog quantity index $\tilde{Q}_T$ satisfies neither. Neither Q_F nor $\tilde{Q}_T$ satisfy BT10. (If we could restrict quantities to be nonnegative, then Q_F would also satisfy BT10.)

All of the above tests seem to be intuitively desirable with the exception of BT10: when the quantity vectors y^i can have negative components, the mean value test is no longer reasonable.

To sum up this section, we have considered ten tests or properties that we would like our bilateral quantity index Q to have. In comparing our two desirable formulae Q_F and $\tilde{Q}_T$ (derived from the viewpoint of economy theory), we found that both satisfied tests BT2 through BT7. In addition, Q_F satisfied BT8 and BT9 while $\tilde{Q}_T$ satisfied BT1. We leave it up to the reader to decide which formula is preferable.

4. Multilateral Comparisons of Real Output: The Test Approach

We shall denote desirable properties for a multilateral system for making real output comparisons by MT1, MT2, etc. As in earlier sections, we denote the N dimensional price and net output vectors for country i by p^i and y^i for $i = 1, \ldots, I$. In order to economize on space, we now define the $N \times I$ matrices of all prices and all quantities by $p \equiv (p^1, p^2, \ldots, p^I)$ and $y \equiv (y^1, y^2, \ldots, y^I)$ respectively. A multilateral system of output indexes is a set of I functions, $S_i(p, y)$, $i = 1, \ldots, I$, where S_i is to be interpreted as country i's share of world private product.

In general, there will be domain restrictions on the $S_i(p, y)$ functions. As a minimal set of restrictions, we assume that $p^i \gg 0_N$ and $p^i \cdot y^i > 0$ for $i = 1, \ldots, I$. At times, we will also want to add additional restrictions such as

$p^i \cdot y^j > 0$ for all i and j, particularly when the share functions S_i are functions of bilateral Fisher quantity indexes.

The first property we want the S_i to satisfy is a fundamental one, since it allows us to interpret the functions S_i as shares.

MT1: *Share Test*: $S_i(p, y) > 0$, $i = 1, \ldots, I$ and $\sum_{i=1}^{I} S_i(p, y) = 1$. Furthermore, the functions $S_i(p, y)$ are continuous in p, y.

The following test is an approximate counterpart to the bilateral identity test BT2:

MT2: *Weak Proportionality*: Let $\alpha_i > 0$, $\beta_i = 0$ for $i = 1, \ldots, I$ with $\sum_{i=1}^{I} \beta_i = 1$. Let $p^1 = p^2 = \cdots = p^I$ and $y^1 = y^2 = \cdots = y^I$. Then $S_i(\alpha_1 p^1, \alpha_2 p^2, \ldots, \alpha_I p^I, \beta_1 y^1, \beta_2 y^2, \ldots, \beta_I y^I) = \beta_i$ for $i = 1, \ldots, I$.

The above property is a very fundamental one and we would expect every reasonable multilateral system to satisfy it. The test says the following: suppose that the price and quantity vectors in each country are (separately) proportional to the price and quantity vectors of all other countries. Then country i's share of world output is equal to its (common) share of world output on all N output markets. This test contains an identity test as a special case: if all output vectors are identical and all price vectors are proportional, we have

$$\begin{aligned} S_i(\alpha_1 p^1, \ldots, \alpha_I p^1, y^1, \ldots, y^I) &= S_i\Big(\alpha_1 p^1, \ldots, \alpha_I p^1, \\ &\qquad I^{-1} \sum_{j=1}^{I} y^1, \ldots, I^{-1} \sum_{j=1}^{I} y^1\Big) \\ &= 1/I \quad \text{for} \quad i = 1, \ldots, I, \end{aligned}$$

so that in this case, each country has an equal share of world output. Test MT2 also enables us to deduce what the functional form for $S_i(p, y)$ must be if $N = 1$ so that there is only one net output. In this case, we have

$$\begin{aligned} S_i(p^1, \ldots, p^I, y^1, \ldots, y^I) &= S_i\Big[1p^1, \ldots, (p^I/p^1)p^1, \beta_1 \sum_{j=1}^{I} y^j, \ldots, \beta_I \sum_{j=1}^{I} y^j\Big] \\ &= \beta_i \\ &= y^i / \sum_{j=1}^{I} y^j \quad \text{for} \quad i = 1, \ldots, I. \end{aligned} \tag{12}$$

The next multilateral test is a counterpart to the bilateral proportionality test, BT3.

MT3: *Proportionality*: Let $\lambda_i > 0$. Then for $i = 1, 2, \ldots, I$:

$$\begin{aligned} &S_j(p, y^1, \ldots, y^{i-1}, \lambda_i y^i, y^{i+1}, \ldots, y^I) \\ &\quad = \lambda_i S_i(p, y)/[(1 + S_i(p, y)(\lambda_i - 1)] \qquad \text{for } j = i \\ &\quad = S_j(p, y)/[1 + S_i(p, y)(\lambda_i - 1)] \qquad \text{for } j \neq i. \end{aligned} \tag{13}$$

On the left hand side of (13), we have the country j share function after the original country i quantity vector y^i has been multiplied by the positive scalar λ_i; on the right hand side of (13), we have some algebraic expressions involving the original (before multiplication by λ_i) share functions $S_i(p, y)$. Equations (13) are supposed to hold for all I choices for the quantity vector y^i which is multiplied by the scalar λ_i.

We derived test MT3 in analogy to the $N = 1$ case when the share functions are defined by (12); in this case, it can be verified that equations (13) will hold.

In order to obtain a better intuitive feeling for the test MT3, one should consider the test in ratio form. Denote the left hand side of (13) by $S_j(\lambda_i)$. Then (13) implies $S_i(\lambda_i)/S_k(\lambda_i) = \lambda_i S_i(p,y)/S_k(p,y)$ for $k \neq i$ and $S_j(\lambda_i)/S_k(\lambda_i) = S_j(p,y)/S_k(p,y)$ for $j \neq i$ and $k \neq i$. These multilateral proportionality properties seem to be just as reasonable as the bilateral proportionality property BT3. Unfortunately, as we shall soon see, it proves to be very difficult to simultaneously satisfy MT3 along with other reasonable multilateral tests.

The following four tests are counterparts to the corresponding bilateral tests, BT4, BT5, BT6 and BT7.

MT4: *Monetary Unit Test*: Let $\alpha_i > 0$, $i = 1, \ldots, I$ and $\beta > 0$. Then $S_i(\alpha_1 p^1, \ldots, \alpha_I p^I, \beta y^1, \ldots, \beta y^I) = S_i(p, y)$, $i = 1, \ldots, I$.

This property says that inflation levels in each country and a uniform growth in outputs over all countries do not affect each country's share of world output.

MT5: *Commensurability* (or *Invariance to Scale Changes in Units*): Let D be a diagonal $N \times N$ matrix with positive elements on its main diagonal. Then $S_i(Dp, D^{-1}y) = S_i(p, y)$, $i = 1, \ldots, I$.

MT6: *Symmetric Treatment of Countries* (*Multilateral Country Reversal Test*): Let Π be an $I \times I$ permutation matrix and let $S(p, y)$ be the row vector of share functions; i.e., $S(p, y) \equiv [S_1(p, y), \ldots, S_I(p, y)]$. Then $S(p, y)\Pi = S(p\Pi, y\Pi)$.

The above property says that if we evaluate our system of share functions using a permutation of the data, then the resulting share functions are the same permutation of the original share functions. This property means that no country can be singled out to play a special role in generating the country share functions. This property is termed *base country invariance* in Kravis, Kenessey, Heston and Summers [1975].

MT7: *Symmetric Treatment of Commodities* (*Commodity Reversal Test*): Let P denote an $N \times N$ permutation matrix. Then $S_i(Pp, Py) = S_i(p, y)$ for $i = 1, \ldots, I$.

The next three tests have no counterparts in the bilateral context. This is due to the fact that they are consistency in aggregation tests; i.e., we ask that the system of share functions changes in a consistent manner as the number of countries I changes. In the bilateral context, it is not natural to think of changing the number of countries (which is two of course).

The following test formalizes the consistency in aggregation property that was mentioned in earlier sections.

MT8: *Country Partitioning Test*: Let $S_j^I \equiv S_j(p^1,\ldots,p^I,y^1,\ldots,y^I)$ for $j=1,\ldots,I$ and let $0<\lambda<1$. Define $S_j^{I+1} \equiv S_j(p^1,\ldots,p^{i-1}, p^i, p^{i+1},\ldots, p^I, p^i, y^1,\ldots,y^{i-1}, \lambda y^i, y^{i+1},\ldots,y^I, (1-\lambda)y^i)$ for $j=1,\ldots,I,I+1$. Then $S_j^{I+1}=S_j^I$ for $j=1,\ldots,i-1,i+1,\ldots,I$, $S_i^{I+1}=\lambda S_i^I$ and $S_{I+1}^{I+1}=(1-\lambda)S_i$. This property is to hold no matter which country i is partitioned.

The functions S_j^I are the share functions for the initial world economy that consists of I countries. The functions S^{I+1} correspond to a new world economy, where the original country i, which had price vector p^i and quantity vector y^i, is partitioned into two countries. The new country i has price vector p^i and quantity vector λy^i where λ is a positive fraction and the separatist offshoot of the old country i is now country $I+1$ with price vector p^i and quantity vector $(1-\lambda)y^i$. Test MT8 says that under these conditions the old country i share S_i^I splits into λS_i^I and $(1-\lambda)S_i^I$ and the remaining shares are unaffected by this partitioning of countries. Note that MT8 implicitly requires that the share functions can be defined for $I+1$ countries as well as I countries.

MT9: *Irrelevance of Tiny Countries Test*: Let $\lambda_i>0$ and define $S_j^I(\lambda_i) \equiv S_j(p^1,\ldots,p^I,y^1,\ldots,y^{i-1},\lambda_i y^i,y^{i+1},\ldots,y^I)$ for $j=1,\ldots,I$. Define $S_j^{I-1} \equiv S_j(p^1,\ldots,p^{i-1},p^{i+1},\ldots,p^I,y^1,\ldots,y^{i-1},y^{i+1},\ldots,y^I)$ for $j=1,\ldots,I-1$. Then $\lim_{\lambda_i\to 0} S_j^I(\lambda_i) = S_j^{I-1}$ for $j=1,\ldots,i-1$ and $\lim_{\lambda_i\to 0} S_{j+1}^I(\lambda_i) = S_j^{I-1}$ for $j=i,i+1,\ldots,I-1$. This property is to hold for all choices of the disappearing country i. Note that MT9 implicitly requires that the share functions can be defined for $I-1$ countries and we also require the existence of the limit.

The above property may be explained in words as follows. Consider an initial world economy with I countries in it. Deflate the quantity vector for country i down to zero in a proportional manner and consider the resulting system of limiting share functions. For all countries except country i, the limiting share is equal to the corresponding share we would get if we simply deleted the data for country i and defined the resulting system of shares for the world economy consisting of only $I-1$ countries. Roughly speaking, MT9 says that the existence of tiny countries should not materially change the output shares of the remaining countries.

The following test is also a consistency in aggregation test, but it is not as compelling as the previous two tests.

MT10: *Strong Dependence on a Bilateral Formula*: Let $Q(p^i,p^j,y^i,y^j)$ be a bilateral index number formula and define the bilateral ratios $Y_{ij} \equiv Q(p^i,p^j,y^i,y^j)$, $i,j=1,2,\ldots,I$. The share functions $S_i(p,y)$ depend only on the bilateral ratios; i.e., we have

$$(14)\qquad S_i(p,y)=F_i(Y_{11},Y_{12},\ldots,Y_{I-1,I},Y_{II}),\quad i=1,\ldots,I,$$

for some functions of I^2 variables, F_i, $i = 1, \ldots, I$. Moreover, as the quantity vectors for all countries except i and j are deflated proportionally to zero, we have for $1 \leq i < j \leq I$,

$$(15) \quad \lim_{\lambda \to 0} \frac{S_i(p, \lambda y^1, \ldots, \lambda y^{i-1}, y^i, \lambda y^{i+1}, \ldots, \lambda y^{j-1}, y^j, \lambda y^{j+1}, \ldots, \lambda y^I)}{S_j(p, \lambda y^1, \ldots, \lambda y^{i-1}, y^i, \lambda y^{i+1}, \ldots, \lambda y^{j-1}, y^j, \lambda y^{j+1}, \ldots, \lambda y^I)} = Q(p^j, p^i, y^j, y^i).$$

A weaker version of MT10 would drop (15). An alternative weaking of MT10 would drop (14) but keep (15). The justification for this test or its weaker versions depends on how strongly one feels about the accuracy of the bilateral formula Q. If one feels strongly that Q is best for making bilateral comparisons, then property (15) is a very natural one. It says that the relative outputs for countries i and j should approach the value given by the bilateral index number formula as the outputs of all other countries shrink to zero.

Many additional multilateral tests could be devised. However, the above ten tests are sufficient to enable us to discriminate partially among the six multilateral systems developed in Section 6 above. We shall now verify which of the above tests each system satisfies.

PROPOSITION 1. *Suppose the bilateral quantity index $Q(p^i, p^j, y^i, y^j)$ satisfies the bilateral tests BT1 through BT7. Then: (i) the multilateral star system defined by (4) using country j as the numeraire country satisfies all of the above ten multilateral tests except MT6, MT9 (actually the tiny country test fails only when the tiny country is chosen to be the numeraire country j) and MT10; (ii) the own share multilateral system defined by (5) and (6) satisfies all tests except MT3; (iii) the EKSCCD system defined by (9) and (10) satisfies all tests except MT8, MT9 and MT10; (iv) the democratic weights system defined by (11) with $s_j \equiv 1/I$ satisfies all tests except MT3, MT8, MT9 and MT10; (v) the plutocratic weights system defined by (11) with $s_j \equiv p^j \cdot y^j / \sum_{k=1}^{I} p^k \cdot y^k$ satisfies all tests except MT3 and MT4, and (vi) the quantity weights system defined by (11) with the shares s_j defined by the own shares (6) satisfies all tests except MT3.*

The above proposition is very useful since it enables us to eliminate three of the six multilateral systems from contention if we want a country symmetric system: (i) is eliminated since it fails MT6; (iv) is eliminated since it is dominated by (ii), (iii) and (vi); and (v) is eliminated since it is dominated by (ii) and (vi). In addition, if we want our multilateral indexes to satisfy the proportionality test MT3, we should choose (iii), the EKSCCD system. If we want our indexes to have good aggregation properties and satisfy MT8 and MT9, we should choose the own share system (ii) or the own share quantity weights system (vi).

The star system is quite a reasonable one to use if there is a natural numeraire country. Thus if a single country is interested in comparing its real output with a possibly varying number of other countries, it is quite natural to give up on the symmetry test MT6 and use the star system, choosing the numeraire country to be the home country.

However, if a group of countries wish to jointly determine their relative real outputs, then it is natural to demand that the symmetric treatment of countries test MT6 hold, and thus the choice is between systems (ii) or (iii) and (vi). Note that from a numerical point of view, systems (ii) and (iii) should yield very similar results, depending on how closely various harmonic means approximate the corresponding geometric means (recall (8)).

Thus far, we have dealt with quantity indexes but not price indexes. However, once we have decided what the "correct" quantity shares $S_i(p, y)$ for each country are, we can obtain consistent country i price indexes $P_i(p, y)$ by dividing the country i nominal private product, $p^i \cdot y^i$, by the share function, $S_i(p, y)$; i.e.,

$$(16) \qquad P_i(p,y) \equiv p^i \cdot y^i / S_i(p,y), \quad i = 1, \ldots, I.$$

The resulting system of price index and share functions will have the following property:

$$\sum_{i=1}^{I} P_i(p,y) S_i(p,y) = \sum_{i=1}^{I} p^i \cdot y^i$$
$$= \text{nominal world private product.}$$

If prices p^i in each country are measured in a common currency unit such as U.S. dollars, then $P_i(p, y)$ may be interpreted as country i's *purchasing power parity*. If $P_i(p, y) > 1$, then this indicates that prices in country i are high relative to average world prices.

We conclude this chapter by noting that our analysis of the multilateral test approach to internationally comparisons is seriously incomplete. We have not shown that our ten multilateral axioms or tests are independent. Furthermore, it would be useful to introduce additional axioms to better discriminate among the six alternatives we have considered here. Finally, we have not answered the following question: Is it possible to find a system of share functions that satisfies all ten multilateral tests?

Essays in Index Number Theory, Volume I
W.E. Diewert and A.O. Nakamura (Editors)

Chapter 13

Fisher Ideal Output, Input and Productivity Indexes Revisited*

W.E. Diewert

1. Introduction

The basic purpose of this paper is to explain how productivity change can be measured for a firm under ideal circumstances.

If a firm produces only one output and utilizes only one input during each accounting period, then it is straightforward to define the productivity change for the firm between two periods. Let $y^t > 0$ denote the quantity of output produced during period t and let $x^t > 0$ denote the quantity of input utilized by the firm during period t for $t = 0, 1$. Then the productivity change going from period 0 to 1 may be defined by:

$$Pr(x^0, x^1, y^0, y^1) \equiv (y^1/y^0)/(x^1/x^0); \tag{1}$$

i.e., the productivity change is the firm's output ratio divided by its input ratio. Thus if output grows faster (slower) than input, Pr will be greater than one (less than one) and we say that the firm has experienced a productivity improvement (decline).

However, all firms utilize more than one input and virtually all firms produce more than one output. A basic research question is: how can definition (1) be generalized to the case of a multiple output, multiple input firm?

We shall take several different approaches to answering the above question. In our first approach, we replace the output ratio by an output quantity index, $Q(p^0, p^1, y^0, y^1)$, and replace the input ratio by an input quantity index, $Q^*(w^0, w^1, x^0, x^1)$, where $p^t \equiv (p_1^t, \ldots, p_M^t)$ and $y^t \equiv (y_1^t, \ldots, y_M^t)$ are the output price and quantity vectors pertaining to period t and $w^t \equiv (w_1^t, \ldots, w_N^t)$ and $x^t \equiv (x_1^t, \ldots, x_N^t)$ are the input price and quantity vectors pertaining to

---

*First published in the *Journal of Productivity Analysis*, Vol. 3, No. 3, 1992, pp. 211–248. This research was supported by a Strategic Grant from the Social Science and Humanities Research Council of Canada. The author thanks M. Denny and A. Nakamura for comments.

period t for $t = 0, 1$. Note that the index number functions Q and Q^* depend on the relevant quantity vectors for the two periods but they also use the corresponding price vectors as a means of weighting the quantities.

The problem with this index number approach is that we must specify concrete functional forms for the quantity indexes Q and Q^*. Four commonly used functional forms for Q are the Laspeyres quantity index Q_L, the Paasche quantity index Q_P, the Fisher ideal quantity index Q_F and the translog or Törnqvist quantity index Q_T defined by (2) – (5) below:[1]

$$Q_L(p^0, p^1, y^0, y^1) \equiv p^0 \cdot y^1 / p^0 \cdot y^0; \tag{2}$$

$$Q_P(p^0, p^1, y^0, y^1) \equiv p^1 \cdot y^1 / p^1 \cdot y^0; \tag{3}$$

$$Q_F(p^0, p^1, y^0, y^1) \equiv (p^0 \cdot y^1 p^1 \cdot y^1 / p^0 \cdot y^0 p^1 \cdot y^0)^{1/2}; \tag{4}$$

$$Q_T(p^0, p^1, y^0, y^1) \equiv \prod\nolimits_{i=1}^{M} (y_i^1 / y_i^0)^{(s_i^0 + s_i^1)/2} \tag{5}$$

where $s_i^t \equiv p_i^t y_i^t / p^t \cdot y^t$ is the period t revenue share of output i. Note that $Q_F = (Q_L Q_P)^{1/2}$; i.e., the Fisher ideal quantity index is the geometric mean of the Laspeyres and Paasche quantity indexes. Input quantity indexes Q_L^*, Q_P^*, Q_F^* and Q_T^* can be defined in a manner analogous to (2) – (5), except that input prices and quantities, w^t and x^t, replace the output prices and quantities, p^t and y^t.

In Sections 2 and 3 below, we use the *test* or *axiomatic approach* to index number theory to determine the functional forms for Q and Q^*. This approach dates back to C.M. Walsh [1901] [1921] and Irving Fisher [1911] [1921] [1922] but in more recent times, the main contributors have been Eichhorn [1976]

---

[1]The Laspeyres quantity index matches up with the Paasche [1874] price index while the Paasche quantity index matches up with the Laspeyres [1871] price index. Bowley [1921; 203] advocated the use of the Laspeyres quantity index to measure output change in the production context. The "ideal" properties of Q_F were stressed by Fisher [1922]. The quantity index Q_T is first mentioned in Fisher [1922; 473] where it is the quantity index which corresponds to price index number 124, but Törnqvist [1936] was the first to stress its favorable properties. Christensen and Jorgenson [1970] used Q_T and Q_T^* as discrete time approximations to the continuous time Divisia [1926] indexes while Solow [1957] used Q_L^* as a discrete approximation to the Divisia index of inputs. Diewert [1976a; 118–120] showed that Q_T corresponded to a translog aggregator function introduced by Christensen, Jorgenson and Lau [1971] and so Jorgenson and Nishimizu [1978] called Q_T the translog quantity index. Notation: $p \cdot y = p^T y \equiv \sum_{m=1}^{M} p_m y_m$.

[1978b] and his co-workers; see Eichhorn and Voeller [1976] and Funke and Voeller [1978] [1979].

In Section 2, we list some twenty tests or mathematical properties that have been suggested as desirable for an output quantity index $Q(p^0, p^1, y^0, y^1)$ and in Section 3, we show that the Fisher ideal quantity index Q_F defined by (4) above is the unique function which satisfies all of these tests. Thus, our first approach to the measurement of productivity change in the multiple input and output case leads to the Fisher index of productivity change Pr_F defined by:

$$(6)\quad Pr_F(p^0, p^1, y^0, y^1, w^0, w^1, x^0, x^1) \equiv Q_F(p^0, p^1, y^0, y^1)/Q_F^*(w^0, w^1, x^0, x^1).$$

In the remainder of the paper, we consider economic approaches to the construction of input, output and productivity indexes. In economic approaches, the assumption of optimizing behavior is always used; i.e., we assume that the firm competitively minimizes costs, maximizes revenues or maximizes profits. In the test or axiomatic approach, no assumption about optimizing behavior is required and this can be an advantage of the approach.

Our first economic approach to the measurement of productivity change is contained in Sections 4 and 5. It is well known that the technology of a firm can be represented in several alternative ways; e.g., by production functions, variable profit functions or by distance functions.[2] In Section 5, we represent the technology of the firm in period t by a specific functional form for its variable profit function. We then show that the shift in the technology can be computed exactly using a function of observable prices and quantities — in fact, the Fisher productivity index defined by (6) does the job. In Section 4, we show that the specific functional forms for the profit functions used in Section 5 are generalizations of a functional form for a variable profit function that is *flexible*;[3] i.e., the functional form can provide a second order approximation to an arbitrary twice continuously differentiable variable profit function. Thus, the Fisher index of productivity change defined by (6) turns out to be a *superlative*[4] index of productivity change.

Sections 6 and 7 of the paper present economic justifications for the use of the Fisher output and input indexes, Q_F and Q_F^* respectively, that are analogous to the economic justifications for the use of the translog output and input indexes, Q_T and Q_T^*, that were presented by Caves, Christensen and Diewert [1982b; 1395–1401].

[2]See Gorman [1968b], Diewert [1973a], Blackorby, Primont and Russell [1978] and McFadden [1978].

[3]The term "flexible" is due to Diewert [1974a; 113].

[4]Diewert [1976a; 117] used the term "superlative" to describe an index number formula that was exact for a flexible functional form.

In Section 8, a similar economic approach to productivity indexes is developed, except that the firm's technology is represented by an output distance or deflation function instead of a variable profit function. Again we obtain a strong economic justification for the use of the Fisher productivity index (6).

The above economic approaches to productivity indexes, relying on the theory of exact and superlative indexes, are analogous to the approaches used by Diewert [1976a; 124–127][5] [1980; 491–493] [1983b; 1077–1083] and Caves, Christensen and Diewert [1982b; 1401–1408]. In fact, under the hypothesis of a constant returns to scale technology, Caves, Christensen and Diewert [1982b; 1406] present a strong economic justification for the use of the following Törnqvist or translog productivity index to measure the shift in the technology:

$$(7)\quad Pr_T(p^0,p^1,y^0,y^1,w^0,w^1,x^0,x^1) \equiv Q_T(p^0,p^1,y^0,y^1)/Q_T^*(w^0,w^1,x^0,x^1)$$

where Q_T is the translog output index defined by (5) and Q_T^* is the analogous translog input index. The productivity change index (7) has been used by Christensen and Jorgenson [1970], Jorgenson and Griliches [1972] and the Bureau of Labor Statistics in their recent work on multifactor productivity; see Mark and Waldorf [1983; 15]. The results in Sections 5 and 8 below present equally strong economic justifications for the use of the Fisher productivity index Pr_F defined by (6).

Since Pr_T and Pr_F have equally valid economic justifications, the results of Section 3 nudge us in the direction of preferring Pr_F over Pr_T, since from the viewpoint of the test approach to index numbers, the Fisher index Q_F seems preferable to the translog index Q_T.

Section 9 concludes.

2. The Axiomatic or Test Approach to Index Number Theory

As we mentioned in the introduction, we want to use the test approach to index number theory in an attempt to determine the functional forms for the quantity indexes Q and Q^*. We shall concentrate our attention on the determination of the output index $Q(p^0,p^1,y^0,y^1)$; the theory for the input quantity index Q^* is analogous.

Instead of trying to directly determine the functional form for the quantity index Q, it turns out to be more convenient to *simultaneously* determine Q

[5]This approach was not satisfactory since it assumed separability between inputs and outputs. For discussions on separability concepts, see Blackorby, Primont and Russell [1978].

and its corresponding price index $P(p^0, p^1, y^0, y^1)$. Thus we want to find two functions of the $4M$ variables, $p^0 \equiv (p_1^0, \ldots, p_M^0)$, $p^1 \equiv (p_1^1, \ldots, p_M^1)$, $y^0 \equiv (y_1^0, \ldots, y_M^0)$ and $y^1 \equiv (y_1^1, \ldots, y_M^1)$ (which are the price and quantity vectors pertaining to periods 0 and 1), $P(p^0, p^1, y^0, y^1)$ and $Q(p^0, p^1, y^0, y^1)$, which decompose the value change between the two periods, $p^1 \cdot y^1 / p^0 \cdot y^0$, into a price change part P and a quantity change part Q; i.e., we want P and Q to satisfy the following equation:[6]

$$(8) \qquad P(p^0, p^1, y^0, y^1) Q(p^0, p^1, y^0, y^1) = p^1 \cdot y^1 / p^0 \cdot y^0.$$

If $M = 1$, so that there is only one output, then a natural candidate for P is p_1^1/p_1^0, the single price ratio, and a natural candidate for Q is y_1^1/y_1^0, the single quantity ratio.

Note that if either P or Q is determined, then the remaining function Q or P may be defined implicitly or residually using equation (8). Since historically researchers first concentrated on the determination of P, we shall also attempt to determine the functional form for P with the understanding that once P has been determined, Q may be determined using (8).

What index number theorists have done over the years is propose properties or tests that P should satisfy. These properties are generally multidimensional analogues to the one good price index formula, p_1^1/p_1^0. Below, we list twenty tests along with the names of the researchers who have proposed the corresponding tests.

We shall assume that every component of each price and quantity vector is positive; i.e., $p^t \gg 0_M$ and $y^t \gg 0_M$ for $t = 0, 1$. If we want to set $y^0 = y^1$, we shall call the common quantity vector y; if we want to set $p^0 = p^1$, we call the common price vector p.

PT1: *Positivity*: $P(p^0, p^1, y^0, y^1) > 0$. Eichhorn and Voeller [1976; 23] suggested this test.

PT2: *Continuity*: $P(p^0, p^1, y^0, y^1)$ is a continuous function of its arguments.

This test does not seem to have been formally suggested in the literature.[7] However, Irving Fisher [1922; 207–215] seems to have informally suggested the essence of this test.[8]

---

[6]Frisch [1930; 399] called (8) the *product test*. The concept of this test was due to Fisher [1911; 418].

[7]Frisch [1930; 400] assumed differentiability which implies continuity. Also Eichhorn [1978b; 165] made a weaker continuity assumption.

[8]"A formula which can be shown to be especially erratic, as compared with other formulae, has been called freakish" (Fisher [1922; 207]). Also Fisher [1922; 114–115] disapproved of medians and modes because of their insensitivity to possibly large changes in the data and their violent change to possibly small changes in the data.

PT3: *Identity* or *Constant Prices Test*: $P(p, p, y^0, y^1) = 1$; i.e., if the price of every good is identical during the two periods, then the price index should equal unity, no matter what the quantity vectors are. Laspeyres [1871; 308], Walsh [1901; 308] and Eichhorn and Voeller [1976; 24] have all suggested this test.

PT4: *Tabular Standard, Basket* or *Constant Quantities Test*: $P(p^0, p^1, y, y) = p^1 \cdot y / p^0 \cdot y$; i.e., if quantities are constant during the two periods so that $y^0 = y^1 = y$, then the price index should equal the expenditure on the constant basket in period 1, $p^1 \cdot y$, divided by the expenditure on the basket in period 0, $p^0 \cdot y$.

The origins of this test go back at least two hundred years to the Massachusetts legislature which used a constant basket of goods to index the pay of Massachusetts soldiers fighting in the American Revolution; see Willard Fisher [1913]. Other researchers who have suggested the test over the years include: Lowe [1823; Appendix, 95], Scrope [1833; 406], Sidgwick [1883; 67–68], Jevons [1884; 122] originally published in 1865, Edgeworth [1925; 215] originally published in 1887, Marshall [1887; 363], Pierson [1895; 332] Walsh [1901; 540] [1921; 544], Bowley [1901; 227], Pigou [1912; 38], Frisch [1936; 6], Vogt [1978; 132] and Funke [1988; 103]. Scrope [1833; 407] was the first to use the term "tabular standard" while Edgeworth [1925; 331] used the term "consumption standard" to describe the test. Vogt [1978; 132] called the test the "Wertindextreue Test" and Funke [1988; 103] translated this German terminology into the "Value Ratio Preserving Test."

PT5: *Proportionality in Current Prices*: $P(p^0, \lambda p^1, y^0, y^1) = \lambda P(p^0, p^1, y^0, y^1)$ for scalars $\lambda > 0$; i.e., if all period 1 prices are multiplied by the positive number λ, then the new price index is λ times the old price index.

This test was proposed by Walsh [1901; 385], Eichhorn and Voeller [1976; 24] and Vogt [1980; 68].

Walsh [1901] and Irving Fisher [1911; 418] [1922; 420] proposed the related proportionality test $P(p, \lambda p, y^0, y^1) = \lambda$. This last test is a combination of PT3 and PT5; in fact Walsh [1901; 385] noted that this last test implies the identity test, PT3.

PT6: *Inverse Proportionality in Base Prices* (Homogeneity of Degree Minus One in Base Prices): $P(\lambda p^0, p^1, y^0, y^1) = \lambda^{-1} P(p^0, p^1, y^0, y^1)$ for all $\lambda > 0$; i.e., if all period 0 prices are multiplied by the positive number λ, then the new price index equals the old price index divided by λ. Eichhorn and Voeller [1976; 28] suggested this test.

The next seven tests are ***invariance*** or ***symmetry*** tests. Fisher [1922; 62–63 and 458–460] and Walsh [1921; 542] seem to have been the first researchers to appreciate the significance of these kinds of tests. Fisher [1922; 62–63] spoke of *fairness*, but it is clear that he had symmetry properties in mind. It is perhaps unfortunate that he did not realize that there were more symmetry and invariance properties than the ones he proposed; if he had realized this, it is likely that he would have been able to characterize axiomatically his ideal price index, as we shall do below in Section 3.

PT7: *Invariance to Proportional Changes in Current Quantities* (Homogeneity of Degree Zero in Current Quantities): $P(p^0, p^1, y^0, \lambda y^1) = P(p^0, p^1, y^0, y^1)$ for all $\lambda > 0$; i.e., if current period quantities are all multiplied by the number λ, then the price index remains unchanged. Vogt [1980; 70] seems to have been the first to propose this test and his derivation of the test is of some interest. Recall the product test, equation (8) above. Suppose the quantity index Q satisfies the quantity analogue to the price test PT5; i.e., suppose Q satisfies $Q(p^0, p^1, y^0, \lambda y^1) = \lambda Q(p^0, p^1, y^0, y^1)$. Then using (8), we see that $P(p^0, p^1, y^0, y^1)$ must satisfy PT7.

Vogt [1980; 70] used the same type of argument to derive PT4 as a consequence of the quantity index satisfying an analogue to the identity test PT3; i.e., suppose Q satisfies $Q(p^0, p^1, y, y) = 1$. Then by (8), $P(p^0, p^1, y, y) = p^1 \cdot y / p^0 \cdot y Q(p^0, p^1, y, y) = p^1 \cdot y / p^0 \cdot y$ which is PT4.[9] Thus if a quantity index Q satisfies a certain test or property, then (8) may be used to deduce the corresponding property or test that the price index P must satisfy. This observation was first made by Irving Fisher [1911; 400–406] in an almost forgotten (but nonetheless brilliant) work.

PT8: *Invariance to Proportional Changes in Base Quantities* (Homogeneity of Degree Zero in Base Quantities): $P(p^0, p^1, \lambda y^0, y^1) = P(p^0, p^1, y^0, y^1)$ for all $\lambda > 0$; i.e., if all base period quantities are multiplied by the number λ, then the price index remains unchanged.

Surprisingly, this test does not seem to have been proposed before.

If the quantity index Q satisfies the following counterpart to PT6: $Q(p^0, p^1, \lambda y^0, y^1) = \lambda^{-1} Q(p^0, p^1, y^0, y^1)$ for all $\lambda > 0$, then using (8), the corresponding price index P must satisfy: $P(p^0, p^1, \lambda y^0, y^1) = p^1 \cdot y^1 / p^0 \cdot \lambda y^0 Q(p^0, p^1, \lambda y^0, y^1) = p^1 \cdot y^1 / p^0 \cdot \lambda y^0 \lambda^{-1} Q(p^0, p^1, y^0, y^1) = P(p^0, p^1, y^0, y^1)$; i.e., P must satisfy

[9]This derivation may be found in Irving Fisher [1911; 401] if we set his $k^1 = 1$. The complete derivation of Fisher's Test 2 proceeded as follows. Suppose the quantity index satisfies the following Proportionality Test: $Q(p^0, p^1, y, k^1 y) = k^1$. Then P must satisfy: $P(p^0, p^1, y, k^1 y) = p^1 \cdot k^1 y / p^0 \cdot y Q(p^0, p^1, y, k^1 y) = p^1 \cdot k^1 y / p^0 \cdot y k^1 = p^1 \cdot y / p^0 \cdot y$. Thus if Q satisfies Fisher's Proportionality Test, then the corresponding P must satisfy $P(p^0, p^1, y, \lambda y) = p^1 \cdot y / p^0 \cdot y$ for all $\lambda > 0$, which is a stronger version of the Basket Test, PT4. Fisher [1911; 406] thought that this price test was the most important of his eight tests for prices because it was the only test that indicated what type of quantity weighting of the prices was required. However, later Fisher [1922; 420–421] no longer seemed to consider that the test was important. Note that we can now interpret PT4 as an implicit identity test.

PT8. This argument provides some justification for assuming the validity of PT8.

PT7 and PT8 together impose the property that the price index P does not depend on the absolute magnitudes of the quantity vectors y^0 and y^1. Of course, in the one good case, PT4 and either PT7 or PT8 implies that $P(p_1^0, p_1^1, y_1^0, y_1^1) = p_1^1/p_1^0$, the price ratio for the single good.

PT9: *Commodity Reversal Test* (Invariance to Changes in the Ordering of Commodities): $P(\tilde{p}^0, \tilde{p}^1, \tilde{y}^0, \tilde{y}^1) = P(p^0, p^1, y^0, y^1)$ where $\tilde{p}^t$ denotes a permutation of the components of the vector p^t and $\tilde{y}^t$ denotes the same permutation of the components of y^t for $t = 0, 1$. Thus if the ordering of the commodities is changed, the numerical value of the price index remains unchanged.

This test is due to Irving Fisher [1922][10] and it is one of his three famous reversal tests. The other two are the time reversal test and the factor reversal test which will be considered below.

PT10: *Invariance to Changes in the Units of Measurement* (Commensurability Test): $P(\alpha_1 p_1^0, \ldots, \alpha_M p_M^0;\ \alpha_1 p_1^1, \ldots, \alpha_M p_M^1;\ \alpha_1^{-1} y_1^0, \ldots, \alpha_M^{-1} y_M^0;\ \alpha_1^{-1} y_1^1, \ldots, \alpha_M^{-1} y_M^1) = P(p_1^0, \ldots, p_M^0;\ p_1^1, \ldots, p_M^1;\ y_1^0, \ldots, y_M^0;\ y_1^1, \ldots, y_M^1)$ for all $\alpha_1 > 0, \ldots, \alpha_M > 0$; i.e., the price index does not change if the units of measurement for each commodity are changed.

The concept of this test was due to Jevons [1884; 23] and the Dutch economist Pierson [1896; 131], who criticized several index number formula for not satisfying this fundamental test. Fisher [1911; 411] first called this test the change of units test and later Fisher [1922; 420] called it the commensurability test.

PT11: *Time Reversal Test*: $P(p^1, p^0, y^1, y^0) = 1/P(p^0, p^1, y^0, y^1)$; i.e., if the data for periods 0 and 1 are interchanged, then the resulting price index should equal the reciprocal of the original price index. Obviously, in the one good case when the price index is simply the single price ratio, this test is satisfied (as are all of the other tests listed in this section).

When the number of goods is greater than one, many commonly used price indexes fail this test; e.g., the Laspeyres [1871] price index, $P_L \equiv p^1 \cdot y^0 / p^0 \cdot y^0$, and the Paasche [1874] price index, $P_P \equiv p^1 \cdot y^1 / p^0 \cdot y^1$, both fail this fundamental test.

The concept of the test was due to Pierson [1896; 128], who was so upset with the fact that many of the commonly used index number formulae did not satisfy this test that he proposed that the entire concept of an index number should be abandoned. More formal statements of the test were made by Walsh [1901; 368] [1921; 541] and Fisher [1911; 534] [1922; 64].

[10]Fisher [1922; 63] comments on this test that: "This is so simple as never to have been formulated."

Our next two tests are more controversial, since they do not appear to be consistent with the economic approach to index number theory.[11]

PT12: *Quantity Reversal Test* (Quantity Weights Symmetry Test): $P(p^0, p^1, y^0, y^1) = P(p^0, p^1, y^1, y^0)$; i.e., if the quantity vectors for the two periods are interchanged, then the price index remains invariant. This property means that if quantities y_m^t are used to weight the prices in the index number formula, then y_m^0 and y_m^1 must enter the formula in a symmetric manner; i.e., quantities from each period enter the formula symmetrically.

Funke and Voeller [1978; 3] introduced this test; they called it the "weight property".

PT13: *Price Reversal Test* (Price Weights Symmetry Test): $p^1 \cdot y^1/p^0 \cdot y^0 P(p^0, p^1, y^0, y^1) = p^0 \cdot y^1/p^1 \cdot y^0 P(p^1, p^0, y^0, y^1)$. This test has not been proposed before[12] but it is the analogue to PT12 applied to quantity indexes; i.e., if we use (8) to define the quantity index Q in terms of the price index P, then it can be seen that PT13 is equivalent to the following property for the associated quantity index Q:

$$Q(p^0, p^1, y^0, y^1) = Q(p^1, p^0, y^0, y^1); \tag{9}$$

i.e., if the price vectors for the two periods are interchanged, then the quantity index remains invariant. Thus if prices for the same good in the two periods are used to weight quantities in the construction of the quantity index, then property PT13 implies that these prices enter the quantity index in a symmetric manner.

The next three tests are mean value tests.

PT14: *Mean Value Test for Prices*:

$$\min_i\{p_i^1/p_i^0 : i = 1, \ldots, M\} \leq P(p^0, p^1, y^0, y^1) \leq \max_i\{p_i^1/p_i^0 : i = 1, \ldots, M\};$$

i.e., the price index lies between the minimum price ratio and the maximum price ratio. This very desirable property seems to have been first proposed by Eichhorn and Voeller [1976; 10].

PT15: *Mean Value Test for Quantities*:

$$\min_i\{y_i^1/y_i^0 : i = 1, \ldots, M\} \leq p^1 \cdot y^1/p^0 \cdot y^0 P(p^0, p^1, y^0, y^1) \leq \max_i\{y_i^1/y_i^0 : i = 1, \ldots, M\}.$$

[11]See the critical discussion by Sato [1980].

[12]Funke and Voeller [1979; 55] proposed the following (different) price reversal test: $P(p^0, p^1, y^0, y^1)P(p^1, p^0, y^0, y^1) = 1$. Sato [1980; 127] showed that this price reversal test along with the time reversal test PT11 implied the quantity reversal test PT12.

Using (8) to define the quantity index Q in terms of the price index P, we see that PT15 is equivalent to the following property on the associated quantity index:

$$\begin{aligned}(10)\qquad \min_i\{y_i^1/y_i^0 : i = 1,\ldots,M\} &\leq Q(p^0,p^1,y^0,y^1) \\ &\leq \max_i\{y_i^1/y_i^0 : i = 1,\ldots,M\};\end{aligned}$$

i.e., the implicit quantity index Q defined by P lies between the minimum and maximum rates of growth of the individual quantities.

This test does not seem to have been proposed before, but it is an obvious quantity analogue to the price test PT14.

PT16: *Paasche and Laspeyres Bounding Test*: The price index P satisfies at least one of the following inequalities (11) or (12):

$$(11)\qquad p^1\cdot y^0/p^0\cdot y^0 \leq P(p^0,p^1,y^0,y^1) \leq p^1\cdot y^1/p^0\cdot y^1;$$

$$(12)\qquad p^1\cdot y^1/p^0\cdot y^1 \leq P(p^0,p^1,y^0,y^1) \leq p^1\cdot y^0/p^0\cdot y^0;$$

i.e., the price index P must lie between the Laspeyres and Paasche price indexes.

A justification for this test can be made by considering the basket test PT4. If $y^0 = y^1$, then by PT4, the correct functional form for the price index is $p^1\cdot y^0/p^0\cdot y^0 = p^1\cdot y^1/p^0\cdot y^1$. In the general case where $y^0 \neq y^1$, it is natural to think of both y^0 and y^1 as being "extreme" baskets and so the Laspeyres and Paasche price indexes should provide bounds to the "best" price index $P(p^0,p^1,y^0,y^1)$ which treats the quantity data in each period in a symmetric manner instead of the extreme manner implied by the Paasche and Laspeyres price indexes, which each use the quantity data for only one period.

In the context of the axiomatic or test approach to index number theory, PT16 has been proposed by both Bowley [1901; 227] and Fisher [1922; 403].[13]

We could propose a test where the implicit quantity index Q that corresponds to P via (8) is to lie between the Laspeyres and Paasche quantity indexes, Q_L and Q_P, defined by (2) and (3) above. However, the resulting test turns out to be equivalent to (11) and (12).[14]

Our final four tests are monotonicity tests.

PT17: *Monotonicity in Current Prices*: $P(p^0,p^1,y^0,y^1) < P(p^0,p,y^0,y^1)$ if $p^1 < p$; i.e., if some period 1 price increases, then the price index must

[13]Of course, the Paasche and Laspeyres price indexes arise repeatedly as bounds to the true index in the economic theory of index numbers; e.g., see Fisher and Shell [1972b; 57–58], Hicks [1940] or Diewert [1983b].

[14]Thus PT16 can be given a quantity index justification.

increase, so that $P(p^0, p^1, y^0, y^1)$ is increasing in the components of p^1. This property was proposed by Eichhorn and Voeller [1976; 23].

PT18: *Monotonicity in Base Prices*: $P(p^0, p^1, y^0, y^1) > P(p, p^1, y^0, y^1)$ if $p^0 < p$; i.e., if any period 0 price increases, then the price index must decrease, so that $P(p^0, p^1, y^0, y^1)$ is decreasing in the components of p^0. This property was also proposed by Eichhorn and Voeller [1976; 23].

PT19: *Monotonicity in Current Quantities*: $p^1 \cdot y^1 / p^0 \cdot y^0 P(p^0, p^1, y^0, y^1) < p^1 \cdot y / p^0 \cdot y^0 P(p^0, p^1, y^0, y)$ if $y^1 < y$.

PT20: *Monotonicity in Base Quantities*: $p^1 \cdot y^1 / p^0 \cdot y^0 P(p^0, p^1, y^0, y^1) > p^1 \cdot y^1 / p^0 \cdot y P(p^0, p^1, y, y^1)$ if $y^0 < y$.

If we define the implicit quantity index Q that corresponds to P using (8), we find that PT19 translates into the following inequality involving Q:

$$Q(p^0, p^1, y^0, y^1) < Q(p^0, p^1, y^0, y) \quad \text{if} \quad y^1 < y; \tag{13}$$

i.e., if any period 1 quantity increases, then the quantity index must increase.

Similarly, we find that PT20 translates into:

$$Q(p^0, p^1, y^0, y^1) > Q(p^0, p^1, y, y^1) \quad \text{if} \quad y^0 < y; \tag{14}$$

i.e., if any period 0 quantity increases, then the quantity index must decrease.

Tests PT19 and PT20 are due to Vogt [1980; 70].

This concludes our listing of tests. In the next section, we ask whether any index number formula $P(p^0, p^1, y^0, y^1)$ exists that can satisfy all twenty tests.

3. The Test Approach and Fisher Ideal Index Numbers

Recall the twenty tests listed in the previous section. Our main result is Theorem 1.

THEOREM 1. *The only index number formula $P(p^0, p^1, y^0, y^1)$ which satisfies tests PT1 – PT20 is the Fisher ideal price index P_F defined below by (16).*

Proof: Using the price reversal test, PT13, as well as the positivity test, PT1, we may rearrange terms to obtain the following equality:

(15)

$$\begin{aligned} p^1 \cdot y^1 p^1 \cdot y^0 / p^0 \cdot y^0 p^0 \cdot y^1 &= P(p^0, p^1, y^0, y^1)/P(p^1, p^0, y^0, y^1) \\ &= P(p^0, p^1, y^0, y^1)/P(p^1, p^0, y^1, y^0) \\ &\quad \text{using PT12, the quantity reversal test,} \\ &= P(p^0, p^1, y^0, y^1)P(p^0, p^1, y^0, y^1) \\ &\quad \text{using PT11, the time reversal test.} \end{aligned}$$

The desired result now follows by taking the positive square root of both sides of (15); i.e., we obtain:

$$(16)\quad P(p^0,p^1,y^0,y^1) = [(p^1\cdot y^0/p^0\cdot y^0)(p^1\cdot y^1/p^0\cdot y^1)]^{1/2} = (P_L P_P)^{1/2} \equiv P_F$$

where $P_L \equiv p^1\cdot y^0/p^0\cdot y^0$ is the Laspeyres price index and $P_P \equiv p^1\cdot y^1/p^0\cdot y^1$ is the Paasche price index.

Note that we established (16) using only PT1 and the three reversal tests PT11, PT12 and PT13.

It can be verified by direct computations that P_F satisfies the remaining 16 tests. Q.E.D.

Of course, the quantity index that corresponds to the Fisher price index using the product test (8) is Q_F, the Fisher quantity index, defined by (4).

It turns out that P_F satisfies yet another test, PT21, which was Irving Fisher's [1921; 534] [1922; 72–81] third reversal test (the other two being PT9 and PT11):

PT21: *Factor Reversal Test* (Functional Form Symmetry Test): $P(p^0,p^1, y^0,y^1)P(y^0,y^1, p^0,p^1) = p^1\cdot y^1/p^0\cdot y^0$. A justification for this test is the following one: if $P(p^0,p^1,y^0,y^1)$ is a good functional form for the price index, then if we reverse the roles of prices and quantities, $P(y^0,y^1,p^0,p^1)$ ought to be a good functional form for a quantity index (which seems to be a correct argument) and thus the product of the price index $P(p^0,p^1,y^0,y^1)$ and the quantity index $P(y^0,y^1,p^0,p^1)$ ought to equal the value ratio, $p^1\cdot y^1/p^0\cdot y^0$. The second part of this argument does not seem to be valid and thus many researchers over the years have objected to the factor reversal test.[15] However, if one is willing to embrace PT21 as a basic test, Funke and Voeller [1978; 180] obtained the following result:

---

[15]Bowley [1923; 93] objected to PT21 on statistical grounds: "For (2), there seems to be no justification on general principles; the mean of a product only equals the product of the means of its factors if there is no correlation between them." The next objector to PT21 was Davies [1924; 187]: "That price and quantity each requires a distinct type of formula is indicated by the simpler problem where only one commodity is involved, as in the case of the bushels of wheat previously discussed. In this case, the measure of quantities for each period is obviously obtained merely by summing up the number of units sold, while the measure of prices is obtained by dividing the aggregate value by the quantity units. It would therefore be expected that the composite indexes derived by an analogous method would be associated with distinctive formulae for prices and quantities. Hence the factor reversal test may be disregarded." Note that Davies attacked PT21 from the perspective of what Frisch [1930; 397] called the theory of *absolute index numbers* which looks for functions P and Q which satisfy the following modified product tests: $P(p^0)Q(y^0) = p^0\cdot y^0$ or $P(p^0,y^0)Q(p^0,y^0) = p^0\cdot y^0$. Eichhorn [1978b; 141–146] calls these absolute index numbers, P and Q, *price levels* and *quantity levels*, and he finds, under

THEOREM 2. *(Funke and Voeller): The only index number function $P(p^0, p^1, y^0, y^1)$ which satisfies PT1 (positivity), PT11 (time reversal test), PT12 (quantity reversal test) and PT21 (factor reversal test) is the Fisher ideal index P_F defined by (16).*

Our proof of Theorem 1 is very similar to Funke and Voeller's proof of Theorem 2.[16]

The Fisher price index P_F satisfies all 20 of the tests listed in the previous section. Which tests do other commonly used price indexes satisfy? Recall Q_L, Q_P and Q_T defined by (2), (3) and (5) above. The corresponding price indexes defined using (8) are:

(17)
$$p^1 \cdot y^1/p^0 \cdot y^0 Q_L = p^1 \cdot y^1/p^0 \cdot y^1 \equiv P_P(p^0, p^1, y^0, y^1),$$

(18)
$$p^1 \cdot y^1/p^0 \cdot y^0 Q_P = p^1 \cdot y^0/p^0 \cdot y^0 \equiv P_L(p^0, p^1, y^0, y^1), \text{ and}$$

(19)
$$p^1 \cdot y^1/p^0 \cdot y^0 Q_T = p^1 \cdot y^1/p^0 \cdot y^0 \prod_{i=1}^{M} (y_i^1/y_i^0)^{(s_i^0+s_i^1)/2} \equiv \widetilde{P}_T,$$

where P_P, P_L and $\widetilde{P}_T$ are the Paasche, Laspeyres and implicit translog price indexes respectively. Two additional price indexes that are often used in productivity studies are P_T, the direct translog price index, and P_W, the Walsh [1901; 373] price index,[17] defined as follows:

(20)
$$P_T(p^0, p^1, y^0, y^1) \equiv \prod_{i=1}^{M} (p_i^1/p_i^0)^{(s_i^0+s_i^1)/2},$$

(21)
$$P_W(p^0, p^1, y^0, y^1) \equiv \sum_{i=1}^{M} p_i^1 (y_i^1 y_i^0)^{1/2} / \sum_{j=1}^{M} p_j^0 (y_j^1 y_j^0)^{1/2},$$

where $s_i^t \equiv p_i^t y_i^t / p^t \cdot y^t$ is the expenditure share of good i in period t as usual.

Straightforward computations show that the Paasche and Laspeyres price indexes, P_P and P_L, fail only the three reversal tests, PT11, PT12 and PT13.

---

certain regularity conditions, that functions P and Q satisfying the modified product tests do not exist. Frisch [1930; 398] calls functions P and Q that satisfy the usual product test (8) *relative index numbers*. Finally, Samuelson and Swamy [1974; 575] criticize PT21 from the perspective of the economic theory of index numbers and conclude: "A man and wife should be properly matched; but that does not mean I should marry my identical twin!"

[16] For another axiomatic characterization of the Fisher ideal price index (and a history of the subject), see Balk [1985].

[17] Walsh [1901; 373] called his price index "Scrope's emended method."

Since the quantity and price reversal tests, PT12 and PT13, are somewhat controversial, the test performance of P_L and P_P seems quite good. However, the failure of the time reversal test, PT11, seems to be a fatal flaw associated with the use of these indexes.

The Walsh price index, P_W, fails only four tests: PT13, the price reversal test; PT16, the Paasche and Laspeyres bounding test; PT19, the monotonicity in current quantities test; and PT20, the monotonicity in base quantities test.

Finally, the translog price index P_T and the implicit translog price index $\widetilde{P}_T$ each fail nine tests. Both indexes fail PT12, PT13, PT16 and the monotonicity tests PT17 to PT20. In addition, P_T fails PT4 (the basket test) and PT15 (the mean value test for quantities), while $\widetilde{P}_T$ fails PT3 (the identity test) and PT14 (the mean value test for prices). Thus the translog indexes are subject to a rather high failure rate.

As we mentioned at the beginning of Section 2, our 20 tests on the price index function $P(p^0, p^1, y^0, y^1)$ are really 20 tests on P and the corresponding implicit quantity index $Q(p^0, p^1, y^0, y^1)$ which can be defined in terms of P using the product test (8).

The conclusion we draw from the results of this section is that from the viewpoint of the test approach to index numbers, the Fisher quantity index Q_F appears to be far superior to the translog quantity index Q_T. Hence from the viewpoint of the test approach, the Fisher productivity index Pr_F defined by (6) appears to be superior to the translog productivity index Pr_T defined by (7).

In the following sections of this paper, we will also provide economic justifications for the use of the Fisher productivity index Pr_F. In the following section, we start our discussion of the economic approach to index number theory by proving a flexibility theorem for a certain functional form. This functional form will play an important role in subsequent sections. However, the reader who is interested in productivity indexes can skip to Section 5.

4. A New Flexible Functional Form for a Revenue Function

Let $y \equiv (y_1, \ldots, y_M)$ be a nonnegative output vector and let $x \equiv (x_1, \ldots, x_N)$ be a nonnegative input vector. Then the technology of a firm that produces these M outputs and uses these N inputs can be represented by a production or transformation function f; i.e., $y_1 = f(y_2, \ldots, y_M, x)$ represents the maximum[18] amount of output 1 that can be produced using the vector of inputs x given that amounts $y_2, \ldots, y_M$ of outputs $2, \ldots, M$ must be produced.

[18]This maximum is conditional on current managerial knowledge and practices. If the output targets $y_2, \ldots, y_M$ are too high relative to the available amount of inputs $x \equiv (x_1, \ldots, x_N)$, then $f(y_2, \ldots, y_M, x) \equiv -\infty$.

Given a positive vector of output prices $p \equiv (p_1, \ldots, p_M) \gg 0_M$, define the firm's *revenue function*[19] as follows:

$$r(p, x) \equiv \max_y \{p \cdot y : y_1 = f(y_2, \ldots, y_M, x)\}; \tag{22}$$

i.e., the firm maximizes revenue $p \cdot y \equiv \sum_{m=1}^{M} p_m y_m$ subject to its production function constraint and the resulting maximized revenue is set equal to $r(p, x)$.

It is obvious that f completely determines r. Diewert [1973a] shows that under certain regularity conditions on f, r also completely determines f. It can also be shown that $r(p, x)$ will always be linearly homogeneous in the components of p for fixed x. If the production function f is linearly homogeneous, so that the technology is subject to constant returns to scale, then it can be shown[20] that $r(p, x)$ is linearly homogenous in the components of x for fixed p.

In what follows, we shall assume a constant returns to scale technology. However, this is not restrictive: if we want to model the nonconstant returns to scale case, we need only add an artificial fixed factor whose quantity is always set equal to 1; i.e., add $x_{N+1} \equiv 1$ to the other N inputs.[21]

In order for a functional form for a revenue function, $r(p, x)$, to be *flexible* at the point p^*, x^*, the functional form must have enough free parameters so that it can approximate an arbitrary twice continuously differentiable revenue function r^* to the second order of p^*, x^*; i.e., we require that r have enough free parameters so that the following equations can be satisfied:

$$r(p^*, x^*) = r^*(p^*, x^*); \tag{23}$$

$$\nabla_p r(p^*, x^*) = \nabla_p r^*(p^*, x^*); \tag{24}$$

$$\nabla_x r(p^*, x^*) = \nabla_x r^*(p^*, x^*); \tag{25}$$

$$\nabla^2_{pp} r(p^*, x^*) = \nabla^2_{pp} r^*(p^*, x^*); \tag{26}$$

$$\nabla^2_{xx} r(p^*, x^*) = \nabla^2_{xx} r^*(p^*, x^*); \tag{27}$$

$$\nabla^2_{px} r(p^*, x^*) = \nabla^2_{px} r^*(p^*, x^*); \tag{28}$$

where $\nabla_p r \equiv (\partial r/\partial p_1, \ldots, \partial r/\partial p_M)$ and $\nabla_x r \equiv (\partial r/\partial x_1, \ldots, \partial r/\partial x_N)$ are vectors of first order partial derivatives of r with respect to the components of

[19] The concept is due to Samuelson [1953–54; 20]. Gorman [1968b] uses the term "gross profit function" while McFadden [1978a] uses the term "conditional profit function" and Diewert [1973a] [1974a] uses the term "variable profit function."

[20] See Diewert [1973a; 291–294].

[21] The "price" of this input will be w_{N+1}, the pure profits (or losses) of the firm.

p and x respectively and $\nabla^2_{pp}r$, $\nabla^2_{xx}r$ and $\nabla^2_{px}r$ are matrices of second order partial derivatives of r with respect to the components of p and x. Taking into account the symmetry of the matrices $\nabla^2_{pp}r$ and $\nabla^2_{xx}r$, it appears that r would require at least $1+M+N+(1/2)(M+1)M+(1/2)(N+1)N+MN$ parameters. However, this computation neglects the assumption that r and r^* are assumed to be linearly homogeneous in p and x separately.

If $r(p,x)$ is linearly homogeneous in p, then using Euler's theorem on homogeneous functions it can be shown that[22] r must satisfy the following $1+M+N$ restrictions:[23]

$$r(p^*,x^*) = p^{*T}\nabla_p r(p^*,x^*); \tag{29}$$

$$\nabla^2_{pp}r(p^*,x^*)p^* = 0_M; \tag{30}$$

$$p^{*T}\nabla^2_{px}r(p^*,x^*) = \nabla^T_x r(p^*,x^*). \tag{31}$$

Similarly, if $r(p,x)$ is linearly homogeneous in x, then it can be shown that r must satisfy the following $1+N+M$ restrictions:

$$r(p^*,x^*) = x^{*T}\nabla_x r(p^*,x^*); \tag{32}$$

$$\nabla^2_{xx}r(p^*,x^*) = 0_N; \tag{33}$$

$$\nabla^2_{px}r(p^*,x^*)x^* = \nabla_p r(p^*,x^*). \tag{34}$$

However, not all of the restrictions (29) to (34) are independent: if we postmultiply both sides of (31) by x^*, premultiply both sides of (34) by p^{*T} and use (29) and (32), we find that the last equation in (34) is implied by the other equations. Hence there are $2M+2N+1$ independent restrictions on the first and second partial derivations of r in (29)–(34). Note that r^* must also satisfy the restrictions (29)–(34).

Taking into account the restrictions (29)–(34), we see that in order for r to be flexible, it must have at least $1+M+N+(1/2)(M+1)M+(1/2)(N+1)N+MN-(2M+2N+1) = (1/2)M(M+1)+(1/2)N(N+1)+(M-1)(N-1)-1$ independent parameters.

Consider the following functional form for r:

$$r(p,x) \equiv (p^TApx^TCx + \alpha^Tp\beta^Txp^TBx)^{1/2}, \quad A = A^T, \; C = C^T \tag{35}$$

where A, B and C are parameter matrices and α and β are parameter vectors. The following theorem shows that r is a flexible functional form.

---

[22]See Diewert [1973a; 308] or Diewert [1974a; 143–145].

[23]Notation: p^* and $\nabla_p\Pi$ are defined as column vectors and $p^{*T}\nabla_p\Pi$ denotes their inner product so p^{*T} is the transpose of p^*; 0_M denotes a vector of zeros of dimension M.

THEOREM 3. *Let r^* be the revenue function that corresponds to a constant returns to scale technology. Suppose r^* is twice continuously differentiable at $p^* \gg 0_M$, $x^* \gg 0_N$ with $r^*(p^*, x^*) > 0$. Then for every α and β such that*

$$\alpha^T p^* \neq 0, \quad \beta^T x^* \neq 0, \tag{36}$$

there exist symmetric matrices A and C and a matrix B such that

$$p^{*T} B = 0_N^T, \quad Bx^* = 0_M \tag{37}$$

and r defined by (35) approximates r^* to the second order at (p^*, x^*). Moreover, the matrix A can be chosen to satisfy the following normalization:

$$p^{*T} A p^* = r^*(p^*, x^*) > 0. \tag{38}$$

Proof: We need to choose A, B and C so that equations (23)–(28) are satisfied. Note that $r(p, x)$ defined by (35) is linearly homogeneous in p for fixed x and linearly homogeneous in x for fixed p, as is $r^*(p, x)$. Thus both r and r^* satisfy the restrictions (29)–(34).

Define the vectors y^* and w^* as follows:

$$y^* \equiv \nabla_p r^*(p^*, x^*); \qquad w^* \equiv \nabla_x r^*(p^*, x^*). \tag{39}$$

Using (29) and (32) applied to r^*, we have

$$r^*(p^*, x^*) = p^{*T} y^* = w^{*T} x^*. \tag{40}$$

Now define A, B and C as follows:

$$A \equiv \nabla_{pp}^2 r^*(p^*, x^*) + (p^{*T} y^*)^{-1} y^* y^{*T}; \tag{41}$$

$$C \equiv \nabla_{xx}^2 r^*(p^*, x^*) + (w^{*T} x^*)^{-1} x^* x^{*T}; \tag{42}$$

$$B \equiv 2(\alpha^T p^* \beta^T x^*)^{-1} [p^{*T} y^* \nabla_{px}^2 r^*(p^*, x^*) - y^* w^{*T}]. \tag{43}$$

Using properties (30) and (33) applied to r^*, we have:

$$Ap^* = y^* \qquad \text{and} \tag{44}$$

$$Cx^* = w^*. \tag{45}$$

Equations (40) and (44) may be used to show that A satisfies (38).

Now use (31) and (34) applied to r^*, (36), (39) and (40) to show that B defined by (43) satisfies the restrictions (37).

Premultiply both sides of (45) by x^{*T} and use (40) to show:

$$x^{*T}Cx^* = r^*(p^*, x^*). \tag{46}$$

Now use (35), (37), (38) and (46) to show that (23) is satisfied. Differentiate r with respect to p and obtain:

$$\begin{aligned}
\nabla_p r(p^*, x^*) &= [r(p^*, x^*)]^{-1}Ap^*x^{*T}Cx^* && \text{using (37)}\\
&= [r^*(p^*, x^*)]^{-1}Ap^*x^{*T}Cx^* && \text{using (23)}\\
&= Ap^* && \text{using (46)}\\
&= y^* && \text{using (44)}\\
&= \nabla_p r^*(p^*, x^*) && \text{using (39).}
\end{aligned}$$

Thus equations (24) are satisfied. Similarly, differentiate r with respect to x and obtain

$$\begin{aligned}
\nabla_x r(p^*, x^*) &= [r(p^*, x^*)]^{-1}p^{*T}Ap^*Cx^* && \text{using (37)}\\
&= Cx^* && \text{using (23) and (38)}\\
&= w^* && \text{using (45)}\\
&= \nabla_x r^*(p^*, x^*) && \text{using (39)}
\end{aligned}$$

and thus equations (25) are also satisfied.

Finally, it can be shown that equations (41)–(43) imply equations (26)–(28). Q.E.D.

Thus for essentially arbitrary α and β vectors, the r defined by (35) is a flexible functional form. Note that given α and β, we have used a minimal number of parameters in the A, B and C matrices to achieve the flexibility result; i.e., taking into account the restriction (38), there are $(1/2)M(M+1)-1$ independent a_{ij} parameters, $(1/2)N(N+1)$ independent c_{ij} parameters and, taking into account the restrictions (37), $(M-1)(N-1)$ independent b_{ij} parameters. We shall use generalizations of this functional form in the next section.

5. An Economic Approach to Productivity Indexes

Assume that we can observe a firm's vector of outputs y^t, inputs x^t, output prices p^t and input prices w^t for periods $t = 0$ and $t = 1$. The firm's production function in period t is f^t and the corresponding dual revenue function is r^t defined by (22) (with r^t and f^t replacing r and f) for $t = 0, 1$.

In the economic approach to productivity measurement, a change in productivity is taken to be a shift in the production function or in one of the dual representations of the production function such as the revenue function.[24]

In this section, we shall identify a productivity change with a shift in the firm's revenue function. More specifically, consider the following two economic productivity change indexes, Pr^0 and Pr^1, defined as follows:[25]

$$(47) \qquad Pr^t \equiv r^1(p^t, x^t)/r^0(p^t, x^t), \qquad t = 0, 1.$$

Thus Pr^t is a measure of the outward shift in the technology going from period 0 to period 1, using the period t output price vector p^t and the period t input quantity vector x^t as reference vectors.

If there has been an efficiency improvement due to a process innovation or improved managerial practices, then Pr^0 and Pr^1 will be greater than 1.

We shall assume competitive profit maximizing behavior on the part of the firm in each period. This implies both revenue maximizing and cost minimizing behavior. Thus we have observed period t revenue, $p^t \cdot y^t$, equal to maximized revenues; i.e., we have

$$(48) \qquad p^t \cdot y^t = r^t(p^t, x^t), \qquad t = 0, 1.$$

If r^t is differentiable at p^t, x^t, then by a result due to Hotelling [1932; 594], we have

$$(49) \qquad y^t = \nabla_p r^t(p^t, x^t), \qquad t = 0, 1.$$

Furthermore, under the assumption of competitive profit maximizing behavior, for $t = 0, 1$, the period t observed input vector x^t must be a solution to the profit maximization problem:

$$(50) \qquad \max_x \{r^t(p^t, x) - w^t \cdot x\}.$$

---

[24]This shift in the production function approach dates back to Tinbergen [1942] and Solow [1957]. Jorgenson and Griliches [1967; 253] seem to have been the first to note that productivity change could also be defined in terms of shifts in the dual profit function.

[25]This type of theoretical productivity index was defined by Diewert [1983b; 1063].

Assuming that $x^t \gg 0_N$, the first order necessary conditions for (50) are:

$$(51) \qquad w^t = \nabla_x r^t(p^t, x^t), \qquad t = 0, 1.$$

We shall assume that the firm's period t revenue function r^t has the following functional form which generalizes (35):

$$(52) \quad r^t(p, x) \equiv \sigma_t(p^T A p x^T C x + \alpha^t \cdot p \beta^t \cdot x p^T B^t x)^{1/2}, \qquad A = A^T,\ C = C^T,\ t = 0, 1,$$

where σ_t is a positive number, A, B^t and C are parameter matrices and α^t and β^t are parameter vectors. Note that the symmetric parameter matrices A and C are constant over the two periods but the other parameters are allowed to vary.

THEOREM 4. *Suppose that the parameters of the revenue functions r^0 and r^1 defined by (52) satisfy restrictions (53) or (54):*

$$(53) \qquad p^{0T} B^0 = 0_N^T; \quad B^0 x^0 = 0_M; \quad \alpha^0 \cdot p^1 = 0 \text{ and } \beta^0 \cdot x^1 = 0 \text{ or}$$

$$(54) \qquad p^{1T} B^1 = 0_N^T; \quad B^1 x^1 = 0_M; \quad \alpha^1 \cdot p^0 = 0 \text{ and } \beta^1 \cdot x^0 = 0.$$

Suppose further that the observed period t input vector x^t is strictly positive for $t = 0, 1$. Then assuming competitive profit maximizing behavior for periods 0 and 1, the theoretical productivity indexes Pr^0 and Pr^1 defined above by (47) are both equal to the Fisher productivity index Pr_F defined by (6); i.e., we have

$$(55) \qquad Pr^0 = Pr_F(p^0, p^1, y^0, y^1, w^0, w^1, x^0, x^1) = Pr^1 = \sigma_1/\sigma_0.$$

Proof: Under assumptions (53) or (54), it can be verified that

$$(56) \qquad r^t(p^t, x^t) = \sigma_t(p^{tT} A p^t x^{tT} C x^t)^{1/2}, \qquad t = 0, 1.$$

Hence using definitions (47), we have

$$(57) \qquad Pr^0 = Pr^1 = \sigma_1/\sigma_0.$$

From (49), we have for $t = 0, 1$:

$$(58) \qquad y^t = \nabla_p r^t(p^t, x^t) = \sigma_t^2 A p^t x^{tT} C x^t / r^t(p^t, x^t),$$

differentiating definitions (52) and using (53) or (54). Similarly, from (51) we have for $t = 0, 1$:

$$(59) \qquad w^t = \nabla_x r^t(p^t, x^t) = \sigma_t^2 C x^t p^{tT} A p^t / r^t(p^t, x^t).$$

Using (59) and (56), we have

(60) $$w^t \cdot x^t = r^t(p^t, x^t), \qquad t = 0, 1.$$

The square of the Fisher output index is:

$$[Q_F(p^0, p^1, y^0, y^1)]^2 = (p^0 \cdot y^1 / p^0 \cdot y^0)/(p^1 \cdot y^0 / p^1 \cdot y^1)$$

$$= [\sigma_1^2 p^{0T} A p^1 x^{1T} C x^1 / r^1(p^1, x^1) p^0 \cdot y^0] / [\sigma_0^2 p^{1T} A p^0 x^{0T} C x^0 / r^0(p^0, x^0) p^1 \cdot y^1]$$

using (58)

(61) $$= (\sigma_1^2 x^{1T} C x^1)/(\sigma_0^2 x^{0T} C x^0) \qquad \text{using } A = A^T \text{ and (48).}$$

The square of the Fisher input index is:

$$[Q_F^*(w^0, w^1, x^0, x^1)]^2 = (w^0 \cdot x^1 / w^0 \cdot x^0)/(w^1 \cdot x^0 / w^1 \cdot x^1)$$

$$= [\sigma_0^2 x^{1T} C x^0 p^{0T} A p^0 / r^0(p^0, x^0) w^0 \cdot x^0] / [\sigma_1^2 x^{0T} C x^1 p^{1T} A p^1 / r^1(p^1, x^1) w^1 \cdot x^1]$$

using (59)

$$= [\sigma_0^2 p^{0T} A p^0 / r^0(p^0, x^0)^2] / [\sigma_1^2 p^{1T} A p^1 / r^1(p^1, x^1)^2]$$

using $C = C^T$ and (60)

$$= (\sigma_0^2 p^{0T} A p^0 / \sigma_0^2 p^{0T} A p^0 x^{0T} C x^0)/(\sigma_1^2 p^{1T} A p^1 / \sigma_1^2 p^{1T} A p^1 x^{1T} C x^1)$$

using (56)

(62) $$= x^{1T} C x^1 / x^{0T} C x^0.$$

Take the ratio of (61) to (62), take the positive square root and we obtain:

(63) $$Pr_F(p^0, p^1, y^0, y^1, w^0, w^1, x^0, x^1) = \sigma_1/\sigma_0.$$

The equalities (57) and (63) imply (55). Q.E.D.

The restrictions (53) are consistent with $r^0(p, x)$ being flexible at p^0, x^0 while the restrictions (54) are consistent with $r^1(p, x)$ being flexible at (p^1, x^1). Thus since the Fisher productivity index Pr_F is exact for the theoretical productivity indexes Pr^0 and Pr^1, Pr_F is a *superlative* measure of productivity change.[26]

---

[26]This terminology is analogous to Diewert's [1976a; 117] definition of a superlative index number formula.

Thus Theorem 4 provides a strong economic justification for the use of the Fisher productivity index.[27] However, the revenue functions r^t in the theorem correspond to constant returns to scale technologies. Thus if we want to apply the theorem to a diminishing returns to scale technology, it is necessary to add an artificial fixed input and set it equal to one (i.e., $x^t_{N+1} \equiv 1$). The corresponding period t price, w^t_{N+1}, must be set equal to the firm's period t pure profits.

In the remainder of this paper, we follow the example of Caves, Christensen and Diewert [1982b] and use the distance function approach for defining theoretical input, output and productivity indexes. The reader who is mainly interested in productivity indexes can skip ahead to Section 8.

6. An Economic Approach to Indexes of Real Input

As usual, let $w^t \equiv (w^t_1, \ldots, w^t_N)$ be the input price vector for a firm in period t and let $x^t \equiv (x^t_1, \ldots, x^t_N)$ be the corresponding positive input quantity vector for $t = 0, 1$.[28]

In order to aggregate x^0 and x^1, Caves, Christensen and Diewert [1982b; 1395–1399] developed the concept of the Malmquist [1953] input index, which was first defined geometrically in the two input case by Moorsteen [1961; 460] and was perhaps verbally stated by Hicks [1961; footnote 4] [1981; 256]. Caves, Christensen and Diewert [1982b; 1398] found that an average of two theoretical Malmquist input indexes could be approximated rather well under certain conditions by a translog or Törnqvist index of inputs Q^*_T defined as:

$$(64) \qquad Q^*_T(w^0, w^1, x^0, x^1) \equiv \prod_{i=1}^{N} (x^1_i/x^0_i)^{(1/2)(s^0_i+s^1_i)}$$

where $s^t_i \equiv p^t_i x^t_i / p^t \cdot x^t$ is the period t cost share for input i.

Our purpose in this section is to provide an analogous justification for the Fisher input index Q^*_F defined by:

$$(65) \qquad Q^*_F(w^0, w^1, x^0, x^1) \equiv (p^1 \cdot x^1 p^0 \cdot x^1 / p^0 \cdot x^1 p^0 \cdot x^0)^{1/2}.$$

[27]Diewert [1976a; 126–130] provided an economic justification for the use of Pr_F under rather strong separability assumptions; i.e., the period 0 transformation function had the form $g(y) = f(x)$ while the period 1 transformation function had the form $g(y) = (1+\tau)f(x)$. See Blackorby, Primont and Russell [1978] on separability concepts.

[28]Alternatively, t could index two different firms in the same industry.

As in Section 4, let f denote a firm's production function, let $x \equiv (x_1, \ldots, x_N)$ denote a positive input vector and let $y \equiv (y_1, \ldots, y_M)$ denote a nonnegative output vector. In order to define our theoretical indexes of real input, it is first necessary to define the firm's *input distance* (or deflation) *function* D as follows:[29]

$$(66) \qquad D(y,x) \equiv \max_{\delta>0}\{\delta : y_1 \leq f(y_2, \ldots, y_M, x_1/\delta, x_2/\delta, \ldots, x_N/\delta)\}.$$

Thus $D(y,x)$ is the maximal deflation factor δ^* which will just put the output vector y and the deflated input vector x/δ^* onto the boundary of the feasible production set.

It is easy to verify that $D(y,x)$ defined by (66) will be linearly homogeneous in the components of x; i.e., for a scalar $\lambda > 0$, we have

$$(67) \qquad D(y, \lambda x) = \lambda D(y,x).$$

Hence if D is twice continuously differentiable with respect to the components of x, Euler's theorem on homogeneous functions may be used to establish the following identities:

$$(68) \qquad D(y,x) = x^T \nabla_x D(y,x);$$

$$(69) \qquad \nabla^2_{xx} D(y,x) x = 0_N;$$

$$(70) \qquad \nabla^2_{yx} D(y,x) x = \nabla_y D(y,x).$$

If the production function f is linearly homogeneous, so that there are constant returns to scale, then it can be shown that $D(y,x)$ defined by (66) is homogeneous of degree minus one in the components of y; i.e., for $\lambda > 0$, we have:

$$(71) \qquad D(\lambda y, x) = \lambda^{-1} D(y,x).$$

Hence if D is twice continuously differentiable, Euler's theorem on homogeneous functions may be used to establish the following identities:

$$(72) \qquad D(y,x) = -y^T \nabla_y D(y,x);$$

$$(73) \qquad \nabla^2_{yy} D(y,x) y = -2\nabla_y D(y,x);$$

$$(74) \qquad y^T \nabla^2_{yx} D(y,x) = -\nabla^T_x D(y,x).$$

---

[29]Some minimal regularity conditions on f will be required to ensure that the maximum in (66) exists. For the one output case ($M = 1$), appropriate regularity conditions and duality theorems may be found in Blackorby, Primont and Russell [1978; 25–26] and in Diewert [1982; 559–561].

In the remainder of this section, we shall assume that production functions are linearly homogeneous so that constant returns to scale prevail.[30]

We first want to find a functional form for $D(y,x)$ which is *flexible* at y^*, x^*; i.e., given an arbitrary twice continuously differentiable input distance function $D^*(y,x)$ which is homogeneous of degree one in x and homogeneous of degree minus one in y, we want to find a functional form for D such that the following equalities hold:

$$D(y^*,x^*) = D^*(y^*,x^*) = 1; \tag{75}$$

$$\nabla_y D(y^*,x^*) = \nabla_y D^*(y^*,x^*); \tag{76}$$

$$\nabla_x D(y^*,x^*) = \nabla_x D^*(y^*,x^*); \tag{77}$$

$$\nabla^2_{yy} D(y^*,x^*) = \nabla^2_{yy} D^*(y^*,x^*); \tag{78}$$

$$\nabla^2_{xx} D(y^*,x^*) = \nabla^2_{xx} D^*(y^*,x^*); \tag{79}$$

$$\nabla^2_{yx} D(y^*,x^*) = \nabla^2_{yx} D^*(y^*,x^*). \tag{80}$$

We have set $D^*(y^*,x^*) = 1$, which means that (y^*,x^*) is on the production surface. The equalities (76)–(80) mean that D is to have the same first and second order partial derivatives as D^* at the point (y^*,x^*).

For a positive output vector $y \equiv (y_1,\ldots,y_M)$, we shall define y^{-1} to be the vector $(y_1^{-1},\ldots,y_M^{-1})$. Now consider the following functional form for D:

$$D(y,x) \equiv [(y^T Ay)^{-1} x^T Cx + \alpha \cdot y^{-1} \beta \cdot x (y^{-1})^T Bx]^{1/2}, \quad A = A^T,\ C = C^T \tag{81}$$

where A, B and C are parameter matrices and α and β are parameter vectors.

It is easy to verify that $D(y,x)$ defined by (81) is homogeneous of degree one in x and homogeneous of degree minus one in y. Thus both D and D^* satisfy the restrictions (68)–(74) at (y^*,x^*).

THEOREM 5. *Let $y^* \gg 0_M$ and $x^* \gg 0_N$. Then for every pair of vectors α and β such that*

$$\alpha \cdot y^{*-1} \neq 0, \qquad \beta \cdot x^* \neq 0, \tag{82}$$

[30] However, the results of this section are still valid for the nonconstant returns to scale case: we need only add an artificial output $M+1$ whose quantity is always equal to one; i.e., define $y_{M+1} \equiv 1$ and redefine the output vector as $y \equiv (y_1,\ldots,y_M,y_{M+1})$. Note that the restriction (72) is implied by (68), (70) and (74).

there exist symmetric matrices A and C such that

$$y^{*T}Ay^* = 1, \qquad x^{*T}Cx^* = 1 \tag{83}$$

and a matrix B such that

$$y^{*-1T}B = 0_N^T, \qquad Bx^* = 0_M \tag{84}$$

and D defined by (81) is flexible at y^*, x^*.

Proof: Define A, B and C in terms of the derivatives of D^* as follows:

$$A \equiv -\nabla^2_{yy}D^*(y^*,x^*) + 3\nabla_y D^*(y^*,x^*)\nabla_y^T D^*(y^*,x^*); \tag{85}$$

$$C \equiv \nabla^2_{xx}D^*(y^*,x^*) + \nabla_x D^*(y^*,x^*)\nabla_x^T D^*(y^*,x^*); \tag{86}$$

$$\begin{aligned} B \equiv (\alpha \cdot y^{*-1})^{-1}(\beta \cdot x^*)^{-1}\hat{y}^{*2}\{&-\nabla^2_{yx}D^*(y^*,x^*) \\ &+ \nabla_y D^*(y^*,x^*)\nabla_x^T D^*(y^*,x^*)\} \end{aligned} \tag{87}$$

where $\hat{y}^*$ is a diagonal matrix with the elements of y^* on the main diagonal and $\hat{y}^{*2} \equiv \hat{y}^*\hat{y}^*$.

We first show that A, B and C defined by (85)–(87) satisfy the restrictions (83) and (84). Postmultiply both sides of (85) by y^* and use (72) and (73) applied to D^* to obtain the following equation:

$$Ay^* = -\nabla_y D^*(y^*,x^*). \tag{88}$$

Now premultiply both sides of (88) by y^{*T}, use (72) applied to D^* and use $D^*(y^*,x^*) = 1$ to obtain the first equality in (83).

Postmultiply both sides of (86) by x^*, use (68) and (69) applied to D^* and obtain:

$$Cx^* = \nabla_x D^*(y^*,x^*). \tag{89}$$

Premultiply both sides of (89) by x^{*T}, use (68) applied to D^* and use $D^*(y^*,x^*) = 1$ to obtain the second equality in (83).

Premultiply both sides of (87) by y^{*-1T}. Note that $y^{*-1T}\hat{y}^{*2} = \hat{y}^{*T}$. Now use (72) and (74) applied to D^* and use $D^*(y^*,x^*) = 1$ to obtain the first set of equations in (84).

Postmultiply both sides of (87) by x^*. Using (68) and (70) applied to D^* and $D^*(y^*,x^*) = 1$, we obtain the second set of equations in (84).

Using the definition of D, (81), and the restrictions (83) and (84), it can be verified that $D(y^*,x^*) = 1$. Hence (75) holds.

Now differentiate the D defined by (81) and evaluate the first order partial derivatives at y^*, x^*. Using the restrictions (83) and (84), we obtain equations (90) and (91) below:

$$\begin{aligned} (90) \qquad \nabla_y D(y^*, x^*) &= -Ay^* \\ &= \nabla_y D^*(y^*, x^*) \quad \text{using (88) and} \end{aligned}$$

$$\begin{aligned} (91) \qquad \nabla_x D(y^*, x^*) &= Cx^* \\ &= \nabla_x D^*(y^*, x^*) \quad \text{using (89).} \end{aligned}$$

Thus (76) and (77) are satisfied.

The matrices of second order derivatives of $D(y^*, x^*)$ are given by equations (92)–(94) below, using the restrictions (83) and (84):

$$(92) \qquad \nabla^2_{yy} D(y^*, x^*) = -A + 3(Ay^*)(Ay^*)^T;$$

$$(93) \qquad \nabla^2_{xx} D(y^*, x^*) = C - (Cx^*)(Cx^*)^T;$$

$$(94) \qquad \nabla^2_{yx} D(y^*, x^*) = -(\alpha \cdot y^{*-1})(\beta \cdot x^*)\hat{y}^{*-2}B - (Ay^*)(Cx^*)^T.$$

If we equate the second order derivatives of D to the corresponding second order derivatives of D^*, then using (76), (77), (90) and (91), it can be seen that the resulting three matrix equations are equivalent to equations (85)–(87), which were used to define A, B and C. Thus equations (78)–(80) are also satisfied and hence D defined by (81) is flexible at y^*, x^*. Q.E.D.

We now allow the firm's production function to depend on time; i.e., in period t, the production function is f^t. We define an *input deflation function* D^t for each production function f^t as follows: for $t = 0, 1$,

$$(95) \qquad D^t(y, x) \equiv \max_{\delta > 0}\{\delta : y_1 \le f^t(y_2, \ldots, y_M, x_1/\delta, x_2/\delta, \ldots, x_N/\delta)\}.$$

In each period t, we assume that the observed output vector y^t and the observed input vector x^t are on the period t production surface, so that we have:

$$(96) \qquad D^t(y^t, x^t) = 1, \qquad t = 0, 1.$$

Following Caves, Christensen and Diewert [1982b; 1396], we define two Malmquist theoretical input indexes Q^{*0} and Q^{*1} as follows:

$$\begin{aligned} (97) \qquad Q^{*0}(x^0, x^1) &\equiv D^0(y^0, x^1)/D^0(y^0, x^0) \\ &= D^0(y^0, x^1) \qquad \text{using (96);} \end{aligned}$$

$$\begin{aligned} (98) \qquad Q^{*1}(x^0, x^1) &\equiv D^1(y^1, x^1)/D^1(y^1, x^0) \\ &= 1/D^1(y^1, x^0) \qquad \text{using (96).} \end{aligned}$$

Let us interpret Q^{*0}. Let $\delta^0 = D^0(y^0, x^1)$. Then using definition (95), we have $y_1^0 = f^0(y_2^0, \ldots, y_M^0, x_1^1/\delta^0, \ldots, x_N^1/\delta^0)$ and by (96), we have $y_1^0 = f^0(y_2^0, \ldots, y_M^0, x_1^0, \ldots, x_N^0)$. Thus the deflated input vector x^1/δ^0 is equivalent to the period 0 input vector x^0 from the viewpoint of the period 0 technology and the deflation factor $\delta^0 = D^0(y^0, x^1) = Q^{*0}(x^0, x^1)$ is a natural measure of the size of x^1 relative to x^0; if $\delta^0 > 1$, then x^1 is bigger than x^0; if $\delta^0 = 1$, then x^0 is equivalent to x^1; and if $\delta^0 < 1$, then x^1 is less than x^0.

The interpretation for (98) is similar but the comparisons are made using the period 1 technology. Let $\delta^1 = 1/D^1(y^1, x^0)$, or $D^1(y^1, x^0) = 1/\delta^1$. Then by definition (95), we have $y_1^1 = f^1(y_2^1, \ldots, y_M^1, \delta^1 x_1^0, \ldots, \delta^1 x_N^0)$ and by (96), we have $y_1^1 = f^1(y_2^1, \ldots, y_M^1, x_1^1, \ldots, x_N^1)$. Thus $\delta^1 x^0$ is equivalent to the period 1 input vector x^1 from the viewpoint of the period 1 technology and $\delta^1 = 1/D^1(y^1, x^0) = Q^{*1}(x^0, x^1)$ is another natural measure for the size of x^1 relative to x^0.

The firm's period t cost function C^t may be defined in terms of the period t production function f^t as follows: for a target output vector $y \equiv (y_1, \ldots, y_M)$ and a given vector of positive input prices w, define

$$(99) \qquad C^t(y, w) \equiv \min_x \{w \cdot x : y_1 \leq f^t(y_2, \ldots, y_M, x)\}, \qquad t = 0, 1.$$

For our next theorem, we will require the hypothesis of cost minimizing behavior on the part of the firm for periods 0 and 1; i.e., if w^t is the observed input price vector for period t and x^t, y^t are the observed input and output vectors respectively for period t, then we assume:

$$(100) \qquad w^t \cdot x^t = C^t(y^t, w^t), \qquad t = 0, 1.$$

Assuming that $D^t(y^t, x^t)$ is differentiable with respect to the components of w, Caves, Christensen and Diewert [1982b; 1397] show that the following equalities hold:

$$(101) \qquad w^t / w^t \cdot x^t = \nabla_x D^t(y^t, x^t), \qquad t = 0, 1.$$

We now assume that the period t input deflation function D^t has the following functional form:

$$(102) \quad D^t(y, x) \equiv [(y^T A^t y)^{-1} x^T C x + \alpha^t \cdot y^{-1} \beta^t \cdot x y^{-1T} B^t x]^{1/2}, \qquad t = 0, 1,$$

where the matrices A^t and C are symmetric.

Note that the D^t defined by (102) are generalizations of the flexible D defined by (81): in (102), the matrices A^t and B^t and the vectors α^t and β^t are now allowed to depend on time, whereas in (81), these parameter matrices and vectors were fixed. However, note that the parameter matrix C in (102) is fixed across the two time periods.

THEOREM 6. *Suppose the parameters of D^0 and D^1 defined by (102) satisfy the following restrictions:*

(103) $$(y^{tT}A^t y^t)^{-1} x^{tT} C x^t = 1, \qquad t = 0, 1;$$

(104) $$(y^0)^{-1T} B^0 = 0_N^T;$$

(105) $$\alpha^1 \cdot (y^1)^{-1} = 0;$$

where the column vectors $(y^t)^{-1} \equiv [(y_1^t)^{-1}, (y_2^t)^{-1}, \ldots, (y_M^t)^{-1}]^T$ for $t = 0, 1$. Suppose also that the firm is engaging in competitive cost minimizing behavior in the two periods so that the relations (96), (100) and (101) hold. Then the Fisher ideal input index Q_F^* is equal to each of the theoretical input indexes Q^{*0} and Q^{*1} defined by (97) and (98); i.e., we have

(106) $$Q_F^*(w^0, w^1, x^0, x^1) = Q^{*0}(x^0, x^1) = Q^{*1}(x^0, x^1).$$

Proof: Note that the restrictions (103)–(105) imply that $D^0(y^0, x^0) = D^1(y^1, x^1) = 1$. Now write the square of the Fisher input index as follows:

$$Q_F^*(w^0, w^1, x^0, x^1)^2 = [(w^0/w^0 \cdot x^0) \cdot x^1]/[(w^1/w^1 \cdot x^1) \cdot x^0]$$

$$= [x^{1T} \nabla_x D^0(y^0, x^0)]/[x^{0T} \nabla_x D^1(y^1, x^1)]$$

using (101)

$$= (x^{1T} C x^0 / y^{0T} A^0 y^0)/(x^{0t} C x^1 / y^{1T} A^1 y^1)$$

using (102)–(105)

(107) $$= (y^{0T} A^0 y^0)^{-1}/(y^{1T} A^1 y^1)^{-1}$$

using $C = C^T$

$$= (y^{0T} A y^0)^{-1} x^{1T} C x^1 / (y^{1T} A^1 y^1)^{-1} x^{1T} C x^1$$

$$= (y^0 A y^0)^{-1} x^{1T} C x^1$$

using (103)

$$= [D^0(y^0, x^1)]^2$$

using (102) and (104)

(108) $$= [Q^{*0}(x^0, x^1)]^2$$

using (97)

$$= (y^{0T} A^0 y^0)^{-1} x^{0T} C x^0 / (y^{1T} A^1 y^1)^{-1} x^{0T} C x^0$$

using (107)

$$= 1/(y^{1T} A^1 y^1)^{-1} x^{0T} C x^0$$

using (103)

$$= 1/[D^1(y^1, x^0)]^2$$

using (102) and (105)

(109) $$= [Q^{*1}(x^0, x^1)]^2$$

using (98).

Taking square roots of (108) and (109) yields (106). Q.E.D.

COROLLARY. *The above theorem holds if the restrictions (104) and (105) are replaced by the following restrictions:*

$$(y^1)^{-1T} B^1 = 0_N^T; \tag{110}$$

$$\alpha^0 \cdot (y^0)^{-1} = 0. \tag{111}$$

The proof follows by a series of computations similar to (107) through (109).

We note that the restrictions (104) are consistent with D^0 being flexible at (y^0, x^0) while the restrictions (110) are consistent with D^1 being flexible at (y^1, x^1).

Note that the above theorem does not require optimizing behavior with respect to outputs.

Theorem 6 is a Fisher input index counterpart to the translog input index justification derived by Caves, Christensen and Diewert [1982b; 1398] who showed that under certain conditions, a geometric mean of Q^{*0} and Q^{*1} was equal to the translog input index Q_T^* defined by (64). Our present result is perhaps marginally better in that we no longer have to take a geometric mean of Q^{*0} and Q^{*1}.

We turn now to a parallel discussion of output indexes.

7. An Economic Approach to Indexes of Real Output

Our theoretical treatment of output indexes follows that of Caves, Christensen and Diewert [1982b; 1399–1401]. We now represent the technology in period t by means of an input requirements function g^t where $x_1 = g^t(y_1, \ldots, y_M, x_2, \ldots, x_N)$ is the minimum amount of input one required to produce the vector of outputs $y \equiv (y_1, \ldots, y_M)$ given that x_n units of input n are available for $n = 2, 3, \ldots, N$.

The period t *output deflation or distance function* d^t is defined as follows:

$$d^t(y, x) \equiv \min_{\delta > 0}\{\delta : g^t(y_1/\delta, \ldots, y_M/\delta, x_2, \ldots, x_N) \leq x_1\}, \quad t = 0, 1. \tag{112}$$

It is straightforward to show that $d^t(y, x)$ must be homogeneous of degree one in the components of y. Moreover, if the period t technology exhibits constant returns to scale, then g^t is homogeneous of degree one in its arguments and it can be shown that $d^t(y, x)$ must be homogeneous of degree minus one in the components of x. Thus in the constant returns to scale case, the regularity conditions on the output deflation functions d^t are the reverse of the regularity conditions on the input deflation function $D(y,x)$ which was defined in the previous section. Thus in the differentiable case, conditions analogous to (67)–(74) hold for d^t except that the roles of x and y are interchanged.

For $x \gg 0$, define the vector $x^{-1} \equiv (x_1^{-1}, \ldots, x_N^{-1})$ and define the following distance function:

$$(113)\quad d(y,x) \equiv [y^T Ay(x^T Cx)^{-1} + \alpha \cdot y\beta \cdot x^{-1} y^T Bx^{-1}]^{1/2}, \quad A = A^T,\ C = C^T$$

where A, B and C are matrices of parameters and α and β are vectors of parameters.

The following theorem is analogous to Theorem 5 and can be proven in the same manner.

THEOREM 7. *Let $y^* \gg 0_M$ and $x^* \gg 0_N$. Then for every α and β such that*

$$(114)\qquad \alpha \cdot y^* \neq 0, \qquad \beta \cdot x^{*-1} \neq 0,$$

there exist symmetric matrices A and C such that

$$(115)\qquad y^{*T} Ay^* = 1, \qquad x^{*T} Cx^* = 1$$

and a matrix B such that

$$(116)\qquad y^{*T} B = 0_N^T, \qquad Bx^{*-1} = 0_M$$

and the output deflation function d defined by (113) is flexible at y^*, x^*.

If we wish to relax the assumption of a constant returns to scale technology, then we need only add an extra input to the list of inputs and fix its level; i.e., define $x_{N+1} \equiv 1$. Then $d(y,x)$ defined by (113) (where x is now an $N+1$ dimensional vector) will be a flexible output deflation function in the class of nonconstant returns to scale technologies.

We shall assume that the observed period t output and input vectors, y^t and x^t respectively, are efficient relative to the firm's period t technology; i.e., we assume that $x_1^t = g^t(y^t, x_2^t, \ldots, x_N^t)$ for $t = 0, 1$. Then by (112), we shall have:

$$(117)\qquad d^t(y^t, x^t) = 1, \qquad t = 0, 1.$$

Following the example of Caves, Christensen and Diewert [1982b; 1400], we use the output deflation functions d^t in order to define the following theoretical Malmquist[31] output indexes Q^0 and Q^1:

$$(118)\qquad Q^0(y^0, y^1) \equiv d^0(y^1, x^0)/d^0(y^0, x^0) = d^0(y^1, x^0) \qquad \text{using (117);}$$

$$(119)\qquad Q^1(y^0, y^1) \equiv d^1(y^1, x^1)/d^1(y^0, x^1) = 1/d^1(y^0, x^1) \qquad \text{using (117).}$$

[31]The basic idea of these indexes appears to be in Hicks [1961] [1981; 256] and they are defined geometrically in the two output case by Moorsteen [1961; 452].

An interpretation for the theoretical index defined by (118) runs as follows. Let $\delta^0 = d^0(y^1, x^0)$. Then by definition (112), we have $x_1^0 = g^0(y^1/\delta^0, x_2^0, \ldots, x_N^0)$. From (117) we have $x_1^0 = g^0(y^0, x_2^0, \ldots, x_N^0)$. Thus the deflated period 1 output vector y^1/δ^0 is "equivalent" using the period 0 technology to the period 0 output vector y^0, and the deflation factor $\delta^0 = d^0(y^1, x^0) = Q^0(y^0, y^1)$ is a natural scalar measure of the size of y^1 relative to y^0.

Similarly, let $1/d^1 = d^1(y^0, x^1)$. Then by definition (112), $x_1^1 = g^1(y^0\delta^1, x_2^1, \ldots, x_N^1)$ and by (117), $x_1^1 = g^1(y^1, x_2^1, \ldots, x_N^1)$. Thus $\delta^1 y^0$ is "equivalent" to y^1 and $\delta^1 = 1/d^1(y^0, x^1) = Q^1(y^0, y^1)$ is a natural scalar measure of the size of y^1 relative to y^0.

The firm's *period t revenue function* $r^t(p, x)$ may be defined in terms of the factor requirements function g^t as follows: given a vector of inputs $x \equiv (x_1, \ldots, x_N)$ and a vector of positive output prices $p \equiv (p_1, \ldots, p_M)$, define

$$r^t(p, x) \equiv \max_y \{p \cdot y : g^t(y, x_2, \ldots, x_N) \leq x_1\}, \qquad t = 0, 1. \tag{120}$$

For our next theorem, we shall require the hypothesis of revenue maximizing behavior on the part of the firm for periods 0 and 1; i.e., if p^t is the observed output price vector for period t and x^t, y^t are the observed input and output vectors respectively for period t, then we shall assume:

$$p^t \cdot y^t = r^t(p^t, x^t), \qquad t = 0, 1. \tag{121}$$

Assuming that $d^t(y^t, x^t)$ was differentiable with respect to the components of y, Caves, Christensen and Diewert [1982b; 1401] showed that the following equalities hold:

$$p^t/p^t \cdot y^t = \nabla_y d^t(y^t, x^t), \qquad t = 0, 1. \tag{122}$$

We may now prove the following theorem which is an exact counterpart to Theorem 6 in the previous section.

THEOREM 8. *Suppose that the firm's output deflation function in period t, d^t, has the following functional form for $t = 0, 1$:*

$$d^t(y, x) \equiv [y^T A y (x^T C^t x)^{-1} + \alpha^t \cdot y \beta^t \cdot x^{-1} y^T B^t x^{-1}]^{1/2} \tag{123}$$

where A and C^t are symmetric matrices and the parameters in (123) satisfy the following restrictions:

$$y^{tT} A y^t (x^{tT} C^t x^t)^{-1} = 1, \qquad t = 0, 1 \tag{124}$$

and <u>either</u> the following restrictions are satisfied:

$$B^0 (x^0)^{-1} = 0_M; \qquad \beta^1 \cdot (x^1)^{-1} = 0, \tag{125}$$

or the following restrictions are satisfied:

$$B^1(x^1)^{-1} = 0_M; \qquad \beta^0 \cdot (x^0)^{-1} = 0. \tag{126}$$

Suppose also that the firm is engaging in competitive revenue maximizing behavior in periods 0 and 1 so that the relations (121), (122) and (117) hold. Then the Fisher output index Q_F defined by (4) is equal to each of the theoretical indexes Q^0 and Q^1 defined by (118) and (119); i.e., we have:

$$Q_F(p^0, p^1, y^0, y^1) = Q^0(y^0, y^1) = Q^1(y^0, y^1). \tag{127}$$

The proof is analogous to the proof of Theorem 6 in the previous section. We note that the restrictions (125) are consistent with d^0 being flexible at (y^0, x^0) while the restrictions (126) are consistent with d^1 being flexible at (y^1, x^1).

Theorem 8 is a Fisher index analogue to Theorem 2 in Caves, Christensen and Diewert [1982b; 1401] which showed that the translog or Törnqvist output index Q_T defined by (5) was equal to the geometric mean of Q^0 and Q^1 provided that the firm's distance functions d^t had the translog functional form with identical quadratic coefficients for the second order terms in $\ln y_1, \ldots, \ln y_M$. Note that our present result does not require us to take a geometric mean of Q^0 and Q^1. Note also that the matrix A which has the quadratic terms in y in $d^0(y, x)$ and $d^1(y, x)$ defined by (123) is constant across time periods. Thus in both Theorem 8 and Theorem 2 of Caves, Christensen and Diewert, the technologies in the two time periods cannot be completely different.

8. An Alternative Economic Approach to Productivity Indexes

We can draw on the material of the previous two sections to define productivity or efficiency indexes.

One approach would be to define a productivity index to be the ratio of the Malmquist output index $Q^0(y^0, y^1)$ or $Q^1(y^0, y^1)$ defined by (118) and (119) divided by a Malmquist input index $Q^{*0}(x^0, x^1)$ or $Q^{*1}(x^0, x^1)$ defined by (97) or (98). Thus there are four possible theoretical productivity indexes that could be defined in this manner. In the two input, two output case, these theoretical efficiency change indexes were suggested by Moorsteen [1961; 462] and perhaps by Hicks [1961; footnote 4] in the general multiple input and output case.

Rather than follow this Hicks-Moorsteen approach to productivity indexes, we shall follow the approach taken by Caves, Christensen and Diewert [1982b; 1401–1408] and use only the output deflation functions $d^t(y, x)$ defined in the previous section by (112) in order to define theoretical productivity indexes.

We assume that in each period, either the firm's technology is subject to constant returns to scale or it is subject to diminishing returns to scale but the firm's pure profits are imputed to an artificial fixed factor.

Let x^t and y^t be the observed input and output vectors for period t and let $d^t(y,x)$ be the firm's period t output deflation function defined by (112) for $t = 0,1$. Following Caves, Christensen and Diewert [1982b; 1402], define the following *theoretical productivity indexes* Π^0 and Π^1:

$$\text{(128)} \qquad \begin{aligned} \Pi^0(x^0,x^1,y^0,y^1) &\equiv d^0(y^1,x^1)/d^0(y^0,x^0) \\ &= d^0(y^1,x^1) \qquad \text{using (117);} \end{aligned}$$

$$\text{(129)} \qquad \begin{aligned} \Pi^1(x^0,x^1,y^0,y^1) &\equiv d^1(y^1,x^1)/d^1(y^0,x^0) \\ &= 1/d^1(y^0,x^0) \qquad \text{using (117).} \end{aligned}$$

We can interpret (128) as follows. Let $\delta^0 = d^0(y^1,x^1)$. Then by definition (112), $x_1^1 = g^0(y^1/\delta^0, x_2^1,\ldots,x_N^1)$. Thus the deflated period 1 output vector g^1/δ^0 and the period 1 input vector x^1 are on the production surface for period 0. Of course, we also have $x_1^0 = g^0(y^0,x_2^0,\ldots,x_N^0)$ so that the period 0 output vector y^0 and the period 0 input vector x^0 are on the period 0 production surface. If there were a productivity improvement going from period 0 to 1, we would expect that the deflation factor δ^0 would be greater than 1 and $\delta^0 = d^0(y^1,x^1) = \Pi^0(x^0,x^1,y^0,y^1)$ can serve as a measure of the magnitude of the productivity improvement. If $\delta^0 = 1$, then the period 1 output-input combination (y^1,x^1) is on the period 0 production surface and there has been no efficiency improvement.

To interpret (129), let $1/\delta^1 = d^1(y^0,x^0)$. Then by definition (112), $x_1^0 = g^1(y^0\delta^1,x_2^0,\ldots,x_N^0)$. This means that the inflated period 0 output vector $\delta^1 y^0 \equiv (\delta^1 y_1^0,\ldots,\delta^1 y_M^0)$ and the period 0 input vector x^0 are on the production surface for period 1. Thus using the period 0 input vector, the period 1 technology can produce δ^1 times the period 0 output vector. Hence the blow up factor $\delta^1 = 1/d^1(y^0,x^0) = \Pi^1(x^0,x^1,y^0,y^1)$ can serve as an index of the productivity improvement.

In Theorem 9 below, we shall again assume revenue maximizing behavior as in the previous section. Thus (121) and (122) must hold if $d^t(y^t,x^t)$ is differentiable. We shall also assume cost minimizing behavior as in Section 6. Under the assumption of cost minimizing behavior and a constant returns to scale technology, Caves, Christensen and Diewert [1982b; 1403–1404] show that the following relations must hold if d^t is differentiable:

$$\text{(130)} \qquad \nabla_x d^t(y^t,x^t) = -w^t/w^t\cdot x^t, \qquad t=0,1.$$

In the following theorem, we shall use more general versions of the output deflation function $d(y,x)$ defined by (81), which was shown to be a flexible functional form in Theorem 7.

THEOREM 9. *Suppose that the firm's output deflation function in period t, d^t, is defined as follows for $t = 0, 1$:*

$$(131)\quad d^t(y,x) \equiv \sigma^t[y^T Ay(x^T Cx)^{-1} + \alpha^t \cdot y\beta^t \cdot x^{-1} y^T B^t x^{-1}]^{1/2}, \quad A = A^T,\ C = C^T$$

where the parameter matrices A, B^t and C, the parameter vectors α^t and β^t and the parameter scalars σ^t satisfy the following restrictions:

$$(132)\quad \sigma^t[y^{tT} Ay^t(x^{tT} Cx^t)^{-1}]^{1/2} = 1, \qquad t = 0, 1$$

and either the following restrictions are satisfied:

$$(133)\quad \begin{array}{c} B^0(x^0)^{-1} = 0_M; \quad y^{0T} B^0 = 0_N^T; \quad \alpha^1 \cdot y^1 = 0; \\ \beta^1 \cdot (x^1)^{-1} = 0; \quad \alpha^0 \cdot y^1 \beta^0 \cdot (x^1)^{-1} = 0 \quad \text{and} \quad y^{0T} B^1 (x^0)^{-1} = 0 \end{array}$$

or the following restrictions are satisfied:

$$(134)\quad \begin{array}{c} B^1(x^1)^{-1} = 0_M; \quad y^{1T} B^1 = 0_N^T; \quad \alpha^0 \cdot y^0 = 0; \\ \beta^0 \cdot (x^0)^{-1} = 0; \quad \alpha^1 \cdot y^0 \beta^1 \cdot (x^0)^{-1} = 0 \quad \text{and} \quad y^{1T} B^0 (x^1)^{-1} = 0. \end{array}$$

Suppose also that the firm competitively maximizes revenues given inputs and competitively minimizes costs given outputs in each period.[32] *Then the Fisher productivity index Pr_F defined by (6) is equal to each of the theoretical productivity indexes Π^0 and Π^1 defined by (128) and (129); i.e., we have*

$$(135)\quad Q_F(p^0,p^1,y^0,y^1)/Q_F^*(w^0,w^1,x^0,x^1) = \Pi^0(x^0,x^1,y^0,y^1) = \Pi^1(x^0,x^1,y^0,y^1).$$

Proof: The restrictions (132) and (133) or (132) and (134) imply that d^t defined by (131) satisfies the following restriction:

$$(136)\quad d^t(y^t,x^t) = 1, \qquad t = 0, 1.$$

The square of the Fisher productivity index is

$$[Q_F(p^0,p^1,y^0,y^1)/Q_F^*(w^0,w^1,x^0,x^1)]^2$$
$$= \Big[[(p^0/p^0 \cdot y^0) \cdot y^1]/[(p^1/p^1 \cdot y^1) \cdot y^0]\Big] \Big/ \Big[[(w^0/w^0 \cdot x^0) \cdot x^1]/[(w^1/w^1 \cdot x^1) \cdot x^0]\Big]$$

[32]These assumptions are implied by the assumption of competitive profit maximizing behavior in each period.

$= [y^1 \cdot \nabla_y d^0(y^0, x^0)/y^0 \cdot \nabla_y d^1(y^1, x^1)]$

$/[(-1)x^1 \cdot \nabla_x d^0(y^0, x^0)/(-1)x^0 \cdot \nabla_x d^1(y^1, x^1)]$

using (122) and (130)

$= [[(\sigma^0)^2 y^{1T} A y^0 / x^{0T} C x^0]/[(\sigma^1)^2 y^{0T} A y^0 / x^{1T} C x^1]]$

$/ \ [[(\sigma^0)^2 x^{1T} C x^0 y^{0T} A y^0 (x^{0T} C x^0)^{-2}]/[(\sigma^1)^2 x^{0T} C x^1 y^{1T} A y^1 (x^{1T} C x^1)^{-2}]]$

differentiating (131) and using (132)–(134)

(137)
$= (\sigma^0)^2/(\sigma^1)^2$ using $A = A^T$, $C = C^T$ and (132)

$= (\sigma^0)^2 y^{1T} A y^1 (x^{1T} C x^1)^{-1} / (\sigma^1)^2 y^{1T} A y^1 (x^{1T} C x^1)^{-1}$

$= (\sigma^0)^2 y^{1T} A y^1 (x^{1T} C x^1)^{-1}$ using (132)

$= (\sigma^0)^2 [y^{1T} A y^1 (x^{1T} C x^1)^{-1} + \alpha^0 \cdot y^1 \beta^0 \cdot (x^1)^{-1} y^{1T} B^0 (x^1)^{-1}]$

using either (133) or (134)

$= [d^0(y^1, x^1)]^2$ by definition (131)

(138)
$= (\Pi^0(x^0, x^1, y^0, y^1)]^2$ using (128)

$= (\sigma^0)^2 y^{0T} A y^0 (x^{0T} C x^0)^{-1} / (\sigma^1)^2 y^{0T} A y^0 (x^{0T} C x^0)^{-1}$ using (137)

$= 1/(\sigma^1)^2 y^{0T} A y^0 (x^{0T} C x^0)^{-1}$ using (132)

$= 1/(\sigma^1)^2 [y^{0T} A y^0 (x^{0T} C x^0)^{-1} + \alpha^1 \cdot y^0 \beta^1 \cdot (x^0)^{-1} y^{0T} B^1 (x^0)^{-1}]$

using (133) or (134)

$= 1/[d^1(y^0, x^0)]^2$ by definition (131)

(139)
$= [\Pi^1(x^0, x^1, y^0, y^1)]^2$ using (129).

Q.E.D.

Taking square roots of (138) and (139) yields (135).

The restrictions (133) are consistent with d^0 being flexible at (y^0, x^0) and the restrictions (134) are consistent with d^1 being flexible at (y^1, x^1). Thus the above Theorem is a fairly close analogue to Theorem 3 of Caves, Christensen and Diewert [1982b; 1404] which showed that the translog productivity index

Pr_T defined by (7) was exactly equal to the geometric mean of the theoretical productivity indexes Π^0 and Π^1 defined by (128) and (129) provided that the distance functions d^0 and d^1 had translog functional forms (with identical second order coefficients). Our present result is perhaps a bit stronger in that we find that Pr_F is equal to Π^0 and Π^1 (i.e., we do not have to take a geometric mean of Π^0 and Π^1).

Theorem 9 provides yet another strong justification for the use of the Fisher productivity index.

9. Conclusion

We have presented a number of justifications for the use of the Fisher output index Q_F, the Fisher input index Q_F^* and the Fisher productivity index $Pr_F \equiv Q_F/Q_F^*$. Theorem 1 presents a strong justification for the use of these indexes from the viewpoint of the test approach.

Using the economic approach to index number theory, Theorem 6 justifies the use of the Fisher input index while Theorem 8 justifies the use of the Fisher output index. Theorems 4 and 9 provide strong justifications for the use of the Fisher productivity index from the viewpoint of the economic approach.[33]

If we compare the test approach to productivity indexes with the economic approach, the following points emerge: (i) the test approach suffers from a lack of consensus on what the appropriate tests or axioms should be; (ii) the economic approach requires the assumption of competitive price taking behavior and the assumption of constant (or diminishing) returns to scale. Thus both approaches have their weaknesses. However, we have presented strong justifications for the use of the Fisher productivity index from *both* viewpoints, which should reduce objections to its use in many contexts.

We conclude by noting that both the economic and test approaches to the measurement of productivity change make assumptions which are often not satisfied in empirical contexts such as: (i) all prices and quantities in the two periods are known with certainty;[34] and (ii) there are no new inputs that are used in period 1 but not in period 0 and there are no new outputs that are produced in period 1. With respect to the new good problem, the economic

[33]Many years ago, Denny [1980; 537] asked whether some of my translog exact aggregation results could be generalized to other functional forms. At the time, I could not answer his question, but now Theorems 4, 6, 8 and 9 show that his question has an affirmative answer.

[34]Students of accounting will know that this assumption is very suspect. For example, consider the problems in estimating user costs for durable capital inputs when there is rapid inflation and tax complications.

approach can at least provide a theoretical solution to the problem[35] but the empirical difficulties remain formidable.

[35]In period 0, the quantity of the new good is obviously zero and its price is set equal to that shadow price which would just cause the firm to demand or supply a zero amount of that good in period 0. The basic idea is due to Hicks [1940; 114]; see also Fisher and Shell [1972b; 101] and Diewert [1980; 498–503].

Essays in Index Number Theory, Volume I
W.E. Diewert and A.O. Nakamura (Editors)

Chapter 14
SYMMETRIC MEANS AND CHOICE UNDER UNCERTAINTY*

W.E. Diewert

1. Introduction

A symmetric mean of N numbers, $x_1, x_2, \ldots, x_N$, is simply an average of the N numbers, where each x_i is treated in the same manner in forming the average. More formally, a symmetric mean is a function of N variables, $M(x_1, x_2, \ldots, x_N)$, which has at least two properties: (i) $M(k, k, \ldots, k) = k$ and (ii) $M(Px) = M(x)$ where $x \equiv (x_1, \ldots, x_N)$ and Px denotes a permutation of the components of the vector x. Property (i) expresses the fundamental property that a mean or average must have: if all of the N numbers being averaged are equal to a common value k, then the mean must also equal this common value. Property (ii) expresses mathematically the symmetry property of a symmetric mean.

Symmetric means occur in the economics literature in at least three distinct areas: (i) index number theory; (ii) welfare economics and the measurement of inequality; and (iii) the specification of a decision maker's preferences over equally likely uncertain alternatives. We shall give a brief discussion of each application in this introduction.

The use of symmetric means in index number theory dates back at least to the pioneering work of Irving Fisher [1922] who used symmetric means to take averages of prices to form price indexes. Fisher [1922; 63] called the symmetry property of the mean "the commodity reversal test." More recently, Samuelson and Swamy [1974; 581–582] showed that a symmetric mean of the Paasche and Laspeyres price indexes approximates the true cost of living index (in the case of homothetic preferences) to the second order.[1] Diewert [1978b; 897] showed

*This research was supported by a Strategic Grant from the Social Science and Humanities Research Council of Canada. Thanks are due Shelley Hey and Louise Hebert for typing a difficult manuscript, and to J. Aczél, W. Bossert, D. Donaldson and L.G. Epstein for valuable comments and to C. Blackorby for valuable discussions.

[1]They also referred to an earlier paper by Samuelson [1953] who showed that a symmetric mean of overcompensated and undercompensated demand functions approximated a consumer's system of Hicksian demand functions to the second order.

that a twice continuously differentiable symmetric mean of the Paasche and Laspeyres price indexes approximates a superlative price index to the second order in a certain sense.[2]

The use of symmetric means in welfare economics and the measurement of inequality can be traced back to the work of Kolm [1969] [1976a] [1976b]. Let x_i denote the income of the ith individual and let $W(x_1, x_2, \ldots, x_N)$ denote the aggregate social welfare of N individuals having incomes $x_1, x_2, \ldots, x_N$. In order to treat each individual fairly, it is necessary that the social welfare function W be symmetric in its variables. Kolm [1969; 186] defined society's *equally distributed equivalent income* as the number $m \equiv M(x_1, \ldots, x_N)$ which satisfies the equation $W(m1_N) = W(x_1, \ldots, x_N)$ where 1_N is a vector of ones of dimension N.[3] The symmetry of the social welfare function W implies the symmetry of the equivalent income function M. In addition, it can be seen that M has the mean property; i.e., $M(\lambda 1_N) = \lambda$. Thus M is a symmetric mean. In fact, M is simply a convenient cardinalization of the (ordinal) social welfare function W.[4]

Given the symmetric mean M, Kolm [1969; 186] [1976a; 438] defined an index of *relative inequality* by

$$I_r(x_1, \ldots, x_N) \equiv 1 - [M(x_1, \ldots, x_N)/\bar{x}] \tag{1}$$

where $\bar{x} \equiv \sum_{i=1}^{N} x_i/N$ is the arithmetic mean of the x_i. Kolm also defined an index of *absolute inequality*[5] by

$$I_a(x_1, \ldots, x_N) \equiv \bar{x} - M(x_1, \ldots, x_N). \tag{2}$$

---

[2]A superlative price index is exact for an aggregator function that can provide a second order approximation to an arbitrary twice continuously differentiable linearly homogeneous aggregator function; see Diewert [1976a] [1978b].

[3]Sufficient conditions that guarantee the existence of a unique solution to the equation are that W be continuous and increasing along the equal incomes ray. Other terms used to describe the equal equivalent income include Atkinson's [1970; 250] "equally distributed equivalent income" and Blackorby and Donaldson's [1980; 118] "representative income".

[4]Kolm [1976a; 437] notes that M is simply an increasing transformation of W as do Blackorby, Donaldson and Auersperg [1981; 668] and Blackorby and Donaldson [1984; 17].

[5]Kolm [1969; 186] initially called I_r an "index of relative injustice" and I_a a "monetary measure of injustice". If M is a mean of order r (see Hardy, Littlewood and Polya [1934; 12] for a definition), then I_r defined by (1) becomes a member of the Kolm [1969; 194]-Atkinson [1970] family of inequality measures. On the other hand, if M is a Nagumo [1930] mean of order s, then I_a defined by (2) becomes a member of what Blackorby and Donaldson [1980; 117] [1982; 257] call the Kolm [1969; 194] [1976a; 419] [1976b; 95]-Pollak [1971b] family of inequality measures. Kolm [1976a; 417-419] identified these special cases of (1)

It can be seen that if M is (positively) ***linearly homogeneous*** (i.e., $M(\lambda x) = \lambda M(x)$ for $x \gg 0_N$ and $\lambda > 0$), then I_r is invariant to proportional shifts in the income distribution (i.e., $I_r(\lambda x) = I_r(x)$ for $x \gg 0_N$ and $\lambda > 0$). On the other hand, if M is ***unit translatable***[6] (i.e., $M(x+\lambda 1_N) = M(x)+\lambda$ for all $x \geq 0_N$ and λ such that $x+\lambda 1_N \geq 0_N$), then I_a is invariant to additive shifts in the income distribution; i.e., $I_a(x + \lambda 1_N) = I_a(x)$. Two other mathematical properties that are often imposed on M in the inequality literature are: (i) continuity and (ii) increasingness; i.e., $M(x_1, x_2, \ldots, x_N)$ is an increasing function of each of its variables.

Other properties that have been imposed on M in the inequality literature are properties that express an aversion to inequality. The first such property was Dalton's [1920; 352] ***principle of transfers***: if a transfer of income takes place from the richer to the poorer without reversing their ranks, inequality must be diminished. Other inequality aversion axioms have been proposed by Kolm [1969; 188–190] [1976a; 441–442] [1976b; 87–93], Atkinson [1970; 247], Dasgupta, Sen and Starrett [1973; 181–184] and Rothschild and Stiglitz [1973; 189–196]. We shall study some of these inequality aversion axioms[7] in more detail in Section 2 below.

Thus far, we have assumed that the x_i are individual incomes. If we are willing to assume that individual utilities can be measured and are interpersonally comparable, then the x_i can be interpreted as "real incomes" or "utilities" and $M(x_1, \ldots, x_N)$ can be interpreted as a *social evaluation function*; see Blackorby and Donaldson [1982] [1984] for applications of this framework and references to the literature.

Turning now to the use of symmetric means in the context of choice under uncertainty, Atkinson [1970; 247] pointed out that Dalton's principle of transfers was identical to the concept of a mean preserving spread introduced by Rothschild and Stiglitz [1970]. The latter concept was introduced in order to

---

and (2) as "rightist" and "leftist" measures of inequality. Means of order r and s will be considered in more detail in Section 4 below. Other terms for I_r and I_a include Bossert and Pfingsten's [1990; 125-126] "relative and absolute waste of income due to inequality." Finally, Blackorby, Donaldson and Auersperg [1981; 669] call I_a a "per capita index of inequality".

[6] The concept is due to Nagumo [1930; 77] and the term "unit translatable" is due to Blackorby and Donaldson [1980; 109] [1982; 254]. Aczél [1966; 151] uses the term "translativity" to describe this property.

[7] It should be noted that much of the inequality literature proposes axioms on I_r or I_a defined by (1) and (2) rather than on the underlying (cardinal) social welfare function M; e.g. Eichhorn and Gehrig [1982; 681–683] list 14 possible axioms that could be imposed on inequality measures. Of course, it is usually straightforward to translate these axioms on I_r or I_a into axioms on M.

determine when a random variable was riskier or had a higher variance than another random variable which had the same mean. Many other authors pointed out that various inequality aversion axioms on a social welfare function had risk aversion counterparts in the context of a decision maker's preferences over uncertain choices; e.g. see Kolm [1976a; 421–422], Machina [1982; 312–313] and Chew and Epstein [1989a; 218–223].

To see more directly how symmetric means can play a role in the modeling of uncertain choices, assume that a decision maker has a *certainty utility function* $f(y)$ where y is a choice vector belonging to some set S. There are N states of nature and each state is equally likely and hence has probability $1/N$. Denote the consumption vector of the decision maker if state i occurs by y^i for $i = 1, 2, \ldots, N$. Then following Arrow [1953; 41] and Debreu [1959; 101], it is natural to assume that the decision maker's preferences over state contingent commodities can be represented by a continuous utility function $F(y^1, y^2, \ldots, y^N)$. Following Samuelson [1952; 674], it is also natural to assume that the state contingent preference function F has the following structure:

$$(3) \qquad F(y^1, y^2, \ldots, y^N) = M[f(y^1), f(y^2), \ldots, f(y^N)]$$

where f is the certainty utility function and M is a function which expresses the decision maker's attitude towards uncertainty. If the state contingent consumption vectors are all equal, so that $y^1 = y^2 = \cdots = y^N \equiv y$, then it does not matter which state of nature occurs. Under these circumstances, it is natural to require that

$$(4) \qquad M[f(y), f(y), \ldots, f(y)] = f(y);$$

i.e., the state contingent preferences collapse down to the certainty preferences when all of the state contingent decision vectors y^i are forced to be the same vector y. But property (4) implies that the *stochastic preference function* M is a mean. Furthermore, since the states of nature are equally probable, it is natural to require M to be a symmetric function. Thus the stochastic preference function M must be a *symmetric mean* under the above hypotheses.

The above (equally probable) Arrow-Debreu state contingent preference model was used by Blackorby, Davidson and Donaldson [1977; 352–354] in their proof of the Expected Utility Theorem. In addition, they placed a separability assumption on F and then applied a representation theorem due to Gorman [1968a] to deduce an additively separable functional form for M; i.e., under Blackorby, Davidson and Donaldson's assumptions, M can be written as follows:[8]

$$(5) \qquad M(x_1, x_2, \ldots, x_N) = \phi^{-1}\left[\sum_{i=1}^{N} (1/N)\phi(x_i)\right]$$

[8]Blackorby, Donaldson and Auersperg [1981; 670] also used separability assumptions and Gorman's [1968a] Theorem in order to derive a social evaluation function of the form (5).

where $x_i \equiv f(y^i)$ for $i = 1, 2, \ldots, N$ and ϕ is a continuous, increasing function of one variable.

The right hand side of (5) defines what we will call a *separable symmetric mean*. Axiomatic characterizations for this class of means were originally derived by Nagumo [1930] and Kolmogoroff [1930]. However, these authors did not provide a list of axioms that characterize a separable symmetric mean involving only N variables; rather, they provided axioms that characterize an entire family of separable symmetric means $M^n(x_1, x_2, \ldots, x_n)$ where n is any positive integer. The key axiom in their characterizations was the following one: for all positive integers m and n,

$$(6)\quad M^{m+n}(x_1, \ldots, x_m, y_1, \ldots, y_n) = M^{m+n}[M^m(x_1, \ldots, x_m)1_m, y_1, \ldots, y_n].$$

Property (6) is a *consistency in aggregation* property. It says that when calculating the mean of $m+n$ numbers $x_1, \ldots, x_m, y_1, \ldots, y_n$, we can aggregate the first m numbers by calculating the mean $M^m(x_1, \ldots, x_m)$ and then each of the numbers x_i can be replaced by $M^m(x_1, \ldots, x_m)$ and $M^{m+n}[M^m(x_1, \ldots, x_m), \ldots, M^m(x_1, \ldots, x_m), y_1, \ldots, y_n]$ will equal $M^{m+n}(x_1, \ldots, x_m, y_1, \ldots, y_n)$. It is interesting to note that Blackorby and Donaldson [1984; 19] independently utilized the above consistency in aggregation axiom in order to determine a family of consistent social evaluation functions over variable populations.[9]

The axiomatic characterizations of separable symmetric means derived by Nagumo [1930] and Kolmogoroff [1930] are not completely satisfactory for all purposes; for example, we may want to characterize this family of means for a fixed number of independent variables N rather than for all possible combinations of variables. Sections 3 and 4 below will be devoted to providing alternative axiomatic characterizations for this class of means.[10]

Returning to the choice under uncertainty problem, it turns out that stochastic preference functions M of the additively separable form given by (5)

---

[9] They called the axiom the "population substitution principle".

[10] Hardy, Littlewood and Polya [1934; 65–69 and 158–163] provided an alternative axiomatic characterization using the Stieltjes integral. Unfortunately, none of the early authors provided a satisfactory name for the class of separable symmetric means: Nagumo had no name at all, Kolmogoroff [1930; 388] used the term "regular mean" while Hardy, Littlewood and Polya [1934; 65] used the awkward phrase "mean values with an arbitrary function". Turning to more recent authors, Eichhorn [1978b; 32] uses the term "quasilinear mean" to describe the function $\phi^{-1}\left[\sum_{i=1}^N a_i\phi(x_i)\right]$ where $a_i > 0$ and $\sum_{i=1}^N a_i = 1$ while Bossert and Pfingsten [1990; 129] use the term "quasiarithmetic mean" to describe (5). These last two terms were also used by Aczél [1966; 279].

cannot provide an adequate empirical description of actual choices made under uncertainty, as the pioneering work of Allais [1953; 527–530] made clear.[11] However, if we simply allow M to be an arbitrary symmetric mean as in (3), the resulting class of state contingent preference functions F is too general in a certain sense. The problem is that (3) does not indicate to us how the decision maker's preferences depend on the probabilities of the N stochastic events. The problem will become clear in Section 7 below when we follow the example of Blackorby, Davidson and Donaldson [1977] and show how the separable model (5) can be generalized to deal with the unequal probabilities case. It does not seem possible to achieve a corresponding generalization of model (3) to cover the unequal probabilities case. Thus we require a class of stochastic preference functions M which are more general than the class of separable symmetric means used in (5) but are less general than the class of symmetric means used in (3). We shall show that an appropriate class of mean functions which will do the job is the class of *implicitly separable*[12] *symmetric means* which will be defined and discussed in Sections 5 and 6 below. The application of this class of functions to choice under uncertainty will be made in Section 7 where we end up with an axiomatic model of uncertain choice which is similar to the *implicit linear utility models* obtained by Dekel [1986], Chew [1989] and Chew and Epstein [1989a; 211].

We can now explain the two main purposes of this paper: (i) to provide axiomatic characterizations for various classes of symmetric means and (ii) to provide an alternative axiomatic foundation for Chew and Epstein's implicit linear utility model.

In Section 2 below, we review various basic properties for symmetric means and provide a few results on the independence of the various axioms.[13] In Section 3, we characterize separable symmetric means and in Section 4, we add the additional axioms of homogeneity and translatability. In Section 5, we characterize implicitly separable symmetric means, and in Section 6 we add the additional axioms of homogeneity and translatability.

---

[11]See the discussion and references to the Allais Paradox in Machina [1982; 287–289] and Chew and Epstein [1989a; 212–213].

[12]Implicitly separable functions were considered briefly in Blackorby, Primont and Russell [1978; 113–114] and more comprehensively in Blackorby, Davidson and Schworm [1991].

[13]In general, we would like to follow the methodological approach to functional equations laid down by Eichhorn [1976; 255] [1978b] and Eichhorn and Gehrig [1982; 680–681]: (i) propose a list of desirable properties or axioms that M should possess; (ii) derive sets of independent and consistent properties from this list, choosing a minimal set to characterize M; (iii) add additional axioms to the basic set and attempt to determine the class of functions which satisfy each set of consistent axioms.

In Section 7, we rework Blackorby, Davidson and Donaldson's [1977] proof of the Expected Utility Theorem and we use their technique of proof to show how an analogue to Chew and Epstein's [1989a; 211] implicit linear utility model can be obtained as a limiting case of implicitly separable stochastic preference functions defined over finite numbers of equally likely alternatives. In the following three sections, we apply this new model of choice under uncertainty to a few simple problems involving insurance (Section 8), gambling (Section 9) and investing (Section 10). The reader who is primarily interested in modeling choice under uncertainty will probably find it easier to read the applications Sections 8, 9 and 10 first, and then read Section 7 which develops the underlying theoretical framework. Section 11 concludes.

2. Basic Properties for Symmetric Means

In this section, we consider various properties that we may wish to impose on a symmetric mean.

A mean is a function of N variables, say $M(x_1, x_2, \ldots, x_N) = M(x)$, which has property (A1) listed below. The domain of definition of M is the closed[14] hypercube $S^N \equiv \{x : a1_N \leq x \leq b1_N\}$ where $a < b$. The number a could equal $-\infty$ and b could equal $+\infty$ unless we explicitly state that a and b are finite. Throughout the paper, we assume that $N \geq 2$.

We take the following four properties as the minimal properties that a symmetric mean should possess:

(A1) $\lambda 1_N \in S^N$ implies $M(\lambda 1_N) = \lambda$ (*Mean Value Property*);

(A2) $M(Px) = M(x)$ where Px is a permutation of the components of the vector $x \equiv (x_1, x_2, \ldots, x_N)$ (*Symmetry*);

(A3) $M(x)$ is a *continuous function* for $x \in S^N$;

(A4) $M(x)$ is a *strictly increasing function* over S^N; i.e., if $x^1 \in S^N$, $x^2 \in S^N$ and $x^1 < x^2$, then $M(x^1) < M(x^2)$.

Axiom (A1) is the defining property for an average or a mean value; if all components of $x \equiv (x_1, \ldots, x_N)$ equal a common value λ, then $M(x)$ should also equal this common value λ.

---

[14]In Proposition 10 where we characterize means of order r, the domain of definition set is the positive orthant, $\{x : x \gg 0_N\}$. Notation: $x \geq 0_N$ means each component of x is nonnegative; $x \gg 0_N$ means each component is positive and $x > 0_N$ means $x \geq 0_N$ but $x \neq 0_N$.

Axiom (A2) ensures that the mean function $M(x_1, x_2, \ldots, x_N)$ is a symmetric function of its variables.[15]

Axiom (A3) ensures that $M(x)$ does not change erratically as the components of x change and (A4) ensures that the mean of the N numbers $x_1, \ldots, x_N$ increases if any one of the x_i increases.

PROPOSITION 1. *The axioms (A1)–(A4) are logically independent in the sense that for each subset of three of the four axioms, an $M(x)$ can be found which satisfies the three properties but not the fourth.*

Proof: (i) $M(x) \equiv (1_N \cdot x)^2$ satisfies (A2)–(A4) but not (A1). (ii) $M(x) \equiv \alpha \cdot x$, $\alpha \gg 0_N$, $\alpha \cdot 1_N = 1$, $\alpha \neq (1/N)1_N$ satisfies (A1), (A3) and (A4) but not (A2). (iii) For $0 < \alpha < 1$ and $x \equiv (x_1, \ldots, x_N) \in S^N$, define

$$M(x) \equiv \begin{cases} (1_N \cdot x)/N & \text{if } (1_N \cdot x) < N \\ [(1_N \cdot x)/N] + \alpha[\max_i\{x_i : i = 1, \ldots, N\} - 1_N \cdot x/N] & \text{elsewhere.} \end{cases}$$

It can be verified that $M(x)$ is discontinuous along the hyperplane $\{x : 1_N \cdot x = N\}$ for $x \neq 1_N$. However the above $M(x)$ satisfies (A1), (A2) and (A4). (iv) $M(x) \equiv \min_i\{x_i : i = 1, \ldots, N\}$ satisfies (A1)–(A3) but not (A4). Q.E.D.

It turns out that the above axioms for a symmetric mean imply an interesting fifth axiom as the following proposition indicates.

PROPOSITION 2. *If $M(x)$ satisfies (A1) and (A4), then it also satisfies:*

(A5) $\alpha \equiv \min_i\{x_i : i = 1, \ldots, N\} \leq M(x) \leq \max_i\{x_i : i = 1, \ldots, N\} \equiv \beta$ *for* $x \equiv (x_1, \ldots, x_N) \in S^N$. *If $x \neq \lambda 1_N$, then the above inequalities are strict (Min-Max Property).*

Proof: Let $x \in S^N$. If $x = \lambda 1_N$, then $\alpha = \beta = \lambda$ and using (A1), we have $M(x) = M(\lambda 1_N) = \lambda$. Thus the inequalities in (A5) hold as equalities if $x = \lambda 1_N$.

If $x \neq \lambda 1_N$, then the definitions of α and β in (A5) imply that

$$\alpha 1_N < x < \beta 1_N. \tag{7}$$

The first inequality in (7) and (A4) imply that $M(x) > M(\alpha 1_N) = \alpha$, using (A1). The second inequality in (7) and (A4) imply that $M(x) < M(\beta 1_N) = \beta$ where the equality follows using (A1). Thus the inequalities in (A5) hold strictly if $x \neq \lambda 1_N$. Q.E.D.

[15] Other terms used to describe the symmetry property include "impartiality" or "non-discrimination" (Kolm [1969; 188]) and "anonymity" (Blackorby and Donaldson [1984; 16]).

The above technique of proof was used by Eichhorn and Voeller [1976; 10] in the context of index number theory. Our axiom (A1) is a combination of the *identity* and *proportionality* axioms of index number theory and our (A4) is the *monotonicity* axiom of index number theory.[16]

Note that (A5) implies that $M(x) \geq 0$ if $a = 0$ where a is the lower bound which occurs in the definition of $S^N \equiv \{x : a1_N \leq x \leq b1_N\}$.

It is sometimes useful to know something about the derivatives of a symmetric mean $M(x)$ evaluated at a point on the equality ray (where $x = \lambda 1_N$); e.g. see Diewert [1978b; 897–898]. The following proposition shows that the first derivatives $M_i(x) \equiv \partial M(x_1, \ldots, x_N)/\partial x_i$ are all equal to $1/N$ and the second derivatives $M_{ij}(x) \equiv \partial^2 M(x_1, \ldots, x_N)/\partial x_i \partial x_j$ depend on only one unknown parameter when evaluated at a point of equality.

PROPOSITION 3. *Let $M(x)$ satisfy (A1) and (A2) and, in addition, let $M(x)$ be twice continuously differentiable at the point $x = \lambda 1_N$ where $\lambda 1_N$ belongs to the interior of S^N. Then the first and second derivatives of M evaluated at $\lambda 1_N$ satisfy the following restrictions:*

$$M_i(\lambda 1_N) = 1/N \quad \text{for } i = 1, \ldots, N; \tag{8}$$

$$M_{ii}(\lambda 1_N) = \gamma \quad \text{for } i = 1, \ldots, N; \tag{9}$$

$$M_{ij}(\lambda 1_N) = -(N-1)^{-1}\gamma \quad \text{for } 1 \leq i, j \leq N,\ i \neq j, \tag{10}$$

where the parameter γ will depend on λ in general.

Proof: From (A1), $M(\lambda 1_N) = \lambda$. Differentiate this equation with respect to λ to get

$$\sum_{i=1}^{N} M_i(\lambda 1_N) = 1. \tag{11}$$

Using (A2), for $j = 2, 3, \ldots, N$:

$$M(x_1, x_2, \ldots, x_{j-1}, x_j, x_{j+1}, \ldots, x_N) = M(x_j, x_2, \ldots, x_{j-1}, x_1, x_{j+1}, \ldots, x_N). \tag{12}$$

Differentiate equation (12) with respect to x_1 and evaluate the resulting partial derivatives at $x = \lambda 1_N$ to get

$$M_1(\lambda 1_N) = M_j(\lambda 1_N), \quad j = 2, \ldots, N. \tag{13}$$

Substitute the $N-1$ equations (13) into (11) to get

$$N M_1(\lambda 1_N) = 1. \tag{14}$$

---

[16]Axiom (A5) is the counterpart to the mean value test in index number theory; see Eichhorn and Voeller [1976; 10].

Equations (13) and (14) imply equations (8).

Now differentiate equation (12) with respect to x_1 twice and evaluate the resulting partial derivatives at $x = \lambda 1_N$. We find that

$$(15) \qquad M_{11}(\lambda 1_N) = M_{jj}(\lambda 1_N) \quad \text{for } j = 2, 3, \ldots, N.$$

Define $\gamma \equiv M_{11}(\lambda 1_N)$ and this definition plus equations (15) imply equations (9).

(A2) implies that for $1 \leq i < j \leq N$:

$$(16) \quad M(x_1, x_2, x_3, \ldots, x_{i-1}, x_i, x_{i+1}, \ldots, x_{j-1}, x_j, x_{j+1}, \ldots, x_N) \\ = M(x_i, x_j, x_3, \ldots, x_{i-1}, x_1, x_{i+1}, \ldots, x_{j-1}, x_2, x_{j+1}, \ldots, x_N).$$

Partially differentiate (16) with respect to x_1 and then x_2 and evaluate the resulting partial derivatives at $x = \lambda 1_N$ to get

$$(17) \qquad M_{12}(\lambda 1_N) = M_{ij}(\lambda 1_N), \quad 1 \leq i < j \leq N,$$

$$(18) \qquad = M_{ji}(\lambda 1_N),$$

where (18) follows from (17) using the assumption that M is twice continuously differentiable at $x = \lambda 1_N$.

Now differentiate (11) with respect to λ and obtain the following equation:

$$(19) \qquad \sum_{i=1}^{N} \sum_{j=1}^{N} M_{ij}(\lambda 1_N) = 0.$$

Substituting (9), (17) and (18) into (19) yields $N\gamma + N(N-1)M_{12}(\lambda 1_N) = 0$ or $M_{12}(\lambda 1_N) = -(N-1)^{-1}\gamma$. This last equation plus (17) and (18) yield equations (10). Q.E.D.

Thus the first and second order partial derivatives of a differentiable symmetric mean are determined up to a single parameter γ provided that the derivatives are evaluated at an equal quantities point.

We now consider some additional basic properties that economic researchers have found convenient to impose on symmetric means: (i) linear homogeneity; (ii) translatability; and (iii) various curvature properties.

For the following axiom, we require that the domain of definition parameter a be nonnegative (recall that $S^N \equiv \{x : a1_N \leq x \leq b1_N\}$ with $a < b$).

(A6) S^N is a subset of the nonnegative orthant, $x \in S^N$, $\lambda \geq 0$, $\lambda x \in S^N$ implies $M(\lambda x) = \lambda M(x)$ (*Positive Linear Homogeneity*).

The linear homogeneity axiom (A6)[17] implies that each indifference surface of $M(x)$ is a radial blowup of a single indifference surface. If $a = 0$ and $b = +\infty$,

[17]This axiom was explicitly introduced into the mathematics literature on means by Nagumo [1930; 79].

then the unit indifference surface, $I(1) \equiv \{x : M(x) = 1,\ x \geq 0_N\}$, determines the entire family of indifference surfaces, $I(u) \equiv \{x : M(x) = u,\ x \geq 0_N\}$, for $u > 0$; i.e., if $x \in I(1)$, then $ux \in I(u)$.

The following axiom implies that each indifference surface is a translation of a base indifference surface.

(A7) $x \in S^N$, $x + \lambda 1_N \in S^N$ implies $M(x + \lambda 1_N) = M(x) + \lambda$ (*Translatability*).

Aczél [1966; 151] uses the term "translativity" to describe axiom (A7) while Blackorby and Donaldson [1980; 109] [1982; 254] use the term "unit-translatability." The concept was introduced into the mathematics literature on means by Nagumo [1930; 77] and into the economics literature on inequality by Kolm [1969; 190]. Axiom (A7) tells us if we have a point $(x_1, \ldots, x_N)$ on the u indifference surface, then adding the same number λ to each x_i gives us a point $(x_1 + \lambda, \ldots, x_N + \lambda)$ which is on the $u + \lambda$ indifference surface.

We now turn our attention to properties that describe the curvature properties of the indifference surfaces. The first two of these axioms limit the behavior of $M(x)$ as x moves away from the ray of equality.

(A8) $\lambda 1_N \in S$, $0 < \delta$, $v \neq 0_N$, $v \cdot 1_N = 0$, $(\lambda 1_N + \delta v) \in S^N$ implies $M(\lambda 1_N) > M(\lambda 1_N + \delta v)$ (*Equality Preference*).

Alternative terms for axiom (A8) might be *inequality aversion* or *dispersion aversion* or *risk aversion*. The meaning of (A8) can be expressed as follows: start at the point $\lambda 1_N \equiv (\lambda, \lambda, \ldots, \lambda)$, a point on the ray of equality, pick a *direction vector* $v \equiv (v_1, \ldots, v_N) \neq 0_N$ which is perpendicular or orthogonal to the ray of equality (so that $v \cdot 1_N = 0$). Then the mean of the vector which has all components equal to $\lambda, M(\lambda 1_N)$, is greater than the mean $M(\lambda 1_N + \delta v)$ of a vector which is equal to the initial vector $\lambda 1_N$ plus a vector $\delta v = (\delta v_1, \delta v_2, \ldots, \delta v_N)$ which is perpendicular to the initial vector (note that $(\delta v) \cdot (\lambda 1_N) = \delta \lambda v \cdot 1_N = \delta \lambda 0 = 0$).

Our axiom (A8) is related to the *globally equality preferring axiom* of Rothschild and Stiglitz [1973; 196], which is $M(x) \leq M[\alpha \bar{x} 1_N + (1 - \alpha) x]$ where $0 \leq \alpha \leq 1$ and $\bar{x} \equiv x \cdot 1_N / N$. Using the notation of Rothschild and Stiglitz, our axiom (A8) can be rephrased as $M(x) < M(\bar{x} 1_N)$, where $\lambda \neq \bar{x} 1_N$, which is a strict version of the Rothschild and Stiglitz property for $\alpha = 1$.

A stronger axiom than (A8) is

(A9) $0 \leq \delta_1 < \delta_2$, $v \neq 0_N$, $v \cdot 1_N = 0$, $(\lambda 1_N + \delta_2 v) \in S^N$ implies $M(\lambda 1_N + \delta_1 v) > M(\lambda 1_N + \delta_2 v)$ (*Increasing Inequality Aversion*).

Alternative terms for axiom (A9) are *increasing dispersion aversion* or *increasing risk aversion*.

Using the definition of S^N, we can show that if $\lambda 1_N + \delta_2 v$ belongs to S^N, then $\lambda 1_N + \delta v$ also belongs to S^N for $0 \leq \delta \leq \delta_2$. Thus (A9) implies (A8) if we set $\delta_1 = 0$ and $\delta_2 = \delta$. Note also that the arithmetic mean of $\lambda 1_N + \delta v$ is λ provided that $v \cdot 1_N = 0$. Using the notation that Rothschild and Stiglitz [1973] used, property (A9) can be rephrased as: $f(\alpha) \equiv M[\alpha \bar{x} 1_N + (1-\alpha)x]$ is an increasing function of α for $0 \leq \alpha \leq 1$, provided that $x \neq \lambda 1_N$.

In words, axiom (A9) says that the mean of $\lambda 1_N + \delta v$ steadily decreases as δ goes from 0 to $\delta_2 > 0$. Thus in the context of welfare or inequality measurement, as we move from an equal incomes point along a line segment that is perpendicular to the ray of equality, welfare steadily decreases. Axiom (A9) can be interpreted as an axiomatization of Dalton's [1920; 352] *principle of transfers.*[18]

There are other ways of axiomatizing the principle of transfers. One such alternative method is due to Kolm [1969; 189], who suggested the following property for M:

(20) $x \in S^N$, $x \neq \lambda 1_N$ and B is a bistochastic matrix such that Bx is not a permutation of x implies $M(Bx) > M(x)$.

Kolm [1969; 189] called (20) a *strict averages preference.* An $N \times N$ matrix B is bistochastic if and only if $B \geq 0_{N\times N}$, $1_N^T B = 1_N^T$ and $B 1_N = 1_N$. If, in addition, each element of B is either 0 or 1, then B is a permutation matrix. It turns out that (20) is generally a stronger property than (A9) as the following proposition shows.

PROPOSITION 4. *Strict averages preference (20) implies increasing inequality aversion (A9) but not vice versa if $N > 2$.*

Proof: Assume (20) holds and let $0 \leq \delta_1 < \delta_2$, $v \neq 0_N$, $v \cdot 1_N = 0$ and $(\lambda 1_N + \delta_2 v) \in S^N$. Define $\delta \equiv \delta_1/\delta_2$ and note that $0 \leq \delta < 1$. Define the bistochastic matrix B by

$$B \equiv \delta I_N + (1-\delta)(1/N) 1_N 1_N^T \tag{21}$$

where I_N is the $N \times N$ identity matrix and 1_N^T is the transpose of the (column) vector 1_N. Now define $x \equiv \lambda 1_N + \delta_2 v$ and note that $x \neq \lambda 1_N$. Compute:

$$\begin{aligned} Bx &= [\delta I_N + (1-\delta)(1/N)1_N 1_N^T](\lambda 1_N + \delta_2 v) \\ &= \delta\lambda 1_N + \delta\delta_2 v + (1-\delta)\lambda 1_N && \text{since } 1_N^T v = 0 \\ &= \lambda 1_N + \delta\delta_2 v \\ &= \lambda 1_N + \delta_1 v && \text{since } \delta = \delta_1/\delta_2. \qquad (22) \end{aligned}$$

[18] "Maintaining the assumptions laid down in Section 2 above, we may safely say that, if there are only two income-receivers, and a transfer of income takes place from the richer to the poorer, inequality is diminished." (Dalton [1920; 352]).

From (20), $M(Bx) > M(x)$ and using the definition of x and (22), this inequality translates to $M(\lambda 1_N + \delta_1 v) > M(\lambda 1_N + \delta_2 v)$. Thus (20) implies (A9).

The above proof shows that property (A9) is in fact property (20), except that the bistochastic matrices B must be chosen to be of the form (21) for $0 < \delta < 1$. Thus if $N > 2$, then (20) is a stronger property than (A9), since there are bistochastic matrices satisfying (20) which are not of the form (21).

Q.E.D.

When $N = 2$, the only bistochastic matrices that satisfy condition (20) are of the form given by (21). Thus when $N = 2$, our increasing inequality aversion axiom (A9) coincides with Kolm's strict averages preference property (20). When N is greater than two, (20) is a stronger property than (A9). Since we do not find the strict averages property as transparent as (A9), we will not consider (20) any further.[19]

We turn now to our last curvature property. $M(x)$ is a *strictly quasiconcave function* over S^N if and only if

(23) $x \in S^N$, $y \in S^N$, $x \neq y$, $0 < \lambda < 1$, $M(x) \leq M(y)$ implies $M(x) < M[\lambda x + (1-\lambda)y]$.

Our final axiom that we consider in this section is:

(A10) M is a *strictly quasiconcave function* over S^N.

Property (A10) implies that the upper level sets of M, $L(u) \equiv \{x : M(x) \geq u, x \in S^N\}$, are strictly convex sets. Debreu [1959; 60–61 and 101] introduced this assumption (following the lead of Arrow [1953] who introduced the related assumption of quasiconcavity where the strict inequality in (23) is relaxed to a weak inequality) in the context of the Arrow-Debreu state contingent commodity approach to modeling consumer demand under uncertainty. Both Arrow and Debreu noted that the assumption of a quasiconcave or strictly quasiconcave preference function implied an attitude of risk aversion on the part of the consumer.

It is not immediately evident whether (A10) implies (A9) (and hence (A8)). However, the following proposition shows that (A10) does imply (A8) and (A9) if M is a symmetric mean.

[19] Kolm [1976b; 90–92 and 104–106] and Blackorby and Donaldson [1980; 110] indicate that if M is symmetric and quasiconcave, then M also satisfies (20), but the converse is not true.

PROPOSITION 5. *Let M satisfy (A2), (A3) and (A10). Then M also satisfies (A8) and (A9).*

Proof: Let $\delta_2 > 0$, $v \neq 0_N$, $v \cdot 1_N = 0$ and $(\lambda 1_N + \delta_2 v) \in S^N$. Let e_i be the ith unit vector of dimension N for $i = 1, 2, \ldots, N$. Define the $N \times N$ permutation matrix $P \equiv (e_N, e_1, e_2, \ldots, e_{N-1})$ and define $P^1 \equiv P$, $P^2 \equiv PP$, $P^3 \equiv PPP$, ... , $P^N = I_N$. Define

$$x^i \equiv P^i(\lambda 1_N + \delta_2 v), \quad i = 1, 2, \ldots, N. \tag{24}$$

Since each x^i is a permutation of the components of $\lambda 1_N + \delta_2 v$, we have by (24) and (A2):

$$M(x^i) = M(\lambda 1_N + \delta_2 P^i v) = M(\lambda 1_N + \delta_2 v), \quad i = 1, \ldots, N. \tag{25}$$

By (A10), $\{x : M(x) \geq M(\lambda 1_N + \delta_2 v),\ x \in S^N\}$ is a convex set. The equalities (25) imply that each x^i belongs to this convex set. Thus the convex combination $\sum_{i=1}(1/N)x^i$ also belongs to this set so that

$$\begin{aligned} M(\lambda 1_N + \delta_2 v) &\leq M\left[\sum\nolimits_{i=1}^{N}(1/N)x^i\right] \\ &= M\left[\lambda 1_N + (1/N)\delta_2\left(\sum\nolimits_{i=1}^{N} P^i\right)v\right] \text{ using (24)} \\ &= M[\lambda 1_N + (1/N)\delta_2(1_N 1_N^T)v] \text{ using the definition of the } P^i \\ &= M(\lambda 1_N) \text{ since } 1_N^T v = 0. \end{aligned} \tag{26}$$

An inequality analogous to (26) can be derived for any δ such that $-\delta_2 \leq \delta \leq \delta_2$; i.e. we have

$$M(\lambda 1_N + \delta v) \leq M(\lambda 1_N) \text{ for } -\delta_2 \leq \delta \leq \delta_2.$$

Thus the function of one variable $f(\delta) \equiv M(\lambda 1_N + \delta v)$ attains a local maximum at $\delta = 0$, and using a result found in Diewert, Avriel and Zang [1981; 406], this local maximum is the unique global maximizer. Moreover, $f(\delta)$ must decrease monotonically as δ goes from 0 to δ_2 (for otherwise, the continuous function f would attain a local minimum, which cannot happen if f is strictly quasiconcave and continuous; see Diewert, Avriel and Zang [1981; 406]). Thus we have established (A9), and (A8) follows from (A9). Q.E.D.

It is easy to construct examples of symmetric means that satisfy (A8) but not (A9) or that satisfy (A10) but not (A9). For example, let $N = 2$, $a = 0$, $b = +\infty$ and let $M(x_1, x_2)$ satisfy (A6). If $u > 0$, then the indifference curve $I(u) \equiv \{(x_1, x_2) : M(x_1, x_2) = u,\ x_1 \geq 0,\ x_2 \geq 0\}$ determines the entire function. In Figure 1 below, OA is the ray of equality and the other dashed lines are perpendicular to OA. If the base indifference curve is given by the "flying bird" indifference curve BCD, then it can be seen that the resulting

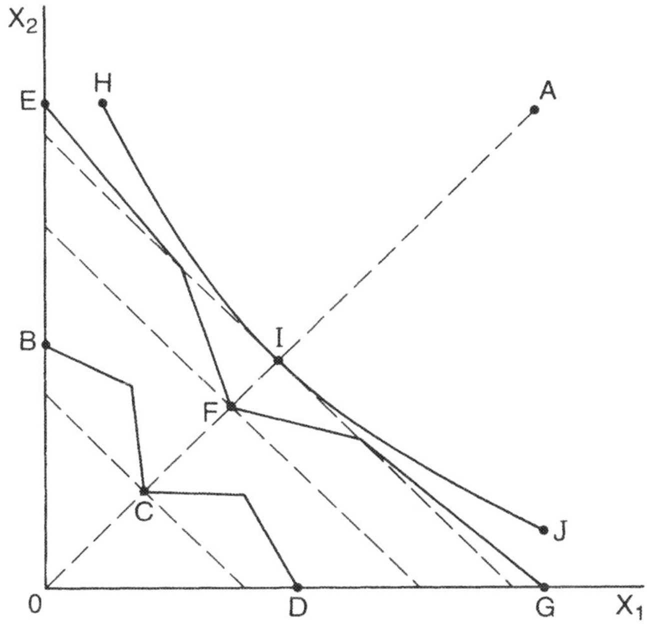

Figure 1

$M(x)$ satisfies (A8) but not (A9). If the base indifference curve is EFG, then the resulting $M(x)$ satisfies (A9) but not (A10) since the corresponding upper level set $L(u) \equiv \{(x_1, x_2) : M(x_1, x_2) \geq u,\ x_1 \geq 0,\ x_2 \geq 0\}$ is not convex and hence M is neither quasiconcave nor strictly quasiconcave. Finally, if the base indifference curve is HIJ, then the corresponding $M(x_1, x_2)$ is strictly quasiconcave.

A stronger curvature property than (A10) is the assumption that M is a strictly concave function over S^N. However, this property is inconsistent with (A1) so we cannot require that a symmetric mean be strictly concave.

In the following section, we consider axioms that will imply the separability of a symmetric mean. We shall also establish counterparts to the results of Kolm [1969; 193] who showed that under the assumption of separability, very weak curvature axioms will imply much stronger ones.

3. Separable Symmetric Means

In this section, we shall provide axiomatic characterizations for additively separable symmetric means of the form (5).

For each positive integer n, define the domain of definition set $S^n \equiv \{(x_1, x_2, \ldots, x_n)\colon a \le x_i \le b,\ i = 1, 2, \ldots, n\}$ where $a < b$. Consider the following property for a symmetric mean function $M(x_1, x_2, \ldots, x_N)$:

(A11) There exists a function of two variables $M_2(x_1, x_2)$ defined over S^2 such that for every $(x_1, x_2, \ldots, x_N) \in S^N$, we have $M(x_1, x_2, x_3, \ldots, x_N) = M[M_2(x_1, x_2),\ M_2(x_1, x_2),\ x_3, \ldots, x_N]$ where M_2 satisfies axioms (A1)–(A4) (*Consistency in Aggregation*).

It can be seen that our consistency in aggregation axiom[20] (A11) is a weakened version of the Nagumo [1930], Kolmogoroff [1930] and Blackorby and Donaldson [1984; 19] consistency in aggregation property (6): the difference between the two properties is that (6) requires the existence of an entire family of aggregator functions, M^n for $n = 1, 2, \ldots$, whereas (A11) requires the existence of only two functions, $M(x_1, \ldots, x_N)$ and $M_2(x_1, x_2)$.

PROPOSITION 6. *Let $N \ge 3$ and let $M(x_1, x_2, \ldots, x_N)$ satisfy (A1)–(A4) and (A11). Then there exists a continuous monotonic function of one variable $\phi(z)$ defined for $a \le z \le b$ such that, for $(x_1, x_2) \in S^2$, we have*

$$M_2(x_1, x_2) = \phi^{-1}[(1/2)\phi(x_1) + (1/2)\phi(x_2)] \tag{27}$$

and for $(x_1, x_2, \ldots, x_N) \in S^N$, we have

$$M(x_1, x_2, \ldots, x_N) = \phi^{-1}\left[\sum\nolimits_{i=1}^{N} (1/N)\phi(x_i)\right]. \tag{28}$$

Proof: Let $x \in S^N$. Then we have

$$\begin{aligned} M(x_1, x_2, x_3 \ldots, x_N) &= M[M_2(x_1, x_2), M_2(x_1, x_2), x_3, \ldots, x_N] \\ &\equiv F[M_2(x_1, x_2), x_3] \end{aligned} \tag{29}$$

where, in (29), it is understood that F also depends on $x_4, x_5, \ldots, x_N$ if $N > 3$, but since we will be holding these variables constant for the time being, we suppressed these variables temporarily. Since, by (A2), M is symmetric in its variables, we have for all $(x_1, x_2, x_3) \in S^3$,

$$F[M_2(x_1, x_2), x_3] = F[M_2(x_1, x_3), x_2]. \tag{30}$$

---

[20]For references to the economics literature on consistency in aggregation, see Vartia [1974] [1976a], Diewert [1978b; 883] and Blackorby and Primont [1980].

Equation (30) is a generalized equation of associativity; see Aczél [1966; 310]. Axioms (A3) and (A4) applied to M_2 and M imply that M_2 and F are continuous and increasing in their variables. Thus the hypotheses of Corollary 1 in Aczél [1966; 312] are satisfied and we conclude that there exist continuous and monotonic (increasing or decreasing) functions of one variable f, g, h, k and m defined over $S^1 \equiv \{z : a \leq z \leq b\}$ such that

(31) $$F(y,z) \equiv h[f(y)+g(z)] \text{ and}$$

(32) $$M_2(x_1,x_2) \equiv f^{-1}[k(x_1)+m(x_2)].$$

For later results (see Proposition 10), we note that Aczél's Theorem is still valid if the domain of definition sets are of the form $S^n \equiv \{(x_1,\ldots,x_n) : a < x_i \leq b,\ i=1,2,\ldots,n\}$.

Since $M_2(x_1,x_2)$ is symmetric in x_1 and x_2,

(33) $$k = m.$$

Using axiom (A1) for M_2, we have for $z \in S^1$:

$$\begin{aligned} z &= M_2(z,z) \\ &= f^{-1}[m(z)+m(z)] && \text{using (32) and (33)} \\ &= f^{-1}[2m(z)] && \text{or} \\ f(z) &= 2m(z) && \text{or} \end{aligned}$$

(34) $$m(z) = (1/2)f(z).$$

Substituting (33) and (34) into (32) yields

(35) $$M_2(x_1,x_2) = f^{-1}[(1/2)f(x_1)+(1/2)f(x_2)].$$

Now substitute (35) into (31) to get:

(36) $$\begin{aligned} F[M_2(x_1,x_2),x_3] &= h\big[f\{f^{-1}[(1/2)f(x_1)+(1/2)f(x_2)]\}+g(x_3)\big] \\ &= h[(1/2)f(x_1)+(1/2)f(x_2)+g(x_3)]. \end{aligned}$$

We now recognize that F depends on $x_4,\ldots,x_N$. Hence h and g could depend on $x_4,\ldots,x_N$ as well. Thus using (29) and (36), M may be written as follows:

(37) $$\begin{aligned} M(x_1,\ldots,x_N) = h[&(1/2)f(x_1)+(1/2)f(x_2) \\ &+ g(x_3,x_4,\ldots,x_N),x_4,\ldots,x_N]. \end{aligned}$$

Since, by (A2), the left hand side of (37) is symmetric in $x_1, \ldots, x_N$, the right hand side of (37) must also be symmetric. There are two ways to achieve this symmetry: (i) the first argument of h is symmetric in $x_1, \ldots, x_N$ and h is constant with respect to its last $N-3$ arguments or (ii) h is symmetric in x_1, x_2, x_3 and independent of $x_4, \ldots, x_N$ in its first argument. We analyze each possibility in turn.

Case (i). We require $(1/2)f(x_1) + (1/2)f(x_2) + g(x_3, \ldots, x_N)$ to be symmetric in $x_1, \ldots, x_N$. This requires that $g(x_3, \ldots, x_N) = \sum_{i=3}^{N}(1/2)f(x_i)$. Substituting this last equation into (37) and noting that, in this case, h is constant with respect to its last $N-3$ arguments yields

$$M(x_1, \ldots, x_N) = h^*\left[(1/2)\sum\nolimits_{i=1}^{N} f(x_i)\right]. \tag{38}$$

Case (ii). In order to make the first argument of h symmetric in x_1, x_2, x_3, we require that $g(x_3, \ldots, x_N) \equiv (1/2)f(x_3)$. Substituting this equation into (37) yields $M(x_1, \ldots, x_N) = h[(1/2)f(x_1)+(1/2)f(x_2)+(1/2)f(x_3), x_4, \ldots, x_N]$. In order to make this last equation symmetric in $x_1, \ldots, x_N$, we again require that M have the representation (38).

Finally, we determine h^* in terms of f. By (A1),

$$\begin{aligned} z = M(z1_N) &= h^*\left[(1/2)\sum\nolimits_{i=1}^{N} f(z)\right] \\ &= h^*[(N/2)f(z)]. \end{aligned} \tag{39}$$

Now let $y = (N/2)f(z)$ and substitute this equation into (39) to obtain

$$z = h^*(y) = f^{-1}[(2/N)y]. \tag{40}$$

Substitute (40) into (38) and obtain $M(x_1, \ldots, x_N) = f^{-1}[\sum_{i=1}^{N}(1/N)f(x_i)]$. Now let $\phi = f$ and we have (28). Equation (27) follows from (35) and $\phi = f$.

Q.E.D.

COROLLARY 1. (Nagumo [1930; 77] and Hardy, Littlewood and Polya [1934; 66]): *The continuous monotonic function $\phi(z)$ which occurs in (27) and (28) can be replaced by $\chi(z) \equiv \alpha + \beta\phi(z)$ where $\beta \neq 0$. Thus, in particular, if ϕ is monotonically decreasing, we can replace it by the monotonically increasing function of one variable $\chi(z) \equiv -\phi(z)$.*

Proof: Let $x \in S^N$ and let $z = M(x) = \phi^{-1}\left[\sum_{i=1}^N (1/N)\phi(x_i)\right]$. Thus

$$\begin{aligned} \phi(z) &= (1/N)\sum_{i=1}^N \phi(x_i) && \text{or} \\ \alpha + \beta\phi(z) &= \alpha + \beta(1/N)\sum_{i=1}^N \phi(x_i) && \text{if } \beta \neq 0 \\ &= (1/N)\sum_{i=1}^N [\alpha + \beta\phi(x_i)] && \text{or} \\ \chi(z) &= (1/N)\sum_{i=1}^N \chi(x_i) && \text{or} \\ z &= \chi^{-1}\left[\sum_{i=1}^N (1/N)\chi(x_i)\right]. \end{aligned}$$

Q.E.D.

The above corollary shows that the generating function ϕ that is used to represent a separable symmetric mean is unique only up to an affine transformation.

COROLLARY 2. *$M(x_1, \ldots, x_N)$ is consistent in aggregation for any subset of m variables where $2 \leq m \leq N-1$; i.e., for each such m, we have the existence of a function of m variables, M_m, such that*

(41) $M(x_1, \ldots, x_m, x_{m+1}, \ldots, x_N) = M[M_m(x_1, \ldots, x_m)1_m, x_{m+1}, \ldots, x_N]$.

The function M_m is defined in terms of ϕ as follows:

(42) $$M_m(x_1, \ldots, x_m) \equiv \phi^{-1}\left[\sum_{i=1}^m (1/m)\phi(x_i)\right].$$

Proof: Substitute (28) and (42) into (41). Q.E.D.

Corollary 2 says that we can deduce the stronger property (6) (for all m less than N) as a consequence of the much weaker property (A11).

The function M_2 can be interpreted as a subgroup aggregator function in the context of social welfare or inequality measurement. In the context of choice under uncertainty when there are N equally likely states of nature, $M_2(x_1, x_2)$ can be interpreted as the certainty equivalent over two states of nature; i.e., the decision maker will be indifferent to a gamble involving equal probabilities with payoffs x_1 and x_2 or receiving $M_2(x_1, x_2)$ with certainty.

Proposition 6 provides an axiomatic characterization for a separable symmetric mean when $N \geq 3$ that is similar in spirit to the axiomatic characterizations of Nagumo [1930] and Kolmogoroff [1930].[21] The question arises: what

[21] Kolmogoroff's axioms were (A1)–(A4) and (6); Nagumo's axioms were the same except that (A5) was substituted for (A4). Both authors assumed that the domain of definition parameters, a and b, were finite. In our Proposition 6, a could equal $-\infty$ and b could equal $+\infty$, since Aczél's [1966; 12] Theorem allows these infinite values.

happens when $N = 2$? It is certainly true that there are symmetric means that are not of the form $M(x_1, x_2) = \phi^{-1}[(1/2)\phi(x_1) + (1/2)\phi(x_2)]$ when $N = 2$; for example, consider the *logarithmic mean*[22] defined for $x_1 > 0$, $x_2 > 0$ as follows:

$$(43) \qquad M_L(x_1, x_2) \equiv \begin{cases} (x_2 - x_1)/(\ln x_2 - \ln x_1) & \text{if } x_1 \neq x_2 \\ x_1 & \text{if } x_1 = x_2. \end{cases}$$

It can be shown that M_L satisfies (A1)–(A4) but it is certainly not an additively separable function. Thus we require an alternative axiom to (A11) to characterize separable symmetric means when $N = 2$. It turns out that the following axiom will do the job for a general $N \geq 2$.

(A12) There exists a function of two variables $M_2(x_1, x_2)$ defined for $(x_1, x_2) \in S^2$ such that, for every $(x_1, \ldots, x_N) \in S^N$, we have $M(x_1, x_2, x_3, \ldots, x_N) = M[M_2(x_1, x_2),\ M_2(x_1, x_2),\ x_3, \ldots, x_N]$ where M_2 satisfies axioms (A1)–(A4) plus the following additional property for all $(x_1, x_2, x_3, x_4) \in S^4$:

$$(44) \quad M_2[M_2(x_1, x_2), M_2(x_3, x_4)] = M_2[M_2(x_1, x_3), M_2(x_2, x_4)]$$

(*Strong Consistency in Aggregation*).

Thus axiom (A12) is the same as (A11) except that the micro aggregator function M_2 satisfies the additional property (44), which is known as the ***bisymmetry equation*** in the functional equations literature; e.g., see Aczél [1966; 278].

Why is it reasonable to ask that the micro or subgroup aggregator function M_2 satisfy property (44)? Let us use M_2 in two stages to construct a mean function for four variables; i.e., define

$$(45) \qquad M_4(x_1, x_2, x_3, x_4) \equiv M_2[M_2(x_1, x_2), M_2(x_3, x_4)].$$

Of course, by axiom (A2), $M_2(x_1, x_2) = M_2(x_2, x_1)$, so we can interchange x_1 and x_2 or x_3 and x_4 in (45) and the overall mean $M_4(x_1, x_2, x_3, x_4)$ will remain unchanged. However, if we interchange x_1 and x_3 or x_2 and x_4, then unless (44) holds, $M_4(x_1, x_2, x_3, x_4)$ may change. Thus it seems reasonable to impose (44) as a property for M_2.

---

[22]See Carlson [1972] and Vartia [1976a; 122].

In fact, if $N = 4$ and $M(x_1, x_2, x_3, x_4)$ satisfies (A1), (A2) and (A11), then (44) *must* hold, since:

$$\begin{aligned}
M(x_1,x_2,x_3,x_4) &= M[M_2(x_1,x_2), M_2(x_1,x_2), x_3, x_4] \text{ by (A11)} \\
&= M[x_3, x_4, M_2(x_1,x_2), M_2(x_1,x_2)] \text{ by (A2)} \\
&= M[M_2(x_3,x_4), M_2(x_3,x_4), M_2(x_1,x_2), M_2(x_1,x_2)] \text{ by (A11)} \\
&= M[M_2(x_1,x_2), M_2(x_3,x_4), M_2(x_1,x_2), M_2(x_3,x_4)] \text{ by (A2)} \\
&= M\Big[M_2[M_2(x_1,x_2), M_2(x_3,x_4)]1_2, M_2(x_1,x_2), M_2(x_3,x_4)\Big] \\
&\quad \text{by (A11)} \\
&= M\Big[M_2(x_1,x_2), M_2(x_3,x_4), M_2[M_2(x_1,x_2), \\
&\quad M_2(x_3,x_4)]1_2\Big] \text{ by (A2)} \\
&= M\Big[M_2[M_2(x_1,x_2), M_2(x_3,x_4)1_2, M_2[M_2(x_1,x_2), \\
&\quad M_2(x_3,x_4)]1_2\Big] \text{ by (A11)} \\
&= M_2[M_2(x_1,x_2), M_2(x_3,x_4)] \text{ by (A1).} \qquad (46)
\end{aligned}$$

Since $M(x_1, x_2, x_3, x_4)$ is symmetric in its variables, the right hand side of (46) must also be symmetric in its variables and so we obtain property (44). In fact, if $N = 2^n$ for any positive integer n and $M(x_1, \ldots, x_N)$ satisfies (A1), (A2) and (A11), then we can modify the above argument and show that M_2 must satisfy (44). Thus axiom (A12) is not all that much stronger than (A11).

We now use (A12) in place of (A11) and establish the following counterpart to Proposition 6.

PROPOSITION 7. *Let $N \geq 2$, let $M(x_1, \ldots, x_N)$ satisfy (A1)–(A4) and (A12), and let the domain of definition sets be $S^n \equiv \{(x_1, \ldots, x_n) : a \leq x_i \leq b,\ i = 1, \ldots, n\}$ where $a < b$ and a and b are finite numbers. Then there exists a continuous increasing function of one variable $\phi(z)$ defined for $a \leq z \leq b$ such that for $(x_1, x_2) \in S^2$, we have (27) and for $(x_1, \ldots, x_N) \in S^N$, we have (28).*

Proof: Since $M_2(x_1, x_2)$ satisfies the bisymmetry equation (44) and (A1)–(A4), a theorem in Aczél [1966; 281] may be applied to deduce the existence of a continuous monotonically increasing function $\phi(z)$ such that (27) holds. Aczél's proof shows that $\phi(z)$ is uniquely determined given arbitrary values for $\phi(a)$ and $\phi(b)$ such that $\phi(a) < \phi(b)$. The function values for ϕ are determined recursively using the given function M_2 as follows:

$$\begin{aligned}
\phi[M_2(a,b)] &\equiv (1/2)\phi(a) + (1/2)\phi(b); \\
\phi\big[M_2[a, M(a,b)]\big] &\equiv (1/2)\phi(a) + (1/2)\phi[M_2(a,b)] \\
&= (3/4)\phi(a) + (1/4)\phi(b);
\end{aligned}$$

$$\begin{aligned}\phi[M_2[M_2(a,b),b]] &\equiv (1/2)\phi[M_2(a,b)] + (1/2)\phi(b)\\ &= (1/4)\phi(a) + (3/4)\phi(b);\\ \phi\Big(M_2[a, M_2[a, M_2(a,b)]]\Big) &\equiv (1/2)\phi(a) + (1/2)\phi[M_2[a, M_2(a,b)]]\\ &= (7/8)\phi(a) + (1/8)\phi(b); \quad \text{etc.}\end{aligned}$$

Nagumo [1930; 73] used the same constructive method of proof.

Given that we have pinned down the functional form of $M_2(x_1, x_2)$, (A12) implies that

$$\begin{aligned}&M(x_1, x_2, x_3, \ldots, x_N)\\ &\quad = M\left[\phi^{-1}[(1/2)\phi(x_1) + (1/2)\phi(x_2)]1_2, x_3, \ldots, x_N\right]\\ (47)\qquad &\quad = h[\phi(x_1) + \phi(x_2), x_3, \ldots, x_N]\\ (48)\qquad &\quad = h^*\left[\sum_{i=1}^{N} \phi(x_i)\right]\end{aligned}$$

where (48) follows from (47) using (A2), the symmetry property for M. Using (A1), we have

$$\begin{aligned}z &= M(z1_N)\\ &= h^*\left[\sum_{i=1}^{N} \phi(z)\right] && \text{using (48)}\\ &= h^*[N\phi(z)] && \text{or}\\ (49)\qquad h^*(y) &= \phi^{-1}[(1/N)y] && \text{letting } y = N\phi(z).\end{aligned}$$

Now substitute (49) into (48) to get (28). Q.E.D.

COROLLARY. *$\phi(z)$ is uniquely determined given arbitrary values for $\phi(a)$ and $\phi(b)$ such that $\phi(a) < \phi(b)$.*

Comparing Propositions 6 and 7, we see that Proposition 7 uses the somewhat stronger consistency in aggregation axiom (A12) in place of A(11), but the representations (27) and (28) are valid for all $N \geq 2$ instead of for $N \geq 3$. Proposition 7 also assumes that the domain of definition set for M is a closed and bounded hypercube whereas Proposition 6 allows for possibly unbounded domain of definition sets.[23]

[23]However this last difference is not essential; Aczél [1966; 291] shows how his bisymmetry equation results can be extended to a ϕ that is defined over an open or an infinite interval. Aczél [1966; 279] also refers to his 1946 paper where he generalized the bisymmetry equation to an arbitrary N greater than 2. These results of Aczél provide another way of characterizing symmetric means.

We note that Corollaries 1 and 2 to Proposition 6 are also applicable to Proposition 7.

In subsequent sections of this paper, we will argue that the approximation properties of separable symmetric means are inferior to the approximation properties of implicitly separable symmetric means. Hence the next proposition is devoted to the approximation properties of separable symmetric means.

We restrict our attention to the case where the number of variables N equals two and we assume that we are given an arbitrary symmetric mean $U(x_1, x_2)$ satisfying (A1)–(A4) over the region $S^2 \equiv \{(x_1, x_2)\colon\ a \leq x_i \leq b,\ i = 1, 2\}$ where $-\infty < a < b < +\infty$. As is usual in consumer demand theory, the equation $u = U(x_1, x_2)$ may be solved for the indifference curve function, $x_2 = I(x_1, u)$. The following proposition shows that any *single* indifference curve of $U(x_1, x_2)$ can be approximated exactly by a separable symmetric mean $M(x_1, x_2)$.

PROPOSITION 8. *Let $U(x_1, x_2)$ satisfy (A1)–(A4) over $S^2 \equiv \{(x_1, x_2)\colon\ a \leq x_i \leq b,\ i = 1, 2,\}$ where $-\infty < a < b < +\infty$. Then there exists a continuous increasing function of one variable ϕ such that $M(x_1, x_2) \equiv \phi^{-1}[(1/2)\phi(x_1) + (1/2)\phi(x_2)]$ has the same u^* indifference curve as U where $a < u^* < b$.*

Proof: Case (i): $U(a, b) \leq u^* < b$. Let $x_2 = I(x_1, u^*)$ be the indifference curve function that corresponds to $\{(x_1, x_2)\colon\ U(x_1, x_2) = u^*,\ a \leq x_i \leq b,\ i = 1, 2\}$. Since U satisfies (A1),

$$I(u^*, u^*) = u^*. \tag{50}$$

Using the symmetry property (A2) for U, we deduce that:

$$x_1 = I(x_2, u^*) \text{ implies } x_2 = I(x_1, u^*). \tag{51}$$

Using (A3) and (A4), we deduce that $I(z, u^*)$ is a continuous, monotonically decreasing function of z for $u^* \leq z \leq b$. Define the function ϕ as follows:

$$\phi(z) \equiv \begin{cases} z & : a \leq z \leq u^*; \\ 2u^* - I(z, u^*) & : u^* < z \leq b. \end{cases} \tag{52}$$

Using (50) and the fact that $I(z, u^*)$ is continuous and decreasing in z, we see that $\phi(z)$ is continuous and increasing in z for $a \leq z \leq b$. The u^* indifference curve for the resulting $M(x_1, x_2) \equiv \phi^{-1}[(1/2)\phi(x_1) + (1/2)\phi(x_2)]$ is given by $u^* = M(x_1, x_2)$ or

$$\phi(u^*) = (1/2)\phi(x_1) + (1/2)\phi(x_2). \tag{53}$$

In order for x_1 and x_2 to satisfy (53), it must be the case that *either* $x_2 < u^*$ and $x_1 > u^*$ (case (a)) *or* $x_2 \geq u^*$ and $x_1 \leq u^*$ (case (b)).

Using definition (52) for case (a), substitution into (53) yields:

$$
\begin{aligned}
& u^* = (1/2)[2u^* - I(x_1, u^*)] + (1/2)x_2 && \text{or} \\
(54) \qquad & x_2 = I(x_1, u^*) && \text{for } u^* < x_1 \leq b.
\end{aligned}
$$

Using definition (52) for case (b), substitution into (53) yields:

$$
\begin{aligned}
& u^* = (1/2)x_1 + (1/2)[2u^* - I(x_2, u^*)] && \text{or} \\
(55) \qquad & x_1 = I(x_2, u^*) && \text{or} \\
(56) \qquad & x_2 = I(x_1, u^*) && \text{for } a \leq x_1 \leq u^*
\end{aligned}
$$

where (56) follows from (55) using (51). Equations (54) and (56) imply that the u^* indifference curve for $M_2(x_1, x_2)$ is precisely the u^* indifference curve for $U(x_1, x_2)$.

Case (ii): $a < u^* < U(a, b)$. In this case, define

$$
(57) \qquad \phi(z) \equiv \begin{cases} 2u^* - I(z, u^*) & : a \leq z < u^*; \\ z & : u^* \leq z \leq b. \end{cases}
$$

Now use the same technique of proof that was used in case (a). Q.E.D.

The ϕ functions which accomplish the approximation job in the above proposition are not uniquely determined, even up to an affine transformation. For example, let f be an arbitrary increasing and continuous function of one variable. Then the ϕ defined by (52) above can be replaced by the following ϕ defined by (58) below and the resulting $M(x_1, x_2)$ will still attain the u^* indifference curve for $U(x_1, x_2)$:

$$
(58) \qquad \phi(z) \equiv \begin{cases} f(z) & : a \leq z \leq u^*; \\ 2f(u^*) - f[I(z, u^*)] & : u^* < z \leq b. \end{cases}
$$

If the symmetric mean $U(u_1, x_2)$ is once differentiable, then there is another way of exhibiting the restrictiveness of a separable approximation to U. If $M(x_1, x_2) = \phi^{-1}[(1/2)\phi(x_1) + (1/2)\phi(x_2)]$ where ϕ is differentiable and increasing with $\phi'(a) > 0$, then the slope of an indifference curve through a point on the line $x_2 = a$ is $-M_1(x_1, a)/M_2(x_1, a)$ where $M_i(x_1, x_2) \equiv \partial M(x_1, x_2)/\partial x_i$ for $i = 1, 2$. These slopes along this single line determine $M(x_1, x_2)$ (up to an affine transformation) over the entire region S^2. Define $f(x_1)$ as the negative of this slope function:

$$
(59) \qquad f(x_1) \equiv M_1(x_1, a)/M_2(x_1, a) = \phi'(x_1)/\phi'(a), \quad a \leq x_1 \leq b.
$$

Then, by elementary integration theory, we have

$$(60)\qquad \phi(x_1) = \alpha + \beta \int_a^{x_1} f(z)dz \quad \text{for } a \leq x_1 \leq b,$$

where α is an arbitrary constant and $\beta \equiv 1/\phi'(a)$ is an arbitrary positive constant. Thus ϕ is uniquely determined up to an affine transformation by the behavior of $M(x_1, x_2)$ along a single line segment in the domain of definition of M. Thus if M approximates U exactly along this line segment, M is essentially determined over the entire region and hence cannot approximate U adequately away from this line segment.

The above results illustrate the relative inflexibility of separable symmetric means. Since separable symmetric means have only one free function of one variable available to approximate arbitrary (symmetric) indifference curves, we should not be too surprised by the above inflexibility results.

We conclude this section with an analysis of our inequality aversion axioms (A8), (A9) and (A10) in the context of separable symmetric means. In the previous section, we showed that (A8) was weaker than (A9) and (A9) was weaker than (A10). It is a somewhat remarkable fact that if M satisfies (A1)–(A4), (A11) or (A12), then the weakest property (A8) implies the strongest property (A10). Before we prove this, it is necessary to give a few definitions.

Let $f(x)$ be a function of N variables defined over a convex set S. Then f is *concave* over S if and only if:

(61) $x^1 \in S,\ x^2 \in S,\ 0 \leq \lambda \leq 1$ implies
$$f[\lambda x^1 + (1-\lambda)x^2] \geq \lambda f(x^1) + (1-\lambda) f(x^2).$$

The function f is *strictly concave* over S if and only if:

(62) $x^1 \in S,\ x^2 \in S,\ x^1 \neq x^2,\ 0 < \lambda < 1$ implies
$$f[\lambda x^1 + (1-\lambda)x^2] > \lambda f(x^1) + (1-\lambda) f(x^2).$$

The function f is *midconcave* over S if and only if:

(63) $x^1 \in S,\ x^2 \in S$ implies
$$f[(1/2)x^1 + (1/2)x^2] \geq (1/2) f(x^1) + (1/2) f(x^2).$$

The function f is *strictly midconcave* over S if and only if:

(64) $x^1 \in S,\ x^2 \in S,\ x^1 \neq x^2$ implies
$$f[(1/2)x^1 + (1/2)x^2)] > (1/2) f(x^1) + (1/2) f(x^2).$$

Midconcave functions were considered in detail in Roberts and Varberg [1973; 211–225] and strictly midconcave functions were briefly considered by Hardy, Littlewood and Polya [1934; 75].[24] If f is continuous over S, then these authors show that midconcavity is actually equivalent to concavity and strict midconcavity is equivalent to strict concavity. Thus if f is continuous, then (63) implies (61) and (64) implies (62).

[24]Actually Roberts and Varberg considered midconvex functions (f is midconvex if and only if $-f$ is midconcave) and Hardy, Littlewood and Polya considered strictly midconvex functions without using the term midconvex.

PROPOSITION 9. *Let $M(x_1, \ldots, x_N)$ satisfy (A1)–(A4), (A8) and (A11) if $N \geq 3$ or (A12) if $N \geq 2$. Then M also satisfies (A9) and (A10). Moreover, the continuous and increasing function of one variable ϕ which is used to represent M and $M_2(x_1, x_2) \equiv \phi^{-1}[(1/2)\phi(x_1) + (1/2)\phi(x_2)]$ must be strictly concave over $S^1 \equiv \{z : a \leq z \leq b\}$.*

Proof: By Proposition 6 (or 7 if $N \geq 2$), there exists a continuous increasing function of one variable $\phi(z)$ defined for $z \in S^1$ such that for $(x_1, x_2) \in S^2$, we have the separable representation (27) for M_2 and for $(x_1, \ldots, x_N) \in S^N$, we have the separable representation (28) for M.

Let $z \in S^1$, $\delta > 0$, $v \equiv (1, -1, 0_{N-2})$ and $z1_N + \delta v \in S^N$ (which implies that $z + \delta$ and $z - \delta$ both belong to S^1). Then by (A8),

$$M(z1_N) > M(z1_N + \delta v) \qquad \text{or}$$

$$z > \phi^{-1}\left[(1/N)\phi(z+\delta) + (1/N)\phi(z-\delta) + \sum\nolimits_{i=3}^{N}(1/N)\phi(z)\right] \qquad \text{or}$$

$$\phi(z) > (1/N)\phi(z+\delta) + (1/N)\phi(z-\delta) + [(N-2)/N]\phi(z) \qquad \text{or}$$

$$\phi(z) > (1/2)\phi(z+\delta) + (1/2)\phi(z-\delta). \tag{65}$$

But (65) implies that ϕ is a strictly midconcave function over S^1. Hence since ϕ is continuous, ϕ is also strictly concave over S^1. Thus $f(x_1, \ldots, x_N) \equiv \sum_{i=1}^{N}(1/N)\phi(x_i)$ is a strictly concave function over S^N. Since $M(x_1, \ldots, x_N) = \phi^{-1}[f(x_1, \ldots, x_N)]$ is an increasing, continuous transformation of $f(x_1, \ldots, x_N)$, it can be shown that $M(x_1, \ldots, x_N)$ is strictly quasiconcave over S^N. Similarly, $M_2(x_1, x_2)$ is strictly quasiconcave over S^2. Thus M and M_2 satisfy (A10). Finally, M and M_2 also satisfy (A9) using Proposition 5. Q.E.D.

If we weaken the strict inequality in (A8) to the weak inequality $\geq$ and replace (A8) by this weakened property in the hypotheses of Proposition 9, then we can rework the proof to show that ϕ must be concave and the functions M and M_2 must be quasiconcave.[25] Propositions closely related to Proposition 9 have been established by Kolm [1969; 192–193] [1976b; 104].

Proposition 9 further illustrates the restrictive nature of axioms (A11) and (A12). Under these separability axioms, the very mild risk aversion axiom (A8) is equivalent to the much stronger risk aversion axiom (A10).

In the following section, we add the homogeneity axiom (A6) or the translatability axiom (A7) to the list of axioms that characterize a separable symmetric mean and we shall exhibit the explicit functional forms that result.

---

[25]This result is related to a result due to Arrow [1953; 45]: an additive, symmetric quasiconcave function must be concave. See also Blackorby, Davidson and Donaldson [1977; 356].

4. Homogeneous or Translatable Separable Symmetric Means

In this section, we follow the example of Nagumo [1930] and add axiom (A6) (homogeneity) or (A7) (translatability) to the list of axioms that characterize a separable symmetric mean.

PROPOSITION 10. *Let $N \geq 3$ and let $M(x_1, \ldots, x_N)$ satisfy (A1)–(A4), (A6) and (A11) where the domain of definition set for M is the positive orthant. Then $M(x_1, \ldots, x_N) = \phi^{-1}\left[\sum_{i=1}^{N}(1/N)\phi(x_i)\right]$ for some function ϕ such that*

$$\phi(z) = \alpha + \beta f_r(z), \quad \beta \neq 0, \tag{66}$$

where f_r is defined for $z > 0$ and each number r by

$$f_r(z) \equiv z^r, \quad r \neq 0 \text{ or} \tag{67}$$

$$f_0(z) \equiv \ln z, \quad \text{if } r = 0. \tag{68}$$

Proof: Since M satisfies (A1)–(A4) and (A11), by Proposition 6, we can represent $M(x_1, \ldots, x_N)$ by $\phi^{-1}\left[\sum_{i=1}^{N}(1/N)\phi(x_i)\right]$ for some continuous monotonic function ϕ. Now apply a result found in Hardy, Littlewood and Polya [1934; 68–69] or Eichhorn [1978b; 32–35] and we can conclude that ϕ must be defined by (66)–(68). Q.E.D.

It should be noted that Nagumo [1930; 79] was the first to establish a version of Proposition 10. However, Nagumo's proof assumed that ϕ was differentiable whereas later proofs established the differentiability of ϕ (and hence of M) as a consequence of the assumptions. Note also that for $r < 0$, $f_r(z) \equiv z^r$ is decreasing in z and hence to obtain an increasing $\phi(z)$, we must take $\beta < 0$. Finally, note that under the hypotheses of Proposition 10, $M(x_1, \ldots, x_N)$ can be represented as a *mean of order r*, M^r. Hardy, Littlewood and Polya [1934; 12–15] defined M^r as follows: for $x \equiv (x_1, \ldots, x_N) \gg 0_N$,

$$M^r(x) \equiv \begin{cases} \left[\sum_{i=1}^{N}(1/N)x_i^r\right]^{1/r}, & \text{for } r \neq 0; \\ \prod_{i=1}^{N} x_i^{1/N}, & \text{for } r = 0. \end{cases} \tag{69}$$

Note that when $r = 1$, $M^1(x)$ is the arithmetic mean and when $r = 0$, $M^0(x)$ is the geometric mean of $x_1, \ldots, x_N$. Thus Proposition 10 serves to provide an axiomatic characterization for the means of order r defined by (69).

There is one degree of freedom left in the definition of $M^r(x)$: the parameter r. It is natural to attempt to determine this parameter by looking at the curvature properties of the "micro" mean of order r function, $M_2^r(x_1, x_2)$, defined for $x_1 > 0$ and $x_2 > 0$ as follows:

$$M_2^r(x_1, x_2) \equiv \begin{cases} [(1/2)x_1^r + (1/2)x_2^r]^{1/r} & \text{for } r \neq 0 \\ x_1^{1/2}x_2^{1/2} & \text{for } r = 0. \end{cases} \tag{70}$$

Given a general micro symmetric mean $M_2(x_1, x_2)$, solving the equation $u = M_2(x_1, x_2)$ yields the equation for the u indifference curve, $x_2 = I(x_1, u)$. The slope of this indifference curve at x_1 is $\partial I(x_1, u)/\partial x_1 \equiv I_1(x_1, u)$. Now calculate how this slope changes as x_1 changes and convert this rate of change into a unit free elasticity. The resulting elasticity is

$$E(x_1, u) = I_{11}(x_1, u)x_1/I_1(x_1, u), \tag{71}$$

where $I_{11}(x_1, u) \equiv \partial^2 I(x_1, u)/\partial x_1^2$. One half of this elasticity evaluated at a point of equality where $x_1 = x_2 = u$ will serve as a convenient measure of the curvature of the two dimensional indifference curve at the point $(x_1, x_2) = (u, u)$ along the ray of equality:[26]

$$C(u) \equiv (1/2)E(u, u) = (1/2)I_{11}(u, u)u/I_1(u, u). \tag{72}$$

We now show that specifying a numerical value for $C(u)$ suffices to determine the parameter r which occurs in the definition of the mean of order r. The family of indifference curves generated by $M_2^r(x_1, x_2)$ for $r \neq 0$ and $u > 0$ is defined by

$$x_2 = I(x_1, u) \equiv (2u^r - x_1^r)^{1/r}; \quad 0 < x_1^r < 2u^r. \tag{73}$$

Evaluating $C(u)$ for this indifference curve function yields

$$C(u) = -(1 - r) = r - 1. \tag{74}$$

For $r = 0$, we have $x_2 = I(x_1, u) \equiv u^2/x_1$ and evaluating (72) for this I yields $C(u) = -1$. Hence (74) is also valid for $r = 0$. Thus specifying a numerical value for the elasticity $C(u)$ suffices to determine the parameter r. In view of Proposition 10, axioms (A1)–(A4), (A6), (A11) and a numerical value for $C(u)$ provide an axiomatic characterization for the mean of order r defined by (69), provided that $N \geq 3$.

We turn our attention to the curvature properties of the means of order r. It is well known that[27] $M^r(x)$ is a concave function over the positive orthant if and only if $r \leq 1$ (and hence, using (74), if and only if $C(u) \leq 0$) and $M^r(x)$ is a convex function if and only if $r \geq 1$ (and hence, equivalently, if and only if $C(u) \geq 0$). Furthermore, $M^r(x)$ is strictly quasiconcave over $x \gg 0_N$ if and only if $r < 1$ (and hence, equivalently, if and only if $C(u) < 0$).

It can be verified that a necessary condition for M^r to satisfy the equality preference axiom (A8) is $C(u) < 0$. Hence we must have $r < 1$ so M^r is strictly quasiconcave and thus axiom (A10) is satisfied. Thus in order for M^r to satisfy the weak curvature axiom (A8), we require that $r < 1$ (or $C(u) < 0$) and then we find that M^r also satisfies the stronger curvature axioms (A9) and (A10).

We now follow the example of Nagumo [1930; 77] and add axiom (A7), ***translatability***, to the other axioms that characterize a symmetric mean.

---

[26]Let $f(z)$ be a one variable utility function. Arrow [1984a; 151] and Pratt [1964] defined a measure of local (absolute) risk aversion as $-f''(z)/f'(z)$ and a measure of local (relative) risk aversion as $-zf''(z)/f'(z)$. Note the resemblance of this last measure to our measure of indifference curve curvature $C(u)$ defined by (72).

[27]See Hardy, Littlewood and Polya [1934; 30].

PROPOSITION 11. *Let $N \geq 3$ and let $M(x)$ satisfy (A1)–(A4), (A7) and (A11). Let the domain of definition set S^N be R^N. Then*

$$(75) \qquad M(x) = \phi^{-1}\left[\sum_{i=1}^{N}(1/N)\phi(x_i)\right], \quad \phi(z) \equiv \alpha + \beta g_s(z), \quad \beta \neq 0,$$

where the function of one variable $g_s(z)$ is defined by

$$(76) \qquad g_s(z) \equiv \begin{cases} e^{sz} & \text{for } s \neq 0; \\ z & \text{for } s = 0. \end{cases}$$

Proof: Axioms (A1)–(A4) and (A11) imply (via Proposition 6) the existence of a continuous monotonic function of one variable ϕ such that $M(x_1, \ldots, x_N) = \phi^{-1}\left[\sum_{i=1}^{N}(1/N)\phi(x_i)\right]$. Applying (A7) to this representation for M, we obtain for $x \in R^N$ and $z \in R^1$:

$$M(x) + z = M(x + z1_N) \quad \text{or}$$

$$(77) \qquad \phi^{-1}\left[\sum_{i=1}^{N}(1/N)\phi(x_i)\right] + z = \phi^{-1}\left[\sum_{i=1}^{N}(1/N)\phi(x_i + z)\right].$$

For $N = 2$, the functional equation (77) is treated in detail by Aczél [1966; 152–153] and he obtains the family of solutions defined by (75) and (76). However, Aczél [1966; 153] stated that his result is valid for a general $N > 2$, and indeed, it is straightforward to modify his proof to cover the general case. Note that if $s < 0$, then in order for $\phi(z)$ to be increasing in z, we will require that β be negative. On the other hand if $s \geq 0$, then in order for $\phi(z)$ to be increasing, we require $\beta > 0$. Q.E.D.

Results similar to Proposition 11 have been established by Nagumo [1930; 77–78] and Kolm [1976a; 426–430); Nagumo used (6) in place of (A11) and assumed differentiability of ϕ (whereas Aczél's proof deduces the differentiability of ϕ as a consequence of the other assumptions), while Kolm [1976a; 427] placed a separability assumption on M in place of our consistency in aggregation axiom (A11).[28]

We note that under the hypotheses of Proposition 11, $M(x)$ can be represented as a *Nagumo mean of order s*,[29] $N^s(x)$, defined for $x \equiv (x_1, \ldots, x_N) \in R^N$ as follows:

$$(78) \qquad N^s(x) \equiv \begin{cases} (1/s)\ln\left[\sum_{i=1}^{N}(1/N)e^{sx_i}\right] & \text{for } s \neq 0; \\ \sum_{i=1}^{N}(1/N)x_i, & \text{for } s = 0. \end{cases}$$

[28]See also Kolm [1969; 194]. Kolm [1969; 191] termed his separability conditions an "independence" property.

[29]Nagumo [1930; 78–79] appears to have been the first to consider this family of means. They were independently discovered by Kolm [1976b; 95].

When $s = 1$, we call $N^1(x) \equiv \ln\left[\sum_{i=1}^N (1/N)e^{x_i}\right]$ the *exponential mean* of the N numbers $x_1, \ldots, x_N$ and when $s = 0$, $N^0(x) \equiv \sum_{i=1}^N (1/N)x_i$ becomes the ordinary *arithmetic mean*.

Putting together the results of Propositions 10 and 11, it can be seen that the only separable symmetric mean that satisfies both (A6) and (A7) (homogeneity and translatability) is the arithmetic mean.[30]

As was done with our characterization of means of order r, it is natural to characterize the parameter s in the Nagumo means of order s by using the local curvature indicator along the ray of equality, $C(u)$, defined by (72). Define the micro mean of order s function N_2^s by

$$(79) \qquad N_2^s(x_1, x_2) \equiv \begin{cases} (1/s)\ln[(1/2)e^{sx_1} + (1/2)e^{sx_2}] & \text{if } s \neq 0; \\ (1/2)x_1 + (1/2)x_2 & \text{if } s = 0. \end{cases}$$

For $s \neq 0$, the indifference curve function for an arbitrary utility level u is given by $x_2 = I(x_1, u) \equiv s^{-1}\ln(2e^{su} - e^{sx_1})$ where x_1 satisfies $e^{sx_1} < 2e^{su}$. Evaluating $C(u)$ defined by (72) for this $I(x_1, u)$ yields

$$(80) \qquad C(u) = s.$$

For the case $s = 0$, the indifference curve function is $I(x_1, u) \equiv 2u - x_1$ and for this function, $C(u) = 0$. Thus (80) is also correct for the case $s = 0$.

A necessary condition for our equality preference axiom to hold for $M = N^s$ is $C(u) \leq 0$ or, using (80), $s \leq 0$. However, if $s = 0$, then $N^0(x)$ is linear in x and (A8) does not hold. Hence a necessary condition for N^s to satisfy (A8) is $s < 0$. However, the following proposition shows that this condition is sufficient to imply that N^s satisfies (A10), and hence (A9) and (A8) using Proposition 5.

PROPOSITION 12. *A necessary and sufficient condition for the Nagumo mean of order s, N^s defined by (78), to be a strictly quasiconcave function over any convex subset S of R^N is $s < 0$.*

Proof: If $s = 0$, $N^0(x)$ is a linear function of x and hence is not strictly quasiconcave. Hence we can restrict our attention to the case where $s \neq 0$. Let $x \equiv (x_1, \ldots, x_N) \in S$ and calculate the first and second derivatives of $N^s(x)$ defined by (78):

$$(81) \qquad \partial N^s(x)/\partial x_i = e^{sx_i}\Big/\left(\sum_{j=1}^N e^{sx_j}\right) \equiv v_i > 0, \quad i = 1, \ldots, N;$$

$$(82) \qquad \partial^2 N^s(x)/\partial x_i^2 = s(v_i - v_i^2), \qquad i = 1, \ldots, N;$$

$$(83) \qquad \partial^2 N^s(x)/\partial x_i \partial x_j = -s v_i v_j, \qquad i \leq i \neq j \leq N,$$

[30] Aczél [1966; 153] noted that the only quasilinear (or separable) mean that satisfied (A6) and (A7) when $N = 2$ was the (weighted) arithmetic mean.

where the numbers v_i are defined in (81). Define the column vector $v \equiv (v_1, v_2, \ldots, v_N)^T$ and define $\hat{v}$ as a diagonal $N \times N$ matrix with the elements of the vector v running down the main diagonal. Then using matrix notation, the vector of first order partial derivatives of $N^s(x)$, $\nabla N^s(x)$, and the matrix of second order partial derivatives, $\nabla^2 N^s(x)$, can be written as follows:

$$(84) \qquad \nabla N^s(x) = v; \quad \nabla^2 N^s(x) = s[\hat{v} - vv^T]$$

where the vector v satisfies

$$(85) \qquad v \gg 0_N, \quad v^T 1_N = 1.$$

From Diewert, Avriel and Zang [1981; 409], a necessary and sufficient condition for a twice continuously differentiable function $N^s(x)$ to be strongly quasiconcave over S is that for each point $x \in S$, we have:

$$(86) \qquad y \neq 0_N, \quad y^T \nabla N^s(x) = 0 \text{ implies } y^T \nabla^2 N^s(x) y < 0.$$

Substituting (84) into (86) yields the condition:

$$(87) \qquad y \neq 0_N, \quad y^T v = 0 \text{ implies } sy^T[\hat{v} - vv^T]y = sy^T \hat{v} y < 0.$$

Now $y^T \hat{v} y = \sum_{i=1}^N v_i y_i^2 > 0$ since each $v_i > 0$ and at least one $y_i^2 > 0$ and so (87) will be satisfied if and only if $s < 0$. Using a result in Diewert, Avriel and Zang [1981; 409], strong quasiconcavity implies strict quasiconcavity. Hence $s < 0$ implies that $N^s(x)$ is strictly quasiconcave. The condition $s < 0$ is also necessary for $N^s(x)$ to be strictly quasiconcave since for $s = 0$, $N^s(x)$ is linear (and hence not strictly quasiconcave) and for $s > 0$, $N^s(x)$ is strongly quasiconvex and strictly quasiconvex (and hence is not strictly quasiconcave). Q.E.D.

COROLLARY. (Kolm [1976; 95]): *The condition $s \leq 0$ ($s \geq 0$) is necessary and sufficient for the concavity (convexity) of $N^s(x)$ over any convex subset of R^N.*

Proof: A well known[31] necessary and sufficient condition for concavity (convexity) of a twice continuously differentiable function $f(x)$ defined over a convex set with a nonempty interior is that $\nabla^2 f(x)$ be negative (positive) semidefinite for all $x \in S$. Hence in view of (84), we need only show that $\hat{v} - vv^T$ is a positive semidefinite matrix, or equivalently, that $y^T(\hat{v} - vv^T)y \geq 0$ for all vectors $y \in R^N$. This last inequality is equivalent to $y^T vv^T y \leq y^T \hat{v} y$ or, since $v = \hat{v} 1_N$, we need to show that:

$$(88) \qquad (y^T \hat{v} 1_N)^2 \leq y^T \hat{v} y.$$

[31]See the references to the literature in Diewert, Avriel and Zang [1981; 410].

From (85), the numbers v_i are positive. Hence the diagonal matrix $\hat{v}$ has a diagonal square root matrix $\hat{v}^{1/2}$ where the positive square roots $v_i^{1/2}$ run down the main diagonal of $\hat{v}^{1/2}$. Applying the Cauchy-Schwarz inequality[32] to the vectors $\hat{v}^{1/2}y$ and $\hat{v}^{1/2}1_N$, we have the following inequality:

$$\begin{aligned}
(y^T\hat{v}^{1/2}\hat{v}^{1/2}1_N)^2 &\leq (y^T\hat{v}^{1/2}\hat{v}^{1/2}y)(1_N^T\hat{v}^{1/2}\hat{v}^{1/2}1_N) && \text{or} \\
(y^T\hat{v}1_N)^2 &\leq (y^T\hat{v}y)(1_N^T\hat{v}1_N) && \text{or} \\
(y^T\hat{v}1_N)^2 &\leq (y^T\hat{v}y)(1_N^Tv) && \text{since } v = \hat{v}1_N \text{ or} \\
(y^T\hat{v}1_N)^2 &\leq y^T\hat{v}y && \text{since } 1_N^Tv = 1 \text{ by (85).}
\end{aligned}$$

Thus we have established (88). Q.E.D.

Blackorby and Donaldson [1984; 20] noted that N^s is concave if $s \leq 0$. Blackorby and Donaldson [1980; 120] also noted that N^s is quasiconcave if $s < 0$. Finally, Blackorby and Donaldson [1980; 116] [1984; 20] called N^s a *Kolm* [1976a]-*Pollak* [1971b] *index*.

We can summarize the results in the second half of this section as follows: if $N \geq 3$ and the mean function satisfies (A1)–(A4), (A7) and (A11) and $C(u) = s$, then $M(x) = N^s(x)$, a Nagumo mean of order s defined by (78). Thus we have provided an axiomatic characterization for separable symmetric means which also satisfy the translatability axiom (A7).

Looking at the class of generating functions $\phi(z)$ that we have obtained via Propositions 10 and 11, we see that we have obtained z, z^r, $\ln z$, e^z and e^{sz} or all of the nontrigonometric elementary functions of one variable.[33] Moreover, there are separate axiomatic characterizations for each of the symmetric means that correspond to these elementary functions, except that we have provided two such characterizations for the arithmetic mean.

We now turn our attention to a very useful generalization of separable symmetric means.

5. Implicitly Separable Symmetric Means

In this section, we will define a class of symmetric means using the concept of implicit separability. As was mentioned in the introduction, implicitly separable functions were considered briefly in Blackorby, Primont and Russell [1978; 113–114] and in more detail in Blackorby, Davidson and Schworm [1991].

We shall define a weaker consistency in aggregation property than (A12) in Section 3. As in Section 3, the domain of definition for $M(x_1, \ldots, x_N)$ is

[32]See Hardy, Littlewood and Polya [1934; 16].

[33]Aczél and Daroézy [1963] obtained axiomatic characterizations of the elementary trigonometric functions using weighted quasilinear means.

$S^N \equiv \{x : a1_N \leq x \leq b1_N\}$ where $-\infty < a < b < +\infty$. The same finite parameters a and b appear in the definitions $S^n \equiv \{(x_1, \ldots, x_n)\colon a \leq x_i \leq b,$ $i = 1, 2, \ldots, n\}$ for $n = 1, 2, \ldots, N-1$.

(A13) *(Conditional Consistency in Aggregation)*: There exists a function of three variables, $M_2(x_1, x_2, u)$, defined for $(x_1, x_2, u) \in S^3$ and such that $x \in S^N$ and $u = M(x)$ implies $u = M[M_2(x_1, x_2, u), M_2(x_1, x_2, u), x_3, \ldots, x_N]$. The function M_2 has the following properties:

(89) for every $u \in S^1$, $M_2(x_1, x_2, u)$ is a *symmetric mean* in $(x_1, x_2) \in S^2$;

(90) $M_2(x_1, x_2, u)$ is *jointly continuous* in (x_1, x_2, u);

(91) $M_2(x_1, x_2, u)$ is bisymmetric in x_1, x_2 for each $u \in S^1$; i.e., for every $(x_1, x_2, x_3, x_4, u) \in S^5$, $M_2[M_2(x_1, x_2, u), M_2(x_3, x_4, u), u] = M_2[M_2(x_1, x_3, u), M_2(x_2, x_4, u), u]$.

Note that (A13) closely parallels (A12) except that now the micro aggregator function $M_2(x_1, x_2, u)$ depends not only on the two numbers being aggregated x_1 and x_2, but it also depends on the overall mean $u = M(x_1, x_2, x_3, \ldots, x_N)$ of all of the numbers being aggregated. Note also that (89) means that $M_2(x_1, x_2, u)$ satisfies axioms (A1)–(A4) with respect to (x_1, x_2) for each $u \in S^1$. In the context of modeling uncertainty, $M_2(x_1, x_2, u)$ can be interpreted as the *certainty equivalent* of x_1 and x_2, *conditional* on the utility level u; see Machina [1982; 288] who defined the conditional certainty equivalent of a prospect and Chew and Epstein [1989a; 212] who used separability axioms to define contingent certainty equivalents.

Note that if $M_2(x_1, x_2, u)$ does not depend on u, then (A13) collapses down to our old consistency in aggregation axiom (A12).

Recall that axiom (A12) led to the separable symmetric mean $M(x_1, \ldots, x_N) \equiv \phi^{-1}[\sum_{i=1}^{N}(1/N)\phi(x_i)]$ where ϕ is a continuous increasing function of one variable. The following proposition is a counterpart to Proposition 7.

PROPOSITION 13. *Let $N \geq 2$ and let $M(x_1, \ldots, x_N)$ satisfy (A1)–(A4) and (A13) over $S^N \equiv \{x : a1_N \leq x \leq b1_N\}$ where $-\infty < a < b < +\infty$. Then there exists a function of two variables $\phi_u(z)$ defined for $(z, u) \in S^2$ such that for $(x_1, x_2, u) \in S^3$, we have*

$$M_2(x_1, x_2, u) = \phi_u^{-1}[(1/2)\phi_u(x_1) + (1/2)\phi_u(x_2)] \tag{92}$$

and for $x \in S^N$ and $u = M(x)$, we have

$$u = M(x_1, \ldots, x_N) = \phi_u^{-1}\left[\sum_{i=1}^{N}(1/N)\phi_u(x_i)\right]. \tag{93}$$

The function $\phi_u(z)$ has the following properties:

(94) $\phi_u(z)$ *is jointly continuous in* $(z,u) \in S^2$;

(95) *for each* $u \in S^1$, $\phi_u(z)$ *is monotonically increasing in* $z \in S^1$;

(96) *for each* $x \in S^N$, *there exists a unique* $u \in S^1$ *such that* $\phi_u(u) = \sum_{i=1}^{N}(1/N)\phi_u(x_i)$;

(97) *for each* $u \in S^1$, $\phi_u(a) = a$ *and* $\phi_u(b) = b$.

Proof: Since for each $u \in S^1$, $M_2(x_1, x_2, u)$ satisfies the bisymmetry equation (91), the results of Nagumo [1930; 73] and Aczél [1966; 281] may be applied and so for each $u \in S^1$, we have the existence of a continuous monotonically increasing function of z, $\phi_u(z)$, such that for $(x_1, x_2) \in S^2$, equation (92) is satisfied. If we set $\phi_u(a) = a$ and $\phi_u(b) = b$, then for each $u \in S^1$, $\phi_u(z)$ is uniquely determined. The function values for ϕ_u are determined recursively using $M_2(x_1, x_2, u)$ as follows:

$$\begin{aligned}\phi_u[M_2(a,b,u)] &\equiv (1/2)\phi_u(a) + (1/2)\phi_u(b)\\ &= (1/2)a + (1/2)b;\end{aligned}$$

$$\begin{aligned}\phi_u\Big[M_2[a, M_2(a,b,u), u]\Big] &\equiv (1/2)\phi_u(a) + (1/2)\phi_u[M_2(a,b,u)]\\ &= (3/4)a + (1/4)b;\end{aligned}$$

$$\begin{aligned}\phi_u\Big[M_2[M_2(a,b,u), b, u]\Big] &\equiv (1/2)\phi_u[M_2(a,b,u)] + (1/2)\phi_u(b)\\ &= (1/4)a + (3/4)b; \text{ etc.}\end{aligned}$$

Since $M_2(x_1, x_2, u)$ is jointly continuous in its variables, it can be seen from the above construction that $\phi_u(z)$ will also be jointly continuous in (z,u). Thus we have established (92), (94), (95) and (97). It remains to establish (93) and (96).

Let $x \in S^N$ and $u \equiv M(x)$. Then by (A13), we have

$$\begin{aligned}u &= M(x_1, x_2, x_3, \ldots, x_N)\\ &= M[M_2(x_1,x_2,u), M_2(x_1,x_2,u), x_3, \ldots, x_N]\\ &= M\Big[\phi_u^{-1}[(1/2)\phi_u(x_1) + (1/2)\phi_u(x_2)],\\ &\qquad \phi_u^{-1}[(1/2)\phi_u(x_1) + (1/2)\phi_u(x_2)], x_3 \ldots, x_N\Big] \text{ using (92)}\\ &\equiv F[\phi_u(x_1) + \phi_u(x_2), x_3, \ldots, x_N, u]\end{aligned}$$

(98) $$= G\Big[\sum_{i=1}^{N}(1/N)\phi_u(x_i), u\Big]$$

where (98) follows from the line above using (A2), the symmetry of M in $x_1, \ldots, x_N$. Now set all $x_i = u$ in (98) and get

$$\begin{aligned} u &= G[\phi_u(u), u] && \text{or, upon transforming by } \phi_u, \\ \phi_u(u) &= \phi_u\{G[\phi_u(u), u]\}, && \text{and letting } z \equiv \phi_u(u) \\ z &= \phi_u[G(z, u)], && \text{or upon transforming by } \phi_u^{-1} \\ \phi_u^{-1}(z) &= G(z, u). && \end{aligned} \tag{99}$$

Now substitute (99) into (98) and we obtain (93).

Given $x \in S^N$, u is uniquely defined as $u = M(x)$. Transforming both sides of (93) by ϕ_u yields (96). Hence the u solution to (96) must be unique.

Q.E.D.

COROLLARY 1. *(Consistency in Aggregation for any Partition): Under the above hypotheses, for $n = 2, 3, \ldots, N-1$, there exists a conditional aggregator function $M_n(x_1, \ldots, x_n, u)$ such that $u = M(x_1, \ldots, x_N)$ implies $u = M[M_n(x_1, \ldots, x_n, u)1_n, x_{n+1}, \ldots, x_N]$. M_n is defined in terms of ϕ_u as follows:*

$$M_n(x_1, \ldots, x_n, u) \equiv \phi_u^{-1}\left[\sum_{i=1}^{n}(1/n)\phi_u(x_i)\right]. \tag{100}$$

Proof: Let $x \in S^N$ and define $u \equiv M(x_1, \ldots, x_N) = \phi_u^{-1}\left[\sum_{i=1}^{N}(1/N)\phi_u(x_i)\right]$. For $n = 2$, axiom (A13) gives the desired result. For $n = 3, 4, \ldots, N-1$, define M_n by (100) and calculate:

$$\begin{aligned} &M[M_n(x_1, \ldots, x_n, u)1_n, x_{n+1}, \ldots, x_N] \\ &= \phi_u^{-1}\left[\sum_{i=1}^{n}(1/N)\phi_u[M_n(x_i, \ldots, x_n, u)] + \sum_{j=n+1}^{N}(1/N)\phi_u(x_j)\right] \\ &= \phi_u^{-1}\left[(n/N)\phi_u\left[\phi_u^{-1}\left(\sum_{i=1}^{n}(1/n)\phi_u(x_i)\right)\right] + \sum_{j=n+1}^{N}(1/N)\phi_u(x_j)\right] \\ &= \phi_u^{-1}\left[\sum_{i=1}^{n}(1/N)\phi_u(x_i) + \sum_{j=n+1}^{N}(1/N)\phi_u(x_j)\right] \\ &= M(x_1, \ldots, x_N). \end{aligned}$$

Q.E.D.

COROLLARY 2. *(Invariance under Utility Dependent Affine Transformations): Under the hypotheses of Proposition 13, the function $\phi_u(z)$ which was used to represent M via (93) can be replaced by*

$$\chi_u(z) \equiv \alpha(u) + \beta(u)\phi_u(z), \quad \beta(u) \neq 0 \tag{101}$$

where $\alpha(u)$ and $\beta(u)$ are continuous functions of u. If we want $\chi_u(z)$ to be increasing in z, we require in addition that

$$(102) \qquad \beta(u) > 0.$$

If (101) is satisfied, then χ can replace ϕ in (93), (94) and (96). If (102) is also satisfied, then χ can replace ϕ in (95) as well.

Proof: Let $x \in S^N$ and $u \equiv M(x)$. Then u satisfies (96). Multiply both sides of (96) by $\beta(u) \neq 0$ and add $\alpha(u)$ to both sides of the resulting equation. We obtain

$$\beta(u)\phi_u(u) + \alpha(u) = \beta(u)\sum_{i=1}^{N}(1/N)\phi_u(x_i) + \alpha(u) \text{ or}$$
$$\alpha(u) + \beta(u)\phi_u(u) = \sum_{i=1}^{N}(1/N)[\alpha(u) + \beta(u)\phi_u(x_i)] \text{ or}$$
$$\chi_u(u) = \sum_{i=1}^{N}(1/N)\chi_u(x_i).$$

The rest of the proof is obvious. Q.E.D.

We say that a function $M(x)$ is an *implicitly separable symmetric mean* if M satisfies (A1)–(A4) and (A13).

Corollary 1 above shows that an implicitly separable symmetric mean has a (conditional) consistency in aggregation property for any subset of its variables.

Corollary 2 above shows that the invariance properties of the bivariate generating function $\phi_u(z)$ for an implicitly separable symmetric mean are more complex than they were in the case of a separable symmetric mean where the univariate generating function $\phi(z)$ could only be replaced by $\alpha + \beta\phi(z)$, $\beta \neq 0$, where α and β were constants. In the present case where we require that the generating function be continuous in (z, u), α and β can be arbitrary continuous functions of u, subject to the restriction $\beta(u) \neq 0$ or $\beta(u) > 0$. The invariance conditions (101) and (102) were first obtained by Dekel [1986; 307] in his axiomatic characterization of preferences under uncertainty.

We now show that the existence of a $\phi_u(z)$ satisfying (94)–(96) is sufficient to generate an implicitly separable symmetric mean M if we use the equation in (96) to implicitly define $u = M(x_1, \ldots, x_N)$.

PROPOSITION 14. *Let the function $\phi_u(z)$ be defined for $(z, u) \in S^2$ and satisfy (94)–(96). For $x \in S^N$, define $u = M(x)$ as the unique solution to the equation in (96). Then M satisfies (A1)–(A4) and (A13). Note that $S^n \equiv \{x : a1_n \leq x \leq b1_n\}$ where $-\infty < a < b < +\infty$.*

Proof: Let $\lambda \in S^1$. Then $u = \lambda$ is a u solution to $\phi_u(u) = \sum_{i=1}^{N}(1/N)\phi_u(\lambda)$. By (96), this solution is unique. Thus M satisfies (A1).

Since the right hand side of (96) is symmetric in $x_1, \ldots, x_N, M(x_1, \ldots, x_N)$ is also symmetric in $x_1, \ldots, x_N$, which establishes (A2).

For $n = 1, 2, \ldots$, let $x^n \in S^N$ with $\lim_n x^n = x^0$, $u^n \equiv M(x^n)$ and $\lim_n u^n = u^0$. We need to show that $u^0 = M(x^0)$. The equality $u^n = M(x^n)$ means that

$$(103) \qquad \sum_{i=1}^{N}(1/N)\phi_{u^n}(x_i^n) - \phi_{u^n}(u^n) = 0 \text{ for } n = 1, 2, \ldots.$$

Since by (94), $\phi_u(x_i)$ is jointly continuous in (x_i, u), taking limits of both sides of (103) yields

$$(104) \qquad \sum_{i=1}^{N}(1/N)\phi_{u^0}(x_i^0) - \phi_{u^0}(u^0) = 0.$$

But (104) shows that u^0 is the u solution to $\phi_u(u) = \sum_{i=1}^{N}(1/N)\phi_u(x_i^0)$ and hence $u^0 = M(x^0)$ which establishes (A3).

Let $x^0 \in S^N$, $x^1 \in S^N$ with $x^0 < x^1$. Define $u^0 \equiv M(x^0)$ and $u^1 \equiv M(x^1)$. We need to show that $u^0 < u^1$. Since $u^0 = M(x^0)$, by (96), we have

$$(105) \qquad \begin{aligned} 0 &= (1/N)\sum_{i=1}^{N} \phi_{u^0}(x_i^0) - \phi_{u^0}(u^0) \\ &< (1/N)\sum_{i=1}^{N} \phi_{u^0}(x_i^1) - \phi_{u^0}(u^0) \text{ using (95) and } x^0 < x^1. \end{aligned}$$

Define $u^* \equiv \max_i\{x_i^1 : i = 1, 2, \ldots, N\}$. Then

$$(106) \qquad \begin{aligned} 0 &= (1/N)\sum_{i=1}^{N} \phi_{u^*}(u^*) - \phi_{u^*}(u^*) \\ &\geq (1/N)\sum_{i=1}^{N} \phi_{u^*}(x_i^1) - \phi_{u^*}(u^*) \text{ using (95) and } x_i^1 \leq u^*. \end{aligned}$$

Define $f(u) \equiv \sum_{i=1}^{N}(1/N)\phi_u(x_i^1) - \phi_u(u)$. The inequality (105) is equivalent to $f(u^0) > 0$ and (106) is equivalent to $f(u^*) \leq 0$. Since by (94), $f(u)$ is continuous in u, the above inequalities imply that a root u^1 to the equation $f(u) = 0$ exists such that $u^0 < u^1 \leq u^*$. Property (96) implies that this root is unique and we have shown $u^0 < u^1$, which establishes (A4).

To establish (A13), let $x \in S^N$ and let $u = M(x_1, \ldots, x_N)$. Define $M_2(x_1, x_2, u)$ by (92). Using (96) and (92), we have:

$$\begin{aligned} u &= M(x_1, \ldots, x_N) \\ &= \phi_u^{-1}\Big[\sum_{i=1}^{N}(1/N)\phi_u(x_i)\Big] \\ &= \phi_u^{-1}\Big[(1/N)\phi_u(x_1) + (1/N)\phi_u(x_2) + \sum_{i=3}^{N}(1/N)\phi_u(x_i)\Big] \\ &= \phi_u^{-1}\Big[(1/N)[(1/2)\phi_u(x_1) + (1/2)\phi_u(x_2)] \\ &\qquad + (1/N)[(1/2)\phi_u(x_1) + (1/2)\phi_u(x_2)] + \sum_{i=3}^{N}(1/N)\phi_u(x_i)\Big] \\ &= \phi_u^{-1}\Big[(1/N)\phi_u[M_2(x_1, x_2, u)] + (1/N)\phi_u[M_2(x_1, x_2, u)] + \sum_{i=3}^{N}(1/N)\phi_u(x_i)\Big] \\ &= M[M_2(x_1, x_2, u), M_2(x_1, x_2, u), x_3, \ldots, x_N] \end{aligned}$$

which establishes the conditional consistency in aggregation property for M and M_2. Using definition (92), it is easy to check that M_2 satisfies (89)–(91).

Q.E.D.

Thus the existence of a function of two variables $\phi_u(z)$ satisfying (94), (95) and (96) is sufficient to imply that the mean function $M(x_1, \ldots, x_N)$ defined via (93) or (96) will satisfy (A1)–(A4) and (A13). Conversely, if M satisfies (A1)–(A4) and (A13), then by Proposition 13, a ϕ satisfying (94), (95) and (96) exists. We call ϕ the generating function for the implicitly separable symmetric mean M.

Given a function $\phi_u(z)$, it is straightforward to verify conditions (94) and (95). However, it will not always be easy to verify whether a given ϕ satisfies (96). The following condition (which was used by Chew and Epstein [1989a; 217] in their implicit linear utility model) is *sufficient* to imply (96):

(107) for every $z \in S^1$, $\tau(z, u) \equiv \phi_u(z) - \phi_u(u)$ is decreasing in u for $u \in S^1$.

Unfortunately, (107) is not a necessary condition for (96). We will return to the problem of obtaining necessary and sufficient conditions for property (96) in Section 7 below.

We now consider the approximation capabilities of implicitly separable symmetric means. If $N = 2$, we shall show that any symmetric mean $U(x_1, x_2)$ can be represented as an implicitly separable symmetric mean.

Let $U(x_1, x_2)$ satisfy (A1)–(A4) over $S^2 \equiv \{(x_1, x_2)\colon a \le x_i \le b,\ i = 1, 2\}$ where $-\infty < a < b < +\infty$. The family of indifference curve functions, $x_2 = I(x_1, u)$, that corresponds to U is defined as follows: for each $u \in S^1$, define $I(x_1, u)$ by the following equation:

(108) $$u = U[x_1, I(x_1, u)].$$

For each $u \in S^1$, we shall require later that $I(x_1, u)$ be defined for all x_1 such that $a \le x_1 \le b$. However, if U is defined only over S^2, the above property will be true only when $u = U(a, b)$. Thus we shall require that $U(x_1, x_2)$ either be defined over the larger set $S^{2*} \equiv \{(x_1, x_2)\colon a \le x_1 \le b;\ a \le U(x_1, x_2) \le b\}$ or we need to artificially extend the domain of definition of U so that the indifference curves of the extended function, $I^*(x_1, u)$, are defined for all $(x_1, u) \in S^2$.

Given that $U(x_1, x_2)$ satisfies (A1)–(A4) over S^{2*}, it can be verified that $I(x_1, u)$ has the following properties:

(109) for each $u \in S^1$, $I(x_1, u)$ is decreasing in x_1;

(110) for each $x_1 \in S^1$, $I(x_1, u)$ is increasing in u;

(111) $I(x_1, u)$ is jointly continuous in (x_1, u);

(112) if $x_2 = I(x_1, u)$ for $(x_1, u) \in S^2$, then $x_1 = I(x_2, u)$;

(113) for each $u \in S^1$, $u = I(u, u)$.

Property (112) reflects the fact that $U(x_1, x_2)$ is symmetric in (x_1, x_2) and property (113) reflects the fact that $u = U(u, u)$.

For each $u \in S^1$, we use the indifference curve function $I(x_1, u)$ to define $\phi_u(z)$ as follows: for $u \in S^1$,

$$\phi_u(z) \equiv \begin{cases} z & \text{for } a \le z \le u \\ 2u - I(z, u) & \text{for } u \le z \le b. \end{cases} \tag{114}$$

Note that (113) implies that $\phi_u(u) = u$.

We need to verify that ϕ defined by (114) satisfies properties (94)–(96). Since $I(z, u)$ is jointly continuous in (z, u), it can be seen that $\phi_u(z)$ will be jointly continuous in (z, u). By (109), $I(z, u)$ is decreasing in z so $-I(z, u)$ is increasing in z and thus $\phi_u(z)$ is increasing in z for each fixed u. Thus $\phi_u(z)$ satisfies (94) and (95).

To verify that property (96) holds, consider first *case* (i), where $a \le x_1 = x_2 \le b$. We need to show that there exists a unique $u \in S^1$ such that

$$\begin{aligned} 0 &= (1/2)\phi_u(x_1) + (1/2)\phi_u(x_1) - \phi_u(u) \\ &= \phi_u(x_1) - \phi_u(u). \end{aligned} \tag{115}$$

Since $\phi_u(z)$ is monotonically increasing in z, the unique solution to (115) is $u = x_1$. Now consider *case* (ii), where $a \le x_2 < x_1 \le b$. It is easy to show that any u solution to (96) when $N = 2$ must lie between x_1 and x_2; i.e., we must have $x_2 < u < x_1$. Thus we require a unique u such that $x_2 < u < x_1$ and

$$\begin{aligned} 0 &= (1/2)\phi_u(x_1) + (1/2)\phi_u(x_2) - \phi_u(u) \\ &= (1/2)[2u - I(x_1, u)] - (1/2)\phi_u(x_2) - \phi_u(u) && \text{using (114) and } u < x_1 \\ &= u - (1/2)I(x_1, u) + (1/2)x_2 - \phi_u(u) && \text{using (114) and } x_2 < u \\ &= u - (1/2)I(x_1, u) + (1/2)x_2 - u && \text{using (114)} \\ &= (1/2)[x_2 - I(x_1, u)]. \end{aligned} \tag{116}$$

From the properties of indifference curve functions, it is clear that the u solution to $x_2 = I(x_1, u)$ is unique. Hence the u solution to (116) is also unique. Finally, consider *case* (iii), where $a \le x_1 < x_2 \le b$. Again, any u solution to (96) must lie between x_1 and x_2 so we must have $x_1 < u < x_2$. We require a unique u solution to

$$\begin{aligned} 0 &= (1/2)\phi_u(x_1) + (1/2)\phi_u(x_2) - \phi_u(u) \\ &= (1/2)x_1 + (1/2)[2u - I(x_2, u)] - u \\ &\qquad \text{using (114) and } x_1 < u < x_2 \\ &= (1/2)[x_1 - I(x_2, u)]. \end{aligned} \tag{117}$$

The u solution to (117) is unique and is equal to the u solution to $x_2 = I(x_1, u)$ using (112). Thus $\phi_u(z)$ satisfies (96).

Now that we have shown that $\phi_u(z)$ defined by (114) satisfies (94)–(96), we can use this function to generate a symmetric mean $M(x_1, x_2)$; i.e., for $a \leq x_1$, $x_2 \leq b$, $u = M(x_1, x_2)$ is the unique u solution to

$$(118) \qquad (1/2)\phi_u(x_1) + (1/2)\phi_u(x_2) - \phi_u(u) = 0.$$

However, equations (115)–(117) show that (118) is equivalent to the equation $x_2 = I(x_1, u)$ and from the definition of I as the indifference curve function which corresponds to U, we see that $x_2 = I(x_1, u)$ is equivalent to $u = U(x_1, x_2)$. Thus for $a \leq x_1$, $x_2 \leq b$, we have $M(x_1, x_2) = U(x_1, x_2)$ and we have represented the arbitrary symmetric mean U as an implicitly separable symmetric mean M over S^2. Comparing this result with the approximation result in Proposition 8 above, we see that in general, only a *single* indifference curve for U can be attained by a separable symmetric mean but the *entire family* of indifference curves for $U(x_1, x_2)$ can be attained by an implicitly separable symmetric mean. Thus the class of implicitly separable symmetric means is far more flexible than the class of separable symmetric means.

Note that definition (114) allows us to convert an arbitrary two dimensional symmetric mean $U(x_1, x_2)$ (which has indifference curves $x_2 = I(x_1, u)$) into a generating function $\phi_u(z)$ which satisfies (94)–(96) and hence we can use the resulting ϕ as the generating function for an N dimensional implicitly symmetric mean $M(x_1, \ldots, x_N)$ via definition (93).

Recall that property (107) was sufficient to imply (96). It is of some interest to determine what conditions the symmetric mean $U(x_1, x_2)$ would have to satisfy in order to have the $\phi_u(z)$ defined by (114) satisfy the stronger condition (107). It can be verified that $\phi_u(z)$ defined by (114) will satisfy (107) if and only if the following condition on $I(x_1, u)$ is satisfied:

$$(119) \qquad \text{for every } z \in S^1,\ u - I(z, u) \text{ is decreasing in } u \text{ for } a \leq u \leq z.$$

Condition (119) translates into the following condition on U:

$$(120) \qquad \text{for every } x_1 \in S^1,\ U(x_1, x_2) - x_2 \text{ is decreasing in } x_2 \text{ for } a \leq x_2 \leq x_1.$$

Condition (120) is somewhat restrictive, which shows that the sufficient condition (107) for (96) is stronger than (96).

We conclude this section with a brief discussion of our curvature axioms (A8)–(A10).

Since we have just shown that an arbitrary two dimensional symmetric mean $U(x_1, x_2)$ can be represented as an implicitly separable symmetric mean, it can be seen that if $M(x_1, \ldots, x_N)$ satisfies (A1)–(A4), (A13) and the weak curvature axiom (A8), it is not the case (as it was with separable symmetric means) that M also necessarily satisfies the stronger curvature axioms (A9) and (A10). Thus for implicitly separable symmetric means, the axioms (A8), (A9) and (A10) lead to distinct classes of functions.

It is difficult (for this author) to obtain simple necessary and sufficient conditions on the generating function $\phi_u(z)$ that will imply that the resulting $M(x)$ satisfies (A8), (A9) or (A10). We will provide only a simple sufficient condition for $M(x)$ to satisfy (A10).

Consider the following condition on ϕ:

(121) for every $u \in S^1$, $\phi_u(z)$ is a strictly concave function of z over S^1.

PROPOSITION 15. *Let $\phi_u(z)$ satisfy (94)–(96) and (121). Define $M(x_1, \ldots, x_N)$ implicitly by (96). Then M satisfies (A10), strict quasiconcavity.*[34]

Proof: Let $x^1 \in S^N$, $x^2 \in S^N$, $x^1 \neq x^2$, $u^1 \equiv M(x^1)$, $u^2 \equiv M(x^2)$, $u^1 \leq u^2$ and $0 < \lambda < 1$. We need to show that

$$u^1 < M[\lambda x^1 + (1-\lambda)x^2]. \tag{122}$$

Define $F(u, x) \equiv \sum_{i=1}^N (1/N)\phi_u(x_i) - \phi_u(u)$. Using (96), $u^i \equiv M(x^i)$ means that

$$F(u^1, x^1) = 0; \tag{123}$$
$$F(u^2, x^2) = 0. \tag{124}$$

We need to establish the inequality

$$F(u^1, x^2) \geq 0. \tag{125}$$

If $u^1 = u^2$, then using (124), (125) holds as an equality. If all of the components of x^2 are equal so that $x^2 = u^2 1_N$, then (125) becomes $\sum_{i=1}^N (1/N)\phi_{u^2}(u^2) - \phi_{u^2}(u^2) = 0$ and thus (125) holds as an equality if $x^2 = u^2 1_N$. Thus in verifying (125), we can assume

$$u^1 < u^2, \quad x^2 \neq u^2 1_N. \tag{126}$$

[34] Chew and Epstein [1989a; 219] claim that the following related result is true: concavity of $\phi_u(z)$ in z for every u is a necessary and sufficient condition for M to be Schur-concave. $M(x)$ is Schur-concave over S^N if and only if it satisfies (A2) and a weakened version of property (20) where the strict inequality $>$ is replaced by $\geq$; see Kolm [1976a; 422] for references to the concept of Schur-concavity.

Define $u^* \equiv \min_i\{x_i^2 : i = 1,\ldots,N\}$. Then since $x^2 \neq u^2 1_N$, $u^* 1_N < x^2$ and $u^* < u^2$. Calculate:

$$\begin{aligned}F(u^*, x^2) &\equiv \sum_{i=1}^{N}(1/N)\phi_{u^*}(x_i^2) - \phi_{u^*}(u^*)\\ &> \sum_{i=1}^{N}(1/N)\phi_{u^*}(u^*) - \phi_{u^*}(u^*) \quad \text{using (95) and } u^* 1_N < x^2\\ &= 0. \end{aligned} \tag{127}$$

Now *suppose* that (125) does not hold when (126) holds. Hence

$$F(u^1, x^2) < 0. \tag{128}$$

Note that (94) implies that $F(u,x)$ is jointly continuous in its variables. Thus (127) and (128) imply that a u root to $F(u,x^2) = 0$ exists *between* u^1 and u^*. But by (96), this u root is unique and hence must equal u^2. But this leads to a contradiction since $u^1 < u^2$ and $u^* < u^2$. Thus our *supposition* is false and (125) holds.

We now calculate:

$$\begin{aligned}F[u^1, \lambda x^1 + (1-\lambda)x^2] &\equiv \sum_{i=1}^{N}(1/N)\phi_{u^1}[\lambda x_i^1 + (1-\lambda)x_i^2] - \phi_{u^1}(u^1)\\ &> \lambda\sum_{i=1}^{N}(1/N)\phi_{u^1}(x_i^1) + (1-\lambda)\sum_{i=1}^{N}(1/N)\phi_{u^1}(x_i^2) - \phi_{u^1}(u^1) \text{ by (121)}\\ &= \lambda F(u^1, x^1) + (1-\lambda)F(u^1, x^2)\\ &\geq \lambda(0) + (1-\lambda)(0) \quad \text{using (123) and (125).}\end{aligned} \tag{129}$$

Now define $\bar{u} \equiv \min_i\{\lambda x_i^1 + (1-\lambda)x_i^2 \colon i = 1,\ldots,N\}$ and $\bar{\bar{u}} \equiv \max_i\{\lambda x_i^1 + (1-\lambda)x_i^2 \colon i = 1,\ldots,N\}$. Since $0 < \lambda < 1$ and $x^1 \neq x^2$, we have

$$\bar{u}1_N < \lambda x^1 + (1-\lambda)x^2 < \bar{\bar{u}}1_N. \tag{130}$$

Finally calculate:

$$\begin{aligned}F[\bar{u}, \lambda x^1 + (1-\lambda)x^2] &\equiv \sum_{i=1}^{N}(1/N)\phi_{\bar{u}}[\lambda x_i^1 + (1-\lambda)x_i^2] - \phi_{\bar{u}}(\bar{u})\\ &> \sum_{i=1}^{N}(1/N)\phi_{\bar{u}}(\bar{u}) - \phi_{\bar{u}}(\bar{u})\\ &\qquad \text{by (95) and } \bar{u}1_N < \lambda x^1 + (1-\lambda)x^2\\ &= 0;\end{aligned} \tag{131}$$

$$\begin{aligned}F[\bar{\bar{u}}, \lambda x^1 + (1-\lambda)x^2] &\equiv \sum_{i=1}^{N}(1/N)\phi_{\bar{\bar{u}}}[\lambda x_i^1 + (1-\lambda)x_i^2] - \phi_{\bar{\bar{u}}}(\bar{\bar{u}})\\ &< \sum_{i=1}^{N}(1/N)\phi_{\bar{\bar{u}}}(\bar{\bar{u}}) - \phi_{\bar{\bar{u}}}(\bar{\bar{u}})\\ &\qquad \text{by (95) and } \lambda x^1 + (1-\lambda)x^2 < \bar{\bar{u}}1_N\\ &= 0.\end{aligned} \tag{132}$$

The inequalities (129) and (132) imply that there exists a $\tilde{u}$ between u^1 and $\bar{\bar{u}}$ which solves $F[\tilde{u}, \lambda x^1 + (1-\lambda)x^2] = 0$. By (96), the root is unique. If $u^1 \geq \bar{\bar{u}}$, then (129), (131), (132) and $\bar{u} < \bar{\bar{u}}$ would imply the existence of two u roots to $F[u, \lambda x^1 + (1-\lambda)x^2] = 0$ which would contradict (96). Hence we must have $u^1 < \bar{\bar{u}}$ and

$$u^1 < \tilde{u} < \bar{\bar{u}}. \tag{133}$$

Since $\tilde{u} = M[\lambda x^1 + (1-\lambda)x^2]$, (133) implies (122). Q.E.D.

COROLLARY. *If the strict concavity condition for ϕ (121) is replaced by the corresponding condition of concavity, then the associated M defined by (96) is quasiconcave.*

Proof: Replace the strict inequality in (129) by the weak inequality $\geq$. Q.E.D.

Remember that throughout this paper, we are assuming $N \geq 2$.

6. Homogeneous and Translatable Implicitly Separable Symmetric Means

In this section, we would like to add the homogeneity axiom (A6) or the translatability axiom (A7) to the list of axioms which characterize an implicitly separable symmetric mean (these axioms are (A1)–(A4) and (A13)) and then work out what extra conditions are implied on the generating function $\phi_u(z)$ in addition to conditions (94)–(96). Unfortunately, we are unable to implement this research program, so we only provide extra conditions on ϕ which are sufficient to imply that the corresponding M generated by ϕ satisfies (A6) or (A7).

Throughout this section, the domain of definition sets S^n are the bounded, closed hypercubes $\{(x_1, \ldots, x_n)\colon\ a \leq x_i \leq b,\ i = 1, \ldots, n\}$ where $-\infty < a < b < +\infty$, except when we explicitly specify more restrictive conditions in Proposition 17.

Consider the following property for $\phi_u(z)$:

$$(z, u) \in S^2, \lambda > 0, (\lambda z, \lambda u) \in S^2 \text{ implies } \phi_{\lambda u}(\lambda z) = h(\lambda)\phi_u(z) \tag{134}$$

for some continuous function h such that $h(\lambda) > 0$ for $\lambda > 0$.

PROPOSITION 16. *Let $\phi_u(z)$ be defined for $(z,u) \in S^2$ and satisfy (94)–(96) and (134). For $x \equiv (x_1,\ldots,x_N) \in S^N$, define $u = M(x)$ as the u solution to the equation in (96). Then M satisfies (A1)–(A4), (A6) and (A13).*

Proof: In view of Proposition 14, we need only show M satisfies (A6). Let $x \in S^N$, $\lambda > 0$ and $\lambda x \in S^N$. Then $u = M(x)$ is the unique u which solves (96). Now calculate, using (134):

$$\begin{aligned}\sum_{i=1}^{N}(1/N)\phi_{\lambda u}(\lambda x_i) - \phi_{\lambda u}(\lambda u) &= h(\lambda)\Big[\sum_{i=1}^{N}(1/N)\phi_u(x_i) - \phi_u(u)\Big] \\ &= h(\lambda)(0) \qquad \text{using (96)} \\ &= 0.\end{aligned}$$

Hence λu solves (96) when x is replaced by λx and thus $\lambda u = M(\lambda x)$ as is required for (A6) to hold. Q.E.D.

Some results presented in Aczél [1966; 304–305] can be utilized to show that the function $h(\lambda)$ which appears in (134) must actually equal λ^k. Hence property (134) essentially implies that ϕ is homogeneous of degree k in its two variables. Thus a homogeneous of degree k ϕ (for any k) leads to a linearly homogeneous implicitly separable mean M.

For the following proposition, we require slightly different domain of definition sets S^n. Define for $n = 1, 2, \ldots, N$:

$$S^n \equiv \{(x_1,\ldots,x_n) : a \le x_i \le b,\ i = 1,\ldots,n\} \quad \text{where } 0 < a \le 1 \le b < +\infty \text{ with } a < b. \tag{135}$$

Now consider the following property for $\phi_u(z)$:

(136) for $(z,u) \in S^2$, $\phi_u(z) = \gamma(z/u)$ where $\gamma(z)$ is a continuous increasing function of z for $a/b \le z \le b/a$.

PROPOSITION 17. *Let $\phi_u(z)$ be defined for $(z,u) \in S^2$ and satisfy (136). For $x \equiv (x_1,\ldots,x_N) \in S^N$, define $u = M(x)$ as the u solution to*

$$\sum_{i=1}^{N}(1/N)\gamma(x_i/u) - \gamma(1) = 0. \tag{137}$$

Then M satisfies (A1)–(A4), (A6) and (A13). The sets S^2 and S^N are defined by (135).

Proof: Since γ is an increasing, continuous function, it can be seen that $\phi_u(z) \equiv \gamma(z/u)$ is increasing in z, decreasing in u and jointly continuous (and finite) for $(z,u) \in S^2$. Hence ϕ satisfies (94), (95) and (107), which in turn implies (96). Note that the equation in (96) reduces to (137) if $\phi_u(z) \equiv \gamma(z/u)$. Thus we may apply Proposition 14 and conclude that the M implicitly defined by (137) satisfies (A1)–(A4) and (A13). Using (137), it is easy to verify that M also satisfies (A6). Q.E.D.

Equation (137) in Proposition 17 shows how a large class of homogeneous implicitly separable symmetric means can be generated: we require only a continuous and increasing function of one variable γ to do the job. This class of functions was considered briefly in Chew [1989; 287]. Note that if we let $\gamma(z) \equiv z^r$ for $r > 0$, $\gamma(z) \equiv -z^r$ for $r < 0$ or $\gamma(z) \equiv \ln z$, then the resulting mean functions M are the means of order r, $M^r(x)$, defined earlier by (69). As we noted earlier, these means of order r are the *only* homogeneous separable symmetric means. We now see that the class of homogeneous implicitly separable symmetric means is much larger, since γ can be a *general* increasing and continuous function of one variable.

We turn now to the derivation of sufficient conditions for $\phi_u(z)$ to generate a translatable implicitly separable symmetric mean.

Consider the following property for $\phi_u(z)$:

$$(138) \qquad (z,u) \in S^2,\ (z+\lambda, u+\lambda) \in S^2 \text{ implies } \phi_{u+\lambda}(z+\lambda) = \phi_u(z) + \lambda.$$

Property (138) is simply the property of translatability for the function of two variables ϕ over the domain of definition set S^2.

PROPOSITION 18. *Let $\phi_u(z)$ be defined for $(z,u) \in S^2$ and satisfy (94)–(96) and (138). For $x \in S^N$, define $u = M(x)$ as the u solution to the equation in (96). Then M satisfies (A1)–(A4), (A7) and (A13); i.e., M is a translatable implicitly separable symmetric mean.*

Proof: In view of Proposition (14), we need only show that M satisfies (A7). Let $x \in S^N$ and $x + \lambda 1_N \in S^N$. Then $u = M(x)$ is the unique solution to (96). Now calculate:

$$\begin{aligned}
&\textstyle\sum_{i=1}^{N}(1/N)\phi_{u+\lambda}(x_i+\lambda) - \phi_{u+\lambda}(u+\lambda) \\
&\quad = \textstyle\sum_{i=1}^{N}(1/N)[\phi_u(x_i)+\lambda] - [\phi_u(u)+\lambda] \text{ using (138)} \\
&\quad = \textstyle\sum_{i=1}^{N}(1/N)\phi_u(x_i) - \phi_u(u) \\
(139) &\quad = 0 \qquad\qquad \text{using } u = M(x) \text{ and (96).}
\end{aligned}$$

Since $u + \lambda$ is a solution to (139), we have $u + \lambda = M(x + \lambda 1_N)$ as required for M to satisfy (A7). Q.E.D.

Thus if the generating function ϕ is translatable, then the corresponding implicitly separable symmetric mean is also translatable.

Now consider the following property for ϕ:

(140) for $(z,u) \in S^2$, $\phi_u(z) \equiv \gamma(z-u)$ where $\gamma(z)$ is a continuous increasing function of z for $a - b \le z \le b - a$.

PROPOSITION 19. *Let ϕ satisfy (140). For $x = (x_1, \ldots, x_N) \in S^N$, define $u = M(x)$ as the u solution to*

$$(141) \qquad \sum_{i=1}^{N} (1/N)\gamma(x_i - u) - \gamma(0) = 0.$$

Then M satisfies (A1)–(A4), (A7) and (A13); i.e., M is a translatable implicitly separable symmetric mean.

Proof: Using the continuity and monotonicity properties of γ, it can be seen that $\phi_u(z) = \gamma(z-u)$ satisfies (94), (95) and (107) and hence (96) as well. Also if $\phi_u(z) = \gamma(z-u)$, then the equation in (96) reduces to (141). Hence by Proposition 14, M satisfies (A1)–(A4) and (A13). It remains to show that M also satisfies (A7). Let $x \in S^N$ and let $u = M(x)$ so that u and $x \equiv (x_1, \ldots, x_N)$ satisfy (141). Let $x + \lambda 1_N \in S^N$. Then $u^* \equiv M(x + \lambda 1_N)$ and u^* solves the following equation:

$$(142) \qquad \sum_{i=1}^{N} (1/N)\gamma(x_i + \lambda - u^*) - \gamma(0) = 0.$$

Using (141), it is easy to verify that $u^* = u + \lambda$ solves (142). Thus we have $u^* = M(x + \lambda 1_N) = u + \lambda = M(x) + \lambda$ which shows that M satisfies (A7).

Q.E.D.

Equation (141) in Proposition 19 shows how a large class of translatable implicitly separable symmetric means can be generated: we require only a continuous increasing function of one variable γ to act as the underlying generator function. Again, this class of functions was considered briefly by Chew [1989; 287]. If we let $\gamma(z) \equiv e^{sz}$ for $s > 0$, $\gamma(z) \equiv -e^{-sz}$ for $s < 0$ or $\gamma(z) \equiv z$ for $s = 0$, then it can be verified that the corresponding symmetric means M are the members of the Nagumo means of order s defined by (78). As we showed earlier, these Nagumo means of order s are the only translatable separable symmetric means. Thus the results of Proposition 19 show that the class of translatable implicitly separable symmetric means is much larger since the generating function γ can be *any* increasing continuous function of one variable (and not just the functions e^{sz}, $-e^{sz}$ and z).

The indifference curves shown in Figure 1 along with the approximation results developed in the previous section (recall the ϕ defined by (114)) can be used to show that the curvature axioms (A8), (A9) and (A10) each give rise to a distinct class of implicitly separable homogeneous or translatable symmetric means. We are unable to provide nice necessary and sufficient conditions for the functions $\phi_u(z)$ or $\gamma(z)$ that will imply that the corresponding M will satisfy (A8), (A9) or (A10). However, Proposition 15 can still be applied to Propositions 16–19 in this section; i.e., if $\phi_u(z)$ is strictly concave in z for every u, then the associated M will be strictly quasiconcave and satisfy (A10). For Propositions 17 and 19, strict concavity of γ over the relevant domain of definition is sufficient to imply that the corresponding M is strictly quasiconcave.

We turn now to a major application of the theory of symmetric means.

7. Separable Symmetric Means and the Expected Utility Theorem

Blackorby, Davidson and Donaldson [1977] used an approach based on separability assumptions to derive an expected utility theorem.[35] We present the essence of their approach here (with a few modifications) and then generalize their approach from the separable preferences case to cover the case of implicitly separable preferences.

Blackorby, Davidson and Donaldson first considered the case where the decision maker has equal (possibly subjective) probabilities over N states of nature.[36] Let there be K goods that matter to the decision maker and denote the quantity of the kth decision variable by y_k for $k = 1, \ldots, K$. Let the certainty preferences of the decision maker be defined by the continuous utility function, $f(y_1, \ldots, y_K) \equiv f(y)$, for $y \in S$ where S is a closed, convex subset of R^K. Denote the "consumption" vector of the decision maker if state i occurs by $y^i \equiv (y_1^i, \ldots, y_K^i)$ for $i = 1, \ldots, N$. Following Arrow [1953; 41] and Debreu [1959; 101], we assume that the decision maker's preferences over state contingent commodities can be represented by a continuous utility function, $F(y^1, y^2, \ldots, y^N)$. It is natural to assume[37] that the state contingent preferences defined by F are related to the certainty preferences defined by f as follows:

$$F(y^1, y^2, \ldots, y^N) = M[f(y^1), f(y^2), \ldots, f(y^N)], \tag{143}$$

---

[35]Their approach was a simplification of an approach due originally to Arrow [1984b]. The first attempts to prove an expected utility theorem were made by Bernoulli [1738] and Ramsey [1926; 178-183]. Early rigorous proofs of expected utility theorems were made by Neumann and Morgenstern [1947], Marshak [1950], Arrow [1951; 424-425] [1984b; 180–193], Samuelson [1952; 672] and Savage [1954; 75]. Chew and Epstein [1989a; 212] make extensive use of separability axioms in their axiomatic approach to non-expected utility theorems and they also note that Savage's [1954; 23] sure-thing principle is in fact a separability axiom. Note that Segal [1992] makes some corrections to Chew and Epstein's [1989a; 227] generalization of Gorman's [1968a] Separability Theorem; see also Chew and Epstein [1992].

[36]The concept of equal subjective probabilities (applied to two alternatives) first appeared in Ramsey [1926; 177–179]. We note that the type of uncertainty that we are modeling is of the ethically neutral type, a concept also due to Ramsey [1926; 177].

[37]This assumption seems to be consistent with the assumptions used by Samuelson [1952; 674]. He noted that his state contingent preference function had to approach a conventional nonstochastic preference function as all but one of his probabilities tended to zero.

where M is the decision maker's *stochastic preference function*; i.e., M expresses his or her attitude towards uncertainty. If we force all outcomes to be equal so that $y^1 = y^2 = \cdots = y^N \equiv y$, then it does not matter which state of nature occurs. Under these circumstances, it is natural to require that the state contingent preferences collapse down to the certainty preferences so that we have $F(y, y, \ldots, y) = M[f(y), f(y), \ldots, f(y)] = f(y)$. Hence M must be a mean; i.e., M satisfies (A1). Since the states of nature are equally probable, F must be symmetric in the vectors $y^1, y^2, \ldots, y^K$ and hence M must be symmetric in its N arguments. Since we are assuming that F and f are continuous functions, it is natural to assume that M is also continuous with respect to its variables. Finally if F and f are increasing in their variables, then M must also be increasing in its variables. Thus it is natural to assume that the stochastic preference function M in (143) is a *symmetric mean*; i.e., M satisfies (A1)–(A4).

In what follows, we simply assume that the certainty utility function is continuous and we concentrate our attention on the properties of the stochastic preference function M.

Suppose that $N \geq 3$ and suppose that M satisfies the consistency in aggregation axiom (A11) in addition to axioms (A1)–(A4). The fact that M satisfies (A11) means that the decision maker can aggregate or average over the uncertainty in the first two states of nature and find the certainty equivalent, $M_2[f(y^1), f(y^2)]$, that is independent of whatever happens in the other states of nature.[38] Applying Proposition 6 above, we have the existence of a continuous increasing function of one variable ϕ such that

$$(144) \qquad M[f(y^1), \ldots, f(y^N)] = \phi^{-1}\left[\sum\nolimits_{i=1}^{N} (1/N)\phi[f(y^i)]\right] \text{ and}$$

$$(145) \qquad M_2[f(y^1), f(y^2) = \phi^{-1}\{(1/2)\phi[f(y^1)] + (1/2)\phi[f(y^2)]\}.$$

Since ϕ is monotonically increasing, maximizing $F(y^1, \ldots, y^N)$ with respect to the components of $y^1, \ldots, y^N$ is equivalent to maximizing the following monotonic transform of F;

$$(146) \qquad \phi[F(y^1, \ldots, y^N)] = \sum\nolimits_{i=1}^{N} (1/N)\phi[f(y^i)].$$

The right hand side of (146) can be interpreted as an *expected utility* for the equal probabilities case, using the transformed certainty utility function, $\phi[f(y)]$.

The above argument which led to the state contingent preference function defined by the right hand side of (146) is essentially due to Blackorby, Davidson and Donaldson [1977; 352–354], except that they use a separability axiom

[38]Thus (A11) can be interpreted as an independence of irrelevant alternatives axiom or as a counterpart to Savage's [1954; 21–23] sure-thing principle.

(and Gorman's [1968a] Separability Theorem) in place of our consistency in aggregation axiom, (A11).

We turn now to the case where the probabilities for the N states are not necessarily equal. Suppose that the probabilities are $p_i \geq 0$ for $i = 1, 2, \ldots, N$ with $\sum_{i=1}^{N} p_i = 1$. In order to model this situation, we require that the decision maker have preferences over choices involving an infinite sequence of equally probable events. Thus we require that the decision maker have a sequence of stochastic preference functions M^n such that $F^n(y^1, y^2, \ldots, y^n) \equiv M^n[f(y^1), f(y^2), \ldots, f(y^n)]$ represents the decision maker's state contingent preferences when there are n equally likely states of nature for $n = 3, 4, \ldots$. We assume that for each n, M^n satisfies (A1)–(A4) and (A11). Thus by axiom (A11), we have for each n, the existence of a micro certainty equivalence function M_2^n such that

$$(147) \quad M^n[f(y^1), f(y^2), f(y^3), \ldots, f(y^n)] \\ = M^n\{M_2^n[f(y^1), f(y^2)], M_2^n[f(y^1), f(y^2)], f(y^3), \ldots, f(y^N)\}.$$

However, as n increases, we want the decision maker's preferences to remain unchanged when facing choices involving fewer than n equally likely alternatives. For example, let $n = 4$ and set $y^1 = y^3$ and $y^2 = y^4$. Then $M^4[f(y^1), f(y^2), f(y^1), f(y^2)]$ should equal $M^2[f(y^1), f(y^2)]$ since we have simply reparameterized the same uncertain situation in two equivalent ways.[39] A way of forcing preferences to be consistent is to require that the micro functions M_2^n be independent of n. Thus to prevent stochastic preferences from changing in an inconsistent manner, we add the following axiom to our list of assumptions:

(A14) $M_2^n(x_1, x_2) = M_2(x_1, x_2)$ for $n = 3, 4, \ldots$. *(Consistency of Stochastic Preferences in the Separable Model).*

Assuming that M^n satisfies (A1)–(A4), (A11) and (A14) for $n = 3, 4, \ldots$, Proposition 6 implies that we can use the *same* continuous increasing function of one variable ϕ in order to represent the functions M^n and M_2^n. Thus for $n = 3, 4, \ldots$, we have, letting $x_i \equiv f(y^i)$:

$$(148) \qquad M^n(x_1, \ldots, x_n) = \phi^{-1}\left[(1/n)\sum_{i=1}^{n} \phi(x_i)\right] \text{ and}$$

$$(149) \qquad M_2^n(x_1, x_2) = \phi^{-1}[(1/2)\phi(x_1) + (1/2)\phi(x_2)].$$

[39]Thus using the arguments associated with (46), we must have $M_2^2(x_1, x_2) = M_2^4(x_1, x_2)$ for consistency.

Now we are ready to deal with the unequal probabilities case. For n large, we approximate each p_i by a rational number:

$$(150) \qquad p_i \cong n_i/n, \quad i = 1, 2, \ldots, N,$$

where each n_i is a nonnegative integer with $\sum_{i=1}^{N} n_i = n$. As in Blackorby, Davidson and Donaldson [1977; 354], we can model the unequal probabilities case by an equal probabilities state contingent utility function F^n, except that the first n_1 vector arguments of F^n are forced to equal y^1, the next n_2 vector arguments are forced to equal y^2, ... , and the last n_N vector arguments are forced to equal y^N. The resulting state contingent utility function, $F^{n*}(y^1, y^2, \ldots, y^N)$ depends only on the N state contingent consumption or decision vectors $y^1, \ldots, y^N$ where y^i indicates the decision maker's "consumption" if state i occurs for $i = 1, 2, \ldots, N$. Thus we have

$$\begin{aligned}(151) \quad F^{n*}(y^1, y^2, \ldots, y^N) &\equiv M^n[f(y^1)1_{n_1}, f(y^2)1_{n_2}, \ldots, f(y^N)1_{n_N}] \\ &= \phi^{-1}\{(n_1/n)\phi[f(y^1)] + (n_2/n)\phi[f(y^2)] + \cdots + (n_N/n)\phi[f(y^N)]\},\end{aligned}$$

where (151) follows from the line above using (148). Now take the limit of the right hand side of (151) as n tends to infinity. Recalling (150), the n_i/n will tend to the probabilities p_i and using the continuity of ϕ and ϕ^{-1}, the right hand side of (151) will tend to the right hand side of (152) below. Defining the limiting state contingent preference function as $F(y^1, y^2, \ldots, y^N)$, we have

$$(152) \qquad F(y^1, y^2, \ldots, y^N) = \phi^{-1}\left[\sum_{i=1}^{N} p_i \phi[f(y^i)]\right].$$

Since ϕ is monotonically increasing, maximizing F with respect to $y^1, \ldots, y^N$ is equivalent to maximizing

$$(153) \qquad \phi[F(y^1, y^2, \ldots, y^N)] = \sum_{i=1}^{N} p_i \phi[f(y^i)],$$

which is the Expected Utility Theorem for the case of unequal probabilities. Thus we have proven:

PROPOSITION 20. (*Expected Utility Theorem for Separable Stochastic Preferences*): *Suppose that the decision maker's certainty utility function is the continuous function f and that the state contingent preferences over n equally likely states of nature can be represented by the function $M^n[f(y^1), \ldots, f(y^n)]$ where the stochastic preference function M^n satisfies (A1)–(A4), (A11) and (A14) for $n = 3, 4, \ldots$. Then there exists a continuous increasing function of one variable ϕ such that the decision maker's preferences over N uncertain* *states of nature with probabilities* $p_1, \ldots, p_N$ *can be represented by the function* $\sum_{i=1}^{N} p_i\phi[f(y^i)]$.

The above proof of the Expected Utility Theorem follows that of Blackorby, Davidson and Donaldson [1977; 353–355] except that these authors did not explicitly make an assumption like (A14), which is necessary so that state contingent preferences over different numbers of equally likely alternatives are consistent as the number of alternatives expands.

The expected utility model represented by (153) has not fared well in a variety of empirical and experimental situations.[40] Thus we need to weaken the axioms on the stochastic preference functions M^n so that we obtain a bigger or more flexible class of state contingent preference functions than those which are represented by the right hand side of (153).

Let us first consider the case of N equally likely states of nature. Then it is natural to replace the consistency in aggregation axiom (A11) by the conditional consistency in aggregation axiom (A13). The micro aggregator function $M_2(x_1, x_2, u)$ which appears in (A13) can now be interpreted as a *conditional* (on u) *certainty equivalence function.*[41] Assuming that the stochastic preference function M which appears in (143) satisfies (A1)–(A4) and (A13), then by applying Proposition 13, we can conclude that $u = F(y^1, \ldots, y^N) \equiv M[f(y^1), \ldots, f(y^N)]$ is the unique u solution to

$$\sum_{i=1}^{N}(1/N)\phi_u[f(y^i)] - \phi_u(u) = 0 \tag{154}$$

where ϕ satisfies (94)–(96).

If we define $x_i \equiv f(y^i)$ for $i = 1, \ldots, N$ and define τ by

$$\tau(z, u) \equiv \phi_u(z) - \phi_u(u), \tag{155}$$

then (154) can be rewritten as follows:

$$\sum_{i=1}^{N}(1/N)\tau(x_i, u) = 0. \tag{156}$$

---

[40]See Sections 8–10 below for some of the difficulties. The first well justified attack on the expected utility model was made by Allais [1953; 527–530]. For more recent references to the Allais Paradox, see Machina [1982; 287–288] and Chew and Epstein [1989a; 212–213].

[41]This terminology follows Machina [1982; 288]. Note that since we are using (A13) in place of (A11), when we apply Proposition 14, we shall require that the decision maker's certainty utility function f have a range which is equal to the closed finite interval (a, b). This will usually require that the decision maker's choice set be a compact connected set.

Equation (156) is a special case (the equal probabilities case) of Dekel's [1986] and Chew and Epstein's [1989a; 211–212] *implicit linear utility model* and of Chew's [1989] implicit weighted utility model.[42]

We now follow the example of Blackorby, Davidson and Donaldson [1977] and generalize the above model to the case where the probabilities of the N states are not all equal, but are equal to $p_i > 0$ for $i = 1, 2, \ldots, N$. Note that we now assume that all probabilities are positive. As was the case with the expected utility model (recall Proposition 20), we need to assume that the decision maker has state contingent preferences of the form (143) over an infinite sequence of equally likely events. Thus we assume that for $n = 2, 3, \ldots$, the decision maker's stochastic preference function $M^n(x_1, \ldots, x_n)$ satisfies (A1)–(A4) and the conditional consistency in aggregation axiom (A13) where $x_i \equiv f(y^i)$. Thus for each n, if $u = M^n(x_1, \ldots, x_n)$, then by Proposition 13, there exists a ϕ^n satisfying (94)–(96) such that M^n and M_2^n have representations that are analogous to (92) and (93) except that ϕ^n replaces ϕ and n replaces N.

In order to ensure that the decision maker's stochastic preferences remain unchanged as n increases, we assume that the following axiom is satisfied by the family of stochastic preference functions $\{M^n : n = 2, 3, \ldots\}$:

(A15) $M_2^n(x_1, x_2, u) = M_2(x_1, x_2, u)$ for $(x_1, x_2, u) \in S^3$ and for $n = 2, 3, \ldots$. *(Consistency of Stochastic Preferences in the Implicitly Separable Model).*

Axiom (A15) says that the conditional certainty equivalence functions M_2^n are always the same no matter how many equally likely alternatives n the decision maker faces.

Assuming that the family of stochastic preference functions $\{M^n\}$ satisfy (A1)–(A4), (A13) and (A15) for $n = 2, 3, \ldots$, Proposition 13 and (A15) imply that we can use the *same* function $\phi_u(z)$ in order to represent the functions M^n and M_2^n. Thus for $n = 2, 3, \ldots$ and $u = M^n[f(y^1), \ldots, f(y^n)]$, we have

$$(157) \qquad M_2^n[f(y^1), f(y^2), u] = \phi_u^{-1}\left[(1/2)\phi_u[f(y^1)] + (1/2)\phi_u[f(y^2)]\right];$$

$$(158) \qquad M^n[f(y^1), \ldots, f(y^n)] = \phi_u^{-1}\left[\sum_{i=1}^{n}(1/n)\phi_u[f(y^i)]\right]$$

where ϕ satisfies (94)–(96).

---

[42] Of course, our state contingent preferences approach is somewhat different from the preferences over lotteries approach used by these authors and our regularity conditions are also somewhat different. Chew and Epstein's [1989a; 211] necessary conditions for τ were that $\tau(z, u)$ be continuous and increasing in z for each u and that the u solution to (156) be unique. Chew and Epstein's [1989a; 217] additional sufficient conditions for τ were that $\tau(z, u)$ be continuous and decreasing in u for each z.

Now we are ready to model the unequal probabilities case. For n large, approximate each of the N probabilities p_i by the rational numbers n_i/n as in (150). Take the state contingent preference function defined over n equally likely states, $F^n(y^1,\ldots,y^n) \equiv M^n[f(y^1),\ldots,f(y^n)]$, and force the first n_1 vector arguments to equal the common vector y^1, force the next n_2 vector arguments to equal y^2, ... , and force the last n_N vector arguments to equal y^N. The resulting state contingent preference function, $F^{n*}(y^1,\ldots,y^N) \equiv M^n[f(y^1)1_{n_1},\ldots,f(y^N)1_{n_N}]$, depends only on the N state contingent consumption vectors $y^1,\ldots,y^N$. Using (158), if $u = M^n[f(y^1)1_{n_1},\ldots,f(y^N)1_{n_N}]$, then we have

$$(159) \qquad u = F^{n*}(y^1,\ldots,y^N) = \phi_u^{-1}\left[\sum\nolimits_{i=1}^{N}(n_i/n)\phi_u[f(y^i)]\right] \text{ or}$$

$$(160) \qquad \sum\nolimits_{i=1}^{N}(n_i/n)\phi_u[f(y^i)] - \phi_u(u) = 0,$$

where we have transformed both sides of (159) by the monotonic function ϕ_u to obtain (160). Now take the limit of (160) as n tends to infinity. Recalling (150), the n_i/n will tend to the p_i and (160) will tend to:

$$(161) \qquad \sum\nolimits_{i=1}^{N} p_i\phi_u[f(y^i)] - \phi_u(u) = 0.$$

Assuming for the moment that the u solution to (161) is unique, denote this u solution as $F(y^1,\ldots,y^N,p_1,\ldots,p_N)$. Then the function F can represent the decision maker's preferences over the N states of nature with probabilities $p_1,\ldots,p_N$. Note that if $\phi_u(z)$ does not depend on u, then F^* collapses down to the F defined by (152) and we obtain our old expected utility model.

If we make use of (155) which defines the function of two variables τ, then using the fact that $\sum_{i=1}^{N} p_i = 1$, (161) can be rewritten as:

$$(162) \qquad \sum\nolimits_{i=1}^{N} p_i\tau[f(y^i),u] = 0$$

which is the state contingent preference counterpart to the fundamental equation in the implicit linear utility model of Chew and Epstein [1989a; 211].

There still remains the problem of finding conditions on the family of stochastic preference functions $\{M^n\}$ that will ensure that the u solution to (161) is unique. It will turn out that the axioms (A1)–(A4) and (A13) for each M^n and axiom (A15) for the family are sufficient to imply that the u solution to (161) is unique. However, the arguments that are required to show this are somewhat technical, so the reader who is interested in applications may want to skip the remainder of this section.

Under our assumptions (A1)–(A4), (A13) and (A15) on the family of stochastic preference functions $\{M^n\}$, we know by Proposition 13 that $\phi_u(z)$

satisfies (94)–(96). If we replace (96) by the Chew and Epstein [1989a; 215] sufficient condition (107), then the u solution to (161) will be unique. Unfortunately, condition (107) for ϕ is not necessary; i.e., it is not implied by our axioms on the stochastic preference functions M^n. Fortunately, we shall show below that our present assumptions on the M^n, (A1)–(A4), (A13) and (A15), imply an additional property on $\phi_u(z)$ that will be sufficient to imply that a unique u solution to (161) exists. Before we introduce this additional property on ϕ, we need to note a few properties of the function $\tau(z,u) \equiv \phi_u(z) - \phi_u(u)$, which was defined earlier by (155).

Note that $\tau(u,u) = 0$ and assuming that $\phi_u(z)$ is increasing in z, we have:

(163) $$\tau(z,u) > 0 \text{ if } z > u;$$

(164) $$\tau(z,u) < 0 \text{ if } z < u.$$

For $a \leq x_1 < u < x_2 \leq b$, define the function r by:

(165) $$r(x_1,x_2,u) \equiv -\tau(x_1,u)/\tau(x_2,u).$$

Since $x_1 < u$, (164) implies $\tau(x_1,u) < 0$ and since $u < x_2$, (163) implies that $\tau(x_2,u) > 0$ and so $r(x_1,x_2,u) > 0$ for all u such that $x_1 < u < x_2$.

Consider the following property for ϕ:

(166) $r(x_1,x_2,u)$ is increasing in u for u such that $a \leq x_1 < u < x_2 \leq b$.

Let the family of stochastic preference functions $\{M^n\}$ satisfy (A1)–(A4), (A13) and (A15). Then by Proposition 13, the M^n have the representation (158) in terms of ϕ, where ϕ satisfies (94)–(96). Now *suppose* that ϕ did not satisfy (166). Then there exist u_1 and u_2, $x_1 < u_1 < u_2 < x_2$, such that

(167) $$r_1 \equiv r(x_1,x_2,u_1) \geq r(x_1,x_2,u_2) \equiv r_2.$$

By Proposition 13, ϕ is continuous and hence so are the functions τ and r defined by (155) and (165). Thus we can assume that the r_1 and r_2 which satisfy (167) are rational numbers. Thus, in particular, we have

(168) $$r_1 = n_1/n_2$$

where n_1 and n_2 are positive integers (since $r_1 = r(x_1,x_2,u_1) = -\tau(x_1,u_1)/\tau(x_2,u_1) > 0$ using (163), (164) and $x_1 < u_1 < x_2$). The equation $n_1/n_2 = -\tau(x_1,u_1)/\tau(x_2,u_1)$ may be rewritten as follows:

$$\begin{aligned} 0 &= n_2\tau(x_1,u_1) + n_1\tau(x_2,u_1) \\ &= n_2[\phi_{u_1}(x_1) - \phi_{u_1}(u_1)] + n_1[\phi_{u_1}(x_2) - \phi_{u_1}(u_1)] \text{ using (155)} \\ &= n_2\phi_{u_1}(x_1) + n_1\phi_{u_1}(x_2) - (n_1+n_2)\phi_{u_1}(u_1) \end{aligned}$$

(169) $$= [n_2/(n_1+n_2)]\phi_{u_1}(x_1) + [n_1/(n_1+n_2)]\phi_{u_1}(x_2) - \phi_{u_1}(u_1).$$

Using the representation (158) of $M^{n_1+n_2}$ in terms of ϕ, it can be seen that (169) implies that

$$u_1 = M^{n_1+n_2}(x_1 1_{n_2}, x_2 1_{n_1}). \tag{170}$$

If $r_2 = r_1$, then we can modify the equalities in (169) and show that

$$u_2 = M^{n_1+n_2}(x_1 1_{n_2}, x_2 1_{n_1}). \tag{171}$$

But (170), (171) and $u_1 < u_2$ show that $M^{n_1+n_2}$ is not a well defined function which contradicts our earlier assumptions on the family of stochastic preference functions $\{M^n\}$.

Now suppose that

$$r_1 > r_2. \tag{172}$$

Using definition (165), we have

$$0 = \tau(x_1, u_2) + r_2\tau(x_2, u_2) < \tau(x_1, u_2) + r_1\tau(x_2, u_2) \tag{173}$$

where the inequality in (173) follows from (172), $u_2 < x_2$ and (163). By (168) and (173), we have

$$\begin{aligned} 0 &< n_2\tau(x_1, u_2) + n_1\tau(x_2, u_2) \\ &= n_2\phi_{u_2}(x_1) + n_1\phi_{u_2}(x_2) - (n_1 + n_2)\phi_{u_2}(u_2) \text{ using (155).} \end{aligned} \tag{174}$$

We also have

$$\begin{aligned} &n_2\phi_{x_2}(x_1) + n_1\phi_{x_2}(x_2) - (n_1 + n_2)\phi_{x_2}(x_2) \\ &\quad < n_2\phi_{x_2}(x_2) + n_1\phi_{x_2}(x_2) - (n_1 + n_2)\phi_{x_2}(x_2) = 0 \end{aligned} \tag{175}$$

where the inequality in (175) follows from the increasing in z property of $\phi_u(z)$, property (95), and $x_1 < x_2$. The inequalities (174) and (175) and the continuity of $\phi_u(z)$ in u (recall property (94)) imply that there exists a u_3 such that $u_2 < u_3 < x_2$ and

$$[n_2/(n_1 + n_2)]\phi_{u_3}(x_1) + [n_1/(n_1 + n)]\phi_{u_3}(x_2) - \phi_{u_3}(u_3) = 0. \tag{176}$$

But (176) implies that $u_3 = M^{n_1+n_2}(x_1 1_{n_2}, x_2 1_{n_1}) > u_2 > u_1$ and this inequality along with (170) implies that $M^{n_1+n_2}$ is not a well defined function, a contradiction.

Thus the supposition (167) is not consistent with assumptions (A1)–(A4), (A13) and (A15) on the family of stochastic preference functions M^n and so $\phi_u(z)$ must satisfy (166) under these conditions on the M^n.

Equation (169) can be used to motivate the intuition behind the property (166): as r increases, n_1 must increase and hence the weight $n_1/(n_1+n_2)$ that we are giving to the largest x_i increases and hence the corresponding mean u should increase. Thus u should increase as r increases and vice versa.

We now show that if $\phi_u(z)$ satisfies (94), (95) and (166), then the u solution to (161) is unique in the case where $N=2$.

Using the continuity and monotonicity properties of ϕ, (94) and (95), it is easy to show that at least one root (call it u_1) to (161) when $N=2$ must exist. Define $x_1 \equiv f(y^1)$ and $x_2 \equiv f(y^2)$. If $x_1 = x_2$, then $u_1 \equiv x_1 = x_2$ is the unique root. Hence we can assume, without loss of generality, that $x_1 < u_1 < x_2$. Thus we have

$$\begin{aligned}
0 &= p_1\phi_{u_1}(x_1) + p_2\phi_{u_1}(x_2) - \phi_{u_1}(u_1) \\
&= p_1[\phi_{u_1}(x_1) - \phi_{u_1}(u_1)] \\
&\qquad + p_2[\phi_{u_1}(x_2) - \phi_{u_1}(u_1)] && \text{using } p_1 + p_2 = 1 \\
&= p_1\tau(x_1,u_1) + p_2\tau(x_2,u_1) && \text{using definition (155)} \\
&= \tau(x_1,u_1) + (p_2/p_1)\tau(x_2,u_1) && \text{using } p_1 > 0 \\
&= \tau(x_1,u_1) + r(x_1,x_2,u_1)\tau(x_2,u_1) && \text{using definition (165).} \qquad (177)
\end{aligned}$$

Suppose there exists another root u_2 to (161) when $N=2$. Then we have $x_1 < u_2 < x_2$ and repeating the arguments which led to (177), we have

$$(178)\quad 0 = \tau(x_1,u_2) + r(x_1,x_2,u_2)\tau(x_2,u_2) > \tau(x_1,u_2) + r(x_1,x_2,u_1)\tau(x_2,u_2)$$

where (178) follows if $u_2 > u_1$, using (166), $u_2 < x_2$ and (163). Since $r(x_1,x_2,u_1) = p_2/p_1$, using definition (155), (178) is equivalent to

$$(179)\qquad 0 > p_1\phi_{u_2}(x_1) + p_2\phi_{u_2}(x_2) - \phi_{u_2}(u_2)$$

which contradicts our supposition that u_2 solves (161). If $u_2 < u_1$, we again obtain (179) with the inequality reversed, which again contradicts our supposition that u_2 solves (161). Hence the root to (161) is unique when $N=2$.

In order to obtain the uniqueness of the u solution to (161) for a general N, we need a condition which is stronger (and more complex) than (166). Hence for $u < x_n$, define the function r^n for $n \geq 2$ as follows:

$$(180)\qquad r^n(x_1,x_2,\ldots,x_n,u) \equiv -\Big[\sum_{i=1}^{n-1}\tau(x_i,u)\Big]/\tau(x_n,u).$$

Note that if $n=2$, $r^2(x_1,x_2,u)$ is equal to our old function $r(x_1,x_2,u)$ defined by (165). The condition which generalizes (166) is the following condition on ϕ:

(181) Let $n = 2,3,\ldots$ and let $a \leq x_1 \leq x_2 \leq \cdots \leq x_n \leq b$ with $x_1 < x_n$. Let u_{n-1} be a u solution to $\sum_{i=1}^{n-1}\tau(x_i,u) = 0$. Then $r^n(x_1,\ldots,x_n,u)$ is increasing in u for $u_{n-1} < u < x_n$.

Assuming that ϕ satisfies (94) and (95) and $n=2$, then the u solution to $\tau(x_1,u)=0$ is unique and is equal to x_1. Thus when $n=2$, (181) collapses down to our old condition, (166).

PROPOSITION 21. *Let the family of stochastic preference functions* $\{M^n\}$ *satisfy (A1)–(A4), (A13) and (A15) where* M^n *is defined over* $S^n \equiv \{x : a1_n \leq x \leq b1_n\}$ *with* $-\infty < a < b < +\infty$. *Then there exists a function of two variables,* $\phi_u(z)$, *defined for* $(z, u) \in S^2$, *such that for* $n = 2, 3, \ldots,$ *and* $x^n \in S^n$ *and* $u = M^n(x^n)$, *we have*

$$u = M^n(x_1^n, \ldots, x_n^n) = \phi_u^{-1}\left[\sum_{i=1}^{n}(1/n)\phi_u(x_i^n)\right]. \tag{182}$$

The function ϕ *can be chosen to have properties (94)–(97) and (181).*

Proof: By Proposition (13), there exists a ϕ^n satisfying (94)–(97) which can represent M^n. By (A15), $M_2^n(x_1, x_2, u) = M_2(x_1, x_2, u)$, which implies $\phi^n = \phi$ for each n. Thus (182) holds for this ϕ for each n, where ϕ satisfies (94)–(97). It remains to show that ϕ satisfies condition (181).

Let $n \geq 2$ and let $a \leq x_1 \leq x_2 \leq \cdots \leq x_n \leq b$ with $x_1 < x_n$. Let u_{n-1} be a solution to $\sum_{i=1}^{n-1} \tau(x_i, u) = 0$. The solution is unique using (96). The fact that $\phi_u(z)$ is increasing in z implies that $x_1 \leq u_{n-1} \leq x_{n-1}$ with strict inequalities unless $x_1 = x_2 = \cdots = x_{n-1}$. Thus we have:

$$0 = \sum_{i=1}^{n-1} \tau(x_i, u_{n-1}) = \sum_{i=1}^{n-1}[\phi_{u_{n-1}}(x_i) - \phi_{u_{n-1}}(u_{n-1})] \text{ using (155).} \tag{183}$$

We also have

$$\begin{aligned}\sum_{i=1}^{n-1} \tau(x_i, x_n) &= \sum_{i=1}^{n-1}[\phi_{x_n}(x_i) - \phi_{x_n}(x_n)] \\ &< \sum_{i=1}^{n-1}[\phi_{x_n}(x_n) - \phi_{x_n}(x_n)] \\ &= 0\end{aligned} \tag{184}$$

where the inequality in (184) follows from $x_1 < x_n$, $x_2 \leq x_n$, ... , $x_{n-1} \leq x_n$ and property (95) for ϕ. The equality (183), the inequality (184) and property (96) applied to M^{n-1} imply that

$$-\left[\sum_{i=1}^{n-1} \tau(x_i, u)\right] > 0 \quad \text{for } u_{n-1} < u \leq x_n. \tag{185}$$

The inequality $u < x_n$ and (163) imply that

$$\tau(x_n, u) > 0 \quad \text{for } u < x_n. \tag{186}$$

The inequalities (185) and (186) imply that

$$r^n(x_1, \ldots, x_n, u) > 0 \quad \text{for } u_{n-1} < u < x_n. \tag{187}$$

To prove that $r^n(x_1,\ldots,x_n,u)$ must be increasing in u for $u_{n-1} < u < x_n$, simply repeat the arguments going from (166)–(176) with obvious modifications.

Q.E.D.

Note that Corollary 2 to Proposition 13 is still valid in the present context; i.e., the uniquely determined $\phi_u(z)$ which occurred in Proposition 21 can be replaced by $\psi_u(z) \equiv \alpha(u)+\beta(u)\phi_u(z)$ where $\alpha(u)$ and $\beta(u) \neq 0$ are continuous functions of u and if we limit ourselves to $\psi_u(z)$ which are increasing in z, then we also require that $\beta(u) > 0$. Note also that conditions (181) are invariant under such utility dependent affine transformations; i.e., if we replace $\phi_u(z)$ by $\alpha(u) + \beta(u)\phi_u(z)$ where $\beta(u) \neq 0$, then the new r^n defined by (180) using the transformed ϕ will coincide with the old r^n defined in terms of the initial ϕ. This utility dependent affine transformation invariance property does *not* hold for (107) in general.

Now we are ready to state the following counterpart to the expected utility theorem, Proposition 20 above.

PROPOSITION 22. (*Implicit Expected Utility Theorem*): *Suppose that a decision maker's preferences over certain alternatives $y \in S$ (where S is a compact and connected subset of R^K) can be represented by the continuous utility function f. Suppose further that the decision maker's preferences over n equally likely choices with utility payoffs $x_i \equiv f(y^i)$ for $i = 1,2,\ldots,n$ can be represented by the stochastic preference function $M^n(x_1,\ldots,x_n)$ where for $n = 2,3,\ldots,M^n$ satisfies (A1)–(A4), (A13) and (A15). Then there exists a function of two variables $\phi_u(z)$ defined over $S^2 \equiv \{(x_1,x_2){:}\ a \leq x_i \leq b,$ $i = 1,2\}$ where $a \equiv \min_y\{f(y){:}\ y \in S\}$ and $b \equiv \max_y\{f(y){:}\ y \in S\}$ such that the decision maker's state contingent preferences over N uncertain events with positive probabilities $p_1,\ldots,p_N$ can be represented by the function F where $u = F(y^1,\ldots,y^N,p_1,\ldots,p_N)$ is the unique u solution to (161). The generating function ϕ for the stochastic preference functions M^n satisfies (94), (95) and (181).*

Proof: Except for the uniqueness of the u solution to (161), the proof of the present proposition follows from the proof of Proposition 21 and the earlier arguments leading to (161).

It remains to prove that the u solution to (161) is unique. If $N = 2$, then the analysis around equations (177)–(179) showed that the u solution to (161) was unique. Hence, in what follows, we can assume that $N \geq 3$.

Let $x_i \equiv f(y^i)$ for $i = 1,2,\ldots,N$ and reorder the x_i if necessary so that

$$x_1 \leq x_2 \leq \cdots \leq x_N. \tag{188}$$

If all of the x_i are equal, then the u solution to (161) is unique (using (95)) and is equal to the common x_i. Hence we can assume that $x_1 < x_N$.

Suppose that the solution to (161) (or equivalently, to (162)) is not unique. Then there exist u_1 and u_2 such that

(189) $$x_1 < u_1 < u_2 < x_N \text{ and}$$

(190) $$\sum_{i=1}^{N} p_i \tau(x_i, u_j) = 0 \quad \text{for } j = 1, 2.$$

Case (i): the probabilities p_i are all rational numbers. In this case, there exist positive integers n_i such that

(191) $$p_i = n_i/n, \quad i = 1, 2, \ldots, N \quad \text{and } n = \sum_{i=1}^{N} n_i.$$

Under these conditions, (190) can be rewritten as

(192) $$\sum_{i=1}^{N} n_i \tau(x_i, u_j) = 0, \quad j = 1, 2.$$

But by Proposition 21, M^n satisfies (96) (with n replacing N) and hence the u solution to (192) is unique, which is a contradiction.

Case (ii): one or more of the probabilities p_i are irrational numbers. Define the vectors $\tau^j(x_1, \ldots, x_N) \equiv [\tau(x_1, u_j),\ \tau(x_2, u_j),\ \ldots\ ,\ \tau(x_N, u_j)]$ for $j = 1, 2$. Define the set of probability vectors $H(x_1, \ldots, x_N) \equiv \{v : v \gg 0_N, v \cdot 1_N = 1$ and $v \cdot \tau^j(x_1, \ldots, x_N) = 0$ for $j = 1, 2\}$, where $v \cdot 1_N \equiv \sum_{i=1}^{N} v_i 1$ is the inner product of the vectors v and 1_N. Note that $(p_1, \ldots, p_N) \in H(x_1, \ldots, x_N)$. Now vary $x_1, \ldots, x_N$, preserving the inequalities (188) and (189). Since $N \geq 3$ and each component of $\tau^j(x_1, \ldots, x_N)$ is continuous and increasing for some x_i for $j = 1, 2$, it can be seen that there exist vectors $x^* \equiv (x_1^*, \ldots, x_N^*)$ and $p^* \equiv (p_1^*, \ldots, p_N^*)$ such that: (i) x^* satisfies (188) and (189); (ii) $p^* \in H(x^*)$; and (iii) the p_i^* are all rational numbers. Now apply case (i) above, replacing x_i by x_i^* and p_i by p_i^* for $i = 1, \ldots, N$. Again we obtain a contradiction to our supposition.

Thus the u solution to (161) must be unique. Q.E.D.

The above proposition shows that under the assumptions of the Implicit Expected Utility Theorem, the generating function ϕ for the stochastic preference functions M^n must satisfy (94), (95) and (181). The following proposition shows that a function of two variables ϕ satisfying these three properties can be used to generate a family of stochastic preference functions M^n via (158) and the M^n will satisfy (A1)–(A4), (A13) and (A15).

PROPOSITION 23. *For each integer $n \geq 1$, define $S^n \equiv \{(x_1, \ldots, x_n)\colon a \leq x_i \leq b,\ i = 1, 2, \ldots, n\}$ where $-\infty < a < b < +\infty$. Let the function of two variables $\phi_u(z)$ be defined for $(z, u) \in S^2$ and have properties (94), (95) and (181). Then for $n \geq 2$ and $(x_1, \ldots, x_n) \in S^n$, the u solution to*

(193) $$\phi_u(u) = \sum_{i=1}^{n} (1/n) \phi_u(x_i)$$

is unique and hence we can define the function $M^n(x_1, \dots, x_n)$ by this solution. For each n, M^n satisfies (A1)–(A4) and (A13) and the family of functions, $\{M^n : n = 2, 3, \dots\}$, satisfies (A15).

Proof: Let $n \geq 2$ and let $(x_1, \dots, x_n) \in S^n$. We need to prove that the u solution to (193) is unique and then the remainder of the proof follows using Proposition 14.

Let us reorder the x_i if necessary so that

$$x_1 \leq x_2 \leq \cdots \leq x_n. \tag{194}$$

If all of the x_i are equal, then property (95) on ϕ will imply that the u solution to (193) is unique and is equal to the common x_i. Hence we can assume $x_1 < x_n$.

Note that if $n = 2$, then the argument around (177)–(179) proves that the u solution to (193) is unique; (set $p_1 = p_2 = 1/2$). We now tackle the case of a general n.

Property (95) on ϕ implies that any u roots to (193) must lie between x_1 and x_n. Property (94) implies that at least one such root will exist. Hence ***suppose*** that two roots, u_1 and u_2, to (193) exist. Then we have:

$$x_1 < u_1 < u_2 < x_n; \tag{195}$$

$$\sum_{i=1}^{n} (1/n)\tau(x_i, u_j) = 0, \quad j = 1, 2, \tag{196}$$

where τ is defined by (155) as usual. Since $u_j < x_n$, $\tau(x_n, u_j) > 0$ for $j = 1, 2$ by (163) and so, using (196), we have $\tau(x_n, u_j) = -\sum_{i=1}^{n-1} \tau(x_i, u_j) > 0$ for $j = 1, 2$ and hence by definition (180),

$$r_n(x_1, x_2, \dots, x_n, u_j) \equiv -\sum_{i=1}^{n-1} \tau(x_i, u_j)/\tau(x_n, u_j) = 1, \quad j = 1, 2. \tag{197}$$

Note that

$$\sum_{i=1}^{n-1} \tau(x_i, u_1) = -\tau(x_n, u_1) < 0. \tag{198}$$

We also have:

$$\begin{aligned} \sum_{i=1}^{n-1} \tau(x_i, x_1) &\geq \sum_{i=1}^{n-1} \tau(x_1, x_1) && \text{using (95)} \\ &= 0 && \text{using (155).} \end{aligned} \tag{199}$$

The inequalities (198) and (199) (using also (94) and (95)) imply that a solution u^* to $\sum_{i=1}^{n-1} \tau(x_i, u) = 0$ exists and satisfies

$$x_1 \leq u^* < u_1. \tag{200}$$

Now rewrite the second equation in (196) as follows:

$$
\begin{aligned}
(201) \qquad 0 &= \sum_{i=1}^{n} \tau(x_i, u_2) \\
&= \sum_{i=1}^{n-1} \tau(x_i, u_2) + r^n(x_1, \ldots, x_n, u_2)\tau(x_n, u_2) \\
&\quad \text{using (197) for } j = 2 \\
&> \sum_{i=1}^{n-1} \tau(x_i, u_2) + r^n(x_1, \ldots, x_n, u_1)\tau(x_n, u_2) \\
&\quad \text{using } u^* < u_1 < u_2,\ \tau(x_n, u_2) > 0 \text{ and (181)} \\
&= \sum_{i=1}^{n-1} \tau(x_i, u_2) + 1\tau(x_n, u_2) \quad \text{using (197) for } j = 1 \\
&= 0 \quad \text{using (196) for } j = 2.
\end{aligned}
$$

But (201) is a contradiction and so our supposition is false and hence the u root to (193) must be unique. Q.E.D.

Thus properties (94), (95) and (181) are necessary and sufficient conditions that the generating function ϕ must possess in order to generate a family of stochastic preference functions $\{M^n\}$ via (193) satisfying (A1)–(A4), (A13) and (A15). These latter axioms were the ones which were used to prove the Implicit Expected Utility Theorem, Proposition 22 above.

We turn now to some applications of the above framework to insurance, gambling and investing.

8. Applications to Insurance

In the applications to be considered in this section, we will not model the consumer's choices over (certain) commodities in any detail: we simply assume the existence of a single composite consumption good, which can be thought of as "income" deflated by a consumer price index, or "real income". In this section, we consider applications of the expected utility and implicit expected utility models presented in the previous section to model the demand for insurance.

Consider the decision to insure some property over a given time period. In the first state of nature, which occurs with probability $p_1 > 0$, the property worth $\bar{\imath} > 0$ consumption units is completely destroyed; in the second state of nature, which occurs with probability $p_2 = 1 - p_1 > 0$, the property remains undamaged over the time period under consideration. Let the consumer's beginning of the period real income be $y > 0$. A unit of insurance (with the face value of the policy equal to one consumption unit) can be purchased at the premium rate r (where $0 < r < 1$). If state 1 occurs during the period, the insurance company pays out the face value of the policy less the premium, and this payout is equal to $1 - r$. If state 2 occurs (so that the property is not destroyed during the period), the company pays nothing to the consumer and keeps the premium. Thus if the consumer purchases $i \geq 0$ units of insurance, the cost to the consumer if state 2 occurs (no damage) is ri while if state 1 occurs (complete destruction), the benefit of the policy is $(1 - r)i$. We assume that the consumer is not allowed to overinsure, so we have the following restrictions on the purchase of insurance:

$$0 \leq i \leq \bar{\imath}. \tag{202}$$

The consumer's certainty utility function in this simple model is $f(x) = x$ where x equals consumption and the consumer's state contingent utility function is $F(x_1, x_2, p_1, p_2)$ where $p_j > 0$ is the probability that state j occurs and x_j is consumption in state j, $j = 1, 2$. We can model the consumer's contingent consumption in state j, x_j, as the function of the amount of insurance purchased i as follows:

$$x_1 = y - \bar{\imath} + (1 - r)i; \quad x_2 = y - ri. \tag{203}$$

If it is costless to run an insurance company and insurance is offered at an actuarially fair price, then the expected value of premium income should equal the expected value of property claims. Thus in this "no loading" case,[43] the premium rate r satisfies $p_1(1 - r)i = p_2 ri = (1 - p_1)ri$ or $r = p_1$; i.e., the premium rate r equals the probability of loss p_1 in the actuarially fair case. Of course, this is only a limiting case and in the real world, we would expect $r > p_1$ (the "loaded" case).

Suppose that the consumer's stochastic preference functions over n equally likely alternatives, M^n, satisfy (A1)–(A4), (A12) and (A14). Then we may apply Proposition 20, the Expected Utility Theorem, and conclude that the consumer's purchase of insurance i^* is the solution to the following constrained utility maximization problem:

$$\max_i \{p_1 \phi(y - \bar{\imath} + (1 - r)i) + p_2 \phi(y - ri) : 0 \leq i \leq \bar{\imath}\} \tag{204}$$

where $\phi(z)$ is a continuous increasing function of one variable, defined for $a \leq z \leq b$.[44]

Now suppose that any one of the stochastic preference functions M^n satisfies any one of the risk aversion axioms (A8), (A9) or (A10). Then since M^n also satisfies (A1)–(A4) and (A12), by Proposition 9, ϕ is a strictly concave

[43] This is the term used in the insurance literature; e.g. see Bowers, Gerber, Hickman, Jones and Nesbitt [1986; 7].

[44] We require that $a \leq y - \bar{\imath}$ and $y \leq b$.

(and hence concave) function of one variable. Thus $\phi^+(z) \equiv \lim_{h\to 0, h>0}[\phi(z+h) - \phi(z)]/h$ exists and $\phi^-(z) \equiv \lim_{h\to 0, h<0}[\phi(z+h) - \phi(z)]/h$ exists, since concave functions always have left and right derivatives.

Define the objective function in (4) as $f(i) \equiv p_1\phi[y - \bar{\imath} + (1-r)i] + (1-p_1)\phi(y - ri)$. Since $\phi(z)$ is a strictly concave function of z, it can be verified that $f(i)$ is a strictly concave function of i. Note that the consumer's optimal insurance problem (204) can be rewritten as:

$$\max_i \{f(i) : 0 \le i \le \bar{\imath}\}. \tag{205}$$

In view of the strict concavity of f, a necessary and sufficient condition for the consumer to fully insure (i.e., for $i = \bar{\imath}$ to solve (205) uniquely) is $f^-(\bar{\imath}) \ge 0$. Using the definition of f, $f^-(\bar{\imath})$ can be written in terms of ϕ as follows:

$$\begin{aligned} f^-(\bar{\imath}) &= p_1\phi^-[y - \bar{\imath} + (1-r)\bar{\imath}](1-r) + (1-p_1)\phi^+(y - r\bar{\imath})(-r) \\ &= p_1(1-r)\phi^-(y - r\bar{\imath}) - r(1-p_1)\phi^+(y - r\bar{\imath}). \end{aligned} \tag{206}$$

Consider the *actuarially fair case* where $r = p_1$. Substitution of this condition into (206) yields

$$f^-(\bar{\imath}) = p_1(1-p_1)[\phi^-(y - r\bar{\imath}) - \phi^+(y - r\bar{\imath})] \ge 0 \tag{207}$$

where the inequality in (207) follows from $p_1 > 0$, $1 - p_1 > 0$ and the fact that $\phi^-(y - r\bar{\imath}) \ge \phi^+(y - r\bar{\imath})$; i.e., if $\phi(z)$ is a concave function, then the left hand derivative $\phi^-(z)$ is always equal to or greater than the right hand derivative $\phi^+(z)$.[45] The inequality (207) implies that $\bar{\imath}$ is the unique solution to (205). Thus under the assumptions of the Expected Utility Theorem (see Proposition 20 above) plus the additional assumption that the consumer's stochastic preference functions M^n satisfy the weak risk aversion axiom (A8), *in the actuarially fair case, the risk averse consumer will always fully insure.* It should be noted that Mossin [1968; 557] established this result assuming a twice differentiable utility function.

Consider now the *differentiable actuarially unfair case* (or the loaded case to use insurance terminology) where the premium rate exceeds the probability of loss ($r > p_1$). As before, we assume that the consumer's stochastic preference functions over equally likely states satisfy the assumptions in the Expected Utility Theorem plus the weak risk aversion axiom (A8). Assuming that any one of the M^n are differentiable will imply that the generating function ϕ is also differentiable. Since $\phi(z)$ is also strictly concave and increasing, we must have $\phi'(z) > 0$ for z in the interior of the domain of definition for ϕ.[46] Recall that

[45]See, for example, Roberts and Varberg [1973; 6].

[46]If we had $\phi'(z^*) = 0$ for z^* such that $a < z^* < b$, then $\phi(z^*) > \phi(z)$ for $z \ne z^*$, which would contradict the increasing property for ϕ.

$f(i)$ is the objective function for the consumer's utility maximization problem (204). Since ϕ is differentiable, so is f and we can compute the derivative of f evaluated at the fully insured point, $\bar{\imath}$:

$$\begin{aligned} f'(\bar{\imath}) &= p_1\phi'[y-\bar{\imath}+(1-r)\bar{\imath}](1-r)+(1-p_1)\phi'(y-r\bar{\imath})(-r) \\ &= (p_1-p_1r)\phi'(y-r\bar{\imath})-(r-p_1r)\phi'(y-r\bar{\imath}) \\ &= (p_1-r)\phi'(y-r\bar{\imath}) \\ &< 0 \end{aligned} \tag{208}$$

where the inequality in (208) follows from $r > p_1$ and $\phi'(y-r\bar{\imath}) > 0$. Since a necessary and sufficient condition for $\bar{\imath}$ to solve (205) when f is differentiable at $\bar{\imath}$ is $f'(\bar{\imath}) \geq 0$, (208) implies that $\bar{\imath}$ will *not* solve (205) or (204). Thus, under the assumptions of the Expected Utility Theorem, the *risk averse decision maker with smooth preferences will never fully insure in the loaded case*. Again, this result was first established by Mossin [1968; 557] using a somewhat more restrictive definition of risk aversion.

In the case where $p_1 = p_2 = 1/2$, the consumer's state contingent preferences can be represented by a symmetric mean $M(x_1, x_2)$. If we assume M satisfies (A1)–(A4) and (A12), then by Proposition 7, $M(x_1,x_2) = \phi^{-1}[(1/2)\phi(x_1) + (1/2)\phi(x_2)]$ where ϕ is a continuous increasing function of one variable. In this equal probabilities case, the indifference curves are symmetric around the equal quantities ray and we can illustrate the consumer's decision to purchase insurance by means of Figure 2.

The point A represents the no insurance point: if state 1 occurs, the consumer has real income $y-\bar{\imath}$ while if state 2 occurs, the consumer's real income is y. The certainty equivalent level of utility where consumption in both states is equalized occurs at the point C, which is on the same indifference curve as point A. The consumer's choice set in the no loading case is given by the line segment AB which has slope equal to -1 in this equal probabilities case. The point B corresponds to the fully insured case where $i = \bar{\imath}$. If the consumer's stochastic preference function $M(x_1, x_2)$ satisfies any of the risk aversion axioms (A8), (A9) or (A10), it can be seen that in the no loading case, the highest level of utility will be attained at the fully insured point B where real income or consumption will equal $y-(1/2)\bar{\imath}$ in both states of nature. If the premium rates are loaded so that $r > p_1 = 1/2$, then the consumer's state contingent consumption possibilities set attainable by different levels of insurance can be represented by the line segment AD: the point A corresponds to the no insurance point ($i = 0$) and the point D corresponds to the fully insured case ($i = \bar{\imath}$). In the smooth risk averse case, where the indifference curves are differentiable and enclose strictly convex sets, it can be seen that the equilibrium point E will always fall short of the fully insured point D.

Since, as Mossin [1968; 558] observed, many consumers fully insure in the loaded case, the above result creates a problem for the expected utility

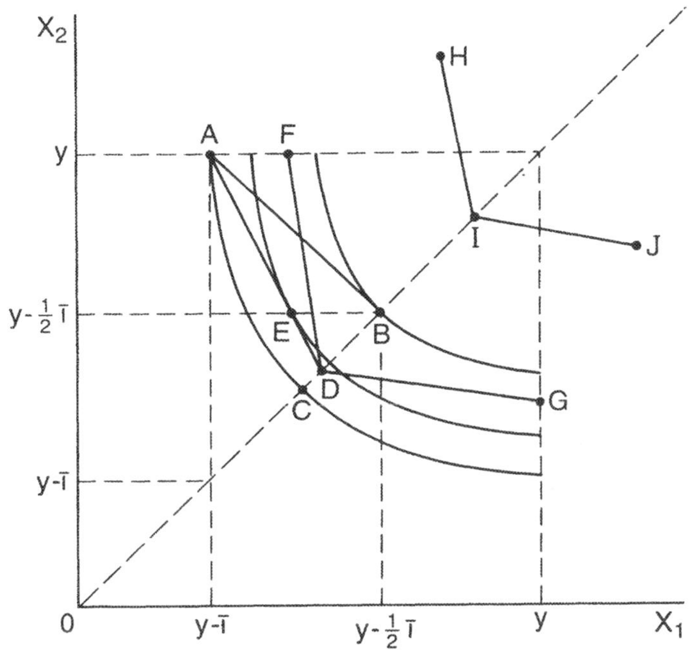

Figure 2

model. We could try to rescue the expected utility model by postulating that the indifference curve through D has a kink at the equal consumption point; for example, the indifference curve through the point D, which has coordinates (x^*, x^*) say, could be the kinked curve FDG. However, in the expected utility model, a kink at D translates into a kink in the generating function ϕ at x^* and so in the risk averse case where ϕ is concave or strictly concave, we must have $\phi^-(x^*) > \phi^+(x^*)$. Now consider another indifference curve $I(u) \equiv \{(x_1, x_2)\colon \phi^{-1}[(1/2)\phi(x_1)+(1/2)\phi(x_2)] = u, a \leq x_i \leq b, i = 1,2\}$ for $u \neq x^*$. If $[x^*, x_2(u)]$ belongs to $I(u)$, then by symmetry, so does $[x_2(u), x^*]$ but there will be kinks at both of these points. Although it is quite reasonable for indifference curves to have kinks at the equal consumption point, it does not seem reasonable to have this systematic pattern of kinks away from the equal consumption line. Moreover, if we attempt to rationalize the decision to fully insure for an interval of real incomes, $y_1 \leq y \leq y_2$, then the expected utility model breaks down altogether, since the underlying generating function $\phi(y)$ cannot have a kink (a point where the left derivative $\phi^-(y)$ is strictly greater than the right derivative $\phi^+(y)$) for every y such that $y_1 \leq y \leq y_2$. This point was first made by Chew [1989; 287] and is repeated in Epstein and Zin [1990; 396].

Having found that the expected utility model is unsatisfactory, we turn now to the implicit expected utility model. Suppose that the consumer's family of stochastic preference functions, $\{M^n : n = 2, 3, \ldots\}$, defined over equally likely alternatives, satisfies (A1)–(A4), (A13) and (A15). Then we may apply Proposition 22, the Implicit Expected Utility Theorem, and conclude that the consumer's state contingent preferences over two states of nature (which have positive probabilities p_1 and p_2) can be represented by the function F where $u = F(x_1, x_2, p_1, p_2)$ is the unique u solution to (209):

$$p_1\phi_u(x_1) + p_2\phi_u(x_2) - \phi_u(u) = 0 \tag{209}$$

where $\phi_u(z)$ is: (i) jointly continuous in (z, u); (ii) increasing in z for each fixed u; and (iii) satisfies property (181).

If we represent consumption x_j in each state of nature $j = 1, 2$ as a function of the amount of insurance i purchased as in equations (203), and we also impose the no over insurance condition (202), then the consumer's decision to purchase insurance i in the implicit expected utility model can be formulated as the following constrained utility maximization problem:

$$\max_{u,i}\{u : p_1\phi_u[y - \bar{\imath} + (1 - r)i] + p_2\phi_u(y - ri) - \phi_u(u) = 0,\ 0 \le i \le \bar{\imath}\}. \tag{210}$$

Note that if $\phi_u(z)$ does not actually depend on u, then (210) is equivalent to (204); i.e., the implicit expected utility model collapses down to the expected utility model in this case.[47]

The mathematics of the constrained maximization problem (210) is somewhat involved and so we will not develop necessary (or sufficient) conditions for (u^*, i^*) to solve (210) in this paper. Our focus here is on demonstrating that the implicit expected utility model is much more flexible than the expected utility model.

In the special case where $p_1 \equiv p_2 = 1/2$, we can represent the consumer's stochastic preferences by means of an arbitrary symmetric mean $U(x_1, x_2)$ where U satisfies (A1)–(A4). Then using the approximation theorem that we developed in Section 5 above (recall (114) which was used to define $\phi_u(z)$ in terms of the indifference curve function $x_2 = I(x_1, u)$ that corresponded to the

[47]If $\phi_u(z) = \phi(z)$, then condition (181) becomes $r^n(x_1, \ldots, x_n, u) = [(n-1)\phi(u) - \sum_{i=1}^{n-1}\phi(x_i)]/[\phi(x_n) - \phi(u)]$ is increasing in u for $u_{n-1} < u < x_n$ and for $n = 2, 3, \ldots$. Note that $r^n(x_1, \ldots, x_n, u) = f^1(u)f^2(u)$ where $f^1(u) \equiv (n-1)\phi(u) - \sum_{i=1}^{n-1}\phi(x_i)$ and $f^2(u) \equiv 1/[\phi(x_n) - \phi(u)]$. If $\phi(u)$ is increasing in u, then obviously so is $f^1(u)$. Also $f^2(u)$ is increasing in u and is positive for $u < x_n$. Moreover, $f^1(u) > 0$ if $u_{n-1} < u$. Under these conditions, $f^1(u)f^2(u)$ will be increasing in u for $u_{n-1} < u < x_n$ and hence (181) will be satisfied.

utility function $u = U(x_1, x_2)$), U can be represented as an implicitly separable symmetric mean and so the consumer's decision to purchase insurance can be modeled by the constrained maximization problem (210) where we set $p_1 = p_2 = 1/2$. Referring to Figure 2, the consumer's family of indifference curves can be an *arbitrary* family of curves (consistent with axioms (A1)–(A4)) in this equal probabilities case such as the family represented by the kinked curves FDG, HIJ, etc. As was the case in the expected utility model, the consumer's state contingent consumption possibilities sets are the line segment AB in the no loading case and the line segment AD in the loaded case. With the family of kinked indifference curves, the consumer will always fully insure until the excess premium rate becomes too large; i.e., when the slope of AD is less than the slope of FD (remember, these slopes are negative).[48]

Returning to the case where the probabilities p_1 and p_2 are arbitrary, we find that we cannot *always* use the approximation theorem developed in Section 5 to define a $\phi_u(z)$ using (114) that can be used as the generating function ϕ in (210) since the resulting ϕ will not necessarily satisfy (181). However, if the symmetric mean $U(x_1, x_2)$ satisfying (A1)–(A4) also satisfies (A6) (with $0 \leq a < 1 \leq b$) so that U is positively linearly homogeneous, then the $\phi_u(z)$ defined by (114) will in fact satisfy (181). To prove this, we need to show that for $n = 2, 3, \ldots, r^n(x_1, \ldots, x_n, u)$ defined by (180) is increasing in u for $u_{n-1} < u < x_n$. If $U(x_1, x_2)$ satisfies (A6), then the corresponding indifference curve function $I(x_i, u)$ will also be positively linearly homogeneous; i.e., $I(\lambda x_i, \lambda u) = \lambda I(x_i, u)$ for $\lambda > 0$. Thus using definitions (114), (130) and (155) (which defined $\tau(x_i, u) \equiv \phi_u(x_i) - \phi_u(u)$) as well as the inequalities $a \leq x_1 \leq x_2 \leq \cdots \leq x_n \leq b$ with $0 \leq a < 1 \leq b$, $x_1 < x_n$, $u_{n-1} < u < x_n$ with u_{n-1} satisfying $\sum_{i=1}^{n-1} \tau(x_i, u_{n-1}) = 0$, we have:

$$\begin{aligned} r^n(x_1, \ldots, x_n, u) &\equiv -\left[\sum\nolimits_{i=1}^{n-1} \tau(x_i, u)\right] / \tau(x_n, u) \\ &= -\left[\sum\nolimits_{i=1}^{n-1} u\tau(x_i/u, 1)\right] / u\tau(x_n/u, 1) \\ &= -\left[\sum\nolimits_{i=1}^{n-1} \tau(x_i/u, 1)\right] / \tau(x_n/u, 1) \end{aligned} \tag{211}$$

where we have used the fact that the linear homogeneity of $I(x_i, u)$ implies that the functions ϕ defined by (114) and τ defined by (155) are also linearly homogeneous in their arguments. Now it is straightforward to prove that $r^n(x_1, \ldots, x_n, u)$ is increasing in u for $u_{n-1} < u < x_n$ using (211) and the

[48]Families of kinked indifference curves seem to have first appeared in Chew [1989; 281] and their application to the problem of fully insuring appears in Segal and Spivak [1990; 116]. The very recent literature on kinked stochastic indifference curves and the related concept of first order risk aversion is reviewed in Section 2.4 of Epstein [1992].

following facts: (i) $\tau(z,1)$ is increasing in z; (ii) $-\sum_{i=1}^{n-1}\tau(x_i/u,1) > 0$; and (iii) $\tau(x_n/u,1) > 0$ for $u_{n-1} < u < x_n$.

Using the result in the previous paragraph, we see that the consumer's stochastic preference function defined over two equally likely stochastic events, $M^2(x_1,x_2) \equiv U(x_1,x_2)$, can be used to define a generating function $\phi_u(z)$ via (114) such that the resulting family of stochastic preference functions, $\{M^n : n = 2,3,\ldots\}$ defined using (158), satisfies the axioms required for the Implicit Expected Utility Theorem, (A1)–(A4), (A13) and (A15), provided that U satisfies (A1)–(A4) and (A6).

Of course the above method for defining a $\phi_u(z)$ which satisfies (94), (95) and (181) is somewhat indirect and more direct methods are available. For example if $\phi_u(z)$ satisfies (94) and (95) and either of the sufficient conditions for the corresponding M to be linearly homogeneous, (134) or (136), then it can be verified that the resulting $\tau(z,u)$ defined by (155) will satisfy (211) (and hence ϕ will satisfy (181)), provided that $0 \le a < 1 \le b$. Thus all three methods lead to a homothetic family of stochastic preference functions $\{M^n : n = 2,3,\ldots\}$; i.e., each of the M^n will satisfy (A6).

It is also possible to use the translatability axiom (A7) to provide sufficient conditions for $\phi_u(z)$ to satisfy (181). For example, suppose the symmetric mean $U(x_1,x_2)$ satisfying (A1)–(A4) also satisfies (A7) (with $a \le 0 \le b$ and $-\infty < a < b < +\infty$). Then the $\phi_u(z)$ defined by (114) will satisfy (94), (95) and (181). To prove that $\phi_u(z)$ satisfies (181), we note that if U is translatable, then so are the corresponding indifference curve function I and the corresponding ϕ defined by (114). Then the corresponding $\tau(z,u)$ defined by (114) has the following property:

$$\tau(z,u) = \tau(z-u,0). \tag{212}$$

Now substitute (212) into the first line of (211) and we obtain:

$$\begin{aligned} r_n(x_1,\ldots,x_n,u) &\equiv -\left[\sum\nolimits_{i=1}^{n-1}\tau(x_i,u)\right]/\tau(x_n,u) \\ &= -\left[\sum\nolimits_{i=1}^{n-1}\tau(x_i-u,0)\right]/\tau(x_n-u,0). \end{aligned} \tag{213}$$

Again it is straightforward to prove that $r_n(x_1,\ldots,x_n,u)$ is increasing in u for $u_{n-1} < u < x_n$ using (213) and the following facts: (i) $\tau(z,0)$ is increasing in z; (ii) $-\sum_{i=1}^{n-1}\tau(x_i-u,0) > 0$; and (iii) $\tau(x_n-u,0) > 0$. Thus the $\phi_u(z)$ defined by (114) satisfies (181) and since we proved in Section 5 that ϕ will also satisfy (94) and (95), we see that the resulting family of stochastic preference functions, $\{M^n : n = 2,3,\ldots\}$ defined using (158), satisfies the axioms required for the Implicit Expected Utility Theorem, (A1)–(A4), (A13) and (A15), provided that $U(x_1,x_2)$ satisfies (A1)–(A4) and (A7).

If $\phi_u(z)$ satisfies (94), (95) and either of the sufficient conditions for the corresponding M to be translatable, (138) or (140), then it can be verified that the resulting $\tau(z,u)$ defined by (155) will satisfy (212) and (213) (and hence ϕ will satisfy (181)), provided that $-\infty < a \leq 0 < b < +\infty$. These two methods for generating a ϕ that satisfies (94), (95) and (181) plus the method outlined in the previous paragraph all lead to a translatable family of stochastic preference functions $\{M^n : n = 2, 3, \dots\}$; i.e., each of the M^n will satisfy (A7).

We conclude that the implicit expected utility model has no difficulty in rationalizing the decision of the consumer to insure fully for all income levels, provided that the loading of the premium rate is not too excessive.

The above model assumes that if disaster strikes, the property is completely destroyed. In many situations, it may be more realistic to assume that the property damage follows a continuous probability distribution $h(\delta)$ for $0 \leq \delta \leq \bar{\imath}$. We assume that this distribution is independent from the discrete distribution (p_1, p_2) which determines whether disaster strikes or not. In this situation, there is the possibility of overinsurance on the part of the consumer; i.e., disaster strikes but the damage δ is less than the amount of insurance coverage i. Thus we need to define the payout function $g(i, \delta)$ as follows:

$$(214)\qquad g(i,\delta) \equiv \begin{cases} i & \text{if } 0 \leq i \leq \delta \\ \delta & \text{if } 0 \leq \delta \leq i. \end{cases}$$

The counterpart to (210) is now the following constrained utility maximization problem:

$$(215)\qquad \max_{u,i}\{u : p_1 \int_0^{\bar{\imath}} \phi_u[y - \delta + (1-r)g(i,\delta)]h(\delta)d\delta + (1-p_1)\phi_u(y - ri) - \phi_u(u) = 0;\ 0 \leq i \leq \bar{\imath}\}.$$

The detailed analysis of this problem is best left to another occasion. The point of the above example is to show that the implicit expected utility model can be extended from the discrete distribution case to cover the case of a continuous distribution.

The insurance demand models presented in this section are only illustrative and indicate possible ways in which the theoretical models presented in Section 7 might be applied. In an actual empirical application, it would probably be necessary to replace the real income variable y which appears in this section with a (real) wealth variable. This replacement would make it necessary to model some of the intertemporal aspects of the consumer's consumption and savings decision.

We turn now to the problem of modeling the decision to gamble.

9. Applications to Gambling

We consider a very simple gambling framework for illustrative purposes. We suppose that a gambler can wager or bet an amount $w \geq 0$ of the composite consumption good up to some positive limit $\overline{w}$.[49] If state 1 occurs (with probability $p_1 > 0$), then the gambler wins Pw where $P > 1$ is the payout ratio. If state 2 occurs (with probability $p_2 = 1 - p_1$), the gambler loses the amount wagered w. Thus if the gambler's real income during the period under consideration is $y > 0$ and we define x_i as consumption in state i for $i = 1, 2$, then the gambler's consumption possibilities set as a function of the amount bet is:

$$x_1 = y + Pw; \quad x_2 = y - w, \quad 0 \leq w \leq \overline{w}. \tag{216}$$

For an actuarially fair lottery, we require that the expected payouts equal the revenue collected; i.e., we require that the payout ratio P satisfy $p_1 Pw = p_2 w = (1 - p_1)w$ or

$$P = (1 - p_1)/p_1. \tag{217}$$

In the "loaded" or actuarially unfair case, we will have:

$$P < (1 - p_1)/p_1. \tag{218}$$

From (216), we see that the gambler's state contingent consumption possibilities set as a function of the amount gambled is a line segment starting at (y, y) and ending at $(y + P\overline{w}, y - \overline{w})$.[50] This line segment has slope equal to $(y - \overline{w} - y)/(y + P\overline{w} - y) = -\overline{w}/P\overline{w} = -1/P$.

Assuming that the gambler's family of stochastic preference functions $\{M^n : n = 3, 4, \ldots\}$ satisfies axioms (A1)–(A4), (A11) and (A14), then by applying the Expected Utility Theorem, Proposition 20 above, to the gambler's choice problem,[51] we can deduce that the gambler's optimal wager w^* is the solution to the following utility maximization problem:

$$\max_w \{p_1 \phi(y + Pw) + p_2 \phi(y - w) : 0 \leq w \leq w^*\}, \tag{219}$$

[49]Let $\bar{c} > 0$ be the minimum amount of the consumption good that is required to survive during the period and let y be the consumer's real income during the period under consideration. Then presumably $\overline{w} \leq y - \bar{c}$ if the consumer cannot borrow or sell assets.

[50]We assume that the domain of definition parameters a and b satisfy the inequalities $a < y - \overline{w}$ and $y + P\overline{w} < b$.

[51]The certainty utility function f which appears in Proposition 20 is simply the identity function in the present context; i.e., $f(x_i) \equiv x_i$ for $i = 1, 2$.

where $\phi(z)$ is an increasing and continuous function of one variable z for $a \leq z \leq b$.

If we also assume that any of the M^n satisfy any of the risk aversion axioms (A8), (A9) or (A10), then as in the previous section, we can deduce that $\phi(z)$ must be a strictly concave function (simply apply Proposition 9). Thus right and left derivatives for ϕ exist in this risk averse case. Define the objective function in (219) as $f(w) \equiv p_1\phi(y + Pw) + (1 - p_1)\phi(y - w)$ and compute the right derivative of f evaluated at $w = 0$:

$$\begin{aligned} f^+(0) &= p_1\phi^+(y)P + (1 - p_1)\phi^-(y)(-1) \\ &\leq p_1\phi^+(y)[(1 - p_1)/p_1] - (1 - p_1)\phi^-(y) \\ &\qquad \text{assuming that (217) or (218) holds} \\ &= (1 - p_1)[\phi^+(y) - \phi^-(y)] \\ &\leq 0, \end{aligned} \tag{220}$$

where the last inequality in (220) follows from $1 - p_1 > 0$ and $\phi^+(y) \leq \phi^-(y)$ which holds when ϕ is concave or strictly concave. In order to derive the first inequality in (220), we also make use of $\phi^+(y) > 0$ which must hold in order for ϕ to be an increasing and concave function of one variable. The inequality (220) is necessary and sufficient for $w^* = 0$ to solve (220) uniquely when ϕ is a strictly concave function. Thus we have just shown that *a risk averse expected utility maximizer will never gamble,*[52] even if risk aversion is defined by the extremely weak risk aversion axiom (A8).

The above risk aversion results for the expected utility model are illustrated in Figure 3, where we have set $p_1 = p_2 = 1/2$ so that the gambler's stochastic indifference curves are symmetric around the 45 degree line. If the consumer's initial real income is y_1, then the no gambling point is at point A. In the fair gambling case, the gambling contingent consumption possibilities set is the line segment AB which has slope equal to -1, since in this case, $P \equiv (1 - p_1)/p_1 = (1/2)/(1/2) = 1$. With risk aversion, the consumer's indifference curves are strictly convex and it can be seen that the highest utility point occurs at the no gambling case A. Even in the unfair case where (218) holds and the consumer's consumption possibilities set is the line segment AC, the highest utility level is attained at the no gambling point A in the risk averse case.

Of course, a large number of consumers gamble even in the unfair case, and it is obvious that the risk averse expected utility model will not be able to model this gambling activity. In order to get a small amount of gambling in the expected utility model, we will have to follow the example of Friedman and Savage [1948; 292] and assume that the generating function ϕ is strictly convex

[52] A version of this result goes back to Friedman and Savage [1948; 280] at least.

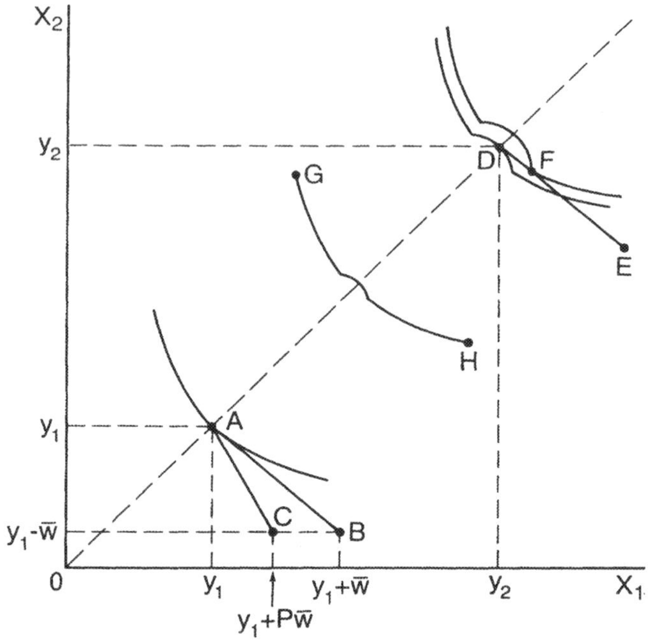

Figure 3

around the consumer's observed real income level. Thus if the consumer's real income is y_2 (see Figure 3) and $\phi(z)$ is strictly convex in a neighborhood of y_2, then there will be a small nonconvexity in the consumer's indifference curve through the initial no gambling point D. In Figure 3, the consumer's fair gambling opportunity set is the straight line segment DE. It can be seen that the consumer attains the highest indifference curve at the point F where the gambling line is tangent to the highest attainable indifference curve. It can be seen that even in the unfair case, the consumer will indulge in a small amount of gambling until the game becomes too unfair.

Note that in this wavy indifference curve case, the consumer will never fully insure, even in the actuarially fair case. However, if the "wave" is small relative to the amount of property at risk, then the consumer will virtually fully insure and the empirical facts of life will not be grossly violated.

The above properties of the expected utility model seem satisfactory thus far. However, a severe problem occurs if we ask that the expected utility model accommodate the propensity to make small gambles for an entire range of real income levels, from say y_1 to y_2. This means that the underlying generator function $\phi(y)$ for the consumer's stochastic preference functions M^n must be strictly convex for all real incomes y between y_1 and y_2. This in turn will imply that the region of nonconvexity in the consumer's indifference curves will be very large for real incomes y around $(y_1+y_2)/2$, so that the consumer will want to gamble huge amounts of his starting real income.[53] This willingness to take large gambles is not consistent with observed behavior for most consumers.

Thus, as in the previous section, we must reject the expected utility model on empirical grounds and turn to the implicit expected utility model.

Suppose that the consumer's family of stochastic preference functions, $\{M^n : n = 2, 3, \ldots\}$, defined over equally likely alternatives satisfies (A1)–(A4), (A13) and (A15). Then we may apply Proposition 22 to our present gambling problem and conclude that the gambler's optimal wager w^* is the solution to the following constrained maximization problem:

$$(221) \qquad \max_{u,w}\{u : p_1\phi_u(y+Pw) + p_2\phi_u(y-w) - \phi_u(u) = 0;\ 0 \leq w \leq \overline{w}\}$$

where $\phi_u(z)$ satisfies (94), (95) and (181).

We will not analyze the mathematics of (221) in any detail in the present paper; we simply note that this implicit expected utility model is perfectly consistent with the consumer having a family of wavy stochastic indifference curves where *each* curve has a small wave around the equal real consumption line. For example in Figure 2, the wavy indifference curve GH can be used to generate an entire family of wavy indifference curves if we are willing to assume that the consumer's stochastic preference functions are all homothetic or are all translatable; see the discussion around equations (211)–(213) in the previous section.

As in the previous section, the models of gambling presented in this section are only illustrative. In an empirical application of these gambling models, it may be necessary to replace the real income variable y which appears in this section with a (real) wealth variable.

We now turn our attention to the consumer's investment decision.

10. Applications to Investing

As usual, we consider a very simple model for an economic agent's investment decision so that we can illustrate the workings of the model by means of a diagram.

The agent has $y > 0$ real income units (dollars divided by a consumer price index) available for investing at the beginning of a period. There is a safe

[53]Friedman and Savage [1948; 294] realized that this problem would occur in their expected utility model.

asset that earns a real rate of return r. There is a single risky asset that earns real rate of return r_1 with probability $p_1 > 0$ and earns real rate $r_2 > r_1$ with probability $p_2 = 1 - p_1 > 0$. State 1 corresponds to the "bad" state of nature where the risky asset earns less than the safe asset (so that $r_1 < r$) and state 2 is the "good" state of nature where the risky asset earns more than the safe asset so we have:

$$r_1 < r < r_2. \tag{222}$$

We let $i \geq 0$ denote the amount of real income that the investor allocates to the risky asset and $y - i \geq 0$ is allocated to the riskless asset. Let x_j denote the investor's end of the period real income if state j occurs for $j = 1, 2$. Then the investor's state contingent consumption possibilities set as a function of the amount invested i is:

$$x_1 = (1+r_1)i + (1+r)(y-i); \; x_2 = (1+r_2)i + (1+r)(y-i); \; 0 \leq i \leq y. \tag{223}$$

If the risky asset earned on average the safe rate of return r, then r_1, r_2, r and p_1 should satisfy the following condition:

$$r = p_1 r_1 + (1 - p_1) r_2. \tag{224}$$

However, typically the risky asset will have a higher expected rate of return than the riskless asset, so in the normal investment situation, we would expect that:

$$r < p_1 r_1 + (1 - p_1) r_2. \tag{225}$$

If condition (225) holds, then the risky asset is said to have a positive risk premium equal to $p_1 r_1 + (1 - p_1) r_2 - r > 0$.

If the decision maker invests nothing in the risky asset, then at the end of the period, his or her real income is equal to $(1 + r)y$ in both states. If the agent invests everything in the risky asset, then the end of the period real income if state 1 occurs is $\bar{x}_1 \equiv (1 + r_1)y$ and the corresponding real income if state 2 occurs is $\bar{x}_2 \equiv (1 + r_2)y$. Thus the investor's state contingent consumption possibilities set is the line segment joining $[(1 + r)y, (1 + r)y]$ to $[(1 + r_1)y, (1 + r_2)y]$. This line segment has slope equal to $[(1 + r_2)y - (1 + r)y]/[(1+r_1)y - (1+r)y] = -(r_2 - r)/(r - r_1) < 0$, where the inequality follows from (222).

Assuming that the investor's family of stochastic preference functions $\{M^n : n = 3, 4, \ldots\}$ over equally likely alternatives satisfies axioms (A1)–(A4), (A11) and (A14), then by applying Proposition 20 above, the Expected

Utility Theorem, we can deduce that the optimal investment i^* is the solution to the following utility maximization problem:[54]

(226) $\max_i \{p_1\phi[(1+r_1)i+(1+r)(y-i)]+p_2\phi[(1+r_2)i+(1+r)(y-i)] : 0 \leq i \leq y\}$

where ϕ is an increasing continuous function of one variable.

If in addition, we assume that the functions M^n satisfy any of the risk aversion axioms (A8), (A9) or (A10), then as in the previous two sections, we can deduce that ϕ must also be a strictly concave function. If the risk premium is zero (so that (224) holds),[55] then we can show that in this risk averse case the expected utility maximizer will not invest in the risky asset. Define $f(i)$ as the objective function in (226). Calculate the right derivative of f evaluated at $i = 0$:

(227)
$$\begin{aligned} f^+(0) &= p_1\phi^-[(1+r)y](r_1 - r) + (1-p_1)\phi^+[(1+r)y](r_2 - r) \\ &\leq \phi^-[(1+r)y]p_1(r_1 - r) + \phi^-[(1+r)y](1-p_1)(r_2 - r) \\ &\quad \text{using } 1 - p_1 > 0,\ r_2 - r > 0 \text{ and } \phi^+(z) \leq \phi^-(z) \\ &= \phi^-[(1+r)y][p_1r_1 + (1-p_1)r_2 - r] \\ &= 0 \quad \text{using (224)} \end{aligned}$$

where $\phi^+(z) \leq \phi^-(z)$ follows from the strict concavity of ϕ. The strict concavity of ϕ implies that $f(i)$ is strictly concave in i. But $f^+(0) \leq 0$ is a necessary and sufficient condition for $i^* = 0$ to be the unique solution to the (strictly) concave programming problem $\max_i\{f(i) : 0 \leq i \leq y\}$. Thus (227) shows that the risk averse expected utility maximizer will not invest in the risky asset if the risk premium is zero, even if the risk aversion is defined by the very weak axiom (A8).

However, if the risk premium is positive so that (225) holds, then it is possible to show that in the case of smooth preferences, the risk averse expected utility maximizer will always invest in the risky asset. In this smooth case, ϕ is differentiable and we may compute the derivative of f evaluated at $i = 0$ as follows:

(228)
$$\begin{aligned} f'(0) &= p_1\phi'[(1+r)y](r_1 - r) + (1-p_1)\phi'[(1+r)y](r_2 - r) \\ &= \phi'[(1+r)y][p_1r_1 + (1-p_1)r_2 - r] \\ &> 0 \end{aligned}$$

---

[54]As usual in our applications, we assume that the investor's certainty utility function f is the identity function; i.e., $f(x_j) = x_j$ for $j = 1, 2$.

[55]The no investment in the risky asset conclusion also holds if the risk premium is negative.

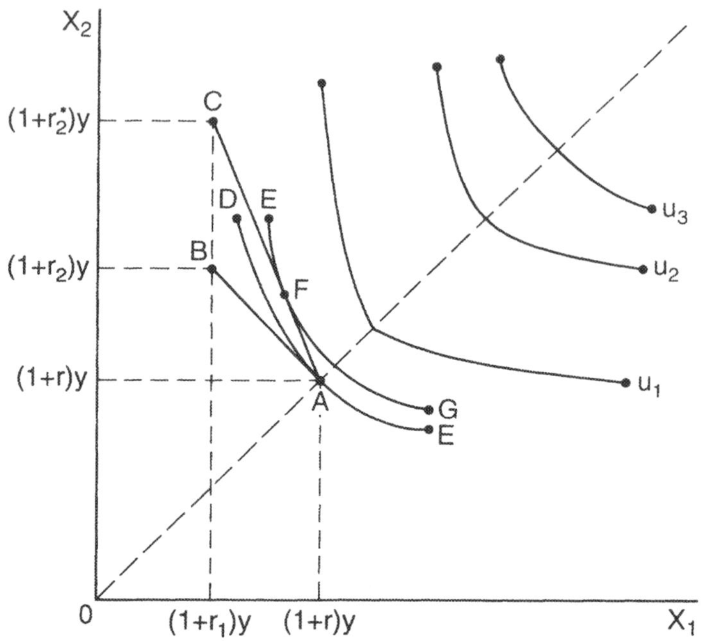

Figure 4

where the inequality in (228) follows using (225) and the fact that since $\phi(z)$ is increasing and strictly concave, we must have $\phi'(z) > 0$ for every z in the interior of the domain of definition for ϕ. The inequality (228) implies that $i^* = 0$ *cannot* solve the (strictly) concave programming problem (226) and hence the solution to (226) must satisfy $i^* > 0$.

The above results are illustrated in Figure 4 for the special case where $p_1 = p_2 = 1/2$, so that the investor's indifference curves are symmetric around the 45 degree line.

Consider first the case where there is no risk premium so that (224) holds (with $p_1 = p_2 = 1/2$). In this case, the investor's state contingent consumption possibilities set is the line segment AB where the point A is the no investment in the risky asset point $[(1+r)y, (1+r)y]$ and the point B is the no investment in the riskless asset point $[(1+r_1)y, (1+r_2)y]$. The slope of this line segment is $-(r_2 - r)/(r - r_1) = -1$, using (224) with $p_1 = p_2 = 1/2$. With risk aversion, the investor's indifference curve through the point A, DAE, will lie above the line segment AB, except for the endpoint A. When the risk premium is negative, the point B shifts downwards and it can be seen that A is still the optimal point for the risk averse investor. However, if the risk premium becomes positive (say r_2 increases to r_2^*), then the point B shifts upwards to the point C and the new consumption possibilities set is the line segment AC. In this case, the risk averse investor with smooth preferences will be induced to invest in the risky asset and the new equilibrium point becomes the point F which is where CA is tangent to the highest attainable indifference curve EFG.

In contrast to our insurance and gambling examples, there appears to be nothing in the above expected utility investing model which is blatantly inconsistent with real life investing behavior. The problems with the expected utility model show up at a more subtle level. For example, suppose that an investor's stochastic preferences over two equally likely alternatives were given by the family of indifference curves indexed by u_1, u_2 and u_3 in Figure 4. Thus for low levels of real income, the investor is quite risk averse and the indifference curves have kinks at the 45 degree line. For higher levels of utility, u_2, the kinks become less pronounced and for higher yet utility levels, u_3, the kinks vanish and the indifference curves become linear (risk neutral) in a neighborhood around the 45 degree line. These preferences seem perfectly reasonable and yet, as we explained in Section 8 above, the expected utility model cannot accommodate them.[56] However, such preferences can easily be accommodated by the implicit expected utility model and we now turn to that model.[57]

Suppose that the investor's family of stochastic preference functions, $\{M^n : n = 2, 3, \dots\}$, defined over equally likely alternatives satisfies (A1)–(A4), (A13) and (A15). Then we may apply Proposition 22 to our present problem and conclude that the optimal investment i^* is the solution to the following constrained maximization problem:

$$(229) \quad \max_{u,i}\{u : p_1\phi_u[(1+r_1)i + (1+r)(y-i)] + p_2\phi_u[(1+r_2)i + (1+r)(y-i)] - \phi_u(u) = 0;\ 0 \le i \le y\}$$

where $\phi_u(z)$ satisfies (94), (95) and (181).

Of course, (229) can readily be generalized to the case where the real return to the risky asset, p, has a continuous distribution rather than a discrete one.

---

[56]Another way of seeing the limitations of the expected utility model is to recall the analysis presented in Section 4 where we showed via (60) that $\phi(z)$ is determined entirely by the behavior of $M^2(x_1, x_2)$ along a single line segment, provided that M^2 is once differentiable. It should also be noted that Chew [1989; 281] has a diagram which is very similar to our diagram of the three indifference curves indexed by u_1, u_2 and u_3; i.e., there are kinks at low levels of income which flatten out as income increases.

[57]For applications of kinked stochastic indifference curve models to problems involving investing behavior, see Epstein and Zin [1990] and Section 2.4 of Epstein [1992].

Thus if we let $g(p)$ be the continuous probability distribution function where the lowest possible real rate of return is r_1 and the highest possible return is r_2, then the continuous distribution counterpart to (229) is:

$$(230)\quad \max_{u,i}\left\{u : \int_{r1}^{r2} \phi_u[(1+p)i+(1+r)(y-i)]g(p)dp-\phi_u(u) = 0;\ 0 \leq i \leq y\right\}.$$

Again we leave the mathematics of the constrained maximization problems (229) and (230) to another occasion.

11. Conclusion

Since the introduction to this paper was rather long, the conclusion will be brief.

We have collected together in one place results on symmetric means that have appeared in rather diverse publications over the past 60 years. At the same time, we established a fair number of new results relating some of the important axioms on symmetric means that have been used in the economics literature.

The discerning reader will have noticed that we studied only properties of symmetric means that were helpful in laying out the two axiomatic approaches to choice under uncertainty that are summarized in Proposition 20 (the Expected Utility Theorem) and Proposition 22 (the Implicit Expected Utility Theorem). We applied these two models to a few problems involving insurance, gambling and investing and, in each case, we found that the implicit expected utility model was more promising from an empirical point of view.

Our implicit expected utility model drew heavily on the implicit linear utility models developed earlier by Dekel [1986] and Chew and Epstein [1989a; 211]. However, our overall approach to modeling choice under uncertainty using the theory of symmetric means also drew heavily on the separability approaches of Arrow [1984b] and Blackorby, Davidson and Donaldson [1977].

There are a number of loose ends that remain to be tied up. For example, in Section 6, we found only sufficient conditions to characterize homogeneous or translatable implicitly separable symmetric means: it would be nice to obtain necessary and sufficient conditions.

As another example of a loose end, in Sections 8, 9 and 10, we did not consider the mathematics of problems (210), (221) and (229) in any detail. We also did not consider specific functional forms for $\phi_u(z)$, the generating function in the implicit models, nor did we derive specific estimating equations that could be used to estimate statistically the unknown parameters that a given $\phi_u(z)$ possesses. The problem of econometric estimation in these implicit models is complicated by the fact that standard duality theory techniques cannot be applied without modification, due to the nonstandard format of these problems and the possibility of corner solutions. We also did not discuss the problems involved in cardinalizing the consumer's certainty utility function when there is more than one consumption good.

A final loose end that we will mention has to do with the various *rank dependent nonexpected utility models* presented in Chew and Epstein [1989a]: can these very general models of choice under uncertainty be axiomatized in terms of axioms on a family of stochastic preference functions M^n defined over n equally likely stochastic events that is similar to our axiomatic treatment of expected utility (Proposition 20) or of the implicit expected utility model (Proposition 22).[58]

However, in spite of the above loose ends, we feel that the implicit expected utility model will prove to be very useful in empirical applications.

[58]Space did not permit us to cover the theory of rank dependent symmetric means; three excellent references on these means are Weymark [1981], Ebert [1988] and Chew and Epstein [1989b].

Essays in Index Number Theory, Volume I
W.E. Diewert and A.O. Nakamura (Editors)

Chapter 15

HICKS' AGGREGATION THEOREM AND THE EXISTENCE OF A REAL VALUE ADDED FUNCTION*

W. Erwin Diewert

1. Introduction

The received theory of consumer demand, which deals with the case of a decision making unit which maximizes a utility function subject to a budget constraint, treats "commodities" as well defined and distinct. However, when we look at empirical applications of this theory, we find that the "commodities" are actually *aggregates* of distinct goods and services, and thus we may ask whether the received theory of consumer demand has any empirical relevance, or alternatively, we must look for conditions which will justify using aggregate commodities in place of micro commodities.

It turns out that this use of aggregates in the theory of consumer demand can be justified, provided that all price changes within an aggregate are proportional. This result is known as *Hicks' Aggregation Theorem.* Somewhat imprecise statements and proofs of this theorem may be found in Hicks [1946; 312–313] and Wold [1953; 109–110], while a rather more precise statement of the theorem may be found in Gorman [1953; 76–77].

In Section 2 of this chapter, we will state a version[1] of Hicks' Aggregation Theorem which is similar to Gorman's version, except that we relax his rather

*First published in *Production Economics: A Dual Approach to Theory and Applications*, Vol. 2, edited by M. Fuss and D. McFadden, Amsterdam: North-Holland, 1978, pp. 17–51. This work was partially supported by the National Science Foundation Grant GS-2874-A1 at the Institute for Mathematical Studies in the Social Sciences, Stanford University, and by a Canada Council Grant S. I am indebted to E.R. Berndt, C. Blackorby, L. Epstein, D. McFadden and A.D. Woodland for helpful discussions. This paper is a revised version of IMSSS Technical Report 84, January 1973.

[1]Our treatment of Hicks' Aggregation Theorem largely parallels that of Pollak [1969; 74–75] and the Arrow-Hahn [1971; 144–145] treatment of the first period derived utility function which arose in their treatment of the temporary equilibrium.

strong regularity conditions on the micro utility function.[2] We then ask the obvious question: what are the properties of the "aggregated" or "macro" utility function when prices within a subgroup do *not* vary in strict proportion? In Section 3, we answer this question and we note that there is a duality between direct "micro" utility functions and "macro" (or "conditional indirect" using Pollak's [1969] terminology or "variable indirect" using Epstein's [1975] terminology) utility functions.

Differentiation of a conditional indirect utility function with respect to prices yields Pollak's [1969] conditional demand functions. We exhibit some convenient functional forms for the conditional indirect utility function and derive the corresponding conditional demand functions. If we econometrically estimate the parameters of these conditional demand functions and if all of the parameters of the conditional indirect utility function appear in these demand functions, then by duality, we will have estimated a representation of the consumer's direct utility function. Thus if we identify the conditional goods as public goods,[3] an appropriate choice of functional form for the conditional indirect utility function can be used to solve one of the most vexing problems in the public finance literature: how can we get people to reveal their preferences about public goods? This use of conditional demand functions to solve the "free rider" problem was suggested by Pollak [1969; 63].

We conclude Section 3 by establishing the implications of concavity of the direct utility function for the conditional indirect utility function, and the implications of convexity in prices of the conditional indirect utility function for the direct utility function. These implications are useful in the analysis of risk aversion in a many commodity world (e.g., see Stiglitz [1969] and Hanoch [1977]).

In Section 4 of this chapter, we develop an analogue to Hicks' Aggregation Theorem in the context of production theory. More specifically, we reconsider the old problem[4] of attempting to represent the "real output" of an industry by using some sort of deflated value added concept. This deflation problem is of some empirical importance, since virtually all (nonfixed coefficients) production function studies use deflated value added (rather than deflated gross output) as their measure of real output. How has this substitution of value added for gross output been justified in the context of production function studies?

---

[2]Gorman's regularity conditions on the micro utility function are: f is a continuous, strictly increasing, strictly quasiconcave, once differentiable function defined over the nonnegative orthant which also has the property that each indifference surface does not intersect the coordinate planes.

[3]Samuelson [1969a; Sec. 5] defines a public good as one that enters two or more persons' utility.

[4]The literature on this problem includes Geary [1944], Berlinguette and Leacy [1961], and David [1966].

Consider the following highly aggregated example. Let Y represent real gross output, K real capital input, L labor input, and M inputs of new materials, energy and other intermediate commodities. Suppose that the technology of the firm or industry can be represented by the production function f where $Y = f(K, L, M)$.

One method of justifying the substitution of value added for gross output has been to assume that the production function f is weakly separable[5] and can thus be written as $f(K, L, M) \equiv g[h(K, L), M]$ in which case we can identify $h(K, L)$ as real value added. This *separability* approach has been studied by Corden [1969], Sims [1969] and Arrow [1974], and we will not pursue it in the present paper.[6]

Another approach to the problem of measuring real value added has been to define nominal value added V as the solution to a profit maximization problem and then to apply Hicks' Aggregation Theorem. That is, we may define nominal value added V as follows, where p_Y is the price of one unit of gross output, p_M is the price of one unit of raw materials, and all other variables have been defined above,

$$V(K, L, p_Y, p_M) \equiv \max_{Y,M}\{p_Y \cdot Y - p_M \cdot M : Y \leq f(K, L, M)\}. \tag{1.1}$$

Now if p_Y and p_M vary in strict proportion, and we let $p \equiv a \cdot p_Y + b \cdot p_M$ be a price index in the prices p_Y and p_M with fixed weights a and b chosen so that p is positive for the base period, then it turns out that the deflated value added function $v(K, L) \equiv V(K, L, p_Y, p_M)/p$ does not depend on the prices p_Y and p_M and, moreover, the function v satisfies the usual neoclassical production function properties in K and L if f satisfies suitable regularity conditions. Thus if producers behave in a profit maximizing manner, and if the prices of outputs and intermediate inputs vary in proportion, then the replacement of gross output Y by deflated value added $v \equiv (p_Y \cdot Y - p_M \cdot M)/p$ can indeed be justified from the viewpoint of production function studies. This second method of justifying the substitution of real value added for gross output is due to Khang [1971] and Bruno [1978].

In Section 4 of the present chapter, we pursue the approach of Khang and Bruno, allowing for joint production and nondifferentiable production functions. We simply note that a nominal or money value added function (such as $V(K, L, p_Y, p_M)$ defined by equation (1.1)) is a special case of a *variable profit*

[5]See Goldman and Uzawa [1964] and Blackorby, Primont, and Russell [1978] for a treatment of weak separability.

[6]Frequently, the separability assumption is implicitly or explicitly strengthened to read $Y = f(K, L, M) \equiv g[h(K, L), M]$ where g is a Leontief or fixed coefficients type production function. In this latter case, either Y, M or $Y - M$ may be used as an index of real value added.

function,[7] and thus known results may be used in order to characterize the properties of the value added function with respect to prices and quantities. We then show that if all output prices and intermediate input prices vary in strict proportion, then the deflated value added function has the properties of a neoclassical production function.

In Section 5, we consider the problem of minimizing the cost of producing one unit of nominal value added and develop a duality theorem between the resulting cost function and the value added function.[8]

In Section 6, we conclude by engaging in some armchair empiricism and we suggest that most postwar production function studies using deflated value added as output are probably somewhat biased.

Section 7 presents proofs for the longer theorems.

In the remainder of the present section, we review some mathematical theorems which will be utilized in the following sections of the chapter.

(1.2) DEFINITION. A function f defined on S, a subset of Euclidean M space, is said to be *continuous from below* at a point $x^0 \in S$ if for every $\varepsilon > 0$, there exists a neighborhood $U(x^0)$ such that $x \in U(x^0)$ implies $f(x) > f(x^0) - \varepsilon$.

(1.3) DEFINITION. A numerical function f defined on S is said to be *continuous from above* at a point $x^0 \in S$ if for every $\varepsilon > 0$ there exists a neighborhood $U(x^0)$ such that $x \in U(x^0)$ implies $f(x) < f(x^0) + \varepsilon$.

Note that f is continuous at x^0 if and only if it is continuous from below and above.[9]

We say that a function f is continuous from above over S if it is continuous from above at each point of S.

(1.4) LEMMA. (Berge [1963; 76], Rockafellar [1970; 51]): *The function f is continuous from above over S if and only if the set $\{x : f(x) \geq \alpha; x \in S\}$ is closed in S for every scalar α.*

(1.5) THEOREM. (Berge [1963; 76]): *If S is a compact (i.e., closed and bounded) subset of Euclidean M space, then a continuous from above function f attains in S the value $M \equiv \sup_x\{f(x) : x \in S\}$.*

---

[7]The variable profit function was introduced into the economics literature by Samuelson [1953–54; 20]. Its properties were more formally studied by Gorman [1968b] who called it a gross profit function, by Diewert [1973a], and McFadden [1978a] who use the terminology restricted profit function, and by Lau [1976].

[8]Denny [1972] has independently studied aspects of this duality relationship in the context of a nonconstant returns to scale production function. Woodland [1977b] has extensively applied this duality in the context of international trade theory.

[9]The property of continuity from above is often called upper semicontinuity (cf. Berge [1963; 74] or Rockafellar [1970; 51]). However, we will use the term upper semicontinuity to describe a property of a *correspondence*.

In the following definitions, let S denote a subset of R^M, T a subset of R^N, $\{x^n\}$ a sequence of points of S and $\{y^n\}$ a sequence of points of T.

(1.6) DEFINITION. ϕ is a *correspondence* from S into T if for every $x \in S$, there exists an image set $\phi(x)$ which is a subset of T.

(1.7) DEFINITION. A correspondence ϕ is *upper semicontinuous* at the point $x^0 \in S$ if $\lim x^n = x^0$; $y^n \in \phi(x^n)$; $\lim y^n = y^0$ implies $y^0 \in \phi(x^0)$. A correspondence ϕ is *lower semicontinuous* at $x^0 \in S$ if $\lim x^n = x^0$; $y^0 \in \phi(x^0)$ implies that there exists a sequence $\{y^n\}$ such that $y^n \in \phi(x^n)$ and $\lim y^n = y^0$. A correspondence ϕ is continuous at $x^0 \in S$ if it is both upper and lower semicontinuous at x^0.

(1.8) UPPER SEMICONTINUITY MAXIMUM THEOREM. (Berge [1963; 116]): *Let f be a continuous from above function defined for (x, y) such that $x \in S$ and $y \in T$ where T is a compact subset of R^N. Suppose that ϕ is a correspondence from S into T and that ϕ is upper semicontinuous at $x^0 \in S$. Then the function g defined by $g(x) \equiv \max_y\{f(x, y) : y \in \phi(x)\}$ is continuous from above at x^0.*

(1.9) MAXIMUM THEOREM. (Debreu [1952; 889–890] [1959; 19], Berge [1963; 116]): *Let f be a continuous real valued function defined for (x, y) such that $x \in S$ and $y \in T$ where T is a compact subset of R^N. Suppose that ϕ is a correspondence from S into T and that ϕ is continuous at $x^0 \in S$. Define the maximum $g(x^0) \equiv \max_y\{f(x^0, y) : y \in \phi(x^0)\}$ and the set of maximizers $\xi(x^0) \equiv \{y : y \in \phi(x^0)$ and $f(x^0, y) = g(x^0)\}$. Then the function g is continuous at x^0 and the correspondence ξ is upper semicontinuous at x^0.*

Now that we have disposed of the mathematical preliminaries, we are in a position to prove a very general version of Hicks' Aggregation Theorem.

2. Hicks' Aggregation Theorem in the Consumer Context

Consider the following micro or *disaggregated utility maximization problem*, where f is the consumer's utility function, $x \geq 0_M$ is a nonnegative M dimensional vector of commodity rentals,[10] $y \geq 0_N$ is a nonnegative N dimensional vector of commodity rentals, $w \gg 0_M$ is an M dimensional vector of positive rental prices, $p \gg 0_N$ is an N dimensional vector of positive rental prices and $Y \geq 0$ is the consumer's "income":

$$(2.1) \qquad \max_{x,y}\{f(x, y) : x \geq 0_M, y \geq 0_N; w^T x + p^T y \leq Y\}.$$

[10] *Notation*: 0_M denotes an M dimensional column vector of zeroes, x^T denotes the transpose of the column vector x, $x \geq 0_M$ means each component of x is nonnegative, and $x \gg 0_M$ means each component is positive.

Observe that if $w \gg 0_M$, $p \gg 0_N$, $Y \geq 0$ and the utility function f is continuous from above, then at least one solution x^*, y^* to the disaggregated utility maximization problem 2.1 will exist, using Theorem 1.5. Thus in what follows, a minimum regularity condition we will impose on the micro utility function f is that it be continuous from above.

Suppose that the prices w, which correspond to the first group of commodities, satisfy the relationship $w = p_0\alpha$ where $\alpha \gg 0_M$ is a fixed vector of constants and $p_0 > 0$ is a scalar. In other words, from period to period, p_0 and p may vary in an arbitrary fashion, but the variation in the price vector w is limited by the equation $w = p_0\alpha$. This vector of fixed constants α is used in order to define the following ***aggregated utility function***:

$$(2.2) \qquad U_\alpha(y_0, y) \equiv \max_x \{f(x, y) : x \geq 0_M, \alpha^T x \leq y_0\},$$

where $y \geq 0_N$, $y_0 \geq 0$, $\alpha \gg 0_M$, and f is the micro utility function. We note that if the micro utility function is continuous from above, then Theorem 1.5 implies that the aggregated (over commodities) utility function defined by equation 2.2 will be well defined as a maximum since the set $\{x : x \geq 0_M, \alpha^T x \leq y_0\}$ is compact if $\alpha \gg 0_M$ and $y_0 \geq 0$.

Now given the macro utility function U_α, $p_0 > 0$, $p \gg 0_N$, $Y \geq 0$, the macro or ***aggregated utility maximization problem*** is defined as follows:

$$(2.3) \qquad \max_{y_0, y} \{U_\alpha(y_0, y) : p_0 y_0 + p^T y \leq Y; y_0 \geq 0, y \geq 0_N\}.$$

(2.4) Hicks' Aggregation Theorem. (Hicks [1946; 312–313], Wold [1953; 109–110], Gorman [1953; 76–77]): *Let the micro utility function f be continuous from above, let $p_0 > 0$, $\alpha \gg 0_M$, $w = p_0\alpha$, $p \gg 0_N$ and $Y \geq 0$. Then (i) the macro utility function U_α defined by equation 2.2 is also continuous from above (and thus the macro utility maximization problem given by equation 2.3 has a solution), (ii) if (x^*, y^*) is any solution to the disaggregated utility maximization problem 2.1, then (y_0^*, y^*) is a solution to the aggregated utility maximization problem 2.3 where $y_0^* \equiv \alpha^T x^* \equiv w^T x^*/p_0$ (and thus the usual procedure of defining an aggregate commodity as expenditure on a group of commodities divided by a price index is justified), and (iii) if the micro utility function f has any of the properties (a) to (e) below in addition to continuity from above, then the macro utility function U_α defined by equation 2.2 also has the corresponding properties: (a) local nonsatiation,*[11] *(b) continuity, (c)*

[11]The utility function $f(z)$ where $z \geq 0_{M+N}$ is subject to local nonsatiation if for every $z^0 \geq 0_{M+N}$ and $\delta > 0$, there exists $z \geq 0_{M+N}$ such that $(z - z^0)^T(z - z^0) \leq \delta^2$ and $f(z) > f(z^0)$.

nondecreasing[12] *in its arguments, (d) quasiconcavity,*[13] *(e) strict quasiconcavity.*

A proof of the above theorem is given in Section 7.[14]

We note that if a micro utility function f has the properties of continuity from above plus local nonsatiation, then data generated as solutions to the micro utility maximization problem 2.1 will satisfy the strong axiom of revealed preference.[15] Hence Theorem 2.4 implies that the aggregated data will also satisfy the strong axiom of revealed preference, provided that prices of the goods in the aggregate vary in strict proportion over time. Thus if the last condition is satisfied, the use of an aggregate commodity in place of the micro commodities can be justified from the viewpoint of the received theory of consumer demand.

If the vector α does not remain constant over time, then the macro utility function $U_\alpha(y_0, y)$ defined by equation 2.2 will be a function of α. In the following section, we determine the properties of U with respect to α.

3. Duality between Direct and Conditional Indirect Utility Functions

We now allow the vector of parameters α found in definition 2.2 to vary and we determine the properties of the macro utility function $U_\alpha(y_0, y)$ with respect to α.

(3.1) THEOREM. *Let the micro utility function $f(x, y)$ be continuous from above with respect to $x > 0_M$ for a fixed $y \geq 0_N$ and let $U_\alpha(y_0, y)$ be defined by equation 2.2 for $\alpha \gg 0_M$, $y_0 \geq 0$, $y \geq 0_N$. Then for fixed y_0 and y, $U_\alpha(y_0, y)$ is (i) continuous from above with respect to α, (ii) nonincreasing with respect to α, and (iii) a quasiconvex function of α over the set $S \equiv \{\alpha : \alpha \gg 0_M\}$. (iv) For fixed y, $U_\alpha(y_0, y)$ is homogeneous of degree zero in (α, y_0); i.e., if $\alpha \gg 0_M$,*

---

[12]The utility function f is nondecreasing if $0_{M+N} \leq z^1 \leq z^2$ implies $f(z^1) \leq f(z^2)$.

[13]The utility function f is quasiconcave if and only if for every scalar k the set $L(k) \equiv \{z : f(z) \geq k : z \geq 0_{M+N}\}$ is convex; f is strictly quasiconcave if and only if, for every k, $L(k)$ is a strictly convex set. A function f is quasiconvex if and only if $-f$ is quasiconcave.

[14]One can also show, using some results due to Danskin [1967; 24], that if the micro utility function f has the properties of local nonsatiation, strict quasiconcavity and in addition is twice continuously differentiable with respect to its arguments, then the macro utility function will also be twice continuously differentiable with respect to its arguments.

[15]See Houthakker [1950; 163] for a statement of the strong axiom of revealed preference.

$\lambda > 0$, $y_0 \geq 0$, then $U_{\lambda\alpha}(\lambda y_0, y) = U_\alpha(y_0, y)$. (v) If the micro utility function f is a continuous function over the nonnegative $(M + N)$ dimensional orthant, then $U_\alpha(y_0, y)$ is jointly continuous with respect to $\alpha \gg 0_M$, $y_0 \geq 0$, and $y \geq 0_N$.

A proof of the above theorem is given in Section 7. Note that Theorem 3.1 yields the properties of the *indirect utility function* as a special case (i.e., let $N = 0$ and the vector y vanishes from definition 2.2, and then y_0 may be interpreted as "income") and thus Theorem 3.1 generalizes somewhat some aspects of the duality theorems between direct and indirect utility (or production) functions due to Newman [1965; 138–172], Lau [1969], Weddepohl [1970; 125] and Shephard [1970; 13–23, 105–111, 301–305].[16] If we look at $U_\alpha(y_0, y)$ as a function of α, Theorem 3.1 tells us that the set of "prices" $\{\alpha : U_\alpha(y_0, y) \leq k;\ \alpha \gg 0_M\}$ will be a convex, closed (in the positive orthant), "nonbackward-bending" set, provided only that the micro utility function is continuous; i.e., the set $\{\alpha : U_\alpha(y_0, y) \leq k;\ \alpha \gg 0_M\}$ will look like an ordinary indifferent-or-preferred-to set. However, note that as k increases, the set $\{\alpha : U_\alpha(y_0, y) \leq k;\ \alpha \gg 0_M\}$ will generally move *towards* the origin instead of away from it; i.e., as the "prices" α become smaller, the set of feasible x's in definition 2.2 becomes larger and thus we would expect $U_\alpha(y_0, y)$ to increase.

The macro utility function $U_\alpha(y_0, y)$ is called a *conditional indirect utility function* by Pollak [1969], or a *variable indirect utility function* by Epstein [1975]. Epstein also showed that the properties of the macro utility function which occurred in Theorems 2.4 and 3.1 completely characterize a certain class of preferences. That is, suppose the direct utility function $f(x, y)$ is a continuous, nondecreasing, quasiconcave function of (x, y) over $x \geq 0_M$, $y \geq 0_N$ and the conditional indirect utility function $U_\alpha(y_0, y)$ is defined by equation 2.2. Then $U_\alpha(y_0, y)$ is: (i) a finite continuous real valued function over the set $S \equiv \{(\alpha, y_0, y) : \alpha \gg 0_M,\ y_0 \geq 0,\ y \geq 0_N\}$, (ii) nonincreasing and quasiconvex in $\alpha \gg 0_M$ for every y_0, y, (iii) homogeneous of degree zero in (α, y_0) for every y, and (iv) nondecreasing and quasiconcave in (y_0, y) for every α. Now extend the domain of definition of $U_\alpha(y_0, y)$ to $\alpha \geq 0_M$ by continuity. (The

[16]The main difference between Theorem 3.1 and the results of Newman, Lau and Shephard lies in the extremely weak regularity conditions we have imposed on the direct utility function f. However, there are other minor differences. Newman [1965; 160] appears to have the curvature of the level sets of the indirect utility function going the wrong way, Lau [1969; 376] asserts that the indirect utility function is a convex function in prices rather than a quasiconvex function and Shephard [1970; 106] is able to prove that the indirect utility function is continuous from below (rather than being continuous from above). However, Shephard's result can be traced to the fact that he defines the indirect utility functions as a sup rather than as a maximum.

resulting function need not be finite if any component of α is zero.) Define the direct utility function f^* using the conditional indirect utility function U_α for $x \geq 0_M$ and $y \gg 0_N$ as follows:

$$f^*(x, y) \equiv \min_\alpha \{U_\alpha(1, y) : \alpha^T x \leq 1, \ \alpha \geq 0_M\}. \tag{3.2}$$

Then Epstein shows that $f = f^*$; i.e., the conditional indirect utility function U_α completely characterizes the direct utility function f.[17] For applications and further extensions of this duality between conditional indirect and direct utility functions, see Blackorby, Primont and Russell [1977a; 1978].

As Hanoch [1978a] has noted in the production theory context, the analogue of the indirect utility function is the indirect production function. Thus we analogously define the conditional indirect production function (while introducing a more traditional notation) as

$$g(v, y) \equiv \max_x \{f(x, y) : v^T x \leq 1, \ x \geq 0_M\}, \tag{3.3}$$

where f is the direct production function, x is an M dimensional vector of variable inputs, $y \geq 0_N$ is a N dimensional vector of fixed inputs, $v \equiv p/Y$ is a vector of *normalized* prices, $p \gg 0_M$ is a vector of variable input rental prices, while $Y > 0$ is expenditure on the variable inputs, and g is the conditional indirect production function. Thus $g(v, y)$ is the analogue to $U_v(1, y)$ and the same duality theorem as held between $f(x, y)$ and $U_\alpha(1, y)$ will hold between f and g, except that it is natural to assume that output is positive if all inputs are positive which in turn will imply that $g(v, y)$ is positive for positive arguments. We denote output as $u \equiv f(x, y)$ since later in this section we will interpret u as utility and f and g as direct and conditional indirect utility functions, respectively.

Looking at equation 3.3, we see that the conditional indirect production function $g(v, y)$ gives the solution to the problem of maximizing output $u = f(x, y)$ given an expenditure constraint on variable inputs x of the form $p^T x \leq Y$ (or $v^T x \leq 1$ where $v \equiv p/Y$) and given the vector of fixed inputs y. On the other hand, given a conditional indirect production function g satisfying the appropriate regularity conditions, the corresponding direct production function f can be calculated as the solution to the following minimization problem (which is the counterpart to equation 3.2):

$$f(x, y) \equiv \min_v \{g(v, y) : v^T x \leq 1, v \geq 0_M\}. \tag{3.4}$$

In the theory of production, it is often assumed that the production function f is positively linearly homogeneous; i.e., for every $x \geq 0_M$, $y \geq 0_N$ and scalar $\lambda > 0$, $f(\lambda x, \lambda y) = \lambda f(x, y)$. The following theorem noted by Epstein [1981a] indicates the implications of linear homogeneity of f on g.

[17]Epstein also notes that the continuity problems which were discussed in Diewert [1974a; 121–123] also occur in the present context but that the same techniques that were used in Diewert can be used in the present context.

(3.5) THEOREM. (Epstein [1981a]): *Let the direct production function f and its dual conditional indirect production function g satisfy Epstein's regularity conditions. Then f is positively linearly homogeneous if and only if for every $v \gg 0_M$, $y \geq 0_N$, $\lambda > 0$, we have*

$$g(\lambda^{-1}v, \lambda y) = \lambda g(v, y). \tag{3.6}$$

If we define a transformed conditional indirect production function g^* for $v \gg 0_M$ by $g^*(v^{-1}, y) \equiv g(v, y)$, where $v^{-1} \equiv (v_1^{-1}, v_2^{-1}, \ldots, v_M^{-1})$ is a vector which has as components the reciprocals of the components of $v \equiv (v_1, v_2, \ldots, v_M)$, then the thrust of Theorem 3.5 is that the direct production function f is positively linearly homogeneous if and only if g^* is positively linearly homogenous in its arguments.

Just as linear homogeneity places certain restrictions on the first and second order partial derivatives of a direct production function, the homogeneity property defined by equation 3.6 places restrictions on the first and second order partial derivatives of g. Assuming that g is twice continuously differentiable at a point $v^* \gg 0_M$, $y^* \gg 0_N$, if we partially differentiate equation 3.6 with respect to λ and set $\lambda = 1$, we obtain the following restriction on the first order partials of g:

$$-v^{*T}\nabla_v g(v^*, y^*) + y^{*T}\nabla_y g(v^*, y^*) = g(v^*, y^*), \tag{3.7}$$

where $\nabla_v g(v^*, y^*)$ denotes the vector of first order partial derivatives of g with respect to v evaluated at v^*, y^* and $\nabla_y g(v^*, y^*)$ denotes the partial derivatives of g with respect to the components of y.

On the other hand, if we partially differentiate the identity equation 3.6 with respect to the components of v and then partially differentiate the resulting M equations with respect to λ, upon setting $\lambda = 1$, we obtain the following M equations:

$$-\nabla^2_{vv} g(v^*, y^*)v^* + \nabla^2_{vy} g(v^*, y^*)y^* = 2\nabla_v g(v^*, y^*), \tag{3.8}$$

where $\nabla^2_{vv} g(v^*, y^*)$ denotes the $M \times M$ matrix of second order partial derivatives of g with respect to v while $\nabla^2_{vy} g(v^*, y^*)$ denotes the $M \times N$ matrix of second order partial derivatives of g with respect to the components of v and y.

Finally, if we partially differentiate equation 3.6 with respect to the components of y, then partially differentiate the resulting N equations with respect to λ, and then set $\lambda = 1$, we obtain the following N equations:

$$-\nabla^2_{yv} g(v^*, y^*)v^* + \nabla^2_{yy} g(v^*, y^*)y^* = 0_N. \tag{3.9}$$

Since g is assumed to be twice continuously differentiable at (v^*, y^*), the matrices of second order partial derivatives $\nabla^2_{vv} g(v^*, y^*)$ and $\nabla^2_{vv} g(v^*, y^*)$ will be symmetric. Thus these symmetry conditions and the $(1+M+N)$ restrictions 3.7 to 3.9 imply that there are $M(M-1)/2+N(N-1)/2+(1+M+N)$ linear restrictions on the first and second order partial derivatives of g (and the level of g), evaluated at the point (v^*, y^*). Thus it can be seen that the following functional form for a conditional indirect production function due to Epstein [1981a] can provide a second order differential[18] approximation to an arbitrary twice continuously differentiable conditional indirect production function which has the homogeneity property 3.6[19] (and thus corresponds to a constant returns direct production function):

$$(3.10) \qquad u = g(v, y) \equiv \sum_{m=1}^{M} \sum_{k=1}^{M} a_{mk} v_m^{-1/2} v_k^{-1/2} + 2\sum_{m=1}^{M} \sum_{n=1}^{N} c_{mn} v_m^{-1/2} y_n^{1/2} + \sum_{n=1}^{N} \sum_{r=1}^{N} b_{nr} y_n^{1/2} y_r^{1/2},$$

where $a_{mk} = a_{km}$ for every m, k, and $b_{nr} = b_{rn}$ for every n, r. It is easy to verify that the functions $y_n^{1/2}$ and $y_n^{1/2} y_r^{1/2}$ are concave in y and that the functions $v_m^{-1/2}$ and $v_m^{-1/2} v_k^{-1/2}$ are convex in v.[20] Thus if all of the coefficients a_{mk}, b_{nr} and c_{mn} are nonnegative, it can be verified that g defined by equation 3.10 is (i) nonnegative, (ii) nonincreasing and convex (and hence quasiconvex) in v for fixed y, (iii) nondecreasing and concave (and hence quasiconcave) in y for fixed v, and (iv) g satisfies the homogeneity property 3.6. Thus g is dual to a well behaved linearly homogeneous direct production function.[21]

Roy's Identity[22] may be used in order to generate the firm's system of M *conditional* or *short run demand functions* $x(v^*, y^*)$ at any point (v^*, y^*) where the conditional indirect production function is differentiable,

$$(3.11) \qquad x(v^*, y^*) \equiv \nabla_v g(v^*, y^*)/v^{*T} \nabla_v g(v^*, y^*).$$

[18]The term is due to Lau [1974], but the concept is discussed by McFadden [1978a] and is termed the second order approximation property.

[19]It is easier to see that the functional form defined by equation 3.10 provides a second order differential approximation to a twice continuously differentiable transformed conditional indirect production function, $g^*(v^{-1}, y) = g^*(z)$, since $g(z)$ defined by equation 3.10 is a generalized Leontief functional form in $z \equiv (v^{-1}, y)$ and thus the "flexibility" theorem of Diewert [1974a; 117] applies.

[20]It is well known that the reciprocal of a positive concave function is convex; e.g., see Zangwill [1969; 60].

[21]If some of the coefficients of g defined by equation 3.10 are negative, then g can still satisfy the appropriate regularity conditions over a region and be a valid representation of technology over that region. See Epstein [1973b] and Diewert [1973a; 305–306].

[22]See Roy [1942], Konüs and Byushgens [1926], or Diewert [1974a; 124].

If the firm's conditional indirect production function is defined by equation 3.10, then the firm's output maximizing demands $x \equiv (x_1, x_2, \ldots, x_M)$ for variable inputs given the expenditure constraint $v^T x \leq 1$ are defined by the following equation (upon application of Roy's Identity 3.11):

$$x_m = \frac{-\sum_{k=1}^{M} a_{mk} v_m^{-3/2} v_k^{-1/2} - \sum_{n=1}^{N} c_{mn} v_m^{-3/2} y_n^{1/2}}{-\sum_{i=1}^{M}\sum_{j=1}^{M} a_{ij} v_i^{-1/2} v_j^{-1/2} - \sum_{h=1}^{M}\sum_{k=1}^{N} c_{hk} v_h^{-1/2} y_k^{1/2}}, \tag{3.12}$$

for $m = 1, 2, \ldots, M$.

Given data on output u, variable inputs x, fixed inputs y and variable input prices p (so that normalized variable input prices $v \equiv p/p^T x$ can be calculated), we may use equation 3.10 and the M variable input demand equations 3.12 (only $M - 1$ of them are independent) in order to estimate the parameters of technology, a_{mk}, b_{nr} and c_{mn}.

Suppose now that the firm can periodically vary its fixed inputs y. For example, we might suppose that output and the short run inputs are chosen every week, but that once a year long run inputs are also changed. Suppose that the firm faces long run input rental prices $q = (q_1, q_2, \ldots, q_N)$ and that it wishes to spend no more than $E > 0$ dollars on fixed inputs during the week when long run inputs are chosen. Then given short run input prices p and a budget of $Y > 0$ dollars to spend on short run inputs, the firm will wish to choose the vector of long run inputs y which solves the following maximization problem:

$$\max_y \{g(p/Y, y) : q^T y \leq E,\ y \geq 0_N\} \tag{3.13}$$
$$= \max_y \{g(v; y) : w^T y \leq 1,\ y \geq 0_N\},$$

where $v \equiv p/Y$ is the vector of short run normalized prices, $w \equiv q/E$ is the vector of long run normalized input prices, and g is the producer's conditional indirect production function.

The producer's system of long run *inverse demand* functions $w(v^*, y^*)$ can be defined as the set of normalized long run prices such that $y^* \geq 0_N$ is a solution to the constrained output maximization problem 2.13 when $v = v^*$ and $w = w(v^*, y^*)$. If the conditional indirect production function g is nondecreasing, quasiconcave and once differentiable in y, then the producer's system of long run inverse demand functions can be calculated by using Wold's Identity,[23]

$$w(v^*, y^*) \equiv \nabla_y g(v^*, y^*)/y^{*T} \nabla_y g(v^*, y^*). \tag{3.14}$$

[23]See Konüs and Byushgens [1926], Hotelling [1935], and Wold [1944; 69–71] for proofs in the context of consumer theory. The proof consists of setting up the Lagrangean for the constrained maximization problem 3.13, partially differentiating and then eliminating the Lagrange multiplier.

If the firm's conditional indirect production function is defined by equation 3.10, then application of Wold's Identity yields the following system of long run inverse demand functions:

$$(3.15)\qquad \frac{q_n}{E} \equiv w_n = \frac{\sum_{r=1}^{N} b_{nr} y_n^{-1/2} y_r^{1/2} + \sum_{m=1}^{M} c_{mn} v_m^{-1/2} y_n^{-1/2}}{\sum_{i=1}^{N}\sum_{j=1}^{N} b_{ij} y_i^{1/2} y_j^{1/2} + \sum_{h=1}^{M}\sum_{k=1}^{N} c_{hk} v_h^{-1/2} y_k^{1/2}}.$$

Given data on output u, variable inputs x, "fixed" inputs y, variable input prices p (recall $v \equiv p/p^T x$) and "fixed" input prices q (recall $w \equiv q/q^T y$), then the "weekly" equations 3.10 and 3.12 along with the N "annual" equations 3.14 (only $N-1$ of them are independent) can be used in order to estimate the unknown parameters a_{mk}, b_{nr}, and c_{mn}.

The idea of producers maximizing output subject to one or two expenditure constraints on inputs is perhaps not intuitively appealing.[24] It is more natural to think of producers as maximizing short run profits with respect to variable outputs and inputs and then choosing fixed inputs to maximize long run profits. In this case, the natural way to describe technology is by means of a variable profit function (a concept which will be discussed in the following section).[25]

However, the real usefulness of the concept of the conditional indirect production function lies not so much in the producer context but in the consumer context where the variable profit concept cannot be used — interpret output u as utility, x and y as short and long run commodity rentals with p and q being the corresponding vectors of rental prices, f as the direct utility function and g as the consumer's conditional indirect utility function. The main difference between the consumer and producer interpretations of the basic model is that while output can be observed utility cannot be observed. Thus in the consumer context, we cannot use equation 3.10 to aid us in the econometric estimation of the unknown parameters of the consumer's preferences.

---

[24]It is, however, logically consistent with competitive profit maximization, since the producer will always want to produce a given amount of output at minimum short (and long) run cost.

[25]Moreover, the use of the variable profit function instead of the conditional indirect production function can lead to a system of short run input demand and output supply functions which are *linear* in the parameters of technology (instead of a nonlinear system like 3.12) if one uses an appropriate functional form and Hotelling's lemma as in Diewert [1974a; 137–139]. Furthermore, use of the variable profit function can lead to systems of long run inverse demand functions which are also linear in the parameters of technology. See Diewert [1974a; 140].

Moreover since equations 3.12 and 3.15 are homogeneous of degree zero in the unknown parameters, in the consumer context when equation 3.10 is unavailable for econometric purposes, it will be necessary to make a normalization on the parameters (in order to determine the scale for utility) such as

$$(3.16)\qquad \sum_{m=1}^{M}\sum_{k=1}^{M} a_{mk} + 2\sum_{m=1}^{M}\sum_{n=1}^{N} c_{mn} + \sum_{n=1}^{N}\sum_{r=1}^{N} b_{rn} = 1.$$

If units of measurement are chosen so that $v_m = 1$, $m = 1, 2, \ldots, M$, and $y_n = 1$, $n = 1, 2, \ldots, N$, during a base period, then the effect of the normalization 3.16 is to make utility equal to unity during the base period.

For an application of the above functional form to the problem of estimating a consumer's preferences under uncertainty, see Epstein [1981a; 12–18]. The above model assumes that the underlying direct production or utility function is linearly homogeneous. This restriction can be relaxed by introducing an additional fixed good, y_0 say, which always takes the value unity. This is equivalent to adding terms of the form $2\sum_{m=1}^{M} c_{m0} v_m^{-1/2} + b_{00} + 2\sum_{n=1}^{N} b_{n0} y_n^{1/2}$ to the conditional indirect function defined by equation 3.10.[26] We leave it to the reader to use Roy's Identity 3.11 and Wold's Identity 3.14 in order to derive the resulting short run and long run inverse demand functions. The parameter b_{00} will not be identifiable in the consumer context.

Let us now suppose that the consumer[27] has preferences defined over market goods x and *public goods* y. The above model can be applied except that the consumer cannot generally optimize his choice of public good inputs, so that the maximization problem 3.13 is not applicable in the present context, and thus we cannot derive equations 3.15 to aid in the econometric estimation of the unknown parameters of preferences. We will be left with only equations 3.12 (or their counterparts in the case of nonhomothetic preferences) and thus the parameters b_{nr} which occur in equation 3.10 cannot be identified econometrically.

The following functional form for a conditional indirect utility function leads to a system of short run derived demand functions which does not suffer

---

[26]The resulting conditional indirect function generally has the second order approximation property.

[27]A similar analysis can be applied in the producer context (toll free roads come to mind as an obvious public good in the producer context), but again it is easier from an econometric point of view to phrase the analysis in terms of the variable profit function (where the fixed inputs include the public goods) in place of the conditional indirect production function.

from the above lack of identifiability problem:

(3.17)

$$\begin{aligned}
g(v,y) \equiv\ & 2\sum_{m=1}^{M}\sum_{k=1}^{M} a_{mk} v_m^{-1/4} v_k^{-1/4}\Big(\sum_{n=1}^{N} y_n^{1/2}\Big) \\
& + 2\sum_{m=1}^{M}\sum_{n=1}^{N} c_{mn} v_m^{-1/2} y_n^{1/2} \\
& + 2\sum_{n=1}^{N}\sum_{r=1}^{N} b_{nr} y_n^{1/4} y_r^{1/4}\Big(\sum_{m=1}^{M} v_m^{-1/2}\Big) \\
& + 4\sum_{m=1}^{M} a_{0m} v_m^{-1/4}\Big(\sum_{n=1}^{N} y_n^{3/4}\Big) \\
& + 4\sum_{n=1}^{N} b_{0n} y_n^{1/4}\Big(\sum_{m=1}^{M} v_m^{-3/4}\Big) + a_0 \\
& + 4\sum_{m=1}^{M} a_m v_m^{-1/4}\Big(\sum_{n=1}^{N} y_n^{1/4}\Big) + 4\sum_{n=1}^{N} b_n y_n^{1/4}\Big(\sum_{m=1}^{M} v_m^{-1/4}\Big),
\end{aligned}$$

where $a_{mk} = a_{km}$, $b_{nr} = b_{rn}$, $a_{mm} = 0$ for $m = 1,2,\ldots,M$, and $b_{nn} = 0$ for $n = 1,2,\ldots,N$. These last two sets of restrictions are necessary in order to identify the parameters c_{mn}. If all of the parameters are nonnegative, then it is easy to verify that g defined by equation 3.17 satisfies the Diewert-Epstein regularity conditions for a conditional indirect function globally. The function defined by equation 3.17 generally has the second order approximation property to an arbitrary twice differentiable conditional indirect function. Moreover, if the parameters a_m, $m = 0,1,\ldots,M$ and b_n, $n = 1,2,\ldots,N$, are set equal to zero, the resulting function has the homogeneity property 2.6 and can generally[28] differentially approximate an arbitrary twice differentiable conditional indirect function which corresponds to a linearly homogeneous direct function.

We leave to the reader the task of applying Roy's Identity 3.11 to the function defined by equation 3.17 in order to derive the variable input demand functions $x(v,y)$. If we make a normalization on the parameters of equation 3.17 which makes utility equal to unity for a base observation (and set $a_0 \equiv 0$), then it will turn out that the remaining parameters of g can be econometrically estimated using just the variable input demand functions $x(v,y)$. Thus a consumer's preferences over market goods x and public goods y can be identified econometrically using only the market goods demand functions, provided that the consumer's preferences can be adequately represented by a conditional indirect utility function of the type defined by equation 3.17.

We conclude this section by discussing the implications of concavity of the direct utility or production function (as opposed to the weaker property

[28]The proof is similar to Lemma 5.9 in Diewert [1973a]. We require that a certain matrix whose coefficients are functions of the initial normalized prices and fixed quantities be nonsingular.

of quasiconcavity) on the conditional indirect function, and the implications of convexity in normalized prices of the conditional indirect function (as opposed to the weaker property of quasiconvexity) on the direct function.[29]

(3.18) THEOREM. *Let f be a continuous, nondecreasing, quasiconcave direct function of (x, y) over the nonnegative orthant in $M+N$ space, and let $g(v, y)$ be the corresponding dual conditional indirect function defined by equations 3.3. Then $f(x, y)$ is concave in (x, y) if and only if $g(p/Y, y)$ is concave in (Y, y) for every $p \gg 0_M$. Moreover, under the above conditions, the conditional or restricted*[30] *cost function defined by*

$$C(u; p, y) \equiv \min_x \{p^T x : f(x, y) \geq u, \ x \geq 0_M\} \tag{3.19}$$

is convex in (u, y) for every $p \gg 0_M$.

A proof of this theorem can be found in Section 7. The same proof with minor modifications can be used to show that under the above conditions on f, $f(x, y)$ is concave in x for a fixed y if and only if $g(p/Y, y)$ is concave in Y for every $p \gg 0_M$ or if and only if $C(u; p, y)$ is convex in u for every $p \gg 0_M$.

(3.20) THEOREM. *Let f be a continuous, nondecreasing quasiconcave direct function of (x, y), and let g be the corresponding dual conditional indirect function. Then $g(v, y)$ is convex in normalized prices $v \gg 0_M$ for a fixed $y \geq 0_N$ if and only if $f(x/Z, y)$ is a convex function of the scalar variable $Z > 0$ for every $x \gg 0_M$.*

A very simple proof of this theorem (due to L. Epstein) can be found in Section 7. If y is taken to be the empty set, then the above two theorems specialize to theorems about the direct and indirect production or utility functions.

We now turn our attention to the theory of production and develop an analogue to Hicks' Aggregation Theorem in the context of value added production functions.

---

[29]As well as being of direct interest in the production theory context where it is natural to assume concavity of the production function, these implications play a role in the analysis of consumer behavior under uncertainty; see Stiglitz [1969], Epstein [1975] [1981a], and Hanoch [1977].
[30]See McFadden [1978a] for a detailed analysis.

4. Hicks' Aggregation Theorem in the Producer Context

Denote outputs by positive and inputs by negative numbers. Let T be the production possibilities set for a firm and let it be a nonempty subset of $(I+J)$ dimensional Euclidean space. We assume that there are I produced goods in the economy and that the firm under consideration produces some of these goods and uses others (such as raw materials, fuel and electricity) as intermediate inputs. We assume that there are J nonproduced or primary factors of production (such as land, different grades of labor and capital services). We suppose that T satisfies the following regularity conditions:

(4.1) CONDITIONS I on the Production Possibilities Set T:

(i) T is a closed, nonempty subset of $I+J$ dimensional space.

(ii) If $(u;v) \in T$, then $v \leq 0_J$ (last J goods are always inputs).

(iii) T is a convex set (nonincreasing marginal rates of transformation).

(iv) T is a cone; i.e., $z \in T$, $\lambda \geq 0$ implies $\lambda z \in T$ (constant returns).

(v) If $z' \in T$, $z'' \leq z'$, then $z'' \in T$ (free disposal).

(vi) If $(u;v) \in T$, then the components of u are bounded from above (for finite fixed inputs, the set of producible outputs is also finite).

If $v \leq 0_J$ is a vector of fixed inputs and $p \gg 0_I$ is a vector of positive produced goods prices, we may define the producer's *variable profit* Π as

$$\Pi(p;v) \equiv \max_u \{p^T u : (u;v) \in T\}. \tag{4.2}$$

We can extend the domain of definition of the variable profit function Π to $p \geq 0_I$ by continuity. Let us change the nonpositive vector of primary inputs v into a nonnegative vector of primary inputs by defining $x \equiv -v$. Now for $p \geq 0_I$, $x \geq 0_J$, we may define the *value added function* V as

$$V(p;x) \equiv \Pi(p;-x), \quad p \geq 0_I, \quad x \geq 0_J. \tag{4.3}$$

If the production possibilities set T satisfies conditions I, it can be shown[31] that V satisfies the following conditions:

(4.4) CONDITIONS II on the Value Added Function V:

(i) V is a *nonnegative*, real value function defined for all $(p;x) \geq 0_{I+J}$.

(ii) V is *linearly homogeneous* in p for fixed x; i.e., if $\lambda > 0$, $V(\lambda p;x) = \lambda V(p;x)$.

(iii) V is a *continuous, convex* function in p for fixed x.

[31]See Diewert [1973a] for a proof. Condition (i) of 4.4 is equivalent to condition (i) of 2.21 in Diewert [1973a] once the domain of definition $\Pi(p;v)$ is extended by continuity from $p \gg 0_I$ to $p \geq 0_I$.

(iv) V is *linearly homogeneous* in x for fixed p; i.e., if $\lambda > 0$, $V(p;\lambda x) = \lambda V(p;x)$.
(v) V is *nondecreasing* in x for fixed p.
(vi) V is a *continuous, concave* function in x for fixed p.

On the other hand, given a value added function V satisfying conditions II above, we may obtain the production possibilities set T satisfying conditions I[32] which corresponds to V by means of the following definition:

$$T \equiv \{(u;-x) : p^T u \leq V(p;x) \text{ for every } p \geq 0_I \text{ and } x \geq 0_J\}. \tag{4.5}$$

Thus the value added function V may be used in order to describe completely a firm's technology, provided that the firm's underlying production possibilities set T satisfies the regularity conditions given by 4.1.

Under certain circumstances, we may use the value added function V in order to define a *real* value added function which has the usual neoclassical production function properties. Suppose the prices of produced goods p vary in strict proportion; i.e., $p = p_0\alpha$ where $\alpha \gg 0_I$ is a fixed vector of constants and p_0 is a scalar which may vary from period to period. Then a *real value added* function f_α may be defined by

$$f_\alpha(x) \equiv V(p_0\alpha;x)/p_0, \quad \alpha \gg 0_I, \quad x \geq 0_J, \quad p_0 > 0. \tag{4.6}$$

Note that $f_\alpha(x)$ is an observable quantity; it is equal to (nominal) value added V divided by the price index p_0. Note also that if V satisfies conditions II, then $V(p_0\alpha;x)/p_0 = V(\alpha;x)$, using (ii) of 4.4.

(4.7) THEOREM. (Khang [1971], Bruno [1978]): *If the production possibilities set T satisfies conditions I and if $V(\alpha;x^*) > 0$ for some $x^* \geq 0_J$ where $\alpha \gg 0_I$ (i.e., nominal value added is positive for at least one vector of fixed inputs), then the real value added function f_α defined by equation 4.6 is: (i) linearly homogeneous in x, (ii) nondecreasing in x, (iii) a continuous concave function of x, and (iv) positive if $x \gg 0_J$.*

Proof: Properties (i), (ii), and (iii) follow directly from properties (iv), (v), and (vi) of the (nominal) value added function V. To prove (iv), let $x \gg 0_J$. Then there exists $\lambda > 0$ such that $\lambda x \geq x^*$. Therefore $0 < f_\alpha(x^*) \equiv V(\alpha;x^*) \leq V(\alpha;\lambda x)$ (using (v) of 4.4) $= \lambda V(\alpha;x)$ (using (iv) of 4.4) $= \lambda f_\alpha(x)$. Since $\lambda > 0$, $f_\alpha(x) > 0$ also. Q.E.D.

Thus if prices of outputs and intermediate inputs vary in strict proportion over time, the deflated nominal value added function $V(p_0\alpha;x)/p_0 \equiv f_\alpha(x)$ is a perfectly well behaved neoclassical production function and the substitution

[32]See Diewert [1973a] for a proof under slightly different but equivalent regularity conditions on the variable profit function.

of deflated value added for real output can be justified from the viewpoint of empirical production function studies.

If prices of outputs and intermediate inputs do not vary in strict proportion, then the parameters of technology can still be estimated, provided that we have information on value added V, on produced goods prices p and on primary inputs x — simply assume a functional form for $V(p;x)$ and use regression techniques in order to estimate the parameters of the functional form.[33]

In the following section, we consider the problem of minimizing the cost of producing a given amount of nominal value added, and we develop a duality between the value added function V and the unit value added cost function. This last function may be used in order to obtain systems of demand functions for inputs which are consistent with cost minimization.

5. The Duality between Value Added "Production" and Cost Functions

Throughout this section, we assume that the value added function $V(p;x)$ satisfies conditions II given by 4.4 in the previous section.

In general, value added will not be positive for all output price vectors $p \geq 0_I$; i.e., if the prices of intermediate inputs are high relative to the prices of goods that the firm produces, then it may not pay the firm to produce anything at all and value added will be zero. These considerations lead us to define the following set of prices where value added is zero:

$$P \equiv \{p : V(p; 1_J) = 0;\ p \geq 0_I\}, \tag{5.1}$$

where 1 is a J dimensional vector of primary inputs which has each component equal to unity.

(5.2) Lemma. *If V satisfies conditions II, then P defined by equation 5.1 is a nonempty, closed convex cone.*

Proof: Since $V(p;x)$ is continuous and linearly homogeneous in p, we have $V(0_I;1_J) = 0$ and thus $0_I \in P$. Let $p', p'' \in P$ and let $0 \leq \lambda \leq 1$. Since V is nonnegative, we have $0 \leq V[\lambda p' + (1-\lambda)p''; 1_J] \leq \lambda V(p';1_J) + (1-\lambda)V(p'';1_J)$ (using the convexity of V in p) $= \lambda \cdot 0 + (1-\lambda) \cdot 0 = 0$ and thus P is a convex set. That P is a cone follows from the linear homogeneity of V in p and the closedness of P follows from the continuity of V in p. Q.E.D.

We note that the set P could reduce to the single point $\{0_I\}$.

[33]See Diewert [1973a] [1974a] for some functional forms for V (or equivalently for Π) which are linear in the unknown parameters.

Given that a firm is producing a certain amount of (nominal) value added, it is natural to suppose that under certain circumstances the firm will attempt to choose a combination of primary inputs which will produce the value added at minimum cost. Thus we define the *unit value added cost function*:

$$(5.3) \qquad c(p;w) \equiv \min_x\{w^T x : V(p;x) \geq 1,\ x \geq 0_J\} \quad \text{if } p \geq 0_J,\ p \notin P,$$
$$\equiv +\infty \qquad \text{if } p \in P,$$

where $w \gg 0_J$ is a positive vector of given primary input prices, V is the value added function and P is the set of prices defined by equation 5.1.

(5.4) LEMMA. *The unit value added cost function is well defined as a minimum, provided that $p \notin P$.*

Proof: Let $p^* \geq 0_I$, $w^* \gg 0_J$ and $p^* \notin P$. Then $V(p^*;1_J) > 0$ and using the linear homogeneity of $V(p;x)$ in x, we have $V(p^*;\lambda^* 1_J) = 1$ where $\lambda^* \equiv 1/V(p^*;1_J)$. Thus the set $\{x : V(p^*;x) \geq 1,\ x \geq 0\}$ is not empty and is closed using the continuity of $V(p;x)$ in x. We have $c(p^*;w^*) \equiv \min_x\{w^{*T}x : V(p^*;x) \geq 1,\ x \geq 0,\ w^{*T}x \leq w^{*T}\lambda^* 1\}$, and the minimum of a linear function over a closed, bounded nonempty set is attained. Q.E.D.

If the prices of produced goods are such that the firm cannot produce a positive amount of value added (i.e., $p \in P$), then it is natural to define the cost of producing one unit of value added to be plus infinity. However, the above lemma shows that as long as $p \notin P$ and $w \gg 0$, the unit value added cost function is well defined as a minimum. Note that since $V(p;x)$ is linearly homogeneous in x, in order to find the minimum cost of producing $\overline{V}$ units of nominal value added, simply multiply $c(p;w)$ by $\overline{V}$.

(5.5) THEOREM. *If the value added function V satisfies conditions II given by 4.4, then the unit value added cost function defined by equation 5.3 satisfies the following conditions:*

(5.6) CONDITIONS III on the Unit Value Added Cost Function c:

(i) $c(p;w)$ is a *positive* extended real valued function defined for $p \geq 0_I$, $w \gg 0_J$, and is infinite for $p \in P$ where P is a closed convex cone of prices.
(ii) c is *continuous from above* in $(p;w)$.
(iii) c is *quasiconcave* in $(p;w)$.
(iv) c is *homogeneous of degree* -1 in p for fixed w; i.e., for $p \geq 0_I$, $w \gg 0_J$ and $\lambda > 0$, we have $c(\lambda p;w) = \lambda^{-1}c(p;w)$.
(v) c is *homogeneous of degree* 1 in w for fixed p; i.e., for $p \geq 0_I$, $w \gg 0_J$ and $\lambda > 0$, we have $c(\lambda p;w) = \lambda^{-1}c(p;w)$.
(vi) c is *nondecreasing* in w for fixed p.

A proof of Theorem 5.5 is given in Section 7.

Let us relate the properties of the unit value added cost function to something more familiar. Let $p = p_0\alpha$ where $p_0 > 0$ and $\alpha \geq 0_I$ is such that $\alpha \notin P$. Then given primary input prices $w \gg 0_J$, the minimum cost of producing one unit of *real* value added may be defined as $c(\alpha; w) \equiv \min_x\{w^T x : V(p_0\alpha; x) \geq p_0 \cdot 1,\ x \geq 0_J\}$ where c is the unit value added cost function. It can be seen that $c(\alpha; w)$ regarded as a function of w alone is a positive, homogeneous of degree one, nondecreasing concave[34] and continuous[35] function of $w \gg 0_J$; i.e., it has the properties of a unit cost function which is dual to a constant returns to scale production function. (In fact, $c(\alpha; w)$ may be interpreted as the unit cost function which is dual to the real value added production function $f_\alpha(x)$ defined by equation 4.6.)

It turns out that the unit value added cost function c may be used in order to generate the value added function of the technology. Let us suppose that c satisfies conditions III. Then we may (uniquely) extend the domain of definition of c from $p \geq 0_I$, $w \gg 0_J$ to $(p, w) \geq 0_{I+J}$ by continuity. Having done this, we may define the function V^* for $p \geq 0_I$ and $x^* \gg 0_J$ as follows:

$$(5.7)\qquad \begin{aligned} V^*(p; x^*) &\equiv \max_\lambda \{\lambda : c(p; w)\lambda \leq w^T x^* \\ &\qquad \text{for every } w \geq 0 \text{ such that } w^T x^* = 1\} \\ &= \max_\lambda \Big\{\lambda : \lambda \leq w^T x^* / c(p; w) \\ &\qquad \text{for every } w \geq 0 \text{ such that } w^T x^* = 1\Big\} \\ &= \min_w \Big\{1/c(p; w) : w \leq 0_J,\ w^T x^* = 1\Big\} \\ &= \min_w \Big\{1/c(p; w) : w \geq 0_J,\ w^T x^* \leq 1\Big\}. \end{aligned}$$

In developing the string of equalities in equations 5.7, we have used the facts that c is nonnegative, homogeneous of degree one in w and nondecreasing in w.

(5.8) THEOREM. *If c satisfies conditions III given by 5.6, the $V^*(p; x)$ defined by equations 5.7 satisfies conditions II given by 4.4 over the domain $p \geq 0_I$, $x \gg 0_J$. Moreover, if c was generated by means of definition 5.3 for some V which satisfied conditions II, then $V^* = V$.*

We note that the domain of definition of the function V^* can be (uniquely) extended to $(p; x) \geq 0_{I+J}$ by continuity. A proof of Theorem 5.8 may be found in Section 7.

[34] A function which is positive, quasiconcave and (positively) homogeneous of degree one over a set is concave over that set. See Berge [1963; 208].

[35] A function which is concave over an open convex set is continuous over that set. See Rockafellar [1970; 82].

Theorems 5.5 and 5.8 establish a duality between a value added function satisfying conditions II and a unit value added cost function satisfying conditions III which is similar to the Shephard [1953] [1970] duality theorem between cost and production functions.

The usefulness of the unit value added cost function may be found in the following result:[36]

(5.9) LEMMA. (Shephard [1953]): *Suppose that c satisfies conditions III, is finite at the prices* $p^* \geq 0_I$, $w^* \gg 0_J$, *and is once continuously differentiable with respect to the input prices* w *at* $w = w^*$. *Then the cost minimizing amount of primary input* j, $x_j(V; p^*; w^*)$, *given that value added* V *is to be produced, is given by* $x_j(V; p^*; w^*) \equiv V \cdot \partial c(p^*; w^*)/\partial w_j$ *for* $j = 1, 2, \ldots, J$.

Thus in order to obtain a system of primary input demand functions which is consistent with the value added being maximized while primary input cost is minimized, we need only assume a functional form for c which is consistent with conditions III and in addition is differentiable with respect to primary input prices, and then apply Lemma 5.9.

Finally, we note that Theorems 5.5 and 5.8 reduce to a variant of the Shephard Duality Theorem[37] between unit cost and constant returns to scale production functions if we assume that only one output is produced (so that $I = 1$). In this case, the vector $p \geq 0_I$ reduces to a scalar and if we set $p = 1$, the value added function $V(1; x) \equiv f(x)$ becomes a linearly homogeneous production function which satisfies the same regularity conditions as the unit cost function $c(1; w)$.

6. Concluding Remarks

In Section 2 above, we saw that the use of aggregate commodities in consumer demand studies could be justified without making any restrictive assumptions on the functional form of the micro utility function if the prices of the

[36]See Diewert [1971a; 495] for a proof (originally due to L. McKenzie) which carries over into the present context and for a more detailed list of historical references.

[37]See (i) Shephard [1953] for a proof which assumed differentiability of the production and unit cost functions, (ii) Samuelson [1953–54; 15] for a statement of the theorem, and (iii) Shephard [1970] for a proof which does not assume differentiability. Duality theorems between nonhomogeneous production functions and total cost functions have been proven by Shephard [1953] [1970], Diewert [1971a], Hanoch [1978a] and McFadden [1978a].

micro commodities within a group varied in strict proportion. This condition is not without empirical relevance, since in practice, the prices of many micro commodities do vary proportionately, at least approximately.[38] Thus the work of national income accountants would be of maximum benefit to the applied econometrician concerned with estimating systems of consumer demand functions if microeconomic consumer budget study data were aggregated in a manner which conformed as closely as possible with the hypotheses of Hick's Aggregation Theorem.[39]

In Section 4 above, we saw that the replacement of real output by deflated value added in production function studies could be justified without making any restrictive assumptions on the functional form of the micro production (or transformation) function, provided that prices of outputs and intermediate inputs varied in strict proportion. However, it is extremely unlikely that this last condition has been satisfied in any western industrial economy during the past twenty-five years. The problem is with the intermediate input, *energy*. The price of energy has generally been constant or growing very slowly during the pre 1973 postwar period, and thus the price of energy relative to the price of produced goods in general has fallen.[40] Also the industrial consumption of energy has risen faster than the rate of growth of output in general,[41] and thus one cannot appeal to Leontief's Aggregation Theorem [1936; 55] in order to justify the substitution of deflated value added for real gross output.

Recall the separability approach of Corden [1969], Sims [1969] and Arrow [1974] where the aggregate production function $f(K,L,M)$ was assumed to be of the form $f(K,L,M) \equiv g[h(K,L),M]$ where $h(K,L)$ was identified as real value added, where we now regard M as a vector of intermediate inputs including energy, K is a vector of capital service inputs and L is a vector of labor inputs. Berndt and Christensen [1973a] have shown that the assumption of separability leads to severe restrictions on partial elasticities of substitution[42] between pairs or inputs. If the above separability assumption

---

[38]The prices of two micro commodities can vary in proportion over time for at least two reasons: the goods may be close substitutes in consumption (e.g., butter and margarine) or the goods may be close substitutes in production (e.g., cars and trucks).

[39]Thus if a major change in the relative prices of micro commodities within an aggregate occurred, then the aggregate should be decomposed into at least two subaggregates. For example, the introduction of medicare would radically change the relative prices of the micro commodities in a household services aggregate.

[40]See the price data on energy, other intermediate inputs and output for U.S. manufacturing tabled in Berndt and Wood [1975] for the period 1950 to 1970.

[41]See Berndt and Wood [1975].

[42]See Allen [1938; 503–509] for a definition.

$f(K, L, M) = g[h(K, L), M]$ is satisfied, then it turns out that the elasticity of substitution between energy and *any* capital or labor input must be the same number. This condition seems highly implausible — we would expect energy and certain types of capital services to be complements (i.e., to have negative partial elasticities of substitution), energy and certain types of maintenance workers to be complements, while energy and unskilled labor would be expected to be substitutes (i.e., to have positive partial elasticities of substitution). Thus the separability approach to justifying the use of "real" value added in place of output in production function studies would also seem to be unacceptable on *a priori* empirical grounds.

Our conclusion is that most postwar production function studies that use deflated value added as a measure of real output are probably somewhat biased due to their inadequate treatment of energy inputs. One is also led to wonder about how much of the "unexplained residual" in growth studies (sometimes called "technical progress") is due to the unjustified use of a real value added framework.[43]

Finally, from Theorem 3.1, it can be seen that if the direct utility function is continuous, then the conditional indirect utility function $U_\alpha(y_0, y)$ is continuous with respect to the vector of price proportionality factors, $\alpha \gg 0_M$. Similarly, from definition 4.6, $f_\beta(x) \equiv V(p_0\beta, x)/p_0$, and condition 4.4(iii) on the nominal value added function V, it can be seen that the real value added function $f_\beta(x)$ is continuous with respect to the vector of price proportionality factors $\beta \gg 0_I$.

The above continuity properties imply that approximate versions of Hicks' Aggregation Theorem will hold in both the consumer and producer contexts. Thus if a group of consumer prices vary over time *approximately* according to the vector of proportionality factors α, then the consumer's preferences can *approximately* be represented by the aggregated utility function $U_\alpha(y_0, y)$. Similarly, if a group of producer prices vary over time *approximately* according to the vector of proportionality factors β, then the producer's technology can *approximately* be represented by the real value added function $f_\beta(x)$.[44] Just how much variation in α or β we can allow before the approximation becomes poor must remain an empirical matter. Much would depend on the variation in prices and quantities of the goods which are not aggregated relative to the variability in the prices of the goods within the aggregate.

---

[43]Two notable studies which use a gross output formulation rather than value added are Parks [1971] and Berndt and Wood [1975].

[44]In Diewert [1974c], we show that an aggregate elasticity of substitution between an aggregated "macro" good and a "micro" good can be written as a weighted sum of micro elasticities of substitution under the hypotheses of Hicks' Aggregation Theorem and we indicate how this aggregate elasticity will change if prices do not vary proportionately.

7. Proofs of Theorems

Proof of Theorem 2.4.

(i) The result follows directly from the Upper Semicontinuity Maximum Theorem 1.8.

(ii) Let $\alpha \gg 0_M$, $w \gg 0_N$, $p_0 > 0$, $Y \geq 0$. We have

$$\begin{aligned} f(x^*, y^*) &= \max_{x,y}\{f(x,y) : p_0\alpha^T x + w^T y \leq Y,\ x \geq 0_M,\ y \geq 0_N\} \\ &= \max_{x,y,y_0}\{f(x,y) : p_0 y_0 + w^T y \leq Y,\ \alpha^T x \leq y_0,\ x \geq 0_M\ y \geq 0_N,\ y_0 \geq 0\} \\ &= \max_{y,y_0}\{U_\alpha(y_0, y) : p_0 y_0 + w^T y \leq Y,\ y \geq 0_N,\ y \geq 0\} \\ &= U_\alpha(\alpha^T x^*, y^*). \end{aligned}$$

(iii-a) Let $\alpha \gg 0_M$, $(y_0^*, y^*) \geq 0_{N+1}$ and let $\varepsilon > 0$,

$$\begin{aligned} U_\alpha(y_0^*, y^*) &\equiv \max_x\{f(x, y^*) : x \geq 0_M,\ \alpha^T x \leq y_0^*\} \\ &= f(x^*, y^*), \end{aligned}$$

say. For every $\delta > 0$, we can find $(x_\delta, y_\delta) \geq 0_{M+N}$ such that $f(x_\delta, y_\delta) > f(x^*, y^*)$ and $(x_\delta - x^*)^T(x_\delta - x^*) + (y_\delta - y^*)^T(y_\delta - y^*) \leq \delta^2$. Define $y_0(\delta) \equiv \alpha^T x_\delta$. By the definition of U_α, we have $U_\alpha[y_0(\delta), y_\delta] \geq f(x_\delta, y_\delta) > f(x^*, y^*) = U_\alpha(y_0^*, y^*)$ and

$$\begin{aligned} &[y_0(\delta) - y_0^*]^2 + (y_\delta - y^*)^T(y_\delta - y^*) \\ &\qquad = (x_\delta - x^*)^T \alpha\alpha^T (x_\delta - x^*) + (y_\delta - y^*)^T(y_\delta - y^*) \leq \varepsilon^2 \end{aligned}$$

for δ small enough.

(iii-b) Continuity of U_α follows directly from the Maximum Theorem 1.9.

(iii-c) Let $y_0'' \geq y_0' \geq 0$, $y'' \geq y' \geq 0_N$ and $\alpha \gg 0_M$. Then

$$\begin{aligned} U_\alpha(y_0', y') &\equiv \max_x\{f(x, y') : \alpha^T x \leq y_0';\ x \geq 0_M\} \\ &= f(x', y') \leq f(x', y'') \text{ (since } y' \leq y'') \\ &\leq \max_x\{f(x, y'') : \alpha^T x \leq y_0';\ x \geq 0_M\} \text{ (since } \alpha^T x' \leq y_0') \\ &\leq \max_x\{f(x, y'') : \alpha^T x \leq y_0'';\ x \geq 0_M \text{ (since } y_0'' \geq y_0') \\ &= U_\alpha(y_0'', y''). \end{aligned}$$

(iii-d) Let $y_0' \geq 0$, $y_0'' \geq 0$, $y' \geq 0_N$, $y'' \geq 0_N$, and $\alpha \gg 0_M$. Suppose that $U_\alpha(y_0', y') \equiv \max_x\{f(x, y') : \alpha^T x \leq y_0';\ x \geq 0_M\} = f(x', y') \geq k$ and $U_\alpha(y_0'', y'') = f(x'', y'') \geq k$ where k is a scalar. Let $0 \leq \lambda \leq 1$. Then

$$\begin{aligned} U_\alpha[\lambda y_0' + (1-\lambda)y_0'',\ \lambda y' + (1-\lambda)y''] \equiv \max_x\{f[x, \lambda y' + (1-\lambda)y''] : \\ \alpha^T x \leq \lambda y_0' + (1-\lambda)y_0'', x \geq 0_M\} \end{aligned}$$

$$\geq f[\lambda x' + (1-\lambda)x'', \lambda y' + (1-\lambda)y'']$$

$$(\text{since } \lambda\alpha^T x' + (1-\lambda)\alpha^T x'' \leq \lambda y_0' + (1-\lambda)y_0'')$$

$$\geq k \text{ (using the quasiconcavity of } f).$$

(iii-e) The proof is similar to (iii-d) above, upon noting that $(y_0', y') \neq (y_0'', y'')$ implies that $(x', y') \neq (x'', y'')$.

Proof of Theorem 3.1.

(i) The result follows directly from the Upper Semicontinuity Maximum Theorem 1.8.

(ii) Let $y_0 \geq 0$, $y \geq 0_N$, $\alpha'' \geq \alpha' \gg 0_M$. Then

$$U_{\alpha''}(y_0, y) \equiv \max_x\{f(x,y) : x \geq 0_M, \alpha''^T x \leq y_0\}$$
$$\leq \max_x\{f(x,y) : x \geq 0_M, \alpha'^T x \leq y_0\}$$

(since $\alpha' \leq \alpha''$ and thus the set of feasible x's has not decreased)

$$= U_{\alpha'}(y_0, y).$$

(iii) Let $y_0 \geq 0$, $y \geq 0_N$, $\alpha'' \gg 0_M$, $\alpha' \gg 0_M$, $0 \leq \lambda \leq 1$, and define the sets $H' \equiv \{x : \alpha'^T x \leq y_0;\ x \geq 0_M\}$, $H'' \equiv \{x : \alpha''^T x \leq y_0;\ x \geq 0_M\}$ and $H^\lambda \equiv \{x : [\lambda\alpha' + (1-\lambda)\alpha'']^T x \leq y_0;\ x \geq 0_M\}$. We first show that H^λ is a subset of the union of H' and H''. Suppose $x^* \in H^\lambda$, but $x^* \notin H'$ and $x^* \notin H''$. Thus $[\lambda\alpha' + (1-\lambda)\alpha'']^T x^* \leq y_0$ and $x^* \geq 0_M$ since $x^* \in H^\lambda$. If $x^* \notin H'$, $x^* \notin H''$, then since $x^* \geq 0_M$, we must have $\alpha'^T x^* > y_0$, $\alpha''^T x^* > y_0$. Thus $[\lambda\alpha' + (1-\lambda)\alpha'']^T x^* > y_0$ which contradicts the supposition $x^* \in H^\lambda$. Thus $H^\lambda \subset H' \cup H''$. Now let $U_{\alpha'}(y_0, y) \equiv \max_x\{f(x,y) : x \in H'\} \leq k$, and $U_{\alpha''}(y_0, y) \equiv \max_x\{f(x,y) : x \in H''\} \leq k$ where k is a scalar. Then

$$U_{\lambda\alpha'+(1-\lambda)\alpha''}(y_0, y) \equiv \max_x\{f(x,y) : x \in H^\lambda\}$$
$$\leq \max_x\{f(x,y) : x \in H' \cup H''\} \text{ (since } H^\lambda \subset H' \cup H'')$$
$$\leq k.$$

(iv) Let $y \geq 0_N$, $\alpha \gg 0_M$, $\lambda > 0$, $y_0 \geq 0$. Then

$$U_{\lambda\alpha}(\lambda y_0, y) \equiv \max_x\{f(x,y) : \lambda\alpha^T x \leq \lambda y_0; x \geq 0_M\}$$
$$= \max_x\{f(x,y) : \alpha^T x \leq y_0; x \geq 0_M\} \text{ (since } \lambda > 0)$$
$$= U_\alpha(y_0, y).$$

(v) The result follows directly from the Maximum Theorem 1.9.

Proof of Theorem 3.5.

Let f be positively linearly homogeneous, $v \gg 0_M$, $y \geq 0_N$ and $\lambda > 0$. Then

$$\begin{aligned} g(\lambda^{-1}v, \lambda y) &\equiv \max_x\{f(x, \lambda y) : \lambda^{-1}v^T x \leq 1, x \geq 0_M\} \\ &= \max_{\lambda z}\{f(\lambda z, \lambda y) : \lambda^{-1}v^T \lambda z \leq 1, \lambda z \geq 0_M\} \\ &= \max_z\{\lambda f(z, y) : v^T z \leq 1, z \geq 0_M\} \\ &= \lambda g(v, y). \end{aligned}$$

Now let g have the stated property and let $x \gg 0_M$, $y \geq 0_N$, $\lambda > 0$. Then

$$\begin{aligned} f(\lambda x, \lambda y) &\equiv \min_v\{g(v, \lambda y) : v^T \lambda x \leq 1, v \geq 0_M\} \\ &= \min_{\lambda^{-1}z}\{f(\lambda^{-1}z, \lambda y) : \lambda^{-1}z^T \lambda x \leq 1, \lambda^{-1}z \geq 0_M\} \\ &= \min_z\{\lambda g(z, y) : z^T x \leq 1, z \geq 0_M\} \\ &= \lambda f(x, y). \end{aligned}$$

Proof of Theorem 3.18.

Part 1. We show that concavity of $f(x, y)$ in x, y implies that $C(u; p, y)$ is convex in u, y for every $p \gg 0_M$. Let $p \gg 0_M$, $y^1 \geq 0_N$, $y^2 \geq 0_N$, $0 \leq \lambda \leq 1$, and define x^i, $i = 1, 2$, by

$$\text{(i)} \qquad C(u^i; p, y^i) \equiv \min_x\{p^T x : f(x, y^i) \geq u^i,\ x \geq 0_m\} \equiv p^T x^i.$$

Then

$$\begin{aligned} \text{(i')} \qquad & C[\lambda u^1 + (1 - \lambda)u^2; p, \lambda y^1 + (1 - \lambda)y^2] \\ & \equiv \min_x\{p^T x : f[x, \lambda y^1 + (1 - \lambda)y^2] \geq \lambda u^1 + (1 - \lambda)u^2\} \\ & \leq p^T[\lambda x^1 + (1 - \lambda)x^2] \\ & = \lambda C(u^1; p, y^1) + (1 - \lambda)C(u^2; p, y^2), \end{aligned}$$

since by concavity of f, $f[\lambda x^1 + (1 - \lambda)x^2, \lambda y^1 + (1 - \lambda)y^2] \geq \lambda f(x^1, y^1) + (1 - \lambda)f(x^2, y^2) \geq \lambda u^1 + (1 - \lambda)u^2$ (using (i)), and thus $\lambda x^1 + (1 - \lambda)x^2$ is feasible for the minimization problem (i). Throughout this proof, u^i's are assumed to belong to the range of the direct utility function.

Part 2. We show that convexity of $C(u;p,y)$ in u,y for every $p \gg 0_M$ implies that $g(p/Y,y) \equiv \max_u\{u : C(u;p/Y,y) \leq 1\}$ is concave in $Y > 0$, $y \geq 0_N$ for every $p \gg 0_M$. Let $p \gg 0_M$, $y^1 \geq 0_N$, $y^2 \geq 0_N$, $Y^1 > 0$, $Y^2 > 0$, $0 \leq \lambda \leq 1$, and define u^i, $i = 1,2$, by

(ii) $$g(p/Y^i, y^i) \equiv \max_u\{u : C(u;p/Y^i,y^i) \leq 1\} \equiv u^i.$$

Then

(ii′) $$\begin{aligned} &g\big[p/[\lambda Y^1 + (1-\lambda)Y^2], \lambda y^1 + (1-\lambda)y^2\big] \\ &\equiv \max_u \big\{u : C\big[u;p/[\lambda Y^1 + (1-\lambda)Y^2], \lambda y^1 + (1-\lambda)y^2\big] \leq 1\big\} \\ &= \max_u \big\{u : C[u;p,\lambda y^1 + (1-\lambda)y^2] \leq \lambda Y^1 + (1-\lambda)Y^2\big\} \\ &\geq \lambda u^1 + (1-\lambda)u^2 \\ &= \lambda g(p/Y^1,y^1) + (1-\lambda)g(p/Y^2,y^2), \end{aligned}$$

since, by the convexity of C, $C[\lambda u^1 + (1-\lambda)u^2; p, \lambda y^1 + (1-\lambda)y^2] \leq \lambda C(u^1;p,y^1) + (1-\lambda)C(u^2;p,y^2) \leq \lambda Y^1 + (1-\lambda)Y^2$ (using (ii)), and thus $\lambda u^1 + (1-\lambda)u^2$ is feasible for the maximization problem (ii′).

Part 3. We show that $g(p/Y,y)$ concave in $Y > 0$, $y \geq 0_N$ for every $p \gg 0_M$ implies that the restricted cost function $C(u;p,y) \equiv \min_Y\{g(p/Y,y) \geq u,\ Y > 0\}$ is convex in u, $y \geq 0_N$ for every $p \gg 0_M$. Let $p \gg 0_M$, $y^1 \geq 0_N$, $y^2 \geq 0_N$, $0 \leq \lambda \leq 1$, and define Y^i, $i = 1,2$, by

(iii) $$C(u^i;p,y^i) \equiv \min_Y\{Y : g(p/Y,y^i) \geq u^i\} \equiv Y^i.$$

Then

(iii′) $$\begin{aligned} &C[\lambda u^1 + (1-\lambda)u^2; p, \lambda y^1 + (1-\lambda)y^2] \\ &\equiv \min_Y\{Y : g[p/Y, \lambda y^1 + (1-\lambda)y^2] \geq \lambda u^1 + (1-\lambda)u^2\} \\ &\leq \lambda Y^1 + (1-\lambda)Y^2 \\ &= \lambda C(u^1;p,y^1) + (1-\lambda)C(u^2;p,y^2), \end{aligned}$$

since by the concavity property of g, $g\big[p/[\lambda Y^1 + (1-\lambda)Y^2], \lambda y^1 + (1-\lambda)y^2\big] \geq [\lambda g(p/Y^1,y^1) + (1-\lambda)g(p/Y^2,y^2)] \geq \lambda u^1 + (1-\lambda)u^2$ (using (iii)), and thus $\lambda Y^1 + (1-\lambda)Y^2$ is feasible for the minimization problem (iii′).

Part 4. We show that $C(u;p,y)$ convex in $u, y \geq 0_N$ for every $p \gg 0_N$ implies that the direct function $f(x,y) \equiv \max_u\{u : C(u;p,y) \leq p^T x$ for every $p \gg 0_M\}$ is concave in $x \geq 0_M$, $y \geq 0_N$. Let $x^1 \geq 0_M$, $x^2 \geq 0_M$, $y^1 \geq 0_N$, $y^2 \geq 0_N$, $0 \leq \lambda \leq 1$, and define u^i, $i = 1, 2$, by

(iv) $$f(x^i, y^i) \equiv \max_u\{u : C(u;p,y^i) \leq p^T x^i \text{ for every } p \gg 0_M\} \equiv u^i.$$

Then

(iv′)
$$\begin{aligned} & f[\lambda x^1 + (1-\lambda)x^2, \lambda y^1 + (1-\lambda)y^2] \\ & \equiv \max_u\{u : C[u;p,\lambda y^1 + (1-\lambda)y^2] \\ & \leq p^T[\lambda x^1 + (1-\lambda)x^2] \text{ for every } p \gg 0_M\} \\ & \geq \lambda u^1 + (1-\lambda)u^2 \\ & = \lambda f(x^1,y^1) + (1-\lambda)f(x^2,y^2), \end{aligned}$$

since by the convexity property of C, for every $p \gg 0_M$, $C[\lambda u^1 + (1-\lambda)u^2; p, \lambda y^1 + (1-\lambda)y^2] \leq \lambda C(u^1;p,y^1) + (1-\lambda)C(u^2;p,y^2) \leq \lambda p^T x^1 + (1-\lambda)p^T x^2$ (using (iv)), which means $\lambda u^1 + (1-\lambda)u^2$ is feasible for the maximization problem (iv′), and the solution to the maximization problem is equal to or greater than any feasible solution.

Note: If $C(u;p,y)$ is monotonically nondecreasing and continuous from below in u, then the set $L_p(x,y) \equiv \{u : C(u;p,y) \leq p^T x\}$ will be closed and bounded from above and thus the intersection $L(x,y) \equiv \bigcap_{p \gg 0_M}\{L_p(x,y)\}$ will also be closed and bounded from above. Thus $f(x,y) \equiv \max_u\{u : u \in L(x,y)\}$ is well defined as a maximum.

Proof of Theorem 3.20.

Part 1. Let $g(v,y)$ be convex in $v \gg 0_M$ for $y \geq 0_N$. Using the Fenchel closure operation (see Rockafellar [1970]), the extension of $g(v,y)$ to $v \geq 0_M$ will also be convex. Let $x \gg 0_M$, $Z^i > 0$, $0 \leq \lambda \leq 1$, and define v^i, $i = 1, 2$, by

(i) $$f(x/Z^i, y) \equiv \min_v\{g(v,y) : v^T x/Z^i \leq 1,\ v \geq 0_M\} \equiv g(v^i, y).$$

Then

$$\begin{aligned} f[x/(\lambda Z^1 + (1-\lambda)Z^2, y] & \equiv \min_v\{g(v,y) : v^T x/[\lambda Z^1 + (1-\lambda)Z^2] \leq 1,\ v \geq 0_M\} \\ & = \min_v\{g(v,y) : v^T x \leq \lambda Z^1 + (1-\lambda)Z^2,\ v \geq 0_M\} \\ & \leq g[\lambda v^1 + (1-\lambda)v^2, y] \end{aligned}$$

since by (i), $[\lambda v^1 + (1-\lambda)v^2]^T x \leq [\lambda Z^1 + (1-\lambda)Z^2]$, and hence is feasible for the minimization problem,

$$\leq \lambda g(v^1, y) + (1-\lambda) g(v^2, y)$$

using the convexity of g in v,

$$= \lambda f(x/Z^1, y) + (1-\lambda) f(x/Z^2, y)$$

using (i).

Part 2. Let $f(x/Z, y)$ be a convex function of the scalar variable $Z > 0$ for every $x \gg 0_M$ for some $y \geq 0_N$. Using the Fenchel closure operation, f will have the same property for every $x \geq 0_M$. Let $v \gg 0_M$. Then

$$g(v, y) \equiv \max_z \{ f(z, y) : v^T z \leq 1,\ z \geq 0_M \}$$

$$= \max_x \{ f(x/v^T x, y) : x \geq 0_M,\ x \neq 0_M \}$$

since f is nondecreasing in its arguments, so $x = 0_M$ is not a maximizer,

$$= \max_x \{ h_{x,y}(v) : x \in S \}$$

where $h_{x,y}(v) \equiv f(x/v^T x, y)$ is a convex function in $v \gg 0_M$ for x, y fixed, and $S \equiv \{x : x \geq 0_M, x \neq 0_M\}$,

$$= \text{a convex function of } v$$

since the pointwise maximum over a set of convex functions is also a convex function — see Rockafellar [1970].

Proof of Theorem 5.5.

(i) In view of Lemmas 5.2 and 5.4, it remains to show that $c(p; w) > 0$ if $p \geq 0_I$ and $w \gg 0_J$. If $p \in P$, then $c(p; w) \equiv +\infty > 0$. If $p \notin P$, then $c(p; w) \equiv \min_x \{ w^T x : V(p; x) \geq 1, x \geq 0_J \} = w^T x^*$, say. Using the linear homogeneity (and continuity) of V in x, we see that $x^* > 0_J$ since $V(p; 0_J) = 0$. Since $w \gg 0_J$, $w^T x^* > 0$.

(ii) We wish to show that for every scalar k, the set $\{(p; w) : c(p; w) \geq k,\ p \geq 0_I,\ w \gg 0_J\}$ is closed in the set $\{(p; w) : p \geq 0_I,\ w \gg 0_J\}$. In view of the positivity of $c(p; w)$, we can restrict k to being nonnegative. Let $k \geq 0$, $p^n \geq 0_I$, $w^n \gg 0_J$ for $n = 1, 2, \ldots$, $\lim p^n = p^0 \geq 0_I$, $\lim w^n = w^0 \gg 0_J$ and $c(p^n, w^n) \geq k$ for each n. *Suppose* that $c(p^0, w^0) = k - \varepsilon$ for some $\varepsilon > 0$. Since $c(p^0; w^0)$ is finite, $p^0 \notin P$. Thus there exists $x^0 \geq 0_J$ such that $c(p^0; w^0) = w^{0T} x^0 = k - \varepsilon$, where $V(p^0, x^0) \geq 1$. In view of the continuity of V in p, we have for every $\delta > 0$, $V(p^n; x^0) \geq 1 - \delta$, for n large enough. Choose $\delta > 0$ to be such that $\delta < \min\{1, \varepsilon/2k\}$. Therefore, for n large enough, $1 \leq V(p^n; x^0)/(1-\delta) = V[p^n; x^0/(1-\delta)]$, using the linear homogeneity of V in x. Since $w^{nT}x^0$ tends to $w^{0T}x^0$, for large enough n, we have $w^{nT}x^0 \leq k - \varepsilon/2$. Thus for n large, $c(p^n; w^n) \equiv \min_x\{w^{nT}x : V(p^n; x) \geq 1,\ x \geq 0_J\} \leq w^{nT}x^0/(1-\delta) \leq (k - \varepsilon/2)/(1-\delta) < k$. This contradicts the hypothesis $c(p^n; w^n) \geq k$, and thus our supposition must be false and therefore we must have $c(p^0; w^0) \geq k$.

(iii) Let $p^1 \geq 0_I$, $p^2 \geq 0_I$, $w^1 \gg 0_J$, $w^2 \gg 0_J$, $k \geq 0$, $0 \leq \lambda \leq 1$. Assume that $c(p^1; w^1) \geq k$ and $c(p^2; w^2) \geq k$. We wish to show that

$$c[\lambda p^1 + (1-\lambda)p^2;\ \lambda w^1 + (1-\lambda)w^2] \geq k.$$

There are three cases to consider. Case (1): $p^1 \in P$, $p^2 \in P$. Since P is a convex set, $\lambda p^1 + (1-\lambda)p^2 \in P$, and thus

$$c[\lambda p^1 + (1-\lambda)p^2;\ \lambda w^1 + (1-\lambda)w^2] = +\infty \geq k.$$

Case (2): $p^1 \notin P$, $p^2 \notin P$. If $\lambda p^1 + (1-\lambda)p^2 \in P$, then

$$c[\lambda p^1 + (1-\lambda)p^2;\ \lambda w^1 + (1-\lambda)w^2] = +\infty \geq k.$$

Assume $\lambda p^1 + (1-\lambda)p^2 \notin P$. If $k = 0$, then the desired result follows using the positivity of c. Thus we assume $k > 0$. We have

$$\begin{aligned} c(p^1; w^1) &\equiv \min_x\{w^{1T}x : V(p^1; x) \geq 1, x \geq 0_J\} \\ &= w^{1T}x^1 \geq k > 0 \quad \text{and} \\ c(p^2; w^2) &\equiv \min_x\{w^{2T}x : V(p^2; x) \geq 1, x \geq 0_J\} \\ &= w^{2T}x^2 \geq k > 0. \end{aligned}$$

Now if $x \gg 0_J$, then there exists $\lambda^* > 0$ such that $0_J \leq \lambda^* 1_J \leq x$. Therefore,

$$\begin{aligned} V(p^1; x) &\geq V(p^1; \lambda^* 1_J) \quad \text{(using 4.4(v))} \\ &= \lambda^* V(p^1; 1_J) \quad \text{(using 4.4(iv))} \\ &> 0 \end{aligned}$$

since $p^1 \notin P$. Similarly, if $x \gg 0_J$, then $V(p^2; x) > 0$. Thus if $x \gg 0_J$, then $x/V(p^1; x) \geq 0_J$ and $V(p^1; x/V(p^1; x)) = 1$. Using the minimum nature of $w^{1T}x^1$, we have for every $x \gg 0_J$, $w^{1T}x^1 \leq w^{1T}x/V(p^1; x)$ or $V(p^1; x)w^{1T}x^1 \leq w^{1T}x$ and using the continuity of V in x and the assumption that $w^{1T}x^1 > 0$

we find that $V(p^1;x) \le w^{1T}x/w^{1T}x^1$ for every $x \geq 0_J$. Similarly, $V(p^2;x) \le w^{2T}x/w^{2T}x^2$ for every $x \geq 0_J$. Let $x \geq 0_J$. Then

$$\begin{aligned} V[\lambda p^1+(1-\lambda)p^2;x] &\le \lambda V(p^1;x)+(1-\lambda)V(p^2;x) \quad \text{(using 4.4(iii))} \\ &\le \lambda w^{1T}x/w^{1T}x^1+(1-\lambda)w^{2T}x/w^{2T}x^2 \\ &\le \lambda w^{1T}x/k+(1-\lambda)w^{2T}x/k, \end{aligned}$$

since $w^{1T}x^1 \geq k$ and $w^{2T}x^2 \geq k$. Thus if $x \geq 0_J$ and $V[\lambda p^1+(1-\lambda)p^2;x] \geq 1$, then $[\lambda w^{1T}x+(1-\lambda)w^{2T}x]/k > 1$. Now

$$\begin{aligned} &c[\lambda p^1+(1-\lambda)p^2;\lambda w^1+(1-\lambda)w^2] \\ &\quad \equiv \min_x\{[\lambda w^1+(1-\lambda)w^2]^T x : V[\lambda p^1+(1-\lambda)p^2;x] \geq 1,\ x \geq 0_J\} \\ &\quad \geq \min_x\{[\lambda w^1+(1-\lambda)w^2]^T x : [\lambda w^{1T}x+(1-\lambda)w^{2T}x]/k \geq 1, x \geq 0_J\} \end{aligned}$$

since in general the set of feasible x's in the second minimization problem is larger,

$$= \min_x\{[\lambda w^1+(1-\lambda)w^2]^T x : [\lambda w^1+(1-\lambda)w^2]^T x \geq k; x \geq 0_J\}$$

since $k > 0$,

$$= k.$$

Case (3): $p^1 \notin P$ and $p^2 \in P$. Without loss of generality, we can take p^2 belonging to the boundary of P. Using the results of case (2) above, we need only show that as λ tends to zero, $c[\lambda p^1+(1-\lambda)p^2;\ \lambda w^1+(1-\lambda)w^2]$ tends to plus infinity. Let $p^1 \geq 0_I$, $p^2 \geq 0_I$, $w^1 \gg 0_J$, $w^2 \gg 0_J$, and let $\beta > 0$ be the minimum component of the vectors w^1 and w^2. For $0 \le \lambda \le 1$, define $f(\lambda) \equiv V[\lambda p^1+(1-\lambda)p^2;1_J]$. By hypothesis, $f(0) = 0$, but $f(\lambda) > 0$ for $0 < \lambda \le 1$. Since by (i) and (iii) of 4.4, $V(p;1_J)$ is a nonnegative, convex function of p, we see that $f(\lambda)$ is a nondecreasing, convex function of λ for $0 \le \lambda \le 1$, such that $\lim_{\lambda\to 0}\{f(\lambda) : \lambda > 0\} = 0$. Now if $0 < \lambda \le 1$, we have $V[\lambda p^1+(1-\lambda)p^2;\ 1_J/f(\lambda)] = 1$. Thus if $0_J \le x \ll 1_J/f(\lambda)$, then using (iv) and (v) of 4.4, we find $V[\lambda p^1+(1-\lambda)p^2;x] < 1$. Therefore, for $0 < \lambda \le 1$,

$$\begin{aligned} &c[\lambda p^1+(1-\lambda)p^2;\ \lambda w^1+(1-\lambda)w^2] \\ &\quad \equiv \min_x\{[\lambda w^1+(1-\lambda)w^2]^T x : V[\lambda p^1+(1-\lambda)p^2;x] \geq 1; x \geq 0_J\} \\ &\quad \geq \beta/f(\lambda), \end{aligned}$$

and the last term tends monotonically to plus infinity as λ tends to zero.

(iv) Let $p \geq 0_I$, $w \gg 0_J$ and $\lambda > 0$. Then

$$\begin{aligned} c(\lambda p; w) &\equiv \min_x \{w^T x : V(\lambda p; x) \geq 1, x \geq 0_J\} \\ &= \min_x \{w^T x : V(p; \lambda x) \geq 1, x \geq 0_J\} \end{aligned}$$

using (ii) and 4.4(iv),

$$\begin{aligned} &= \lambda^{-1} \min_x \{w^T \lambda x : V(p; \lambda x) \geq 1, \lambda x \geq 0_J\} \\ &= \lambda^{-1} c(p; w). \end{aligned}$$

(v) Let $p \geq 0_I$, $w \gg 0_J$, $\lambda > 0$. Then $c(p; \lambda w) \equiv \min_x \{\lambda w^T x : V(p; x) \geq 1, x \geq 0_J\} = \lambda c(p; w)$.

(vi) Let $p \geq 0_I$, $w^2 \geq w^1 \gg 0_J$. Then

$$\begin{aligned} c(p; w^2) &\equiv \min_x \{w^{2T} x : V(p; x) \geq 1, \ x \geq 0_J\} \\ &= w^{2T} x^2 = w^{1T} x^2 + (w^2 - w^1)^T x^2 \\ &\geq w^{1T} x^2 \text{ (since } w^2 - w^1 \geq 0_J \text{ and } x^2 \geq 0_J) \\ &\geq \min_x \{w^{1T} x : V(p; x) \geq 1, x \geq 0_J\} \\ &= c(p; w^1). \end{aligned}$$

Proof of Theorem 5.8.

(i) $V^*(p; x) \equiv \min_w \{1/c(p; w) : w \geq 0_J, w^T x \leq 1\}$ for $p \geq 0_I$ and $x \gg 0_J$ is well defined as a minimum, since $1/c(p; w)$ is continuous from below in w, nonnegative and finite for at least one value of w and the minimum of such a function over the closed, bounded set $\{w : w \geq 0_J, w^T x \leq 1\}$ exists. (Recall Theorem 1.5.) We note that since $c(p; w)$ is nonnegative, continuous from above, and quasiconcave for $p \geq 0_I$, $w \geq 0_J$, $1/c(p; w)$ will be nonnegative, continuous from below, and quasiconvex over the same domain.

(ii) Let $p \geq 0_I$, $x \gg 0_J$ and $\lambda > 0$. Then

$$\begin{aligned} V^*(\lambda p; x) &\equiv \min_w \{1/c(\lambda p; w) : w \geq 0_J, w^T x \leq 1\} \\ &= \min_w \{1/\lambda^{-1} c(p; w) : w \geq 0_J, w^T x \leq 1\} \text{ (using 5.6(iv))} \\ &= \lambda V^*(p; x). \end{aligned}$$

(iii) We first show $V^*(p; x)$ is quasiconvex in p for fixed x. Let $x \gg 0_J$, $p^1 \geq 0_I$, $p^2 \geq 0_I$, $0 \leq \lambda \leq 1$ and $k \geq 0$. Let $k \geq V^*(p^i; x) \equiv \min_w \{1/c(p^i; w) :$

$w \geq 0,\ w^T x \leq 1\} = 1/c(p^i; w^i)$, say. Therefore $c(p^1; w^1) \geq 1/k$, $c(p^2; w^2) \geq 1/k$, and by 5.6(iii), $c[\lambda p^1 + (1-\lambda)p^2; \lambda w^1 + (1-\lambda)w^2] \geq 1/k$ also. Thus

$$\begin{aligned} V^*[\lambda p^1 + (1-\lambda)p^2; x] &\equiv \min_w \{1/c[\lambda p^1 + (1-\lambda)p^2; w] : w \geq 0, w^T x \leq 1\} \\ &\leq 1/c[\lambda p^1 + (1-\lambda)p^2; \lambda w^1 + (1-\lambda)w^2] \end{aligned}$$

since $\lambda w^1 + (1-\lambda)w^2 \geq 0$ and $[\lambda w^1 + (1-\lambda)w^2]^T x \leq 1$

$$\leq k.$$

Finally since $V^*(p; x)$ is nonnegative, quasiconvex and (positively) homogeneous of degree one over the set $p \geq 0$ for a fixed x, it can be seen (see Berge [1963; 208]) that $V^*(p; x)$ is also a convex function in p for fixed x. With respect to the continuity of $V^*(p; x)$ in p, we note that the convexity of $V^*(p; x)$ in p will imply that $V^*(p; x)$ is continuous from above in $p \geq 0_I$ for fixed x (see Rockafellar [1970; 84]). Now

$$\begin{aligned} -V^*(p; x) &= -\min_w \{1/c(p; w) : w \geq 0_J, w^T x \leq 1\} \\ &= \max_w \{-1/c(p; w) : w \geq 0, w^T x \leq 1\} \end{aligned}$$

and $-V^*(p; x)$ will be continuous from above by the Upper Semicontinuity Maximum Theorem 1.8 and thus $V^*(p; x)$ will be continuous from below in p for fixed $x \gg 0_J$. Thus $V^*(p; x)$ will be continuous in p.

(iv) Let $p \geq 0_I$, $x \gg 0_J$ and $\lambda > 0$. Then

$$\begin{aligned} V^*(p; \lambda x) &\equiv \min_w \{1/c(p; w) : w \geq 0_J, w^T \lambda x \geq 1\} \\ &= \min_w \{\lambda/c(p; \lambda w) : \lambda w \geq 0, (\lambda w)^T x \leq 1\} \text{ (using 5.6(v))} \\ &= \min_{w^*} \{1/c(p; w^*) : w^* \geq 0, w^{*T} x \leq 1\} \\ &= \lambda V^*(p; x). \end{aligned}$$

(v) Let $x^2 \geq x^1 \gg 0_J$, $p \geq 0_I$. Then

$$\begin{aligned} V^*(p; x^2) &\equiv \min_w \{1/c(p; w) : w \geq 0_J, w^T x^2 \leq 1\} \\ &\geq \min_w \{1/c(p; w) : w \geq 0_J, w^T x^1 \leq 1\} \end{aligned}$$

since the set of feasible w's has not decreased so the minimum cannot increase,

$$= V^*(p; x^1).$$

(vi) If $p \in P$, then $c(p;w) = +\infty$ for all $w \geq 0_J$, and thus $V^*(p;x) = 0$ for all $x \gg 0_J$, and $V^*(p;x)$ is a concave function in x. Let $p^* \geq 0_I$, $p^* \notin P$. Then $V^*(p^*;x)$ will be a concave function in $x \gg 0_J$ if we can show that $V^*(p^*;x)$ is a positive, linearly homogeneous, quasiconcave function in $x \gg 0_J$. (See Berge [1963; 208] or Newman [1969; 300].) Since $p^* \notin P$, we have $c(p^*;w) > 0$ and finite for $w \gg 0_J$. Thus if $x \gg 0_J$, $V^*(p^*;x) > 0$. Linear homogeneity of $V^*(p^*;x)$ in x follows from (iv) above, and thus it remains to show that $V^*(p^*;x)$ is quasiconcave in x. Let $x^1 \gg 0_J$, $x^2 \gg 0_J$, $0 \leq \lambda \leq 1$, $k > 0$, and suppose that

$$\begin{aligned} V^*(p^*;x^1) &\equiv \min_w \{1/c(p^*;w) : w \geq 0_J, w^T x^1 \leq 1\} \\ &= 1/c(p^*;w^1) \geq k, \text{ and} \\ V^*(p^*;x^2) &\equiv \min_w \{1/c(p^*;w) : w \geq 0_J, w^T x^2 \leq 1\} \\ &= 1/c(p^*;w^2) \geq k. \end{aligned}$$

By the definition of V^*, we have

$$\begin{aligned} V^*[p^*;\lambda x^1 + (1-\lambda)x^2] &\equiv \min_w \{1/c(p^*;w) : w \geq 0_J,\ w^T[\lambda x^1 + (1-\lambda)x^2] \leq 1\} \\ &= 1/c(p^*;w^*) \qquad \text{for some } w^*. \end{aligned}$$

Suppose $1/c(p^*;w^*) < k$. Then since $w^{*T}[\lambda x^1 + (1-\lambda)x^2] \leq 1$, we must have either $w^{*T}x^1 \leq 1$ or $w^{*T}x^2 \leq 1$ (or both). If $w^{*T}x^1 \leq 1$, then $V^*(p^*;x^1) \leq 1/c(p^*;w^*) < k$, which contradicts $V^*(p^*;x^1) \geq k$. Similarly, if $w^{*T}x^2 \leq 1$, then we get a contradiction to $V^*(p^*;x^2) \geq k$. Thus our supposition $1/c(p^*;w^*) < k$ must be false, and $V^*[p^*;\lambda x^1 + (1-\lambda)x^2] \geq k$. We have established that for a fixed $p \geq 0_I$, $V^*(p;x)$ is a concave function in x for $x \gg 0_J$, and thus $V^*(p;x)$ is also continuous in $x \gg 0_J$ for fixed p.

To complete the proof of Theorem 5.8, assume that a value added function V satisfied conditions II given by 4.4, that c was defined by equation 5.3 and that V^* was defined by equation 5.7. We wish to show that for every $p \geq 0_I$, $x^* \gg 0_J$, $V^*(p;x^*) = V(p;x^*)$. If $p \in P$ defined by equation 5.1, it is easy to see that $V^*(p;x^*) = V(p;x^*) = 0$. Assume $p \notin P$. Since $x^* \gg 0_J$, $V(p;x^*) = v > 0$. Define the value added production possibilities set $L(v;p) \equiv \{x : V(p;x) \geq v; x \geq 0_J\}$. By properties (iv) and (vi) of 4.4, we see that x^* is a boundary point of the closed convex set $L(v;p)$ and thus $L(v;p)$ will have at least one supporting hyperplane $w^* \neq 0$ at the point x^*; i.e., $w^{*T}x^* \leq w^{*T}x$ for every $x \in L(v;p)$. Property 4.4(v) will imply that $w^* > 0_J$. Thus $w^{*T}(x^*/v) = c(p;w^*) \equiv \min_x\{w^{*T}x : V(p;x) \geq 1, x \geq 0_J\}$ and

$c(p; w) \leq w^T(x^*/v)$ for every $w \geq 0$ since $V(p; x^*/v) \geq 1$. Now

$$\begin{aligned} V^*(p; x^*) &\equiv \min_w \{1/c(p; w) : w \geq 0_J, w^T x^* \leq 1\} \\ &= 1/\max_w \{c(p; w) : w \geq 0_J, w^T x^* \leq 1\} \\ &= v/\max_w \{c(p; w)v : w \geq 0_J, w^T x^* \leq 1\} \text{ (since } v > 0) \\ &= v, \end{aligned}$$

since $c(p; w)v \leq w^T x^*$ for every $w \geq 0_J$ with equality for $w = w^*$

$$= V(p; x^*)$$

by the definition of v. Q.E.D.

Essays in Index Number Theory, Volume I
W.E. Diewert and A.O. Nakamura (Editors)

Chapter 16
A NOTE ON AGGREGATION AND ELASTICITIES OF SUBSTITUTION*

W.E. Diewert

One of the most pressing problems facing an applied economist attempting to model the behavior of certain specific markets (such as labor markets) is to determine to what extent other markets, which are of less interest, can be aggregated. Presumably the benefits of more disaggregation of other markets are more accurate predictions of behavior on the markets of primary concern. However, one is led to ask, at what point will the increased benefits of disaggregation be outweighed by the costs of providing the disaggregated information?[1] The purpose of this note is to provide a framework for evaluating the benefits of increased disaggregation of other markets from the viewpoint of modeling behavior on the markets of primary concern.

Suppose that we are given a production function[2] f where $Y = f(L_1, L_2, \ldots, L_M;\ X_1, X_2, \ldots, X_N)$, Y being the output produced during a given time period, L_m the quantity of the mth type of labor services used for $m = 1, 2, \ldots, M$, and X_n the quantity of the nth category of 'other' inputs used for $n = 1, 2, \ldots, N$. Let W_m be the wage rate for one unit of L_m and let P_n be the rental price for one unit of X_n.

---

*First published in *The Canadian Journal of Economics*, Vol. 7, No. 1, 1974, pp. 12–20. Discussions on this topic with E.R. Berndt, E. Dianne Cummings, K.F.N. Scott, and A.D. Woodland have been very helpful.

[1]In addition to the cost of constructing an extra data series, there will generally be the increased cost of estimating additional unknown parameters. For example, if we are attempting to estimate the parameters of an N factor production function and we introduce an additional input series, then in general, an additional $N + 1$ parameters will have to be estimated if we do not wish *a priori* to constrain elasticities of substitution. Thus as N increases, the number of parameters to be estimated grows as the square of N.

[2]The analysis which follows can also be applied with minor modifications to: (i) multiple output production functions and (ii) systems of consumer demand and labor supply functions.

Suppose the producer attempts to minimize the cost of producing a given amount of output Y; that is, the producer attempts to solve the following minimization problem:

$$(1) \qquad \text{Minimize } \sum_{m=1}^{M} W_m L_m + \sum_{n=1}^{N} P_n X_n$$

with respect to $L_1, \ldots, L_M$ and $X_1, \ldots, X_N$, subject to $Y = f(L_1, \ldots, L_M; X_1, \ldots, X_N)$.

Define the optimal value of the objective function for (1) to be $C(Y; W_1, \ldots, W_M; P_1, \ldots, P_N)$, the total cost function.

A solution to the cost minimization problem given by (1) is the producer's system of labor demand functions

$$L_m(Y; W_1, \ldots, W_M;\ P_1, \ldots, P_N) \text{ for } m = 1, 2, \ldots, M,$$

and the producer's demand for other input functions

$$X_n(Y; W_1, \ldots, W_M;\ P_1, \ldots, P_N) \text{ for } n = 1, 2, \ldots, N.$$

Suppose that we are primarily interested in the labor demand functions and the response of labor demand to changes in the prices of other inputs; that is, we are interested in the price elasticities:

$$(\partial L_m/\partial P_n)(P_n/L_m) \text{ for } m = 1, 2, \ldots, M \text{ and } n = 1, 2, \ldots, N.$$

It turns out that we can express the above price elasticities in terms of the partial derivatives of the total cost function defined by (1) above. In fact, by Shephard's Lemma:[3]

$$(2) \qquad L_m(Y^*; W^*; P^*) = \partial C(Y^*; W^*; P^*)/\partial W_m \text{ for } m = 1, 2, \ldots, M,$$

where Y^* is a given output level, $W^* \equiv (W_1^*, W_2^*, \ldots, W_M^*)$ is a given vector of wages, and $P^* \equiv (P_1^*, P_2^*, \ldots, P_N^*)$ is a given vector of prices for other inputs. Equation (2) says that, given a functional form for the total cost function (satisfying the appropriate regularity conditions), we can obtain the demand for labor functions simply by partially differentiating the total cost function with respect to wage rates.

Define $L_m^* \equiv L_m(Y^*; W^*; P^*)$ for $m = 1, 2, \ldots, M$ and $X_n^* \equiv X_n(Y^*; W^*; P^*)$ for $n = 1, 2, \ldots, N$. If we partially differentiate both sides of (2) with respect to P_n, we can express the price elasticity of demand for the mth type

[3]See Shephard [1953] or Diewert [1971a; 495–496] for a proof and a statement of the appropriate regularity conditions on the production function f and on the total cost function C which is dual to f.

of labor with respect to a change in P_n, the price of the nth type of other input, as:

$$\frac{\partial L_m(Y^*;W^*;P^*)}{\partial P_n}\frac{P_n^*}{L_m^*} = \left[\frac{\partial^2 C(Y^*;W^*;P^*)}{\partial P_n \partial W_m}\right]\left[\frac{P_n^*}{L_m^*}\right]$$

$$= \left[\frac{\partial^2 C(Y^*;W^*;P^*)}{\partial P_n \partial W_m}\frac{C(Y^*;W^*;P^*)}{L_m^* X_n^*}\right]\left[\frac{P_n^* X_n^*}{C(Y^*;W^*;P^*)}\right]$$

$$= \sigma_m^n(Y^*;W^*;P^*)\left[\frac{P_n^* X_n^*}{C(Y^*;W^*;P^*)}\right] \tag{3}$$

$$\text{for } m = 1,2,\ldots,M; \quad n = 1,2,\ldots,N,$$

where $\sigma_m^n(Y^*;W^*;P^*)$ is the partial elasticity of substitution[4] between L_m and X_n evaluated at the initial point $(Y^*;W^*;P^*)$.[5]

Define $\sigma_m^{n^*} \equiv \sigma_m^n(Y^*;W^*;P^*)$ and define the share in total cost of the nth other input as $s^{n^*} \equiv P_n^* X_n^*/C(Y^*;W^*;P^*)$. Now equation (3) can be rewritten as:

$$[\partial L_m(Y^*;W^*;P^*)/\partial P_n](P_n^*/L_m^*) = \sigma_m^{n^*} s^{n^*} \tag{4}$$

$$\text{for } \quad m = 1,2,\ldots,M; \; n = 1,2,\ldots,N.$$

In words, equation (4) says that the percentage change in the demand for the mth type of labor due to a one percent change in the price of the nth type of other input is equal to the (partial) elasticity of substitution between L_m and X_n times the share in total cost of X_n. Thus partial elasticities of substitution are intimately related to ordinary price elasticities.

We now suppose that the prices of inputs $P \equiv (P_1, P_2, \ldots, P_N)$ vary in strict proportion over time; i.e., that the relative prices of other inputs are constant. If this is the case, it can be shown that, from the viewpoint of modeling labor market behavior, there will be no loss in predictive power due to

---

[4]The partial elasticity of substitution concept was originally defined in terms of the partial derivatives of the production function (see Allen [1938; 504]). However, Uzawa [1962] showed that

$$\sigma_m^n(Y^*;w^*;p^*) \equiv [\partial^2 C(Y^*;w^*;p^*)/\partial p_n \partial w_m][C(Y^*;w^*;p^*)/L_m^* X_n^*].$$

[5]In general, the elasticity of substitution σ_m^n will change as outputs, wages, and prices change. Equation 3 is usually derived using inverses of bordered Hessian matrices; see Allen [1938].

the effects of aggregation. This result is part of Hicks' [1946; 312–313] Aggregation Theorem.[6] The theorem states that, under the appropriate regularity conditions, if the prices of a group of goods change proportionately, then that group of goods behaves just as if it were a single commodity.

Let the proportional variation in the prices of other inputs be expressed as:

$$P_1 = \alpha_1 P_0, \quad P_2 = \alpha_2 P_0, \; \ldots, \; P_N = \alpha_N P_0 \tag{5}$$

where P_0 is the price of the aggregate and the positive constant α_n expresses how the price P_n varies as P_0 varies over time for $n = 1, 2, \ldots, N$ (note that we are regarding the 'micro' prices P_1, P_2, ... , P_N as functions of the 'macro' price P_0). We assume that P_0 is initially equal to unity (i.e., $P_0^* = 1$) and thus $P_n^* = \alpha_n \cdot 1$ is the initial price of the nth other input.

Given that the prices of other inputs vary in proportion, we may calculate what percentage change in the demand for the mth type of labor will be induced by a one percent change in the aggregate price index for other inputs; i.e., we may calculate (where $\alpha \equiv (\alpha_1, \alpha_2, \ldots, \alpha_N)$):

$$\frac{\partial L_m(Y^*; W^*; \alpha P_0^*)}{\partial P_0} \frac{P_0^*}{L_m^*}\bigg|_\alpha = \frac{\partial}{\partial P_0}\left[\frac{\partial C(Y^*; W^*; \alpha P_0^*)}{\partial W_m}\right]\frac{P_0^*}{L_m^*}$$

$$= \sum\nolimits_{n=1}^{N}\left[\frac{\partial^2 C(Y^*; W^*; \alpha P_0^*)}{\partial P_n \partial W_m}\right]\frac{\partial P_n}{\partial P_0}\frac{P_0^*}{L_m^*}$$

$$= \sum\nolimits_{n=1}^{N}\left[\frac{\partial^2 C(Y^*; W^*; \alpha P_0^*)}{\partial P_n \partial W_m}\right]\alpha_n \frac{P_0^*}{L_m^*} \quad \text{(using (5))}$$

$$= \sum\nolimits_{n=1}^{N}\left[\frac{\partial^2 C(Y^*; W^*; P^*)}{\partial P_n \partial W_m}\frac{C(Y^*; W^*; P^*)}{L_m^* X_n^*}\right]\left[\frac{P_n^* X_n^*}{C(Y^*; W^*; P^*)}\right]$$

$$= \sum\nolimits_{n=1}^{N} \sigma_m^{n*} s^{n*} \tag{6}$$

where the last equality follows from the fact that equation (3) above is equivalent to equation (4). Thus the response of L_m to a one percent change in P_0, the aggregate price index for other inputs, is a share weighted average of the partial elasticities of substitution of the mth type of labor with the other inputs.

[6] For alternative treatments of Hicks' Aggregation Theorem, see Wold [1953; 109–110], Gorman [1953; 76–77] and Diewert [1978a].

Using equation (3) as a model, we may define the *aggregate elasticity of substitution* of L_m with other inputs as:

$$\sigma_m^{A^*} \equiv \left[\frac{\partial L_m P_0^*}{\partial P_0 L_m^*}\right]\left[\frac{C(Y^*;W^*;P^*)}{\sum_{n=1}^{N} P_n^* X_n^*}\right] \quad \text{for } m = 1,2,\ldots,M$$

$$(7) \qquad = \frac{\sum_{n=1}^{N} \sigma_m^{n^*} s^{n^*}}{\sum_{n=1}^{N} s^{n^*}} \quad \text{using (6) and the definition of } s^{n^*}.$$

Thus the aggregate elasticity of substitution of the mth type of labor with other inputs is equal to a share in total cost weighted average of the micro elasticities of substitution of L_m with X_n divided by the share of other inputs in total cost.[7]

Recall that $\sigma_m^{n^*}$ will be positive if L_m and X_n are substitutes in production and $\sigma_m^{n^*}$ will be negative if the two inputs are complements. Taking a weighted average of both positive and negative[8] micro elasticities of substitution, $\sigma_m^{n^*}$ will tend to give rise to aggregate elasticities of substitution which are considerably *smaller* in magnitude than an average of the absolute values of the micro elasticities of substitution.

Conversely, as we disaggregate, we can expect to encounter increasingly large elasticities of substitution. Two recent empirical papers confirm this statement. Berndt and Christensen [1974] in their 'two types of labor, one type of capital' disaggregation of US manufacturing industries found that the mean partial elasticities of substitution were 7.88 (between blue and white collar workers), 3.72 (between blue collar workers and capital) and −3.77 (between white collar workers and capital). However, when they fitted a model which aggregated the two types of labor into a single labor factor, they found that the aggregate labor capital elasticity of substitution was approximately 1.42, which is considerably smaller than an average of the absolute values of the three

[7]The expression for the aggregate elasticity of substitution given by (7) has been derived by Woodland [1972a; 25, 40] and by Sato and Koizumi [1973]. They did not use Hicks' Aggregation Theorem in their derivations.

[8]It is theoretically possible for all $N(N-1)/2$ partial elasticities of substitution in the N good case to be positive (Hicks [1946; 311]). It is also theoretically possible for all $N(N-1)/2$, save $N-1$, partial elasticities of substitution to be negative (Ibid., 47). From an empirical point of view, it appears that partial elasticities of substitution are more or less randomly distributed with a positive mean and a variance which becomes larger as N increases. See Berndt and Christensen [1974], Gussman [1972] and Woodland [1972a] for empirical tabulations of elasticities of substitution when there are three, eleven, and five goods respectively.

'micro' elasticities of substitution. Similarly, Woodland [1972a] found partial elasticities of substitution in Canadian manufacturing range from −11.16 to 2.18 in his 'four types of capital, one type of labor' disaggregated results, but he found that the aggregate capital-labor elasticity of substitution was only 0.39.

Let us summarize our results thus far. If the prices of other inputs vary in strict proportion, then we may aggregate all the other inputs into a single composite good whose price is represented by a price index P_0 and we can predict what the percentage change in L_m due to a one percent change in the aggregate price of other inputs is by using the following expression (which is obtained by substituting (7) into (6)):

$$\frac{\partial L_m}{\partial P_0}\frac{P_0^*}{L_m^*} = \sigma_m^{A^*}\left(\sum_{n=1}^{N} s^{n^*}\right). \tag{8}$$

That is, the price elasticity of the mth type of labor with respect to a change in the price of other inputs is equal to the aggregate elasticity of substitution of L_m with other inputs times the share in total cost of other inputs. Moreover, we can use standard econometric techniques in order to estimate the aggregate elasticities of substitution $\sigma_m^{A^*}$ for $m = 1, 2, \ldots, M$ using only aggregate data on other inputs. We note that our estimated aggregate elasticity of substitution $\sigma_m^{A^*}$ will be a weighted average of micro elasticities of substitution $\sigma_m^{n^*}$ of unknown magnitudes, but for purposes of predicting labor market behavior, a knowledge of the micro parameters $\sigma_m^{n^*}$ is not necessary — provided that the prices of other inputs vary in strict proportion.

Since prices are unlikely to vary *precisely* in proportion over a period of years, one is led to ask whether the results summarized in the previous paragraph are robust; that is, do they carry over *approximately* provided that prices of other inputs vary approximately in proportion. If this is not the case, then Hicks' Aggregation Theorem would be of little empirical significance. Fortunately, the above results do carry over to the approximate case, as we shall now attempt to show.[9]

Suppose that the prices of other inputs have varied in strict proportion according to the relationship given by (5) and that we have initial prices of other inputs given by:

$$P_1^* = \alpha_1, \quad P_2^* = \alpha_2, \; \ldots, \; P_N^* = \alpha_N. \tag{9}$$

[9]The aggregate production function $\hat{f}_\alpha(L_1, \ldots, L_M; X_0)$ can be defined as the dual to the cost function $\widehat{C}_\alpha(Y; W_1, \ldots, W_M; P_0) \equiv C(Y; W; \alpha P_0)$. If α remains constant, $\hat{f}_\alpha$ will be a well defined production function in $(L; X_0)$. Since $\hat{f}_\alpha$ can be shown to be a continuous function of α, small changes in α will lead to small shifts in the aggregate production function $\hat{f}_\alpha$.

Now suppose that, starting from the initial prices given by (9), prices of other inputs vary according to a different proportionality vector; that is, we assume that as P_0 varies from an initial level of unity, the prices of other inputs vary according to the following relationships:

$$P_1 = (\alpha_1 - \beta_1) + \beta_1 P_0,\ P_2 = (\alpha_2 - \beta_2) + \beta_2 P_0,\ \ldots,\ P_N = (\alpha_N - \beta_N) + \beta_N P_0. \tag{10}$$

Thus when $P_0^* = 1$ initially, the initial prices are given by $P_n^* = \alpha_n$ for $n = 1, \ldots, N$. However, when P_0 increases beyond 1, the old proportionality factors α_n are replaced by the new proportionality factors β_n. We now calculate the response of L_m to a change in the price index of other goods P_0 assuming that the prices of other inputs satisfy the relation (10) rather than (5).[10]

$$\begin{aligned}
&\frac{\partial L_m(Y^*; W^*; \alpha - \beta + \beta P_0^*)}{\partial P_0} \frac{P_0^*}{L_m^*}\bigg|_\beta \\
&= \frac{\partial}{\partial P_0}\left[\frac{\partial C(Y^*; W^*; \alpha - \beta + \beta P_0^*)}{\partial W_m}\right]\frac{P_0^*}{L_m^*} \\
&= \sum_{n=1}^N \left[\frac{\partial^2 C(Y^*; W^*; \alpha - \beta + \beta P_0^*)}{\partial W_m \partial P_n}\right]\frac{\beta_n}{L_m^*} \\
&= \sum_{n=1}^N \sigma_m^{n^*}\left[\frac{\beta_n X_n^*}{C(Y^*; W^*; P^*)}\right]
\end{aligned}$$

or

$$\frac{\partial L_m}{\partial P_0}\frac{P_0^*}{L_m^*}\bigg|_\beta = \sum_{n=1}^N \sigma_m^{n^*} s^{n^*}\left[\frac{\beta_n}{\alpha_n}\right] \tag{11}$$

where $\sigma_m^{n^*}$ and s^{n^*} are defined below equation (3), α_n represents the *old* proportionality factor for P_n (the price of one unit of the nth type of other input), and β_n represents the *new* proportionality factor for P_n.

Let us suppose that we have estimated the elasticity $(\partial L_m/\partial P_0)P_0^*/L_m^*$ based on the assumption that the prices of other inputs vary according to (5). Then we will measure (incorrectly) the percentage change in labor demand due to a one percent change in the aggregate price of other inputs as:

$$\left[\frac{\partial L_m}{\partial P_0}\right]\frac{P_0^*}{L_m^*}\bigg|_\alpha = \sum_{n=1}^N \sigma_m^{n^*} s^{n^*}. \tag{3}$$

[10]Recall that $P_0^* \equiv 1$ and $\alpha \equiv (\alpha_1, \alpha_2, \ldots, \alpha_N)$, and define $\beta \equiv (\beta_1, \beta_2, \ldots, \beta_N)$.

The correct percentage change in labor demand is given by (11). There are several points to note in comparing (3) with (11):

(i) Note that (11) coincides with (3) if $\alpha_n = \beta_n$ for $n = 1, 2, \ldots, N$ and, moreover, the closer the numbers β_n are to α_n then the closer will be the incorrectly measured elasticity (3) to the true elasticity given by (11). Thus an approximate version of Hicks' Aggregation Theorem will hold in the sense that prediction errors due to aggregation will be small provided that deviations from the assumption that the prices of other inputs vary in proportion are small.

(ii) A given percentage change in β_n relative to α_n will generate a bigger prediction error the bigger is the share s^{n^*} of the nth other input in total cost and the bigger the magnitude of the elasticity of substitution $\sigma_m^{n^*}$.

(iii) If we pursue point (ii) a bit further, we see that a major change, say, in the price of the first other input, relative to the general trend in the price level of other inputs, *will generally have a very uneven effect* on the prediction errors for the various labor markets, since micro elasticities of substitution tend to be more or less randomly distributed with a considerable variance.[11] More specifically, let $\beta_n = \alpha_n$ for $n = 2, 3, \ldots, N$ but let the price of the first other input change according to the relation $\beta_1 = \alpha_1(1+\delta)$. Then the prediction error on the mth labor market due to aggregation may be written as the difference between (11) and (3) and is equal to

$$\sigma_m^{1^*} s^{1^*} \delta \tag{12}$$

when there is a nonproportional variation in the price of the first other input. Thus the error will vary in proportion to the percentage change δ in the price of the first other input relative to the general trend in the price of other inputs, the share in total cost of X_1, and the elasticity of substitution between X_1 and L_m, $\sigma_m^{1^*}$. Since it is unlikely that the elasticities of substitution $\sigma_m^{1^*}$, $m = 1, 2, \ldots, M$, are all equal, we see that the prediction errors generated by aggregating other inputs will not be equal in magnitude for every labor market.

(iv) These points suggest that, in order to minimize prediction errors on disaggregated labor markets, we will have to disaggregate other markets as well. Ideally, we would like each of the other input series X_n to consist of inputs whose prices vary in fixed proportions, at least approximately. In practice, it

[11]Actually, this point may be used to explain why Woodland's [1972a] production function results were somewhat unsatisfactory. Woodland used four types of capital and one type of labor. Regard the four types of capital as L_1, L_2, L_3 and L_4 and labor as an aggregate of $X_1, X_2, \ldots, X_N$. Since relative wage rates have not been constant over time, and since the share of labor in value added is large, it can be seen that the prediction errors due to labor aggregation on the capital markets could be fairly large, depending on the magnitudes of the unknown elasticities of substitution.

appears that the more finely we disaggregate labor markets, the more finely we will have to disaggregate other markets in order to keep prediction errors on the disaggregated labor markets down to a satisfactory level.[12] This in turn suggests that it is unlikely that we will be able to predict behavior on very finely disaggregated labor markets.

We conclude that the benefits of disaggregating other markets from the viewpoint of predicting behavior on a group of markets that we are primarily interested in will be substantial provided that the prices of the 'other' goods do not move in fixed proportions. For example, if we take Canadian labor markets to be the markets we are primarily interested in, then Woodland's [1972b] data indicate that the rental prices of machinery and equipment, engineering structures, and operating capital do not all vary in strict proportion (even approximately). Furthermore, Woodland's work was based on the first approximation that the prices of outputs and intermediate inputs varied in proportion.[13] One suspects that this first approximation has not been satisfied - the prices of intermediate inputs such as imports and energy have not been varying in proportion with the prices of various outputs (or with the prices of certain other primary inputs such as land and working inventories). Thus it would appear that we will be unable to adequately predict the demand for Canadian labor on a disaggregated basis without disaggregating the 'other inputs' series into 'reasonably' homogeneous groups.[14]

---

[12]This follows from our expression for the prediction error given by (12), $\sigma_m^{1^*} s^{1^*} \delta$. As we disaggregate labor markets (i.e., increase M), then the numbers $\sigma_m^{1^*}$ will tend to increase in absolute value, while s^{1^*} and δ remain fixed.

[13]It turns out that real value added may be used as a proxy for real gross output of an industry (without any prediction errors on labor markets due to the effects of aggregating outputs minus intermediate inputs) provided that the prices of outputs and intermediate inputs vary in strict proportion. This is another version of Hicks' Aggregation Theorem (see Diewert [1978a]).

[14]Just what is a 'reasonable' degree of disaggregation will depend on the application at hand. For example, if we wished to determine the effects on the demand for labor of a *selective* change in tariffs, then it would be necessary to disaggregate an imports inputs series into two imports series: tariff affected imports and nonaffected imports.

CONSOLIDATED REFERENCES FOR VOLUME I

Aczél, J., 1966. *Lectures on Functional Equations and their Applications*, New York: Academic Press.

Aczél, J. and Z. Daroézy, 1963. "Über verallgemeinerte quasilineare Mittelwerte, die mit Gewichtsfunktionen gebildat sind," *Publicationes Mathematicae Debrecen* 10, 171–190.

Afriat, S.N., 1967. "The Construction of Utility Functions from Expenditure Data," *International Economic Review* 8, 67–77.

Afriat, S.N., 1972a. "Efficiency Estimation of Production Function," *International Economic Review* 13, 568–598.

Afriat, S.N., 1972b. "The Theory of International Comparisons of Real Income and Prices." In *International Comparisons of Prices and Output*, D.J. Daly (ed.), National Bureau of Economic Research, New York: Columbia University Press, 13–69.

Afriat, S.N., 1972c. "The Case of the Vanishing Slutsky Matrix," *Journal of Economic Theory* 5, 208–223.

Afriat, S.N., 1973a. "The Maximum Hypothesis in Demand Analysis," Discussion Paper #79, Department of Economics, University of Waterloo, Canada.

Afriat, S.N., 1973b. "On Integrability Conditions for Demand Functions," Discussion Paper #82, Department of Economics, University of Waterloo, Canada.

Afriat, S.N., 1973c. "Direct and Indirect Utility," Discussion Paper #83, Department of Economics, University of Waterloo, Canada.

Afriat, S.N., 1973d. "On a System of Inequalities in Demand Analysis: An Extension of the Classical Method," *International Economic Review* 14, 460–472.

Afriat, S.N., 1977. *The Price Index*, London: Cambridge University Press.

Allais, M., 1953. "Le comportement de l'homme rationnel devant le risque: critique de postulats et axiomes de l'école américaine," *Econometrica* 21, 503–546.

Allen, R.C., 1979. "International Competition in Iron and Steel: 1850–1913," *Journal of Economic History* 29 (4), 911–937.

Allen, R.C., 1981. "Accounting for Price Changes: American Steel Rails, 1870–1913," *Journal of Political Economy* 89 (3), 512–528.

Allen, R.G.D., 1938. *Mathematical Analysis for Economists*, London: Macmillan.

Allen, R.G.D., 1949. "The Economic Theory of Index Numbers," *Economica N.S.* 16, 197–203.

Allen, R.G.D., 1975. *Index Numbers in Theory and Practice*, London: Macmillan.

Antonelli, G.B., 1971. "On the Mathematical Theory of Political Economy." In *Preferences, Utility and Demand*, J.S. Chipman, L. Hurwicz, M.K. Richter and H.F. Sonnenschein (eds.), New York: Harcourt Brace Jovanovich, 333–364.

Appelbaum, E., 1975. *Essays in the Theory and Application of Duality in Economics*, Ph.D. Thesis, University of British Columbia, Canada.

Appelbaum, E., 1979. "Testing Price Taking Behavior," *Journal of Econometrics* 9, 283–294.

Appelbaum, E. and U.J.R. Kohli, 1979. "Canada-U.S. Trade: Tests for the Small Open Economy Hypothesis," *Canadian Journal of Economics* 12, 1–14.

Archibald, R.B., 1977. "On the Theory of Industrial Price Measurement: Output Price Indexes," *Annals of Economic and Social Measurement* 6, 57–72.

Arrow, K.J., 1951. "Alternative Approaches to the Theory of Choice in Risk-Taking Situations," *Econometrica* 19, 404–437.

Arrow, K.J., 1953. "Le rôle des valeurs boursières pour la répartition la meilleure des risques." In *Econométrie*, Paris: Centre National de la Recherche

Scientifique, 41–48. Reprinted in English translation in 1964 as "The Role of Securities in the Optimal Allocation of Risk-Bearing," *The Review of Economic Studies* 31, 91–96.

Arrow, K.J., 1974. "The Measurement of Real Value Added." In *Nations and Households in Economic Growth: Essays in Honor of Moses Abramovitz*, P. David and M. Reder (eds.), New York: Academic Press.

Arrow, K.J., 1984a. "The Theory of Risk Aversion." In *Individual Choice under Certainty and Uncertainty*, K.J. Arrow (ed.), Cambridge, Mass.: Harvard University Press, 147–171. (Originally given in lecture form in 1963.)

Arrow, K.J., 1984b. "Exposition of the Theory of Choice under Uncertainty." In *Individual Choice under Certainty and Uncertainty*, K.J. Arrow (ed.), Cambridge, Mass.: Harvard University Press, 172–208. (An expanded version of a lecture originally given in 1963.)

Arrow, K.J., H.B. Chenery, B.S. Minhas and R.M. Solow, 1961. "Capital-Labor Substitution and Economic Efficiency," *Review of Economics and Statistics* 63, 225–250.

Arrow, K.J. and F.H. Hahn, 1971. *General Competitive Analysis*, San Francisco: Holden-Day.

Atkinson, A.B., 1970. "On the Measurement of Inequality," *Journal of Economic Theory* 2, 244–263.

Avriel, M., W.E. Diewert, S. Schaible and I. Zang, 1988. *Generalized Concavity*, New York: Plenum Press.

Balk, B.M., 1980. "A Method for Constructing Price Indices for Seasonal Commodities," *Journal of the Royal Statistical Society*, Series A 143, 68–75.

Balk, B.M., 1981. "Second Thoughts on Wald's Cost-of-Living Index and Frisch's Double Expenditure Method," *Econometrica* 49, 1553–1558.

Balk, B.M., 1985. "A Simple Characterization of Fisher's Price Index," *Statistische Hefte* 26, 59–63.

Balk, B.M., 1991. "On the First Step in the Calculation of a Consumer Price Index," Discussion Paper, Netherlands Central Bureau of Statisitcs, P.O. Box 959, 2270 AZ Voorgerg, The Netherlands.

Banerjee, K.S., 1975. *Cost of Living Index Numbers: Practice, Precision and Theory*, New York: Marcel Dekker.

Barnett, W.A., J. Geweke and M. Wolfe, 1991. "Seminonparametric Baysian Estimation of the Asymptotically Ideal Production Model," *Journal of Econometrics* 49, 5–50.

Barnett, W.A. and A.B. Jonas, 1983. "The Müntz-Szatz Demand System: An Application of Globally Well Behaved Series Expansion," *Economics Letters* 11, 337–342.

Barnett, W.A. and P. Yue, 1988. "Semiparametric Estimation of the Asymptotically Ideal Model: The AIM Demand System." In *Advances in Econometrics*, Vol. 7, G.F. Rhodes Jr. and T.B. Fomby (eds.), Greenwich, Conn.: JAI Press, 229–251.

Becker, G.S., 1965. "A Theory of the Allocation of Time," *Economic Journal* 75, 493–517.

Bennet, T.L., 1920. "The Theory of Measurement of Changes in Cost of Living," *Journal of the Royal Statistical Society* 83, 455–462.

Berge, C., 1963. *Topological Spaces*, New York: Macmillan.

Bergson (Burk), A., 1936. "Real Income, Expenditure Proportionality and Frisch's 'New Methods of Measuring Marginal Utility'," *Review of Economic Studies* 4, 33–52.

Bergson (Burk), A., 1961. *National Income of the Soviet Union since 1928*, Cambridge, Mass.: Harvard University Press.

Berlinguette, V.R. and F.H. Leacy, 1961. "The Estimation of Real Domestic Product by Final Expenditure Categories and by Industry of Origin in Canada." In *Output, Input and Productivity Measurement*, Studies in Income and Wealth, Vol. 25, National Bureau of Economic Research, Princeton, N.J.: Princeton University Press, 303–354.

Berndt, E.R. and L.R. Christensen, 1973a. "The Internal Structure of Functional Relationships: Separability, Substitution, and Aggregation," *Review of Economic Studies* 40, 403–410.

Berndt, E.R. and L.R. Christensen, 1973b. "The Translog Function and the Substitution of Equipment, Structures, and Labor in U.S. Manufacturing, 1929–1968," *Journal of Econometrics* 1, 81–114.

Berndt, E.R. and L.R. Christensen, 1974. "Testing for the Existence of a Consistent Aggregate Index of Labor Inputs," *American Economic Review* 64, 391–404.

Berndt, E.R., M.N. Darrough and W.E. Diewert, 1977. "Flexible Functional Forms and Expenditure Distributions: An Application to Canadian Consumer Demand Functions," *International Economic Review* 18, 651–676.

Berndt, E.R. and D.O. Wood, 1975. "Technology, Prices and the Derived Demand for Energy," *Review of Economics and Statistics*, 57, 259–268.

Bernoulli, D., 1738. "Specimen Theoriae Novae de Mensura Sortis," *Commentarii Academiae Scientiarum Imperialis Petropolitanae* 5, 175–192. Reprinted in English translation by L. Sommer in 1954 as "Exposition of a New Theory on the Measurement of Risk," *Econometrica* 22, 23–36.

Binswanger, H.P., 1974. "The Measurement of Technical Change Biases with Many Factors of Production," *American Economic Review* 64, 964–976.

Blackorby, C., R. Boyce and R.R. Russell, 1978. "Estimation of Demand Systems Generated by the Gorman Polar Form: A Generalization of the S-Branch Utility Tree," *Econometrica* 46, 345–364.

Blackorby, C., R. Davidson and D. Donaldson, 1977. "A Homiletic Exposition of the Expected Utility Hypothesis," *Economica* 44, 351–358.

Blackorby, C., R. Davidson and W. Schworm, 1991. "Implicit Separability: Characterisation and Implications for Consumer Demands," *Journal of Economic Theory* 55, 364–399.

Blackorby, C. and W.E. Diewert, 1979. "Expenditure Functions, Local Duality, and Second Order Approximations," *Econometrica* 47, 579–601.

Blackorby, C. and D. Donaldson, 1978. "Measures of Relative Equality and Their Meaning in Terms of Social Welfare," *Journal of Economic Theory* 18, 59–80.

Blackorby, C. and D. Donaldson, 1980. "A Theoretical Treatment of Indices of Absolute Inequality," *International Economic Review* 21, 107–136.

Blackorby, C. and D. Donaldson, 1981. "Ethical Indices for the Measurement of Poverty," *Econometrica* 22, 1053–1060.

Blackorby, C. and D. Donaldson, 1982. "Ratio-Scale and Translation-Scale Full Interpersonal Comparability without Domain Restrictions: Admissible Social Evaluation Functions," *International Economic Review* 23, 249–268.

Blackorby, C. and D. Donaldson, 1984. "Social Criteria for Evaluating Population Change," *Journal of Public Economics* 25, 13–33.

Blackorby, C., D. Donaldson and M. Auersperg, 1981. "A New Procedure for the Measurement of Inequality Within and Among Population Subgroups," *Canadian Journal of Economics* 14, 664–685.

Blackorby, C., C.A.K. Lovell and M. Thursby, 1976. "Extended Hicks Neutral Technical Change," *Economic Journal* 86, 845–852.

Blackorby, C. and D. Primont, 1980. "Index Numbers and Consistency in Aggregation," *Journal of Economic Theory* 22, 87–98.

Blackorby, C., D. Primont and R.R. Russell, 1975a. "Budgeting, Decentralization, and Aggregation," *Annals of Economic and Social Measurement* 4, 23–44.

Blackorby, C., D. Primont and R.R. Russell, 1975b. "Some Simple Remarks on Duality and the Structure of Utility Functions," *Journal of Economic Theory* 11, 155–160.

Blackorby, C., D. Primont and R.R. Russell, 1977a. "Dual Price and Quantity Aggregation," *Journal of Economic Theory* 14, 130–148.

Blackorby, C., D. Primont and R.R. Russell, 1977b. "An Extension and Alternative Proof of Gorman's Price Aggregation Theorem." In *Theory and Applications of Economic Indices*, W. Eichhorn (ed.), Würstburg-Wien: Physica.

Blackorby, C., D. Primont and R.R. Russell, 1977c. "On Testing Separability Restrictions with Flexible Functional Forms," *Journal of Econometrics* 5, 195–209.

Blackorby, C., D. Primont and R.R. Russell, 1977d. "Separability vs. Functional Structure: A Characterization of Their Differences," *Journal of Economic Theory* 15, 135–144.

Blackorby, C., D. Primont and R.R. Russell, 1978. *Duality, Separability and Functional Structure: Theory and Economic Applications*, New York: North-Holland.

Blackorby, C. and R.R. Russell, 1976. "Functional Structure and the Allen Partial Elasticities of Substitution: An Application of Duality Theory," *Review of Economic Studies* 43, 285–292.

Blackorby, C. and R.R. Russell, 1978. "Indices and Subindices of the Cost of Living and the Standard of Living," *International Economic Review* 19, 229–240.

Blackorby, C. and W. Schworm, 1984. "The Structure of Economics with Aggregate Measures of Capital: A Complete Characterization," *Review of Economic Studies* 51, 633–650.

Blackorby, C., W. Schworm and T. Fisher, 1986. "Testing for the Existence of Input Aggregates in an Economy Production Function," CORE Discussion Paper 8646, Université Catholique de Louvain, November.

Bonnesen, T. and W. Fenchel, 1934. *Theorie der Konvexen Körper*, Berlin: Springer.

Bossert, W. and A. Pfingsten, 1990. "Intermediate Inequality: Concepts, Indices and Welfare Implications," *Mathematical Social Sciences* 19, 117–134.

Bowers, N.L., H.U. Gerber, J.C. Hickman, D.A. Jones and C.J. Nesbitt, 1986. *Actuarial Mathematics*, Itasca, Illinois: The Society of Actuaries.

Bowley, A.L., 1899. "Wages, Nominal and Real." In *Dictionary of Political Economy*, Vol. 3, R.H.I. Palgrave (ed.), London: Macmillan, 640–651.

Bowley, A.L., 1901. *Elements of Statistics*, Westminster, England: P.S. King and Son.

Bowley, A.L., 1919. "The Measurement of Changes in the Cost of Living," *Journal of the Royal Statistical Society* 82, 343–372.

Bowley, A.L., 1921. "An Index of the Physical Volume of Production," *Economic Journal* 31, 196–205.

Bowley, A.L., 1923. "Review of the Making of Index Numbers," *Economic Journal* 33, 90–94.

Bowley, A.L., 1928. "Notes on Index Numbers," *Economic Journal* 38, 216–237.

Bowley, A.L., 1938. "Notes on Professor Frisch's 'The Problem of Index Numbers ,'" *Econometrica* 6, 83–84.

Boyce, R., 1975. "Estimation of Dynamic Gorman Polar Form Utility Functions," *Annals of Economic and Social Measurement* 4, 103–116.

Boyce, R. and D. Primont, 1976. "An Econometric Test of the Representative Consumer Hypothesis," Discussion Paper 76-31, Department of Economics, University of British Columbia, Canada.

Braithwait, S.D., 1980. "The Substitution Bias of the Laspeyres Price Index," *American Economic Review* 70, 64–77.

Brown, R.S., D.W. Caves and L.R. Christensen, 1979. "Modelling the Structure of Cost and Production for Multiproduct Firms," *Southern Economic Journal* 46, 256–270.

Bruno, M., 1978. "Duality, Intermediate Inputs and Value-Added." In *Production Economics: A Dual Approach to Theory and Applications*, Vol. 2, M. Fuss and D. McFadden (eds.), Amsterdam: North-Holland, 3–16.

Bureau of Labor Statistics, 1983. *Trends in Multifactor Productivity 1948–81*, Bulletin 2178, Washington, D.C.

Burgess, D.F., 1974. "Production Theory and the Derived Demand for Imports," *Journal of International Economics* 4, 103–117.

Burgess, D.F., 1975. "Duality Theory and Pitfalls in the Specification of Technologies," *Journal of Econometrics* 3, 105–121.

Burgess, D.F., 1976a. "Tariffs and Income Distribution: Some Empirical Evidence for the United States," *Journal of Political Economy* 84, 17–45.

Burgess, D.F., 1976b. "The Income Distributional Effects of Processing Incentives: A General Equilibrium Analysis," *Canadian Journal of Economics* 9, 595–612.

Byron, R.P., 1977. "Some Monte Carlo Experiments Using the Translog Approximation," Discussion Paper, Development Research Center, World Bank, Washington, D.C., 20433.

Byushgens (Buscheguennce), S.S., 1925. "Ob odnom klasse giperpoverkhnostey: po povodu 'idealnovo indeksa' Irving Fischer, a pokupatelnoi sili deneg"

(French translation of Russian title: "Sur une classe des hypersurfaces: A propos de 'l'index ideal' de M. Irving Fischer"), *Mathematischeskii Sbornik* 32, 625–631.

Carli, G.R., 1764. "Del valore e della proporzione de'metalli monetati con i generi in Italia prima delle scoperte dell'Indie colonfronto del valore e della proporzione de'tempi nostri." In *Opere scelte di Carli*, Vol. 1, Milan, 299–366.

Carlson, B.C., 1972. "The Logarithmic Mean," *The American Mathematical Monthly* 79, 615–618.

Caves, D.W., L.R. Christensen, and W.E. Diewert, 1982a. "Multilateral Comparisons of Output, Input and Productivity Using Superlative Index Numbers," *Economic Journal* 92, 73–86.

Caves, D.W., L.R. Christensen, and W.E. Diewert, 1982b. "The Economic Theory of Index Numbers and the Measurement of Input, Output and Productivity," *Econometrica* 50, 1393–1414.

Chew, S.H., 1989. "Axiomatic Utility Theories with the Betweenness Property," *Annals of Operations Research* 19, 273–298.

Chew, S.H. and L.G. Epstein, 1989a. "A Unifying Approach to Axiomatic Non-Expected Utility Theories," *Journal of Economic Theory* 49, 207–240.

Chew, S.H. and L.G. Epstein, 1989b. "Axiomatic Rank-Dependent Means," *Annals of Operations Research* 19, 299–309.

Chew, S.H. and L.G. Epstein, 1992. "A Unifying Approach to Axiomatic Non-Expected Utility Theories: Corrigenda," *Journal of Economic Theory*, forthcoming.

Chipman, J.S., 1966. "A Survey of the Theory of International Trade: Part 3, The Modern Theory," *Econometrica* 34, 18–76.

Chipman, J.S., 1970. "Lectures on the Mathematical Foundations of International Trade Theory: I Duality of Cost and Production Functions," Discussion Paper, Institute of Advanced Studies, April-May, Vienna.

Chipman, J.S., 1972. "The Theory of Exploitative Trade and Investment Policies: A Reformulation and Synthesis." In *International Economics and Development: Essays in Honor of Raul Prebisch*, L. Eugenio Di Marco (ed.), New York: Academic Press.

Chipman, J.S., 1974a. "Homothetic Preferences and Aggregation," *Journal of Economic Theory* 8, 26–38.

Chipman, J.S., 1974b. "The Transfer Problem Once Again." In *Trade, Stability and Macroeconomics: Essays in Honor of Lloyd A. Mezler*, New York: Academic Press, 19–78.

Chipman, J.S. and J.C. Moore, 1976. "The Scope of Consumer's Surplus Arguments." In *Evolution, Welfare and Time in Economics*, A.M. Tang, F.M. Westfield, and J.S. Worley (eds.), Lexington, Mass.: D.C. Heath.

Christensen, L.R., D. Cummings and D.W. Jorgenson, 1980. "Economic Growth, 1947–1973: An International Comparison." In *New Developments in Productivity Measurement and Analysis*, J.W. Kendrick and B.N. Vaccara (eds.), Studies in Income and Wealth, Vol. 44, National Bureau of Economic Research, Chicago: University of Chicago Press, 595–691.

Christensen, L.R. and W.H. Greene, 1976. "Economies of Scale in U.S. Electric Power Generation," *Journal of Political Economy* 84, 655–676.

Christensen, L.R. and D.W. Jorgenson, 1969. "The Measurement of U.S. Real Capital Input, 1929–1967," *Review of Income and Wealth* 15, 293–320.

Christensen, L.R. and D.W. Jorgenson, 1970. "U.S. Real Product and Real Factor Input, 1929–1967," *Review of Income and Wealth* 16, 19–50.

Christensen, L.R., D.W. Jorgenson and L.J. Lau, 1971. "Conjugate Duality and the Transcendental Logarithmic Production Function," *Econometrica* 39, 255–256.

Christensen, L.R., D.W. Jorgenson and L.J. Lau, 1975. "Transcendental Logarithmic Utility Functions," *American Economic Review* 65, 367–383.

Christensen, L.R. and M. Manser, 1975. "Cost-of-Living Indexes and Price Indexes for U.S. Meat and Produce, 1947–1971." In *Household Production and Consumption*, N. Terleckyj (ed.), Studies in Income and Wealth, Vol. 40, National Bureau of Economic Research, New York: Columbia University Press, 399–446.

Christensen, L.R. and M. Manser, 1977. "Estimating U.S. Consumer Preferences for Meat with a Flexible Utility Function," *Journal of Econometrics* 6, 37–53.

Cobb, C. and P.H. Douglas, 1928. "A Theory of Production," *American Economic Review* 18, 139–165.

Cooper, R.J. and K.R. McLaren, 1977. "Intertemporal Duality: Application to Consumer Theory," Working Paper 10, Department of Econometrics and Operations Research, Monash University, Australia.

Corden, W.M., 1969. "Effective Protective Rates in the General Equilibrium Model: A Geometrical Note," Oxford Economic Papers 21(2), 135–141.

Court, L.M. and H.E. Lewis, 1942–43. "Production Cost Indices," *Review of Economic Studies* 10, 28–42.

Cummings, E.D. and L. Meduna, 1973. "The Canadian Consumer Accounts," Department of Manpower and Immigration, Ottawa.

Dalton, H., 1920. "The Measurement of Inequality of Incomes," *Economic Journal* 30, 348–361.

Danskin, J.M., 1967. *The Theory of Max-Min*, New York: Springer.

Darrough, M.N., 1977. "A Model of Consumption and Leisure in an Intertemporal Framework: A Systematic Treatment Using Japanese Data," *International Economic Review* 18, 677–696.

Darrough, M.N. and C. Southey, 1977. "Duality in Consumer Theory Made Simple: The Revealing of Roy's Identity," *Canadian Journal of Economics* 10, 307–317.

Dasgupta, P., A. Sen and D. Starrett, 1973. "Notes on the Measurement of Inequality," *Journal of Economic Theory* 6, 180–204.

David, P.A., 1966. "Measuring Real Net Output: A Proposed Index," *Review of Economics and Statistics* 48(4), 419–425.

Davies, G.R., 1924. "The Problem of a Standard Index Number Formula," *Journal of the American Statistical Association* 19, 180–188.

Deaton, A., 1979. "The Distance Function in Consumer Behavior with Applications to Index Numbers and Optimal Taxation," *Review of Economic Studies* 46, 391–405.

Deaton, A. and J. Muellbauer, 1980. *Economics and Consumer Behavior*, Cambridge, England: Cambridge University Press.

Debreu, G., 1951. "The Coefficient of Resource Utilization," *Econometrica* 19, 273–292.

Debreu, G., 1952. "A Social Equilibrium Existence Theorem," *Proceedings of the National Academy of Sciences* 38, 886–893.

Debreu, G., 1959. *Theory of Value*, New York: John Wiley and Sons.

Dekel, E., 1986. "An Axiomatic Characterization of Preferences under Uncertainty: Weakening the Independence Axiom," *Journal of Economic Theory* 40, 304–318.

Denny, M., 1972. *Trade and the Production Sector: An Exploration of Models of Multi-Product Technologies*, Ph.D. Dissertation, University of California at Berkeley.

Denny, M., 1974. "The Relationship Between Functional Forms for the Production System," *Canadian Journal of Economics* 7, 21–31.

Denny, M., 1980. "Comment on Aggregation Problems in the Measurement of Capital." In *The Measurement of Capital*, D. Usher (ed.), Studies in Income and Wealth, Vol. 45, National Bureau of Economic Research, Chicago: University of Chicago Press, 528–538.

Denny, M. and M. Fuss, 1977. "The Use of Approximation Analysis to Test for Separability and the Existence of Consistent Aggregates," *American Economic Review* 67, 404–418.

Denny, M. and M. Fuss, 1983. "A General Approach to Intertemporal and Interspatial Productivity Comparisons," *Journal of Econometrics* 23, 315–330.

Diamond, P.A. and D.L. McFadden, 1974. "Some Uses of the Expenditure Function in Public Finance," *Journal of Public Economics* 3, 3–22.

Diewert, W.E., 1969. *Functional Form in the Theory of Production and Consumer Demand*, Ph.D. Dissertation, University of California at Berkeley.

Diewert, W.E., 1971a. "An Application of the Shephard Duality Theorem: A Generalized Leontief Production Function," *Journal of Political Economy* 79, 481–507.

Diewert, W.E., 1971b. "Choice on Labour Markets and the Theory of the Allocation of Time," Research Branch, Department of Manpower and Immigration, Ottawa.

Diewert, W.E., 1973a. "Functional Forms for Profit and Transformation Functions," *Journal of Economic Theory* 6, 284–316.

Diewert, W.E., 1973b. "Afriat and Revealed Preference Theory," *Review of Economic Studies* 40, 419–426.

Diewert, W.E., 1974a. "Applications of Duality Theory." In *Frontiers of Quantitative Economics*, Vol. II, M.D. Intriligator and D.A. Kendrick (eds.), Amsterdam: North-Holland, 106–171.

Diewert, W.E., 1974b. "Functional Forms for Revenue and Factor Requirements Functions," *International Economic Review* 15, 119–130.

Diewert, W.E., 1974c. "A Note on Aggregation and Elasticities of Substitution," *Canadian Journal of Economics* 7, 12–20, and reprinted as Ch. 16 in THIS VOLUME, 471–479.

Diewert, W.E., 1974d. "Intertemporal Consumer Theory and the Demand for Durables," *Econometrica* 42, 497–516.

Diewert, W.E., 1974e. "The Effects of Unionization on Wages and Employment: A General Equilibrium Analysis," *Journal of Economic Inquiry* 12, 319–339.

Diewert, W.E., 1974f. "Unions in a General Equilibrium Model," *Canadian Journal of Economics* 7, 475–495.

Diewert, W.E., 1976a. "Exact and Superlative Index Numbers," *Journal of Econometrics* 4, 114–145, and reprinted as Ch. 8 in THIS VOLUME, 223–252.

Diewert, W.E., 1976b. "Harberger's Welfare Indicator and Revealed Preference Theory," *American Economic Review* 66, 143–152, and reprinted in Diewert and Nakamura [1993].

Diewert, W.E., 1976c. "On Symmetry Conditions for Market Demand Functions: A Review and Some Extensions," Discussion Paper, Department of Economics, University of British Columbia, January.

Diewert, W.E., 1977. "Generalized Slutsky Conditions for Aggregate Consumer Demand Functions", *Journal of Economic Theory* 15, 353–362.

Diewert, W.E., 1978a. "Hicks' Aggregation Theorem and the Existence of a Real Value-Added Function." In *Production Economics: A Dual Approach to Theory and Applications* Vol. 2, M. Fuss and D. McFadden (eds.), Amsterdam: North-Holland, 17–51, and reprinted as Ch. 15 in THIS VOLUME, 435–470.

Diewert, W.E., 1978b. "Superlative Index Numbers and Consistency in Aggregation," *Econometrica* 46, 883–900, and reprinted as Ch. 9 in THIS VOLUME, 253–273.

Diewert, W.E., 1978c. "Duality Approaches to Microeconomic Theory," IMSSS Technical Report No. 281, Stanford University. An abridged version was published as Diewert [1982]. A more complete abridged version appears as Ch. 6 in THIS VOLUME, 105–175.

Diewert, W.E., 1978d. "Optimal Tax Perturbations," *Journal of Public Economics* 10, 139–177.

Diewert, W.E., 1980. "Aggregation Problems in the Measurement of Capital." In *The Measurement of Capital*, D. Usher (ed.), Studies in Income and Wealth, Vol. 45, National Bureau of Economic Research, Chicago: University of Chicago Press, 433–528, and reprinted in Diewert and Nakamura [1993].

Diewert, W.E., 1981a. "The Economic Theory of Index Numbers: A Survey." In *Essays in the Theory and Measurement of Consumer Behaviour in Honour of Sir Richard Stone*, A. Deaton (ed.), London: Cambridge University Press, 163–208, and reprinted as Ch. 7 in THIS VOLUME, 177–221.

Diewert, W.E., 1981b. "Symmetry Conditions for Market Demand Functions," *Review of Economic Studies* 47, 595–601.

Diewert, W.E., 1982. "Duality Approaches to Microeconomic Theory." In *Handbook of Mathematical Economics*, Vol. II, K.J. Arrow and M.D. Intriligator (eds.), Amsterdam: North-Holland, 535–599. This is an abridged version of Diewert [1978c]. A more complete abridged version is published as Ch. 6 in THIS VOLUME, 105–175.

Diewert, W.E., 1983a. "The Theory of the Cost-of-Living Index and the Measurement of Welfare Change." In Diewert and Montmarquette [1983; 163–233], reprinted in Diewert [1990; 79–147] and in Diewert and Nakamura [1993].

Diewert, W.E., 1983b. "The Theory of the Output Price Index and the Measurement of Real Output Change." In Diewert and Montmarquette [1983; 1049–1113], and reprinted in Diewert and Nakamura [1993].

Diewert, W.E., 1983c. "The Treatment of Seasonality in a Cost-of-Living Index." In Diewert and Montmarquette [1983; 1019–1045], and reprinted in Diewert and Nakamura [1993].

Diewert, W.E., 1986. Microeconomic Approaches to the Theory of International Comparisons. Technical Working Paper 53, National Bureau of Economic Research, Cambridge, Massachusetts.

Diewert, W.E., 1987. "Index Numbers." In *The New Palgrave: A Dictionary of Economics*, Vol. 2, J. Eatwell, M. Milgate and P. Newman (eds.), London: Macmillan, 767–780, and reprinted as Ch. 5 in THIS VOLUME, 71–104.

Diewert, W.E., 1988. "Test Approaches to International Comparisons." In *Measurement in Economics*, W. Eichhorn (ed.), Heidelberg: Physica-Verlag 67–86, and reprinted as Ch. 12 in THIS VOLUME, 305–316.

Diewert, W.E. (ed.), 1990. *Price Level Measurement*, Amsterdam: North-Holland.

Diewert, W.E., 1992a. "The Measurement of Productivity," *Bulletin of Economic Research* 44 (3), 163–198, and reprinted in Diewert and Nakamura [1993].

Diewert, W.E., 1992b. "Exact and Superlative Welfare Change Indicators," *Economic Inquiry* 30, 1–18, and reprinted in Diewert and Nakamura [1993].

Diewert, W.E., M. Avriel and I. Zang, 1981. "Nine Kinds of Quasiconcavity and Concavity," *Journal of Economic Theory* 25, 397–420.

Diewert, W.E. and J. Bossons, 1992. "Adjusting the Consumer Price Index for Changes in Taxes." In Diewert and Nakamura [1993].

Diewert, W.E. and C. Montmarquette, 1983. *Price Level Measurement: Proceedings from a conference sponsored by Statistics Canada*, Ottawa: Statistics Canada.

Diewert, W.E. and A.O. Nakamura, 1993. *Essays in Index Number Theory*, Vol. II, Amsterdam: North-Holland, forthcoming.

Diewert, W.E. and C. Parkan, 1978. "Tests for the Consistency of Consumer Data and Nonparametric Index Numbers," Discussion Paper 78-27, Department of Economics, University of British Columbia, Canada.

Diewert, W.E. and C. Parkan, 1983. "Linear Programming Tests of Regularity Conditions for Production Functions." In *Quantitative Studies on Production and Prices*, W. Eichhorn, R. Henn, K. Neumann and R.W. Shephard (eds.), Vienna: Physica-Verlag, 131–158.

Diewert, W.E. and T.J. Wales, 1988. "A Normalized Quadratic Semiflexible Functional Form," *Journal of Econometrics* 37, 327–342.

Diewert, W.E. and T.J. Wales, 1991. "Flexible Functional Forms and Tests of Homogeneous Separability," Discussion Paper 91-12, Department of Economics, University of British Columbia, Vancouver, Canada, April.

Diewert, W.E. and T.J. Wales, 1992a. "Linear and Quadratic Spline Models for Consumer Demand Functions," *Canadian Journal of Economics*, forthcoming.

Diewert, W.E. and T.J. Wales, 1992b. "Quadratic Spline Models for Producer Supply and Demand Functions," *International Economic Review*, forthcoming.

Diewert, W.E. and A.D. Woodland, 1977. "Frank Knight's Theorem in Linear Programming Revisited," *Econometrica* 45, 375–398.

Divisia, F., 1925. "L'indice monétaire et la théorie de la monnaie," published in three parts in *Revue d'Économie Politique* 39, 842–861, 980–1008, and 1121–1151.

Divisia, F., 1926. "L'indice monétaire et la théorie de la monnaie," Paris: Sociéte anonyme du Recueil Sirey. Published as Divisia [1925].

Domar, E.D., 1961. "On the Measurement of Technological Change," *Economic Journal* 71, 709–729.

Donovan, D.J., 1977. *Consumption, Leisure, and the Demand for Money and Money Substitutes*, Ph.D. Thesis, University of British Columbia, Canada.

Drobisch, M.W., 1871. "Uber die Berechnung der Veranderungen der Waarenpreise und des Geldswerths," Jahrbucher für Nationalökonomie und Statistik 16, 143–156.

Dutot, C., 1738. *Réflexions politiques sur les finances et le commerce*, The Hague.

Ebert, U., 1988. "Measurement of Inequality: An Attempt at Unification and Generalization," *Social Choice and Welfare* 5, 147–169.

Edgeworth, F.Y., 1888. "Some New Methods of Measuring Variation in General Prices," *Journal of the Royal Statistical Society* 51, 346–368.

Edgeworth, F.Y., 1896. "A Defence of Index-Numbers," *Economic Journal* 6, 132–142.

Edgeworth, F.Y., 1901. "Mr. Walsh on the Measurement of General Exchange Value," *Economic Journal* 11, 404–416.

Edgeworth, F.Y., 1923. "The Doctrine of Index-Numbers According to Mr. Correa Walsh," *Economic Journal* 33, 343–351.

Edgeworth, F.Y., 1925. *Papers Relating to Political Economy*, Vol. 1, New York: Burt Franklin.

Eichhorn, W., 1973. "Zur Axiomatischen Theorie des Preisindex," *Demostratio Matematica* 6, 561–573.

Eichhorn, W., 1976. "Fisher's Tests Revisited," *Econometrica* 44, 247–256.

Eichhorn, W., 1978a. "What is an Economic Index? An Attempt of an Answer." In *Theory and Economic Applications of Economic Indices*, W. Eichhorn, R. Henn, O. Optiz and R.W. Shephard (eds.), Würzburg: Physica-Verlag, 3–42.

Eichhorn, W., 1978b. *Functional Equations in Economics*, Reading, Mass.: Addison-Wesley.

Eichhorn, W. and W. Gehrig, 1982. "Measurement of Inequality in Economics." In *Modern Applied Mathematics: Optimization and Operations Research*, B. Korte (ed.), Amsterdam: North-Holland, 657–693.

Eichhorn, W. and J. Voeller, 1976. *Theory of the Price Index: Fisher's Test Approach and Generalizations*, Lecture Notes in Economics and Mathematical Systems, Vol. 140, Berlin: Springer-Verlag.

Eichhorn, W. and J. Voeller, 1983. "The Axiomatic Foundations of Price Indexes and Purchasing Power Parities." In Diewert and Montmarquette [1983; 411–450], and reprinted in Diewert [1990; 321–358].

Eltetö, O. and P. Köves, 1964. "On a Problem of Index Number Computation Relating to International Comparison," *Statisztikai Szemle* 42, 507–518.

Epstein, L.G., 1974. "Some Economic Effects of Immigration: A General Equilibrium Analysis," *Canadian Journal of Economics* 7, 174–190.

Epstein, L.G., 1975. "A Disaggregate Analysis of Consumer Choice under Uncertainty," *Econometrica* 43, 877–892.

Epstein, L.G., 1977. *Essays in the Economics of Uncertainty*, Ph.D. Thesis, University of British Columbia, Canada.

Epstein, L.G., 1978. "The Le Chatelier Principle in Optimal Control Problems," *Journal of Economic Theory* 19, 103–122.

Epstein, L.G., 1981a. "Generalized Duality and Integrability," *Econometrica* 49, 655–678.

Epstein, L.G., 1981b. "Duality Theory and Functional Forms for Dynamic Factor Demand," *Review of Economic Studies* 48, 81–95.

Epstein, L.G., 1992. "Behavior under Risk: Recent Developments in Theory and Applications." In *Advances in Economic Theory*, J.J. Laffont (ed.), New York: Cambridge University Press.

Epstein, L.G. and S.E. Zin, 1990. "First-Order Risk Aversion and the Equity Premium Puzzle," *Journal of Monetary Economics* 26, 387–407.

Ethier, W., 1974. "Some of the Theorems of International Trade with Many Goods and Factors," *Journal of International Economics* 4, 199–206.

Farrell, M.J., 1957. "The Measurement of Productive Efficiency," *Journal of the Royal Statistical Society* (Series A) 120, 253–281.

Fenchel, W., 1953. "Convex Cones, Sets and Functions," Lecture Notes, Department of Mathematics, Princeton University.

Ferger, W.F., 1946. "Historical Note on the Purchasing Power Concept and Index Numbers," *Journal of the American Statistical Association* 41, 53–57.

Fisher, F.M., 1965. "Embodied Technical Change and the Existence of an Aggregate Capital Stock," *Review of Economic Studies* 32, 263–288.

Fisher, F.M. and K. Shell, 1972a. "The Pure Theory of the National Output Deflator." In *The Economic Theory of Price Indices*, F.M. Fisher and K. Shell (eds.), New York: Academic Press, 49–113.

Fisher, F.M. and K. Shell, 1972b. *The Economic Theory of Price Indexes*, New York: Academic Press.

Fisher, I., 1911. *The Purchasing Power of Money*, London: Macmillan.

Fisher, I., 1921. "The Best Form of Index Number," *Journal of the American Statistical Association* 17, 535–537.

Fisher, I., 1922. *The Making of Index Numbers*, Boston: Houghton Mifflin.

Fisher, W.C., 1913. "The Tabular Standard in Massachusetts History," *Quarterly Journal of Economics* 27, 417–451.

Friedman, J.W., 1972. "Duality Principles in the Theory of Cost and Production Revisited," *International Economic Review* 13, 167–170.

Friedman, M., 1938. "Discussion." In *Studies in Income and Wealth*, Vol. 2, Conference on Research in National Income and Wealth, National Bureau of Economic Research, M. Friedman (ed.), Baltimore: Waverly Press, 123–130.

Friedman, M. and L.J. Savage, 1948. "The Utility Analysis of Choices Involving Risk," *Journal of Political Economy* 56, 279–304.

Frisch, R., 1930. "Necessary and Sufficient Conditions Regarding the Form of an Index Number Which Shall Meet Certain of Fisher's Tests," *American Statistical Association Journal* 25, 397–406.

Frisch, R., 1936. "Annual Survey of General Economic Theory: The Problem of Index Numbers," *Econometrica* 4, 1–39.

Funke, H., 1988. "Mean Value Properties of the Weights of Linear Price Indices." In *Measurement in Economics: Theory and Applications of Economic Indices*, W. Eichhorn (ed.), Heidelberg: Physica-Verlag, 99–115.

Funke, H. and J. Voeller, 1978. "A Note on the Characterization of Fisher's Ideal Index." In *Theory and Applications of Economic Indices*, W. Eichhorn,

R. Henn, O. Opitz and R.W. Shephard (eds.), Würzburg: Physica-Verlag, 177–181.

Funke, H. and J. Voeller, 1979. "Characterization of Fisher's 'Ideal Index' by Three Reversal Tests," *Statistiche Hefte* 20, 54–60.

Fuss, M. and D. McFadden, 1978. *Production Economics: A Dual Approach to Theory and Applications*, Amsterdam: North-Holland.

Gallant, A.R., 1981. "On the Bias in Flexible Functional Forms and an Essentially Unbiased Form: The Fourier Flexible Form," *Journal of Econometrics* 15, 211–245.

Gallant, A.R., 1982. "Unbiased Determination of Production Technologies," *Journal of Econometrics* 20, 285–323.

Geary, P.T. and M. Morishima, 1973. "Demand and Supply Under Separability." In *Theory of Demand: Real and Monetary*, M. Morishima (ed.), Oxford: Clarendon, 87–147.

Geary, R.C., 1944. "The Concept of Net Volume of Output, with Special Reference to Irish Data," *Journal of the Royal Statistical Society, N.S.* 107, 251–259.

Geary, R.C., 1958. "A Note on Comparisons of Exchange Rates and Purchasing Power Between Countries," *Journal of the Royal Statistical Society*, Series A 121, 97–99.

Gini, C., 1931. "On the Circular Test of Index Numbers," *Metron* 9 (2), 3–24.

Goldberger, A.S. and T. Gamaletsos, 1970. "A Cross-Country Comparison of Consumer Expenditure Patterns," *European Economic Review* 1, 357–400.

Goldman, S.M. and H. Uzawa, 1964. "A Note on Separability in Demand Analysis," *Econometrica* 32, 387–398.

Gorman, W.M., 1953. "Community Preference Fields," *Econometrica* 21, 63–80.

Gorman, W.M., 1959. "Separable Utility and Aggregation," *Econometrica* 27, 469–481.

Gorman, W.M., 1968a. "The Structure of Utility Functions," *Review of Economic Studies* 35, 367–390.

Gorman, W.M., 1968b. "Measuring the Quantities of Fixed Factors." In *Value, Capital and Growth: Papers in Honour of Sir John Hicks*, J.N. Wolfe (ed.), Chicago: Aldine, 141–172.

Gorman, W.M., 1970. "Notes on Divisia Indices," Discussion Paper, Department of Economics, University of North Carolina, Chapel Hill, October.

Gorman, W.M., 1976. "Tricks with Utility Functions." In *Essays in Economic Analysis*, M. Artis and R. Nobay (eds.), Cambridge: Cambridge University Press.

Green, H.A.J., 1964. *Aggregation in Economic Analysis: An Introductory Survey*, Princeton, N.J.: Princeton University Press.

Green, J. and W.P. Heller, 1981. "Mathematical Analysis and Convexity with Applications to Economics." In *Handbook of Mathematical Economics*, Vol. I, K.J. Arrow and M.D. Intriligator (eds.), Amsterdam: North-Holland, 15–52.

Griliches, Z., 1990. "Hedonic Price Indexes and the Measurement of Capital and Productivity: Some Historical Reflections." In *Fifty Years in Economic Measurement* 54, E.R. Berndt and J.E. Triplett (eds.), Chicago: University of Chicago Press, 185–202.

Griliches, Z. and V. Ringstad, 1971. *Economies of Scale and the Form of the Production Function*, Amsterdam: North-Holland.

Gussman, T., 1972. "The Demand for Durables, Nondurables, Services and the Supply of Labour in Canada: 1946–1969," Department of Manpower and Immigration, Strategic Planning and Research, Ottawa.

Haberler, G. von, 1927. *Der Sinn der Indexzahlen*, Tübingen: J.C.B. Mohr.

Hall, R.E., 1973. "The Specification of Technology with Several Kinds of Output," *Journal of Political Economy* 81, 878–892.

Hanoch, G., 1970. "Generation of New Production Functions through Duality," Discussion Paper, Harvard Institute of Economic Research.

Hanoch, G., 1975. "Production or Demand Models with Direct or Indirect Implicit Additivity," *Econometrica* 43, 395–420.

Hanoch, G., 1977. "Risk Aversion and Consumer Preferences," *Econometrica* 45(2), 413–426.

Hanoch, G., 1978a. "Symmetric Duality and Polar Production Functions." In *Production Economics: A Dual Approach to Theory and Applications*, Vol. 1, M. Fuss and D. McFadden (eds.), Amsterdam: North-Holland, 111–131.

Hanoch, G., 1978b. "Generation of New Production Functions through Duality." In *Production Economics: A Dual Approach to Theory and Application*, M. Fuss and D. McFadden (eds.), Amsterdam: North-Holland.

Hanoch, G. and M. Rothschild, 1972. "Testing the Assumptions of Production Theory: A Nonparametric Approach," *Journal of Political Economy* 80, 256–275.

Harberger, A.C., 1971. "Three Basic Postulates for Applied Welfare Economics: An Interpretive Essay," *Journal of Economic Literature* 9, 785–797.

Hardy, G.H., J.E. Littlewood, and G. Polya, 1934. *Inequalities*, Cambridge: Cambridge University Press.

Harper, M.J., E.R. Berndt and D.O. Wood, 1989. "Rates of Return and Capital Aggregation Using Alternative Rental Prices." In *Technology and Capital Formation*, D.W. Jorgenson and R. Landau (eds.), Cambridge, Mass.: MIT Press, 331–372.

Harris, R. and E. Appelbaum, 1977. "Estimating Technology in an Intertemporal Framework: A Neo-Austrian Approach," *Review of Economics and Statistics* 59, 161–170.

Hasenkamp, G., 1973. *Specification and Estimation of Multiple Output Production Functions*, Ph.D. Dissertation, University of Wisconsin at Madison.

Hause, J.C., 1975. "The Theory of Welfare Cost Measurement," *Journal of Political Economy* 83, 1145–1182.

Hicks, J.R., 1940. "The Valuation of the Social Income," *Economica* 7, 105–124.

Hicks, J.R., 1941–42. "Consumers' Surplus and Index-Numbers," *Review of Economic Studies* 9, 126–137.

Hicks, J.R., 1946. *Value and Capital*, 2nd ed., Oxford: Clarendon Press.

Hicks, J.R., 1961. "Measurement of Capital in Relation to the Measurement of Other Economic Aggregates." In *The Theory of Capital*, F.A. Lutz and D.C. Hague (eds.), London: Macmillan.

Hicks, J.R., 1981. *Wealth and Welfare*, Cambridge, Mass.: Harvard University Press.

Hill, T.P., 1984. "Introduction: The Special Conference on Purchasing Power Parities," *Review of Income and Wealth* 30, 125–133.

Hill, T.P., 1988. "Recent Developments in Index Number Theory and Practice," OECD Economic Studies No. 10, Spring, 123–148.

Hofsten, E. von, 1952. *Price Indexes and Quality Change*, London: George Allen and Unwin.

Hotelling, H., 1932. "Edgeworth's Taxation Paradox and the Nature of Demand and Supply Functions," *Journal of Political Economy* 40, 577–616.

Hotelling, H., 1935. "Demand Functions with Limited Budgets," *Econometrica* 3, 66–78.

Hotelling, H., 1938. "The General Welfare in Relation to Problems of Taxation and of Railway and Utility Rates," *Econometrica* 6, 242–269.

Houthakker, H.S., 1950. "Revealed Preference and the Utility Function," *Economica* N.S. 17, 159–174.

Houthakker, H.S., 1951–52. "Compensated Changes in Quantities and Qualities Consumed," *Review of Economic Studies* 19, 155–164.

Howe, H., R.A. Pollak and T.J. Wales, 1979. "Theory and Time Series Estimation of the Quadratic Expenditure System," *Econometrica* 47, 1231–1247.

Hudson, E.A. and D.W. Jorgenson, 1974. "U.S. Energy Policy and Economic Growth, 1975–2000," *Bell Journal of Economics and Management Science* 5, 461–514.

Hulten, C.R., 1973. "Divisia Index Numbers", *Econometrica* 41, 1017–1026.

Hurwicz, L., 1971. "On the Problem of Integrability of Demand Functions." In *Preferences, Utility and Demand*, J.S. Chipman, L. Hurwicz, H.K. Richter and H.F. Sonnenschein (eds.), New York: Harcourt Brace Jovanovich, 174–214.

Hurwicz, L. and H. Uzawa, 1971. "On the Integrability of Demand Functions." In *Preferences, Utility and Demand*, J.S. Chipman, L. Hurwicz, M.K. Richter and H.F. Sonnenschein (eds.), New York: Harcourt Brace Jovanovich, 114–148.

Intriligator, M.D., 1981. "Mathematical Programming with Applications to Economics." In *Handbook of Mathematical Economics*, K.J. Arrow and M.D. Intriligator (eds.), Amsterdam: North-Holland, 53–91.

Jacobsen, S.E., 1970. "Production Correspondences," *Econometrica* 38, 754–771.

Jacobsen, S.E., 1972. "On Shephard's Duality Theorem," *Journal of Economic Theory* 4, 458–464.

Jacobsen, S.E., 1974. "Applications of Duality Theory: Comments," In *Frontiers of Quantitative Economics*, Vol. II, M.D. Intriligator and D.A. Kendrick (eds.), Amsterdam: North-Holland, 171–176.

Jastram, R.W., 1951. "Willard Phillips, A Predecessor of Paasche in Index Number Formulation," *Journal of the American Statistical Association* 46, 124–126.

Jevons, W.S., 1865. "Variations of Prices and the Value of Currency since 1762," *Journal of the Royal Statistical Society* 28 (June), 294–325.

Jevons, W.S., 1884. *Investigations in Currency and Finance*, London: Macmillan.

Jones, R.W., 1965. "The Structure of Simple General Equilibrium Models," *Journal of Political Economy* 73, 557–572.

Jones, R.W., 1972. "Activity Analysis and Real Incomes: Analogies with Production Models," *Journal of International Economics* 2, 277–302.

Jones, R.W. and J.A. Scheinkman, 1977. "The Relevance of the Two-Sector Production Model in Trade Theory," *Journal of Political Economy* 85, 909–936.

Jorgenson, D.W. and Z. Griliches, 1967. "The Explanation of Productivity Change," *Review of Economic Studies* 34, 249–283.

Jorgenson, D.W. and Z. Griliches, 1972. "Issues in Growth Accounting: A Reply to Edward F. Denison," *Survey of Current Business* 52, No. 5, Part II, 65–94.

Jorgenson, D.W. and L.J. Lau, 1970. "The 'Transcendental Logarithmic' Utility Function and Demand Analysis," Discussion Paper.

Jorgenson, D.W. and L.J. Lau, 1974a. "The Duality of Technology and Economic Behavior," *Review of Economic Studies* 41, 181–200.

Jorgenson, D.W. and L.J. Lau, 1974b. "Duality and Differentiability in Production," *Journal of Economic Theory* 9, 23–42.

Jorgenson, D.W. and L.J. Lau, 1975. "The Structure of Consumer Preferences," *Annals of Economic and Social Measurement* 4, 49–101.

Jorgenson, D.W. and M. Nishimizu, 1978. "U.S. and Japanese Economic Growth, 1952–1974: An International Comparison," *Economic Journal* 88, 707–726.

Jorgenson, D.W. and D.T. Slesnick, 1983. "Individual and Social Cost-of-Living Indexes." In Diewert and Montmarquette [1983; 241–323], and reprinted in Diewert [1990; 155–234].

Joseph, M.F.W., 1935–36. "Mr. Lerner's Supplementary Limits for Price Index Numbers," *Review of Economic Studies* 3, 155–157.

Juster, F.T. and F.P. Stafford (eds.), 1985. *Time, Goods and Well-Being*, Ann Arbor: Institute for Social Research, University of Michigan.

Kadiyala, K.R., 1971–72. "Production Functions and the Elasticity of Substitution," *Southern Economic Journal* 38, 281–284.

Karlin, S., 1959. *Mathematical Methods and Theory in Games, Programming and Economics*, Vol. I, Palo Alto, California: Addison-Wesley.

Katzner, D.W., 1970. *Static Demand Theory*, New York: Macmillan.

Keynes, J.M., 1930. *Treatise on Money*, Vol. 1, London: Macmillan.

Khaled, M.S., 1978. *Productivity Analysis and Functional Specification: A Parametric Approach*, Ph.D. thesis, University of British Columbia, Canada.

Khamis, S.H., 1972. "A New System of Index Numbers for National and International Purposes," *Journal of the Royal Statistical Society*, Series A 135, 96–121.

Khang, C., 1971. "An Isovalue Locus Involving Intermediate Goods and Its Applications to the Pure Theory of International Trade," *Journal of International Economics* 1, 315–325.

Kloek, T., 1966. *Indexcijfers: enige methodologisch aspecten*, The Hague: Pasmans.

Kloek, T., 1967. "On Quadratic Approximations of Cost of Living and Real Income Index Numbers," Report 6710, Econometric Institute, Netherlands School of Economics, Rotterdam.

Kloek, T. and G.M. de Wit, 1961. "Best Linear Unbiased Index Numbers," *Econometrica* 29, 602–616.

Kohli, U.J.R., 1978. "A Gross National Product Function and the Derived Demand for Imports and Supply of Exports," *Canadian Journal of Economics* 11, 167–182.

Kolm, S.-C., 1969. "The Optimal Production of Social Justice." In *Public Economics*, J. Margolis and H. Guitton (eds.), London: Macmillan, 145–200.

Kolm, S.-C., 1976a. "Unequal Inequalities I," *Journal of Economic Theory* 12, 416–442.

Kolm, S.-C., 1976b. "Unequal Inequalities II," *Journal of Economic Theory* 13, 82–111.

Kolmogoroff, A., 1930. "Sur la notion de la moyenne," *Atti della Reale Accamedia Nazionale dei Lincei, Rendiconti* 12, 388–391.

Konüs, A.A., 1924. English translation, titled "The Problem of the True Index of the Cost of Living," published in 1939 in *Econometrica* 7, 10–29.

Konüs, A.A., 1968. "The Theory of the Consumer Price Indexes and the Problem of the Comparison of the Cost of Living in Time and Space." In *The Social Sciences: Problems and Orientations*, Paris: UNESCO, 93–107.

Konüs, A.A. and S.S. Byushgens, 1926. "K probleme pokupatelnoi cili deneg" (English translation of Russian title: "On the Problem of the Purchasing Power of Money"), *Voprosi Konyunkturi* II(1) (supplement to the Economic Bulletin of the Conjuncture Institute), 151–172.

Kravis, I.B., 1984. "Comparative Studies of National Incomes and Prices," *Journal of Economic Literature* 22, 1–39.

Kravis, I.B., Z. Kenessey, A. Heston, and R. Summers, 1975. *A System of International Comparisons of Cross Product and Purchasing Power*, Baltimore: Johns Hopkins University Press.

Laspeyres, E., 1863. Geschichte der Volkswirtschaftlichen Anschauungen der Niederländer und ihrer Literatur zur Zeit der Republik, Leipzig.

Laspeyres, E., 1871. "Die Berechnung einer mittleren Waarenpreissteigerung," *Jahrbücher für Nationalökonomie und Statistik* 16, 296–315.

Lau, L.J., 1969. "Duality and the Structure of Utility Functions," *Journal of Economic Theory* 1, 374–396.

Lau, L.J., 1973. Unpublished memorandum.

Lau, L.J., 1974 . "Applications of Duality Theory: Comments." In *Frontiers of Quantitative Economics*, Vol. II, M.D. Intriligator and D.A. Kendrick (eds.), Amsterdam: North-Holland, 176–199.

Lau, L.J., 1976. "A Characterization of the Normalized Restricted Profit Function," *Journal of Economic Theory* 12, 131–163.

Lau, L.J., 1977a. "Existence Conditions for Aggregate Demand Functions: The Case of a Single Index," IMSSS Technical Report 248, Stanford University, October.

Lau, L.J., 1977b. "Existence Conditions for Aggregate Demand Functions: The Case of Multiple Indexes," IMSSS Technical Report 249, Stanford University, October.

Lau, L.J., 1977c. "Complete Systems of Consumer Demand Functions Through Duality," In *Frontiers of Quantitative Economics*, Vol. IIIA, M.D. Intriligator (ed.), Amsterdam: North-Holland, 59–85.

Lau, L.J., 1978a. "Applications of Profit Functions." In *Production Economics: A Dual Approach to Theory and Applications*, Vol. 1, M. Fuss and D. McFadden (eds.), Amsterdam: North-Holland, 133–216.

Lau, L.J., 1978b. "Testing and Imposing Monotonicity, Convexity and Quasiconvexity Constraints." In *Production Economics: A Dual Approach to*

Theory and Applications, Vol. 1, M. Fuss and D. McFadden (eds.), Amsterdam: North-Holland, 409–453.

Lau, L.J., 1979. "On Exact Index Numbers," *Review of Economics and Statistics* 61, 73–82.

Lau, L.J., 1982. "A Note on the Fundamental Theorem of Exact Aggregation," *Economic Letters* 9 (1), 119–126.

Lau, L.J. and B.M. Mitchell, 1970. "A Linear Logarithmic Expenditure System: An Application to U.S. Data," paper presented at the Second World Congress of the Econometric Society, Cambridge, England, September.

Lehr, J., 1885. *Beiträge zur Statistik der Preise*, Frankfurt: J.D. Sauerländer.

Leontief, W., 1936. "Composite Commodities and the Problem of Index Numbers," *Econometrica* 4, 39–59.

Leontief, W., 1941. *The Structure of the American Economy 1919–1929*, Cambridge, Mass.: Harvard University Press.

Leontief, W., 1947. "Introduction to a Theory of the Internal Structure of Functional Relationships," *Econometrica* 15, 361–373.

Lerner, A.P., 1935–36. "A Note on the Theory of Price Index Numbers," *Review of Economic Studies* 3, 50–56.

Lowe, J., 1823. *The Present State of England in Regard to Agriculture, Trade and Finance*, second edition, London: Longman, Hurst, Rees, Orme and Brown.

Machina, M., 1982. "'Expected Utility' Analysis without the Independence Axiom," *Econometrica* 50, 277–323.

Malmquist, S., 1953. "Index Numbers and Indifference Surfaces," *Trabajos de Estadistica* 4, 209–242.

Manser, M.E. and R.J. McDonald, 1988. "An Analysis of Substitution Bias in Measuring Inflation, 1959–85," *Econometrica* 56, 909–930.

Mark, J.A. and W.H. Waldorf, 1983. "Multifactor Productivity: A New BLS Measure," Monthly Labor Review, December, 3–15.

Marshak, J., 1950. "Rational Behavior, Uncertain Prospects and Measurable Utility," *Econometrica* 18, 111–141.

Marshall, A., 1887. "Remedies for Fluctuations of General Prices," *Contemporary Review* 51, 355–375. Reprinted as Ch. 8 in *Memorials of Alfred Marshall*, A.C. Pigou (ed.), London: Macmillan, 1925.

McCarthy, M.D., 1967. "Approximation of the CES Production Function: A Comment," *International Economic Review* 8, 190–192.

McFadden, D., 1962. *Factor Substitutability in the Economic Analysis of Production*, Ph.D. Thesis, University of Minnesota.

McFadden, D., 1966. "Cost, Revenue and Profit Functions: A Cursory Review," IBER Working Paper 86, University of California at Berkeley, March.

McFadden, D., 1970. "Cost, Revenue and Profit Functions," Discussion Paper, Department of Economics, University of California at Berkeley.

McFadden, D., 1978a. "Cost, Revenue and Profit Functions." In *Production Economics: A Dual Approach to Theory and Applications*, Vol. 1, M. Fuss and D. McFadden (eds.), Amsterdam: North-Holland, 3–109.

McFadden, D., 1978b. "The General Linear Profit Function." In *Production Economics: A Dual Approach to Theory and Applications*, Vol. 1, M. Fuss and D. McFadden, (eds.), Amsterdam: North-Holland, 269–286.

McKenzie, L.W., 1956–57. "Demand Theory Without a Utility Index," *Review of Economic Studies* 24, 185–189.

McLaren, K.R. and R.J. Cooper, 1980. "Intertemporal Duality: Application to the Theory of the Firm," *Econometrica* 48, 1755–1776.

Minkowski, H., 1911. *Theorie der konvexen Körper*, Gesammelte Abhandlungen II, Leipzig and Berlin: B.G. Teubner.

Mirrlees, J.A., 1981. "The Theory of Optimal Taxation," In *Handbook of Mathematical Economics*, K.J. Arrow and M.D. Intriligator (eds.), Amsterdam: North-Holland.

Montgomery, J.K., 1929. *Is There a Theoretically Correct Price Index of a Group of Commodities?*, Rome: Roma L'Universale Tipogr, Poliglotta.

Montgomery, J.K., 1937. *The Mathematical Problem of the Price Index*, London: P.S. King.

Moorsteen, R.H., 1961. "On Measuring Productive Potential and Relative Efficiency," *Quarterly Journal of Economics* 75, 451–467.

Mossin, J., 1968. "Aspects of Rational Insurance Purchasing," *Journal of Political Economy* 76, 553–568.

Muellbauer, J., 1974. "The Political Economy of Price Indices," Discussion Paper 22, Birkbeck, England, March.

Muellbauer, J., 1975. "Aggregation, Income Distribution and Consumer Demand," *Review of Economic Studies* 42, 525–543.

Muellbauer, J., 1976. "Community Preferences and the Representative Consumer," *Econometrica* 44, 979–999.

Nagumo, M., 1930. "Über eine Klasse der Mittelwerte," *Japanese Journal of Mathematics* 7, 71–79.

Neumann, J. von and O. Morgenstern, 1947. *Theory of Games and Economic Behavior*, Second Edition, Princeton: Princeton University Press.

Newman, P., 1965. *The Theory of Exchange*, Englewood Cliffs: Prentice-Hall.

Newman, P., 1969. *The Theory of Exchange*, Englewood Cliffs: Prentice-Hall.

Ohta, M., 1974. "A Note on Duality between Production and Cost Functions: Rate of Returns to Scale and Rate of Technical Progress," *Economic Studies Quarterly* 25, 63–65.

Paasche, H., 1874. "Über die Preisentwicklung der letzten Jahre nach den Hamburger Börsennotirungen," *Jahrbücher für Nationalökonomie und Statistik* 23, 168–178.

Parkan, C., 1975a. "Nonparametric Index Numbers and Tests for the Consistency of Consumer Data," Department of Manpower and Immigration, Research Projects Group, Ottawa.

Parks, R.W., 1971. "Price Responsiveness of Factor Utilization in Swedish Manufacturing 1870–1950," *Review of Economics and Statistics* 53, 129–139.

Pearce, I.F., 1961. "An Exact Method of Consumer Demand Analysis," *Econometrica* 29, 499–516.

Pearce, I.F., 1964. *A Contribution to Demand Analysis*, Oxford: Clarendon Press.

Pierson, N.G., 1895. "Index Numbers and Appreciation of Gold," *Economic Journal* 5, 329–335.

Pierson, N.G., 1896. "Further Considerations on Index-Numbers," *Economic Journal* 6, 127–131.

Pigou, A.C., 1912. *Wealth and Welfare*, London: Macmillan.

Pigou, A.C., 1932. *The Economics of Welfare*, fourth edition, London: Macmillan.

Pollak, R.A., 1969. "Conditional Demand Functions and Consumption Theory," *Quarterly Journal of Economics* 83, 60–78.

Pollak, R.A., 1971a. "The Theory of the Cost of Living Index," Research Discussion Paper 11, Office of Prices and Living Conditions, Bureau of Labor Statistics, Washington, D.C. In Diewert and Montmarquette [1983; 87–161], and reprinted in Diewert [1990; 5–77] and Pollak [1989; 3–52].

Pollak, R.A., 1971b. "Additive Utility Functions and Linear Engel Curves," *The Review of Economic Studies* 37, 401–413.

Pollak, R.A., 1972. "Generalized Separability," *Econometrica* 40, 431–453.

Pollak, R.A., 1975. "Subindexes of the Cost of Living," *International Economic Review* 16, 135–150, and reprinted in Pollak [1989; 53–69].

Pollak, R.A., 1981. "The Social Cost of Living Index," *Journal of Public Economics* 15, 311–336, and reprinted in Pollak [1989; 128–152].

Pollak, R.A., 1989. *The Theory of the Cost-of-Living Index*, Oxford: Oxford University Press.

Prais, S., 1959. "Whose Cost of Living?" *Review of Economic Studies* 26, 126–134.

Pratt, J.W., 1964. "Risk Aversion in the Small and in the Large," *Econometrica* 32, 122–136.

Primont, D., 1977. "Necessary and Sufficient Conditions for the Consistent Aggregation of Cost-of-Living Subindexes," Discussion Paper 77-08, Department of Economics, University of British Columbia, Canada.

Rader, T., 1976. "Equivalence of Consumer Surplus, the Divisia Index of Output and Eisenberg's Addilog Social Utility Function," *Journal of Economic Theory* 13, 58–66.

Ramsey, F.P., 1926. "Truth and Probability." Reprinted in *The Foundations of Mathematics and other Logical Essays*, R.B. Braithwaite (ed.), New York: The Humanities Press, 1950, 156–198.

Richter, M.K., 1966. "Invariance Axioms and Economic Indexes," *Econometrica* 34, 739–755.

Riddell, W.C., 1983. "Leisure Time and the Measurement of Economic Welfare." In Diewert and Montmarquette [1983; 337–367], and reprinted in Diewert [1990; 249-277].

Rinne, H., 1981. "Ernst Louis Etienne Laspeyres 1834–1913," *Jahrbucher für Nationalökonomie und Statistik* 198, 194–216.

Roberts, A.W. and D.E. Varberg, 1973. *Convex Functions*, New York: Academic Press.

Rockafellar, R.T., 1970. *Convex Analysis*, Princeton: Princeton University Press.

Rothschild, M. and J.E. Stiglitz, 1970. "Increasing Risk I: A Definition," *Journal of Economic Theory* 2, 225–243.

Rothschild, M. and J.E. Stiglitz, 1973. "Some Further Results on the Measurement of Inequality," *Journal of Economic Theory* 6, 188–204.

Roy, R., 1942. *De l'utilité*, Paris: Hermann.

Roy, R., 1947. "La distribution du revenu entre les divers biens," *Econometrica* 15, 205–225.

Rudin, W., 1953. *Principles of Mathematical Analysis*, New York: McGraw-Hill.

Ruggles, R., 1967. "Price Indexes and International Price Comparisons." In *Ten Economic Studies in the Tradition of Irving Fisher*, W. Fellner et al. (ed.), New York: John Wiley, 171–205.

Russell, R.R., 1975. "Functional Separability and Partial Elasticities of Substitution," *Review of Economic Studies* 42, 79–86.

Russell, R.R. and R. Boyce, 1974. "A Multilateral Model of International Trade Flows: A Theoretical Framework and Specification of Functional Forms," Institute of Policy Analysis, La Jolla, California.

Sakai, Y., 1973. "An Axiomatic Approach to Input Demand Theory," *International Economic Review* 14, 735–752.

Sakai, Y., 1974. "Substitution and Expansion Effects in Production Theory: The Case of Joint Production," *Journal of Economic Theory* 9, 255–274.

Samuelson, P.A., 1938. "The Empirical Implications of Utility Analysis," *Econometrica* 6, 344–356.

Samuelson, P.A., 1942. "Constancy of the Marginal Utility of Income." In *Studies in Mathematical Economics and Econometrics*, O. Lange et al. (eds.), in Memory of Henry Schultz, Chicago: University of Chicago Press, 75–91.

Samuelson, P.A., 1947. *Foundations of Economic Analysis*, Cambridge, Mass.: Harvard University Press.

Samuelson, P.A., 1950a. "Evaluation of Real National Income," *Oxford Economic Papers* 2, 1–29.

Samuelson, P.A., 1950b. "The Problem of Integrability in Utility Theory", *Economica* 17, 355–385.

Samuelson, P.A., 1952. "Probability Utility and the Independence Axiom," *Econometrica* 20, 670–678.

Samuelson, P.A., 1953. "Consumption Theorems in Terms of Overcompensation rather than Indifference Comparisons," *Economica* 20, 1–9.

Samuelson, P.A., 1953–54. "Prices of Factors and Goods in General Equilibrium," *Review of Economic Studies* 21, 1–20.

Samuelson, P.A., 1960. "Structure of a Minimum Equilibrium System." In *Essays in Economics and Econometrics: A Volume in Honor of Harold Hotelling*, Ralph W. Pfouts (ed.), University of North Carolina Press.

Samuelson, P.A., 1965. "Using Full Duality to Show that Simultaneously Additive Direct and Indirect Utilities Implies Unitary Price Elasticity of Demand," *Econometrica* 33, 781–796.

Samuelson, P.A., 1966. "The Fundamental Singularity Theorem for Non-Joint Production," *International Economic Review* 7, 34–41.

Samuelson, P.A., 1969a. "Pure Theory of Public Expenditure and Taxation." In *Public Economics*, J. Margolis and H. Guitton (eds.), New York: St. Martin, 98–123.

Samuelson, P.A., 1969b. "Corrected Formulation of Direct and Indirect Additivity," *Econometrica* 37, 355–359.

Samuelson, P.A., 1972. "Unification Theorem for the Two Basic Dualities of Homothetic Demand Theory," *Proceedings of the National Academy of Sciences, U.S.A.* 69, 2673–2674.

Samuelson, P.A., 1974a. "Remembrances of Frisch," *European Economic Review* 5, 7–23.

Samuelson, P.A., 1974b. "Complementarity — An Essay on the 40th Anniversary of the Hicks-Allen Revolution in Demand Theory," *Journal of Economic Literature* 12, 1255–1289.

Samuelson, P.A. and S. Swamy, 1974. "Invariant Economic Index Numbers and Canonical Duality: Survey and Synthesis," *American Economic Review* 64, 566–593.

Sargan, J.D., 1971. "Production Functions." In *Qualified Manpower and Economic Performance*, Part V, P.R.G. Layard, J.D. Sargan, M.E. Ager and D.J. Jones (eds.), London: Penguin.

Sato, K., 1975. *Production Functions and Aggregation*, Amsterdam: North-Holland.

Sato, K., 1976a. "The Ideal Log-Change Index Number," *Review of Economics and Statistics* 58, 223–228.

Sato, K., 1976b. "The Meaning and Measurement of the Real Value Added Index," *Review of Economics and Statistics* 58, 434–442.

Sato, K., 1980. "The Price Reversal Test and Economic Indexes," *Statistiche Hefte* 21, 127–130.

Sato, R. and T. Koizumi, 1973. "The Production Function and the Theory of Distributive Shares," *American Economic Review* 63, 484–489.

Savage, L.J., 1954. *The Foundations of Statistics*, New York: John Wiley.

Schultz, H., 1939. "A Misunderstanding in Index-Number Theory: The True Konüs Condition on Cost-of-Living Index Numbers and Its Limitations," *Econometrica* 7, 1–9.

Schworm, W.E., 1980. "Financial Constraints and Capital Theory," *International Economic Review* 21, 643–660.

Scrope, G.P., 1833. *Principles of Political Economy*, London: Longman, Rees, Orme, Brown, Green and Longman.

Segal, U., 1992. "Additively Separable Representations on Non-Convex Sets," *Journal of Economic Theory* 56, 89–99.

Segal, U. and A. Spivak, 1990. "First Order versus Second Order Risk Aversion," *Journal of Economic Theory* 51, 111–125.

Shafer, W. and H. Sonnenschein, 1980. "Aggregate Demand as a Function of Prices and Income." In *Handbook of Mathamatical Economics*, K.J. Arrow and M.D. Intriligator (eds.), Amsterdam: North Holland.

Shephard, R.W., 1953. *Cost and Production Functions*, Princeton: Princeton University Press.

Shephard, R.W., 1970. *Theory of Cost and Production Functions*, Princeton: Princeton University Press.

Shephard, R.W., 1974. "Applications of Duality Theory: Comments." In *Frontiers of Quantitative Economics*, Vol. II, M.D. Intriligator and D.A. Kendrick (eds.), Amsterdam: North Holland, 200–206.

Sidgwick, H., 1883. *The Principles of Political Economy*, London: Macmillan.

Silberberg, E., 1972. "Duality and the Many Consumer's Surpluses," *American Economic Review* 62, 942–952.

Sims, C.A., 1969. "Theoretical Basis for a Double Deflated Index of Real Value Added," *Review of Economics and Statistics* 51(4), 470–471.

Slutsky, E., 1915. "Sulla teoria del bilancio del consumatore," *Giornale degli Economisti* 51, 1–26.

Solow, R.M., 1955–56. "The Production Function and the Theory of Capital," *Review of Economic Studies* 23, 101–108.

Solow, R.M., 1957. "Technical Change and the Aggregate Production Function," *Review of Economics and Statistics* 39, 312–320.

Sono, M., 1945. English translation titled "The Effect of Price Changes on the Demand and Supply of Separable Goods" published in 1961 in *International Economic Review* 2, 239–275.

Star, S., 1974. "Accounting for the Growth of Output," *American Economic Review* 64, 123–135.

Star, S. and R.E. Hall, 1976. "An Approximate Divisia Index of Total Factor Productivity," *Econometrica*, 44, 257–263.

Stiglitz, J.E., 1969. "Behavior Toward Risk with Many Commodities," *Econometrica* 37, 660–667.

Stone, R., 1956. *Quantity and Price Indexes in National Accounts*, Paris: Organization for European Cooperation.

Stuvel, G., 1957. "A New Index Number Formula," *Econometrica* 25, 123–131.

Szulc (Schultz), B.J., 1964. "Indices for Multiregional Comparisons," *Przeglad Statystyczny* (*Statistical Review*) 3, 239–254.

Szulc (Schultz), B.J., 1983. "Linking Price Index Numbers." In Diewert and Montmarquette [1983; 537-566].

Szulc (Schultz), B.J., 1989. "Price Indices below the Basic Aggregation Level." Appendix 7 in *Consumer Price Indices, An ILO Manual*, Ralph Turvey (ed.), Geneva: International Labour Office.

Theil, H., 1960. "Best Linear Index Numbers of Prices and Quantities," *Econometrica* 28, 464–480.

Theil, H., 1965. "The Information Approach to Demand Analysis," *Econometrica* 33, 67–87.

Theil, H., 1967. *Economics and Information Theory*, Amsterdam: North-Holland.

Theil, H., 1968. "On the Geometry and the Numerical Approximation of Cost of Living and Real Income Indices," *De Economist* 116, 677–689.

Theil, H., 1973. "A New Index Number Formula," *Review of Economics and Statistics* 55, 498–502.

Tinbergen, J., 1942. "Zur Theorie der lanfristigen Wirtschaftsentwicklung," *Weltwirtschaftliches Archiv* 55, 511–549.

Törnqvist, L., 1936. "The Bank of Finland's Consumption Price Index," *Bank of Finland Monthly Bulletin* 10, 1–8.

Törnqvist, L., 1937. "Finlands Banks Consumptionsprisindex," *Nordisk tidskift for Teknisk Okonomic*, 73–95.

Törnqvist, L., 1971. *The Economic Development of the Post and Telegraph Office Until 1970*, Appendix of the Annual Report of the Administration of Posts and Telegraphs for the year 1970, Helsinki.

Triplett, J.E., 1983. "Concepts of Quality in Input and Output Price Measures: A Resolution of the User-Value Resource — Cost Debate." In *The U.S. National Income and Product Accounts: Selected Topics*, M.F. Foss (ed.), Chicago: University of Chicago Press.

Triplett, J.E., 1988. "Price Index Research and its Influence on Data: A Historical Review." Paper presented at the 50th Anniversary Conference of the Conference on Income and Wealth, Washington, D.C., May 12.

Triplett, J.E., 1990a. "The Theory of Industrial and Occupational Classifications and Related Phenomena." In *Proceedings, Bureau of the Census 1990 Annual Research Conference*, Washington, D.C.: U.S. Department of Commerce, Bureau of the Census, 9–25.

Triplett, J.E., 1990b. "Hedonic Methods in Statistical Agency Environments: An Intellectual Biopsy." In *Fifty Years of Economic Measurement*, Studies in Income and Wealth 54, E.R. Berndt and J.E. Triplett (eds.), Chicago: University of Chicago Press, 207–233.

Triplett, J.E., 1991. "Perspectives on the S.I.C.: Conceptual Issues in Economic Classification." In *Proceedings, 1991 International Conference on the Classification of Economic Activities*, Washington, D.C.: U.S. Department of Commerce, Bureau of the Census, 24–37.

Triplett, J.E., 1992. "Economic Theory and BEA's Alternative Quantity and Price Indexes," *Survey of Current Business* 72, 49–52.

Turvey, R., 1979. "The Treatment of Seasonal Items in Consumer Price Indices," *Bulletin of Labour Statistics*, 4th quarter, Geneva: International Labour Office, 13–33.

Usher, D., 1974. "The Suitability of the Divisia Index for the Measurement of Economic Aggregates," *Review of Income and Wealth* 20, 273–288.

Uzawa, H., 1962. "Production Functions with Constant Elasticities of Substitution," *Review of Economic Studies* 29, 291–299.

Uzawa, H., 1964. "Duality Principles in the Theory of Cost and Production," *International Economic Review* 5, 216–220.

van Yzeren (van Ijzerin), J., 1956. *Three Methods of Comparing the Purchasing Power of Currencies*, Statistical Studies No. 7, The Hague: The Netherlands Central Bureau of Statistics.

van Yzeren (van Ijzerin), J., 1983. *Index Numbers for Binary and Multilateral Comparison*, Statistical Studies No. 34, The Hague: The Netherlands Central Bureau of Statistics.

Vartia, Y.O., 1974. *Relative Changes and Economic Indices*, Licensiate Thesis in Statistics, University of Helsinki, June.

Vartia, Y.O., 1976a. "Ideal Log-Change Index Numbers," *Scandinavian Journal of Statistics* 3, 121–126.

Vartia, Y.O., 1976b. *Relative Changes and Index Numbers*, Helsinki: The Research Institute of the Finnish Economy.

Vartia, Y.O., 1978. "Fisher's Five-Tined Fork and Other Quantum Theories of Index Numbers." In *Theory and Applications of Economic Indices*,

W. Eichhorn, R. Henn, O. Opitz and R.W. Shephard (eds.), Würzburg: Physica-Verlag, 271–295.

Vartia, Y.O., 1985. "Defining Descriptive Price and Quantity Index Numbers: An Axiomatic Approach." Paper presented at the Fourth Karlsruhe Symposium on Measurement in Economics, University of Karlsruhe, July.

Ville, J., 1946. "Sur les conditions d'existence d'une ophélimité totale et d'un indice du niveau des prix," *Annales de l'Université de Lyon* 9, 32–39. Reprinted in English translation as Ville [1951–52].

Ville, J., 1951–52. "The Existence-Conditions of a Total Utility Function," *Review of Economic Studies* 19, 123–128. This is an English translation of Ville [1946].

Voeller, J., 1981. "Purchasing Power Parities for International Comparisons." Discussion Paper 161, Institut für Wirtschaftstheorie und Operations Research, University of Karlsruhe, Karlsruhe, West Germany.

Vogt, A., 1978. Der Wertindextreue-Test und eine Vereinfachung des Index Problems," *Statistiche Hefte* 19, 131–139.

Vogt, A., 1980. Der Zeit und der Faktorumkehrtest als 'Finders of Test'," *Statistiche Hefte* 21, 66–71.

Wald, A., 1937. "Zur Theorie der Preisindexziffern," *Zeitschrift für Nationalökonomie* 8, 179–219.

Wald, A., 1939. "A New Formula for the Index of Cost of Living," *Econometrica* 7, 319–331.

Wales, T.J., 1973. "Estimation of a Labour Supply Curve for Self-Employed Business Proprietors," *International Economic Review* 14, 69–80.

Wales, T.J., 1977. "On the Flexibility of Flexible Functional Forms: An Empirical Approach," *Journal of Econometrics* 5, 183–193.

Wales, T.J. and A.D. Woodland, 1976. "Estimation of Household Utility Functions and Labor Supply Response," *International Economic Review* 17, 397–410.

Wales, T.J. and A.D. Woodland, 1977. "Estimation of the Allocation of Time for Work, Leisure and Housework," *Econometrica* 45, 115–132.

Wales, T.J. and A.D. Woodland, 1979. "Labour Supply and Progressive Taxes," *Review of Economic Studies* 46, 83–95.

Walsh, C.M., 1901. *The Measurement of General Exchange Value*, New York: Macmillan.

Walsh, C.M., 1921. "The Best Form of Index Number: Discussion," *Quarterly Publication of the American Statistical Association* 17 (March), 537–544.

Walsh, C.M., 1924. "Professor Edgeworth's Views on Index-Numbers," *Quarterly Journal of Economics* 38, 500–519.

Walters, A.A., 1961. "Production and Cost Functions: An Econometric Survey," *Econometrica* 31, 1–66.

Weddepohl, H.N., 1970. *Axiomatic Choice Models and Duality*, Groningen: Rotterdam University Press.

Westergaard, H., 1890. *Die Grundzüge der Theorie der Statistik*, Jena: Fischer.

Weymark, J.A., 1980. "Duality Results in Demand Theory," *European Economic Review* 14, 377–395.

Weymark, J.A., 1981. "Generalized Gini Inequality Indices," *Mathematical Social Sciences* 1, 409–430.

Wicksell, K., 1958. *Selected Papers on Economic Theory*, Cambridge, Mass.: Harvard University Press.

Wold, H., 1943, 1944. "A Synthesis of Pure Demand Analysis," published in three parts in *Skandinavisk Aktuarietidskrift* 26, 85–144 and 220–75; 27, 69–120.

Wold, H., 1953. *Demand Analysis*, New York: John Wiley and Sons.

Woodland, A.D., 1972a. "Factor Demand Functions for Canadian Industries, 1946-1969," Department of Manpower and Immigration, Strategic Planning and Research, Ottawa.

Woodland, A.D., 1972b. "The Construction of Price and Quantity Components of Inputs for Canadian Industries, 1927–1969," Department of Manpower and Immigration, Strategic Planning and Research, Ottawa.

Woodland, A.D., 1974. "Demand Conditions in International Trade Theory," *Australian Economic Papers* 13, 209–224.

Woodland, A.D., 1975. "Substitution of Structures, Equipment and Labor in Canadian Production," *International Economic Review* 16, 171–187.

Woodland, A.D., 1977a. "A Dual Approach to Equilibrium in the Production Sector in International Trade Theory," *Canadian Journal of Economics* 10, 50–68.

Woodland, A.D., 1977b. "Joint Outputs, Intermediate Inputs and International Trade Theory," *International Economic Review* 18, 517–534.

Woodland, A.D., 1977c. "Estimation of a Variable Profit and Planning Price Functions for Canadian Manufacturing, 1947–1970," *Canadian Journal of Economics* 10, 355–377.

Woodland, A.D., 1978. "On Testing for Weak Separability," *Journal of Econometrics* 8, 383–398.

Zangwill, W.I., 1969. *Nonlinear Programming: A Unified Approach*, Englewood Cliffs, N.J.: Prentice-Hall.

NAME INDEX

SUBJECT INDEX

Lightning Source UK Ltd.
Milton Keynes UK
UKOW06n2109271015

261503UK00001B/16/P